Dielectrics and Waves

For a complete listing of the *Artech House Microwave Library*,
turn to the back of this book

Dielectrics and Waves

Arthur von Hippel

Artech House
Boston • London

Library of Congress Cataloging-in-Publication Data
Von Hippel, Arthur R. (Arthur Robert), 1898-
Dielectrics and waves / Arthur von Hippel.
Originally published: New York : Wiley, 1954
ISBN 0-89006-803-8
1. Dielectrics. 2. Electric waves. I. Title
QC585.V63 1995 94-33549
537'.24–dc20 CIP

British Library Cataloguing in Publication Data
Hippel, Arthur von
Dielectrics and Waves. New ed.
I. Title
537.24

ISBN 0-89006-803-8

International Standard Book Number: 0-89006-803-8
Library of Congress Catalog Card Number: 94-33549

10 9 8 7 6 5 4 3 2

To

NIELS BOHR and JAMES FRANCK

in friendship and gratitude

PREFACE TO THE SECOND EDITION

As a practicing engineer, I have used in my technical work the two-volume von Hippel set as an indispensable reference for dielectric properties of materials. When I realized that they were out of print, I sent copies of selected pages to Artech House for evaluation with the recommendation that both volumes be republished for the benefit of other interested engineers. Several months later, Artech House notified me that a decision had been made to republish, that the rights to republish had been secured from the retired author, and that, as implausible as it may seem, neither the previous publishers nor the author have any remaining copies of this set, as would be required for the republication process. Therefore, I supplied the original editions to Artech House for reproduction. I would like to thank the entire staff at Artech House for their professionalism in the handling of this republication.

Huntington Beach, California
August 1994

ALEXANDER S. LABOUNSKY
Principal engineer/scientist
McDonnell Douglas Aerospace

PREFACE TO THE FIRST EDITION

A treatise intended for physicists, chemists, and electrical engineers is likely to disappoint three groups of readers. If an author, in spite of this danger, embarks on such an adventure, intense compulsion must drive him. For a number of years my demon has urged me to oppose the trend of specialization by helping to develop a field of knowledge that belongs not only to physics and chemistry but is also of vital importance for modern electrical engineering. We may call this subject "dielectrics" by identifying with the name not a narrow class of so-called insulators, but any nonmetal, and even metals as a boundary case, if their interaction with electric, magnetic, or electromagnetic fields is under consideration. *Dielectrics and Waves* has been chosen as the title of this book because wave phenomena play a dominant part in our story, whether electromagnetic waves, probability waves of quantum mechanics, or the elastic waves of crystal lattices.

The phenomena "polarization," "magnetization," and "conduction" are the properties of matter at issue. For macroscopic physics and electrical engineering this is a familiar subject when viewed from the standpoint of Maxwell's theory. Matter appears here as a storage medium and wave guide of electric and magnetic energy and as a dissipator of such energy by conduction and other irreversible processes. These properties are considered as given quantities that may be tabulated by introducing some descriptive parameters, for example, the complex permittivity and permeability used in this book. At this point the subject "materials" is dismissed by the present-day electrical engineer in favor of his private world of field vectors and equivalent circuits.

The physicist and chemist have pushed on to the task of unraveling the molecular phenomena behind these macroscopic parameters. The structure and behavior of isolated atoms and molecules and the behavior of electrons and ions in gases of low pressure are now relatively well understood; the dielectric properties of gases at high pressures and of liquids and solids are still known only in rough outlines. However, this task of "dielectric analysis" has progressed far enough to allow a successful beginning of "dielectric syntheses" in which the properties of materials are tailored to order by combining the proper atoms and molecules into specified arrangements.

This subject of dielectric synthesis is of vital importance to the electrical engineer, promising him a variety of new tools and a release from shackling limitations. Nobody, however, can leave his problems to others without losing control over his destiny. The electrical engineer has to remember that he is an applied scientist and join his colleagues of physics and chemistry in a co-operative venture of "molecular electrical engineering."

Seen from this point of view, I would want the book to be a trumpet of Jericho; alas it may only loosen some bricks that will fall on the author's head. This is a survey book which cannot go into many important details and has to leave unmentioned many significant contributions. Space does not permit to give the molecular aspects of conduction more than a cursory glance. I hope to make "Electric Conduction and Breakdown" the subject of a later volume.

To remedy some of these shortcomings a representative list of books covering special fields has been added. Much additional information may also be found in the companion book, *Dielectric Materials and Applications*, published simultaneously.

It is too much to hope that any reader will follow the unfolding of this biography of dielectrics with the puzzled attention normally reserved for a detective story. But it may be of some help in bringing physicists, chemists, and electrical engineers closer together and provide a better understanding between the mode of thinking of the theorist and the experimentalist on dielectric problems.

In dedicating this book to Niels Bohr and James Franck I am fulfilling a simple duty of gratitude to two masters of science who became my friends at decisive junctures of my life and who have set, with their scientific genius and humanity, an ideal for our generation.

Cambridge, Massachusetts
June 1954

A. VON HIPPEL

ACKNOWLEDGMENTS

In writing a book, one accumulates a great amount of indebtedness. I owe much to discussions with Professor J. C. Slater and with former and present co-workers of our laboratory, especially Dr. S. C. Abrahams, Professor H. B. Callen, M. E. Caspari, Professor D. J. Epstein, Dr. E. P. Gross and W. B. Westphal. My co-workers Caspari and Epstein, together with our excellent draftsman, J. J. Maccarone, translated rough sketches into finished illustrations. Dr. L. G. Wesson gave invaluable help, assisted by Mrs. H. B. Armstrong, Miss E. J. Busby, and later by Miss A. Sils, in getting the book to the printer and seeing it through the press. Dr. E. P. Gross worked with me in the preparation of the problem section; Professor R. H. Cole read the major part of the manuscript in behalf of the publisher and made valuable comments. Dean F. G. Fasset, Jr., the director of the Technology Press, which publishes the companion volume *Dielectric Materials and Applications*, gave helpful counsel.

The scientific work of the Laboratory for Insulation Research is supported jointly by the ONR, the Army Signal Corps, and the Air Force under ONR Contracts N5ori-07801 and N5ori-07858. It is obvious that the book could not have been written without this background of sponsored free research; and it is a pleasant duty to acknowledge gratefully the unfailing support of our sponsors in keeping our research without restrictions and directed towards fundamental issues.

A number of figures have been adapted from publications of other authors as indicated in the references. I am grateful to these colleagues and their publishers for permission to use such material.

A. von Hippel

CONTENTS

I · MACROSCOPIC APPROACH

0 · Survey

In discussing electromagnetic waves and their interaction with dielectric materials, a comprehensive view will be given without prejudice to the entrenched interests of the electrical engineer in the low-frequency range or of the physicist in the optical spectrum. We may pretend that a dielectric can be exposed to electric or magnetic fields of any frequencies by filling a capacitor or a coil with the material in question and connecting it to a voltage source ranging in frequency from zero (direct current) to X-rays ($\simeq 10^{18}$ cps). This Munchausen device requires, of course, a number of equipments for its practical realization (Fig. 0.1).

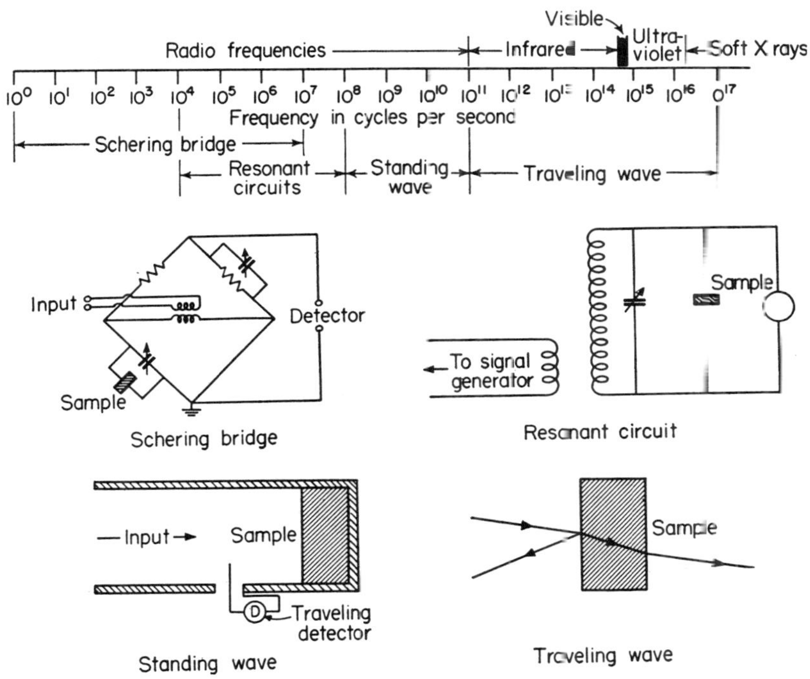

Fig. 0.1. Frequency ranges and measuring techniques.

From direct current to about 10^8 cps a condenser or a coil represents a lumped circuit element; its properties may be measured in the lower half of this range by bridge arrangements and at the higher frequencies in resonant circuits. Between meter and millimeter waves the dimensions of the dielectric become comparable to those of the wavelength, the physical distinction between coil and condenser begins to disappear, and the material is customarily inserted into wave guides and its characteristics are determined by standing-wave patterns. As the wavelength shrinks further, we move out of this range of interference optics into that of geometrical optics and determine the dielectric properties of materials from infrared to ultraviolet by reflection and transmission measurements. Finally, in the X-ray region, we return once more to interference techniques as the size of the atoms and the molecules and their separation become comparable to the incident wave-length.

This whole frequency spectrum is ruled by Maxwell's two field equations which describe quantitatively how a time-varying electric field is accompanied by a time-varying magnetic field, and vice versa. Since both fields may cause energy storage and energy dissipation in matter, two pairs of parameters are required to characterize a dielectric as a carrier of electromagnetic energy.

Various sets of such parameters have been introduced, depending on the specific interests of the observer. The scientist and the engineer concerned with the low-frequency range refer, in general, to permittivity (dielectric constant) and permeability, and the communication engineer to the complex propagation factor; in optics the complex index of refraction is used. Physicist and electrical engineer, allied in language through the advent of radar, may, in addition, characterize a dielectric by its complex impedance, and the chemist may side with the power engineer in referring to dielectric constant and power factor.

To establish order in this confusion of tongues, we introduce, by lumped-circuit considerations, the complex permittivity ϵ^* and permeability μ^* as the fundamental parameters for the macroscopic description of a dielectric exposed to sinusoidal fields (Sec. 1) and then proceed systematically in the development of the electric and the magnetic field concepts.

The electrostatic field is due to the existence of positive and negative charges. The total or true charge stored in a capacitor is only in part free to create an external electric field; the remainder is bound by the countercharges of the dielectric, that is, by polarization. The true-, free-, and bound-charge densities are considered to give rise to three vector fields, characterized by the electric displacement **D**, the electric field strength **E**, and the polarization **P**, respectively. The field concept is extended throughout space by the definition that the electric field strength is equivalent in magnitude and direction to the force per unit charge acting on a detector charge. This dual role of **E** determines the dimensions of the dielectric constant. After introduction of the electric dipole concept, the stage is set for a discussion of magnetic fields created by magnetic dipoles. The magnetic field vectors, the magnetic flux density **B**, field strength **H**, and magnetization **M** become defined, and the dimensions of the magnetic dipole moment are determined by the torque equation (Sec. 2).

Formulating the potential concept, we arrive at Coulomb's law as a special case of the general force law for spherical symmetry and at the dipole field by the superposition of the Coulomb fields of two opposite point charges. The differential operator ∇ is introduced for the discussion of the potential gradient of the dipole field and of the general forces acting on a dipole in an external field (Sec. 3). Next, space charge fields are analyzed in integral and differential formulation, and the divergence ($\nabla \cdot E$) and the Laplacian (∇^2) are formulated (Sec. 4).

Thus far, electricity and magnetism appear as two new phenomena independent of each other and consequently requiring for their description the introduction of two new quantities, an electric and a magnetic one, for example, *electric charge* and *magnetic dipole moment*. Actually, they are interlinked by Ampère's circuital and Faraday's induction law. The equivalence between the fields of magnetic dipoles and of circular currents allows us to attribute the existence of magnetic dipole moments to molecular currents and to reduce the dimensions of the magnetic quantities to electrical and mechanical dimensions (Sec. 5). By extending Ampère's and Faraday's laws to all space, we obtain Maxwell's two field equations; they can be written in complete symmetry (save for a negative sign) by the use of the complex permittivity and permeability (Sec. 6).

From these field equations, the wave equations of the electromagnetic field are derived and solved for plane waves in unbounded space. A discussion of these waves leads to the concepts complex propagation factor, wavelength, phase velocity, and intrinsic impedance (Sec. 7). Herewith the prerequisites are at hand for a quantitative formulation of dimensions and units (Sec. 8) and for the description of dielectrics by various sets of parameters, alternative to ϵ^* and μ^*. Conversion formulas and nomographic charts are given in Sec. 9.

After a short digression concerned with the forces operative in electric and magnetic fields (Sec. 10), the energy of the electromagnetic field and its flow are considered (Sec. 11). The state of polarization of a field is defined (Sec. 12), and the radiation emitted by an electric dipole is calculated (Sec. 13). This extension of the theory to light sources paves the way for a later discussion of the dipole radiation of atoms and molecules.

Electromagnetic fields are usually not observed in free space, but within the confinement of boundaries. In fact, reflection and refraction of waves at boundaries provide the essential means for measuring the interaction between electromagnetic waves and dielectric materials. Boundary conditions are therefore introduced for static fields as well as for plane waves striking the interface between two media (Sec. 14). Snell's laws of reflection and refraction follow from the continuity of the tangential components of the **E** and **H** fields and determine the direction of propagation of the reflected and the refracted beams. Simultaneously, Fresnel's equations are obtained which prescribe the amplitudes of the reflected and the transmitted waves and their states of polarization (Sec. 15). For dielectrics without loss, the phenomena of special interest are the disappearance of one pair of field components when the wave is reflected at Brewster's angle, and the occurrence of total reflection accompanied by the formation of a guided surface wave with longitudinal components (Sec. 16). At normal incidence, standing waves (Sec. 17) and useful methods of measuring dielectric properties by interference (Sec. 18) result.

When electromagnetic waves strike a metal, nearly total reflection ensues, and the weak transmitted beam is rapidly attenuated (Sec. 19). The general case of oblique incidence on media with loss is of great complexity. The index of refraction becomes a function of the angle of incidence, and complicated phase shifts arise because of the appearance of longitudinal field components (Sec. 20).

Viewed from an alternative standpoint, the incident wave, striking a boundary obliquely, forms, with the reflected wave, an interference pattern which glides along the interface with a phase velocity greater than that of the incident wave; the boundary acts as a wave guide. To make such a guide more effective, a second boundary may be placed parallel to the first one in one of the dark interference fringes, and a parallel-plane wave guide with cut-off properties results (Sec. 21). By completing the enclosure we obtain hollow wave guides which behave like highly dispersive dielectrics and propagate characteristic wave types (Sec. 22). Dielectrics can be measured in such wave guides with great precision, whether by mapping with a traveling detector the standing-wave pattern formed in front of the sample (Sec. 23) or by evaluating the effect of the material on the impedance of a cavity resonator (Sec. 24).

Frequently it proves convenient to handle field phenomena by an equivalence approach, in which the role of the electric and the magnetic fields is assumed by voltages and currents in electric circuits. By introducing this complementarity we return to distributed and lumped-circuit concepts (Sec. 25) and close the macroscopic discussion with a formalistic representation of dielectrics by lumped equivalent circuits (Sec. 26).

Summarizing: Part I of this book introduces the complex permittivity and permeability as the fundamental parameters develops in rapid succession the essential field concepts, considers the propagation of electromagnetic waves in unbounded space and under successively more stringent boundary conditions. In the course of this treatment we are logically led to alternative ways of describing the interaction between fields and matter and to conversion formulas interlinking the various parameters. In addition to this quantitative description of fields and dielectrics, the macroscopic theory provides a quantitative basis for measuring ϵ^* and μ^*.

1 · Complex Permittivity and Permeability

A capacitor, connected to a sinusoidal voltage source

$$\mathcal{V} = \mathcal{V}_0 e^{j\omega t} \dagger \qquad (1.1)$$

of the *angular frequency*

$$\omega = 2\pi\nu \qquad (1.2)$$

stores, when vacuum is its dielectric, a charge

$$Q = C_0\mathcal{V}, \qquad (1.3)$$

and draws a *charging current*

$$I_c = \frac{dQ}{dt} = j\omega C_0 \mathcal{V} = I_0 e^{j\left(\omega t + \frac{\pi}{2}\right)}, \qquad (1.4)$$

leading the voltage by a temporal phase angle of 90°

† Throughout this book we will use complex quantities in treating periodic phenomena and represent them in the complex plane. Here the x-axis corresponds to the *axis of reals* and the y-axis to the *axis of imaginaries.* The factor $j = \sqrt{-1}$ in front of a real quantity signifies an imaginary component oriented in the $+y$-axis or $+j$-axis direction. A complex quantity $z = x + jy$ plotted in the complex plane corresponds in polar co-ordinates to a radius vector $\rho = \sqrt{x^2 + y^2}$ inclined by an angle $\theta = \tan^{-1}(y/x)$ towards the real axis: $z = \rho e^{j\theta}$.

The complex function $\mathcal{V} = \mathcal{V}_0 e^{j\omega t} = \mathcal{V}_0(\cos\omega t + j\sin\omega t)$ consequently can be plotted in the complex plane as a radius vector of length $\mathcal{V}_0$, the voltage amplitude, making an angle of ωt radians with the axis of reals. As long as the voltage and current vectors rotate at the same angular velocity of ω radians per second, we can forget this rotation in discussing their relative positions in the complex plane.

We return from the complex functions to actual currents and voltages by taking the real or the imaginary part: Re $(\mathcal{V}) = \mathcal{V}_0 \cos\omega t$, Im $(\mathcal{V}) = \mathcal{V}_0 \sin\omega t$. In dealing with products of complex functions it has to be kept in mind that the product of the real parts of two complex quantities A_1 and A_2 is not equal to the real part of their product, but

$$\text{Re}(A_1)\,\text{Re}(A_2) = \tfrac{1}{4}(A_1 + \bar{A}_1)(A_2 + \bar{A}_2).$$

The symbol $\bar{A}_1$ signifies the conjugate of A_1; for example, if $A_1 = (x + jy)e^{j\omega t}$, then $\bar{A}_1 = (x - jy)e^{-j\omega t}$.

The time average of a periodic function A is $\bar{A} = \frac{1}{T}\int_0^T A\,dt$, where T is the period of the function. If A_1 and A_2 are such functions, the product of the averages of their real parts is $\overline{\text{Re}(A_1)\,\text{Re}(A_2)} = \frac{1}{2}\,\text{Re}(A_1\bar{A}_2)$.

(Fig. 1.1). C_0 is the *vacuum* (or geometrical) *capacitance* of the condenser.

When filled with some substance, the condenser increases its capacitance to

$$C = C_0 \frac{\epsilon'}{\epsilon_0} = C_0 \kappa', \tag{1.5}$$

where ϵ' and ϵ_0 designate the real *permittivities* or *dielectric constants* of the dielectric and of vacuum, respectively, and their ratio κ' the *relative dielectric constant* of

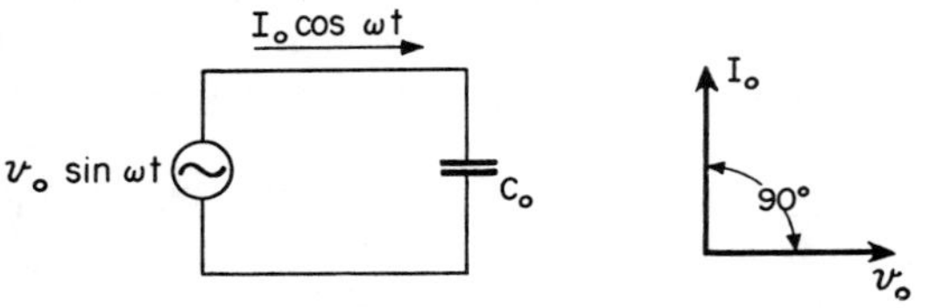

Fig. 1.1. Current-voltage relation in ideal capacitor.

the material. Simultaneously, there may appear, in addition to the charging current component I_c, a *loss current* component

$$I_l = G\mathcal{V} \tag{1.6}$$

in phase with the voltage; G represents the conductance of the dielectric. The total current traversing the condenser,

$$I = I_c + I_l = (j\omega C + G)\mathcal{V}, \tag{1.7}$$

is inclined by a *power factor angle* $\theta < 90°$ against the applied voltage $\mathcal{V}$, that is, by a *loss angle* δ against the $+j$-axis (Fig. 1.2).

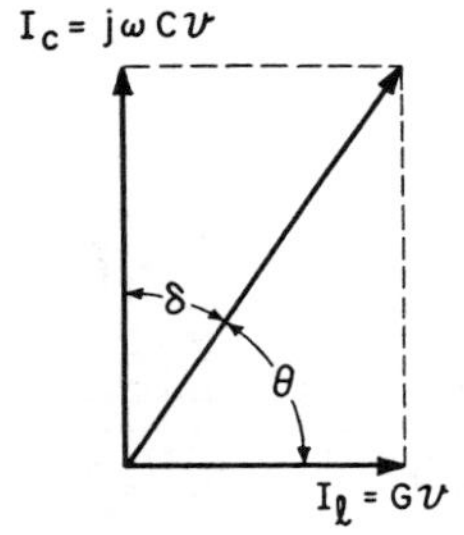

Fig. 1.2. Capacitor containing dielectric with loss.

It would be premature to conclude that the dielectric material corresponds in its electrical behavior to a capacitor paralleled by a resistor (*RC* circuit) (Fig. 1.3). The frequency response of this circuit, which can be expressed by the ratio of loss current to charging current, that is, the *dissipation factor* D or loss tangent $\tan \delta$, as

$$D \equiv \tan \delta = \frac{I_l}{I_c} = \frac{1}{\omega RC}, \tag{1.8}$$

may not at all agree with that actually observed because the conductance term need not stem from a migration of charge carriers, but can represent any other energy-consuming process. It has therefore become customary to refer to the existence of a loss current in addition to a charging current noncommittally by the introduction of a *complex permittivity*

$$\epsilon^* = \epsilon' - j\epsilon''. \tag{1.9}$$

The total current I of Eq. 1.7 may thus be rewritten

$$I = (j\omega\epsilon' + \omega\epsilon'')\frac{C_0}{\epsilon_0}\mathcal{V} = j\omega C_0 \kappa^* \mathcal{V}, \tag{1.10}$$

where

$$\kappa^* \equiv \frac{\epsilon^*}{\epsilon_0} = \kappa' - j\kappa'' \tag{1.11}$$

is the *complex relative permittivity* of the material and

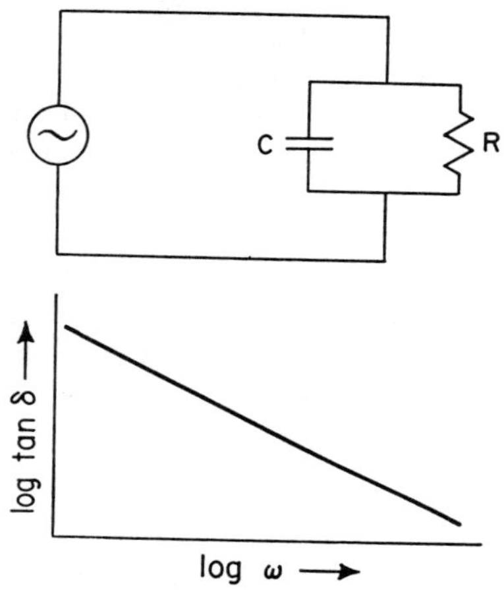

Fig. 1.3. *RC* circuit and its frequency response.

ϵ'' and κ'' the *loss factor* and *relative loss factor*, respectively. The loss tangent becomes

$$\tan \delta = \frac{\epsilon''}{\epsilon'} = \frac{\kappa''}{\kappa'}. \tag{1.12}$$

Since a parallel-plate condenser of the area A and the plate separation d, fringing effects neglected, has the vacuum capacitance

$$C_0 = \frac{A}{d}\epsilon_0, \tag{1.13}$$

the current density J traversing a condenser under the applied field strength

$$E = \mathcal{V}/d \tag{1.14}$$

becomes, according to Eq. 1.10,

$$J = (j\omega\epsilon' + \omega\epsilon'')E = \epsilon^* \frac{dE}{dt} \tag{1.15}$$

(Fig. 1.4). The product of angular frequency and loss factor is equivalent to a *dielectric conductivity*

$$\sigma = \omega\epsilon''. \tag{1.16}$$

This dielectric conductivity sums over all dissipative effects and may represent as well an actual ohmic conductivity caused by migrating charge carriers as refer

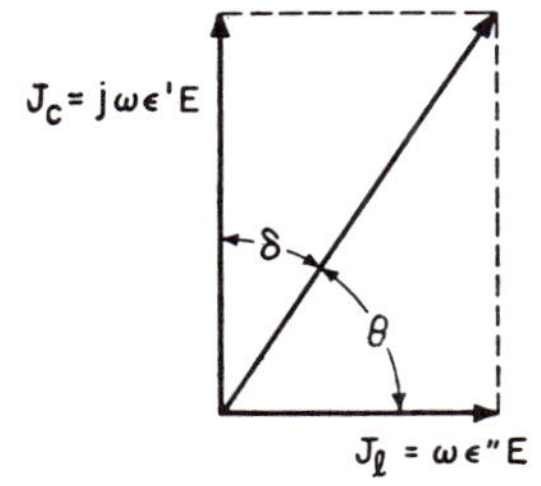

Fig. 1.4. Charging and loss current density.

to an energy loss associated with a frequency dependence (dispersion) of ϵ', for example, to the friction accompanying the orientation of dipoles.

If the dielectric material is transferred from the electric field of the capacitor into the magnetic field of a coil, the voltage $\mathcal{V}$ drives through the coil a magnetization current I_m according to Faraday's inductance law $\left(\mathcal{V} = L\dfrac{dI}{dt}\right)$ as

$$I_m = \frac{\mathcal{V}}{j\omega L_0 \dfrac{\mu'}{\mu_0}} = -j\frac{\mathcal{V}}{\omega L_0 \kappa_m'}. \tag{1.17}$$

L represents the *inductance* and L_0 the *vacuum* (or geometrical) *inductance* of the coil. This magnetization current lags behind the applied voltage by 90° (Fig. 1.5). The *permeabilities* μ' and μ_0 designate the mag-

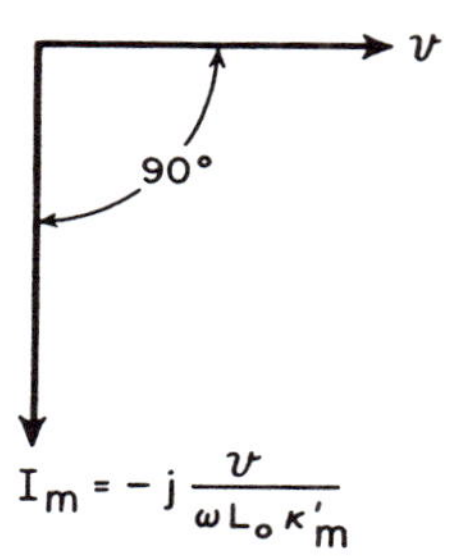

Fig. 1.5. Current-voltage relation in ideal inductor.

netization of the material and of vacuum, respectively, and their ratio

$$\kappa_m' \equiv \mu'/\mu_0 \tag{1.18}$$

the *relative permeability* of the material in which the magnetic field of the coil resides.

Because of the resistance R of the coil windings, an ohmic current component $\mathcal{V}/R$ exists. In addition, there may appear, in phase with $\mathcal{V}$, a magnetic loss current I_l caused by energy dissipation during the magnetization cycle. We shall allow for this magnetic loss by introducing a *complex permeability*

$$\mu^* = \mu' - j\mu'' \tag{1.19}$$

and a *complex relative permeability*

$$\kappa_m^* = \frac{\mu^*}{\mu_0} = \kappa_m' - j\kappa_m'' \tag{1.20}$$

in complete analogy to the electric case. Thus we obtain the total magnetization current

$$I = I_m + I_l = \frac{\mathcal{V}}{j\omega L_0 \kappa_m^*}$$

$$= -\frac{j\mathcal{V}(\mu' + j\mu'')}{\omega \dfrac{L_0}{\mu_0}(\mu'^2 + \mu''^2)}. \tag{1.21}$$

According to these lumped-circuit considerations the macroscopic electric and magnetic behavior of a dielectric material in sinusoidal fields is determined by the two complex parameters ϵ^* and μ^*.

The real and imaginary parts of these complex variables (ϵ' and ϵ'' or μ' and μ'') are even and odd functions, respectively, of the variable ω, that is, conjugate functions, and therefore not entirely independent of each other. Physically speaking, the mechanisms of energy storage and energy dissipation are two aspects of the same phenomenon; hence if one of them is given over the whole frequency spectrum, the other one is prescribed (see II, Fig. 19.1). Mathematically expressed, the calculation of an imaginary part (conjugate function) from a given real part [an arbitrary function $f(\omega)$] and, vice versa, is prescribed by the *Hilbert transforms*, known to physicists as *Kramers' theorem*.[1] This interrelation between the frequency response characteristics of dielectric constant and loss can sometimes prove helpful in checking the reliability of measurements but, in general, various polarization and conduction phenomena superpose and the available frequency range is insufficient for this type of unscrambling (see also Appendix A, I, 1).

[1] See, for example, E. A. Guillemin, *The Mathematics of Circuit Analysis*, John Wiley and Sons, New York, 1949, p. 339; H. Fröhlich, *Theory of Dielectrics*, Clarendon Press, Oxford, 1949, pp. 6 ff.

2 · Polarization and Magnetization

A dielectric material increases the storage capacity of a condenser by neutralizing charges at the electrode surfaces which otherwise would contribute to the external field. Faraday[1] was the first to recognize this phenomenon of *dielectric polarization.* We may visualize it as the action of dipole chains which form under the influence of the applied field and bind countercharges with their free ends on the metal surfaces (Fig. 2.1).

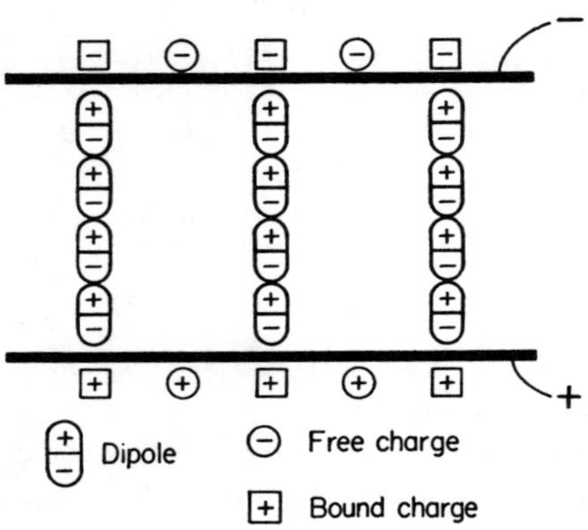

Fig. 2.1. Schematic representation of dielectric polarization.

By writing the voltage of the capacitor according to Eqs. 1.3 and 1.5 as

$$\mathcal{V} = \frac{Q}{\kappa'} \cdot \frac{1}{C_0}, \tag{2.1}$$

we may interpret this equation as stating that only a fraction of the *total charge* Q, the *free charge* Q/κ', contributes to the voltage whereas the remainder, the *bound charge* $Q\left(1 - \frac{1}{\kappa'}\right)$, is neutralized by the polarization of the dielectric.

To obtain a clearer conception of the charge distribution and its effect in space, we represent charge densities by field vectors. The *total* (or *true*) *charge* Q concentrated in the capacitor is distributed over the surface area A of the metal electrodes with a density s as

$$Q = \int_A s\, dA. \tag{2.2}$$

We represent this true charge density s by a vector $\mathbf{D}$, the *electric flux density* (or *dielectric displacement*), such that the surface charge density shall be equal to the normal component of $\mathbf{D}$, or

$$s\, da \equiv D \cos \alpha\, dA \equiv \mathbf{D} \cdot \mathbf{n}\, dA = D_n\, dA. \tag{2.3}$$

A positive value of the *scalar* or *dot product* of the vector $\mathbf{D}$ and the unit normal vector $\mathbf{n}$ indicates a positive charge.

Similarly, we allocate to the free charge density s/κ' a vector $\mathbf{E}$, the *electric field strength* or *field intensity,* by defining

$$\frac{s}{\kappa'}\, da \equiv \epsilon_0 \mathbf{E} \cdot \mathbf{n}\, dA = \epsilon_0 E_n\, dA \tag{2.4}$$

and to the bound charge density a vector $\mathbf{P}$, called the *polarization,* as

$$s\left(1 - \frac{1}{\kappa'}\right) da \equiv \mathbf{P} \cdot \mathbf{n}\, dA = P_n\, dA \tag{2.5}$$

(Fig. 2.2).† From Eqs. 2.3 and 2.4 we obtain the relation between dielectric flux density and field strength

$$\mathbf{D} = \epsilon' \mathbf{E}, \tag{2.6}$$

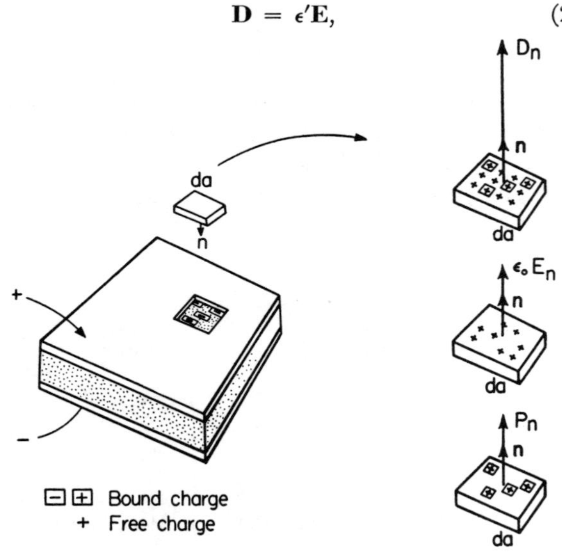

Fig. 2.2. Representation of total, free and bound charge densities by field vectors.

[1] M. Faraday, *Phil. Trans.*, 1837–1838.

† By postulating that the surface charge densities are equal to the normal components of the field vectors instead of 4π times their magnitude, we have decided on a rationalized system of units.

and from Eqs. 2.3 to 2.5 the interrelation between the three field vectors

$$\mathbf{P} = \mathbf{D} - \epsilon_0\mathbf{E} = (\epsilon' - \epsilon_0)\mathbf{E} \equiv \chi\epsilon_0\mathbf{E}. \tag{2.7}$$

The factor

$$\chi = \frac{\mathbf{P}}{\epsilon_0\mathbf{E}} = \kappa' - 1 = \frac{\text{bound charge density}}{\text{free charge density}} \tag{2.8}$$

is known as the *electric susceptibility* of the dielectric material.

Electric flux density **D** and polarization **P** have, according to their defining equations, the dimension "charge per unit area," whereas the electric field strength **E** can have a different physical meaning because the dimensions of the dielectric constant may yet be chosen. We take advantage of this possibility and extend the concept of the electric field into space by postulating that an electric probe charge Q', placed in an electrostatic field of the intensity **E**, is subjected to a force

$$\mathbf{F} \equiv Q'\mathbf{E}. \tag{2.9}$$

Thus the electric field strength **E** becomes equivalent in magnitude and direction to the force per unit charge acting on a detector charge, and the dielectric constant obtains the dimensions

$$[\epsilon] = \left[\frac{\text{charge per unit area}}{\text{force per unit charge}}\right] \tag{2.10}$$

Two electric charges of opposite polarity, $\pm Q$, separated by a distance d, represent a *dipole* of the moment

$$\boldsymbol{\mu} = Q\mathbf{d}; \tag{2.11}$$

this *electric dipole moment* is symbolized by a vector of

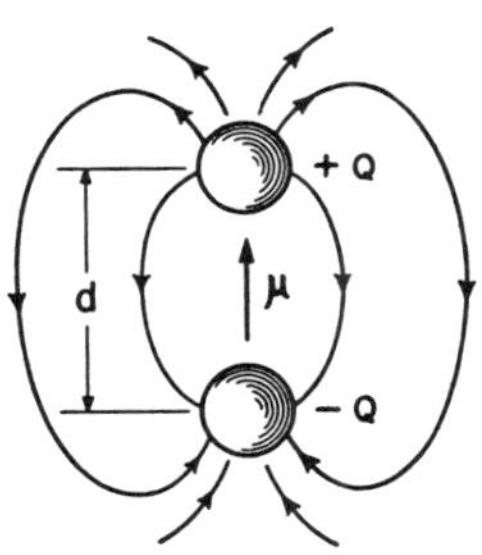

Fig. 2.3. Electric dipole of the moment $\mu = Qd$.

the magnitude $|\boldsymbol{\mu}|$ pointing from the negative to the positive pole (Fig. 2.3).† The polarization vector **P** corresponds in magnitude to the surface charge density bound at the electrodes by the polarized dielectric, and it points in the direction of the applied field. The polarization **P** is therefore obviously identical with the *electric dipole moment per unit volume* of the dielectric material (Fig. 2.4).

† Chemists frequently represent dipole moments by vectors pointing from the positive to the negative charge (+ → −); this convention is incompatible with the definition of P and should be abolished.

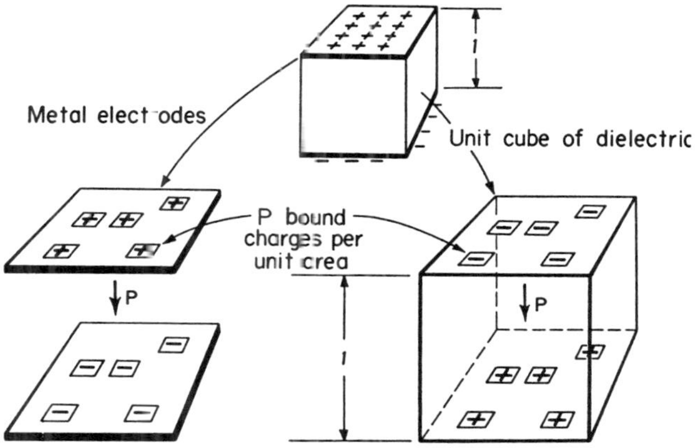

Fig. 2.4. **P** designating both bound charge density and dipole moment per unit volume.

The electrostatic field in space obtains physical meaning because the field strength at any point can be measured by the force acting on a detector charge (Eq. 2.9). Alternatively, it could be measured by the torque **T** exercised by the electric field **E** on an electric dipole **μ** as

$$\mathbf{T} = |\boldsymbol{\mu}|\,|\mathbf{E}|\sin\theta = \boldsymbol{\mu}\times\mathbf{E}, \tag{2.12}$$

which tends to align this dipole in field direction (Fig. 2.5).

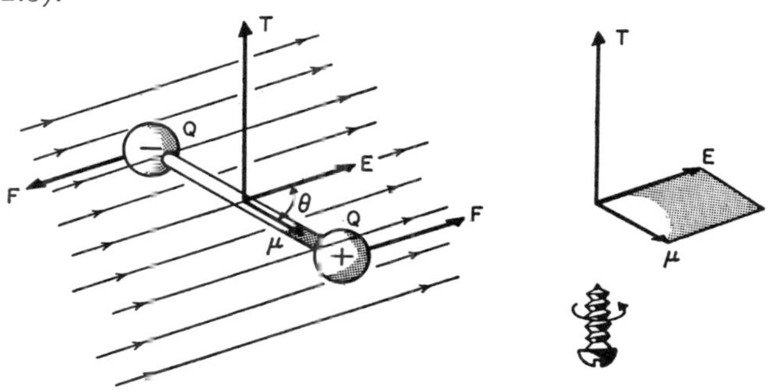

Fig. 2.5. Torque acting on electric dipole.

The concepts developed for the electrostatic field apply for the magnetostatic field with the restriction that individual magnetic point charges of north and south polarity are not known to exist in nature. Hence the *magnetic dipole moment*

$$\mathbf{m} = p\mathbf{d} \qquad (p = \text{pole strength}) \tag{2.13}$$

(Fig. 2.6) is the starting point of the theory. Visualizing that under the influence of a magnetic field **H** magnetic dipole chains form in a dielectric in analogy to the electric polarization of Fig. 2.1, we can introduce a *magnetization* vector **M**, which represents the *magnetic dipole moment per unit volume* of the material. This magnetization **M**, together with the *magnetic field* strength **H** in the dielectric, determines the total *magnetic flux density* (*magnetic induction*) **B**.

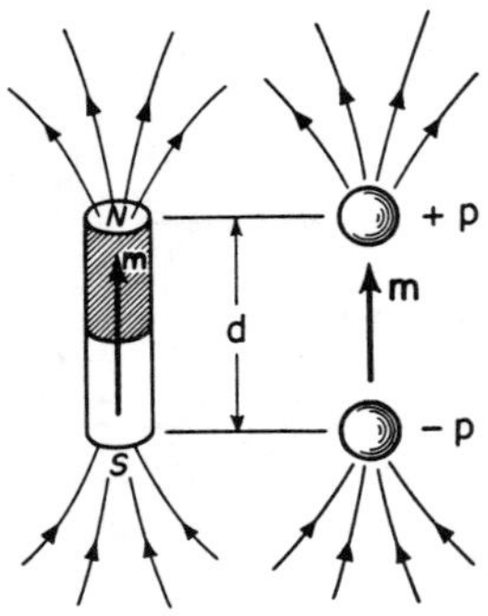

Fig. 2.6. Magnetic dipole.

Unfortunately, at this point the magnetic field theory deviates in its mathematical formulation from the electric theory, causing a great deal of confusion. Electric flux density **D** and polarization **P** have identical dimensions ("surface charge density" or "electric moment per unit volume"), and the interrelating equation

$$\mathbf{D} = \epsilon_0 \mathbf{E} + \mathbf{P} = \epsilon' \mathbf{E} \tag{2.14}$$

allows us to define the electric field strength **E** as "force per unit charge" or "torque per unit dipole moment" by the proper choice of the dimensions of the permittivity (see Eq. 2.10). The magnetic flux density **B**, however, is defined by the equation

$$\mathbf{B} \equiv \mu_0 \mathbf{H} + \mu_0 \mathbf{M} = \mu' \mathbf{H}, \tag{2.15}$$

where the factors μ' and μ_0 represent the *permeability* (*inductive capacity*) of the material and of vacuum, respectively. Thus the magnetic field strength **H** obtains the same dimension ("magnetic moment per unit volume") as the magnetization **M**, and only the magnetic induction **B** can acquire, by a proper choice of the dimensions of the permeability, the meaning of "torque per unit dipole moment." Thus **B** appears in the force and torque equations of the magnetic field, but not **H**, and the magnetic analogue to Eq. 2.12 is

$$\mathbf{T} = |\mathbf{m}|\,|\mathbf{B}| \sin\theta = \mathbf{m} \times \mathbf{B}. \tag{2.16}$$

Rewriting Eq. 2.15 for the magnetization, we obtain

$$\mathbf{M} = \frac{1}{\mu_0}\mathbf{B} - \mathbf{H} \equiv \chi_m \mathbf{H}, \tag{2.17}$$

and define, in analogy to the electric susceptibility of Eq. 2.8, a *magnetic susceptibility*

$$\chi_m \equiv \frac{\mathbf{M}}{\mathbf{H}} = \frac{\mu'}{\mu_0} - 1 \equiv \kappa_m' - 1; \tag{2.18}$$

κ_m' is the *relative permeability*.

3 · Coulomb and Dipole Fields

The electrostatic **E**-field and force **F** originate in free electric charges and are introduced as pointing from the positive to the negative polarity. Once created by the separation of positive and negative charge carriers, an electrostatic field can exist without further energy supply; it has a *conservative* character like the gravitational field of mechanics. When a probe charge Q' is moved from a point 1 to a point 2 in space, the field performs the work

$$W = \int_1^2 \mathbf{F} \cdot d\mathbf{l} = Q' \int_1^2 \mathbf{E} \cdot d\mathbf{l}. \tag{3.1}$$

The work per unit charge done *by* the field is called the *electromotive force* (emf), whereas the work per unit charge done *against* the field is designated as the *voltage* $\mathcal{V}$ between the two points, or

$$\text{emf} = \int_1^2 \mathbf{E} \cdot d\mathbf{l} = -\mathcal{V}. \tag{3.2}$$

Work done by the field results in dissipation, work done against the field, in storage of potential energy.

By definition, the potential energy of a charge at infinity is zero. The work per unit charge done against the field in moving a positive detector charge from infinity to any point in space is called the *potential* ϕ of that point. In consequence,

$$\left.\begin{aligned} \phi_1 &\equiv -\int_\infty^1 \mathbf{E} \cdot d\mathbf{l} \\ \phi_2 &\equiv -\int_\infty^2 \mathbf{E} \cdot d\mathbf{l} \end{aligned}\right\} \quad \phi_1 - \phi_2 = \int_1^2 \mathbf{E} \cdot d\mathbf{l}. \tag{3.3}$$

The *potential difference* $\phi_1 - \phi_2$ between two points is equal to the emf (Fig. 3.1). A surface normal to the field lines ($\mathbf{E} \cdot d\mathbf{l} = 0$) is an *equipotential surface.* A charge can be moved along such a surface without energy expense, and it describes an *equipotential line.*

The work done in an electrostatic field depends only on the initial and final position of the probe charge, but not on the path traversed, because sections of equipotential lines can be inserted at will without changing

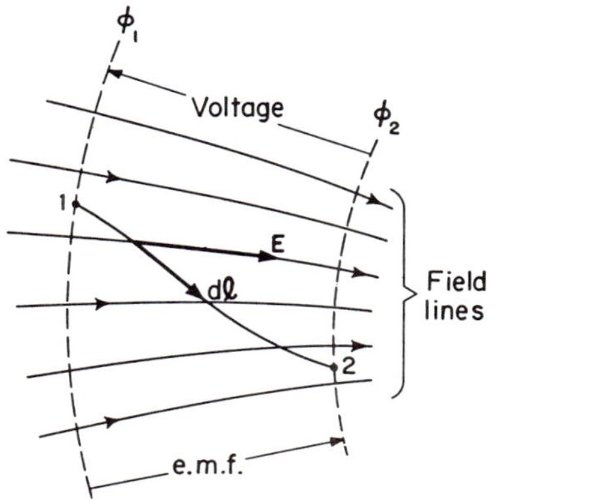

Fig. 3.1. Potential difference, emf, and voltage.

the result. The total work done over a closed path is zero, or

$$\oint \mathbf{E}\cdot d\boldsymbol{l} = 0. \tag{3.4}$$

Equation 3.4 defines the electrostatic field as *conservative* in contrast to the whirlpool fields encountered in the electromagnetic theory.

In an electrostatic field the volume of a conductor must remain field-free and its surface must remain an equipotential surface in order that its mobile charge carriers stay at rest. This is the principle of the *Faraday cage.* A charge Q placed on a conducting sphere of radius r in field-free space therefore spreads uniformly over the surface with a charge density

$$s = Q/4\pi r^2. \tag{3.5}$$

This surface charge density creates an electric field strength at the surface (see Eq. 2.4)

$$\mathbf{E} = \frac{Q}{\epsilon' 4\pi r^2}\,\mathbf{r}^0, \tag{3.6}$$

where ϵ' is the dielectric constant of the surrounding medium and $\mathbf{r}^0$ a unit vector. The field extends radially into space as may be verified by the force acting on a detector charge (see Eq. 2.9). Hence the potential of the sphere is (see Eq. 3.3)

$$\phi = -\int_\infty^r \mathbf{E}\cdot d\mathbf{r} = Q/\epsilon' 4\pi r, \tag{3.7}$$

and the force acting at the surface on a detector charge Q' equals

$$\mathbf{F} = \frac{Q'Q}{\epsilon' 4\pi r^2}\,\mathbf{r}^0. \tag{3.8}$$

Equation 3.8 is *Coulomb's law.*[1] It serves frequently as the starting point of the electrostatic field theory. When written without the factor 4π in the denominator, it commits us to the unrationalized system of units. Coulomb's law is a special case of the general force law (Eq. 2.9) written for spherical symmetry. The intensity of a *Coulomb field* varies therefore inversely with the square of the distance, and a *Coulomb potential* decreases inversely with the distance. Electrostatic fields produced by any arbitrary charge distribution may be considered to originate from the superposition of the Coulomb fields of point charges.

A simple application of this principle is the derivation of the field of an electric dipole (see Fig. 2.3) from the Coulomb field of its opposite charges. Obviously, we may add either the field vectors of the two elemental fields vectorially or the potentials arithmetically and derive the field strength by differentiation. The latter is the simpler procedure and gives directly for any point

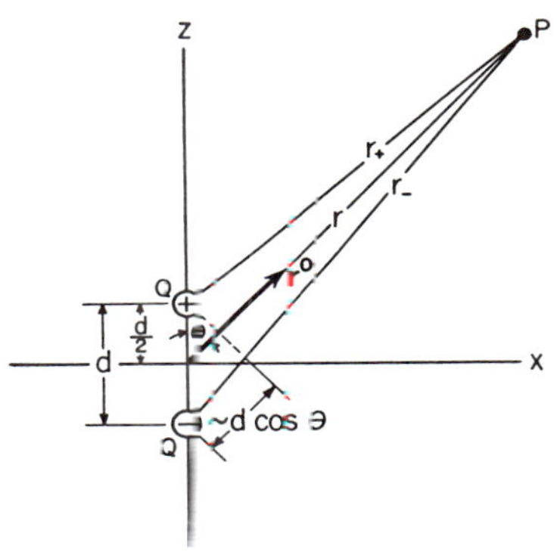

Fig. 3.2. Dipole at origin producing potential at point P.

P, located in space at distances r_+ and r_- from the two charges (Fig. 3.2), the potential

$$\phi_P = \frac{Q}{\epsilon' 4\pi}\left(\frac{1}{r_+} - \frac{1}{r_-}\right). \tag{3.9}$$

For $r \gg d$, $r_- - r_+ \simeq d\cos\theta$ and $r_- r_+ \simeq r^2$, hence

$$\phi_P = \frac{Qd}{\epsilon' 4\pi r^2}\cos\theta = \frac{\boldsymbol{\mu}\cdot\mathbf{r}^0}{\epsilon' 4\pi r^2}. \tag{3.10}$$

The unit vector $\mathbf{r}^0$ points from the center of the dipole towards P.

The field of a dipole has cylindrical symmetry and may therefore be conveniently described either by the

[1] C. A. Coulomb, *Histoire de l'académie royale des sciences,* Paris, 1785.

polar co-ordinates r and θ or by the cylindrical co-ordinates ρ and z, where

$$\cos\theta = z/r \qquad r = \sqrt{\rho^2 + z^2} \tag{3.11}$$

(Fig. 3.3). By introducing the latter co-ordinates, Eq. 3.10 of the dipole potential may be rewritten

$$\phi_P = \frac{|\boldsymbol{\mu}|z}{\epsilon' 4\pi(\rho^2 + z^2)^{3/2}}. \tag{3.12}$$

The electric field strength vector **E**, according to the defining Eq. 3.3 for the potential, points normal to the

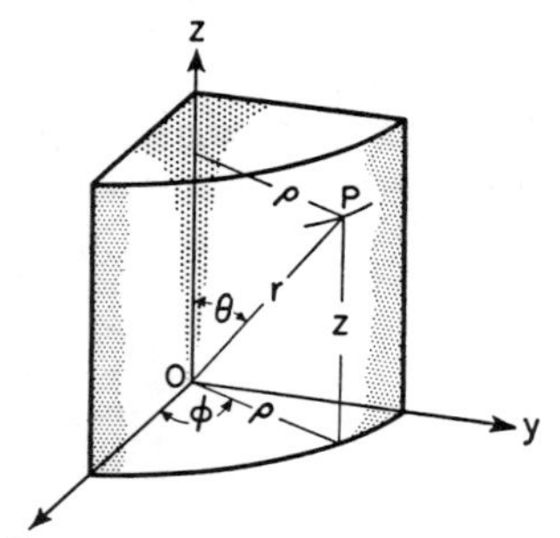

Fig. 3.3. Spherical and cylindrical co-ordinates.

equipotential surfaces in the direction of the steepest potential fall. To determine **E** from ϕ is a special case of a general mathematical problem: to find the maximum rate of increase of some function at any point of space (its *gradient*). On the slope of a mountain, for example, the gradient of the altitude at any point indicates the direction of steepest ascent, and its magnitude equals the maximum rate of change in altitude over an elemental length of distance at that point. If ∂n denotes an element of length normal to an equipotential surface and $\mathbf{n}^0$ a unit vector pointing in the direction of steepest potential rise, the electric field strength is

$$\mathbf{E} = -\mathbf{n}^0\left(\frac{\partial\phi}{\partial n}\right) \equiv -\nabla\phi. \tag{3.13}$$

The differential operator ∇, called the *del operator* because its symbol is an inverted Greek capital delta, denotes the gradient and may be expressed in various ways, depending on the system of co-ordinates chosen (see also Appendix B).

In addition to the normal Cartesian co-ordinates we shall have to refer at times to cylindrical or spherical co-ordinates or to general co-ordinates which may correspond to any one of these orthogonal co-ordinate systems (Table 3.1).

Table 3.1. Orthogonal co-ordinate systems and their geometrical relations

(See Fig. 3.3)

General co-ordinates	u	v	w
Cartesian	x	y	z
Circular cylindrical	ρ	ϕ	z
Spherical	r	ϕ	θ

Geometrical relations

Cylindrical	*Spherical*
$x = \rho\cos\phi$	$x = r\sin\theta\cos\phi$
$y = \rho\sin\phi$	$y = r\sin\theta\sin\phi$
$z = z$	$z = r\cos\theta$

In Cartesian co-ordinates a vector increment of length dl is expressed by

$$dl = \mathbf{i}\,dx + \mathbf{j}\,dy + \mathbf{k}\,dz$$

of the magnitude

$$|dl| = \sqrt{(dx)^2 + (dy)^2 + (dz)^2}; \tag{3.14}$$

i, **j**, and **k** are unit vectors pointing in the positive x, y, and z directions, respectively. In general co-ordinates u, v, w, we have to write instead

$$dl = \mathbf{u}^0\alpha\,du + \mathbf{v}^0\beta\,dv + \mathbf{w}^0\gamma\,dw,$$

$$|dl| = \sqrt{\alpha^2(du)^2 + \beta^2(dv)^2 + \gamma^2(dw)^2}. \tag{3.15}$$

The vectors $\mathbf{u}^0$, $\mathbf{v}^0$, $\mathbf{w}^0$ are unit vectors of any one of the systems in Table 3.1, and α, β, γ are scale factors converting the length of the step taken in the u, v, and w directions to the uniform scale of a local Cartesian system. Table 3.2 shows the interrelations derived from Table 3.1 and these equations.

Table 3.2. Interrelation of unit vectors and scale factors of the co-ordinate systems

General Symbol	Cartesian	Cylindrical	Spherical
$\mathbf{u}^0$	$\mathbf{i}$	$\boldsymbol{\rho}^0$	$\mathbf{r}^0$
$\mathbf{v}^0$	$\mathbf{j}$	$\boldsymbol{\phi}^0$	$\boldsymbol{\phi}^0$
$\mathbf{w}^0$	$\mathbf{k}$	$\mathbf{k}$	$\boldsymbol{\theta}^0$
α	1	1	1
β	1	ρ	$r\sin\theta$
γ	1	1	r
$\|dl\|$	$\sqrt{(dx)^2 + (dy)^2 + (dz)^2}$	$\sqrt{(d\rho)^2 + \rho^2(d\phi)^2 + (dz)^2}$	$\sqrt{(dr)^2 + (r\sin\theta)^2(d\phi)^2 + r^2(d\theta)^2}$

As we move from a point of the potential field comparable distances in the directions of the co-ordinates u, v, w, the potential changes by the amounts

$$\frac{\partial \phi}{\partial l_u} = \frac{1}{\alpha}\frac{\partial \phi}{\partial u}, \quad \frac{\partial \phi}{\partial l_v} = \frac{1}{\beta}\frac{\partial \phi}{\partial v}, \quad \frac{\partial \phi}{\partial l_w} = \frac{1}{\gamma}\frac{\partial \phi}{\partial w}. \tag{3.16}$$

The gradient is the vector sum of these three quantities

$$\text{grad}\ \phi = \nabla\phi = \frac{\mathbf{u}^0}{\alpha}\frac{\partial \phi}{\partial u} + \frac{\mathbf{v}^0}{\beta}\frac{\partial \phi}{\partial v} + \frac{\mathbf{w}^0}{\gamma}\frac{\partial \phi}{\partial w}. \tag{3.17}$$

Hence the differential operator ∇, the del operator, becomes, in the three co-ordinate systems,

$$\nabla \begin{cases} \text{Cartesian} & = \mathbf{i}\dfrac{\partial}{\partial x} + \mathbf{j}\dfrac{\partial}{\partial y} + \mathbf{k}\dfrac{\partial}{\partial z}, \\ \text{Cylindrical} & = \boldsymbol{\rho}^0\dfrac{\partial}{\partial \rho} + \dfrac{\boldsymbol{\phi}^0}{\rho}\dfrac{\partial}{\partial \phi} + \mathbf{k}\dfrac{\partial}{\partial z}, \\ \text{Spherical} & = \mathbf{r}^0\dfrac{\partial}{\partial r} + \dfrac{\boldsymbol{\phi}^0}{r\sin\theta}\dfrac{\partial}{\partial \phi} + \dfrac{\boldsymbol{\theta}^0}{r}\dfrac{\partial}{\partial \theta}. \end{cases} \tag{3.18}$$

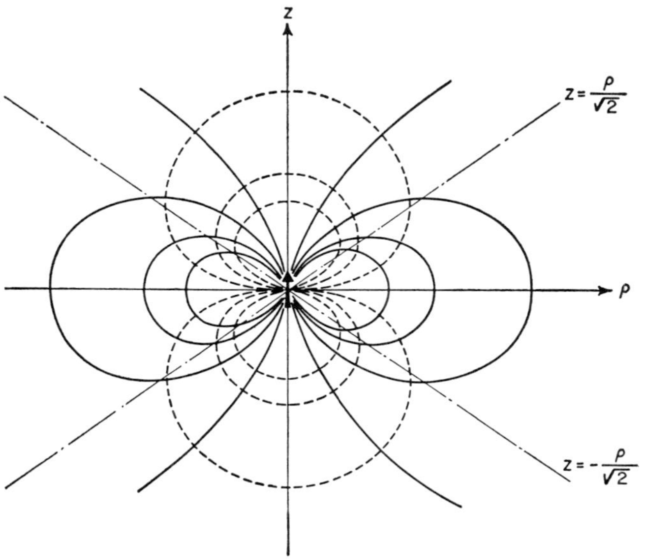

Fig. 3.4. Dipole field (—— field lines, ------ equipotential lines). (After Ollendorff.[2])

Making use of the del operation, we obtain for the field strength of the dipole field in spherical co-ordinates

$$E_r = -\frac{\partial \phi_P}{\partial r} = \frac{|\boldsymbol{\mu}|\cos\theta}{\epsilon' 2\pi r^3} = \frac{\boldsymbol{\mu}\cdot\mathbf{r}^0}{\epsilon' 2\pi r^3},$$

$$E_\theta = -\frac{1}{r}\frac{\partial \phi_P}{\partial \theta} = \frac{|\boldsymbol{\mu}|\sin\theta}{\epsilon' 4\pi r^3}, \tag{3.19}$$

and, in cylindrical co-ordinates,

$$E_z = -\frac{\partial \phi_P}{\partial z} = -\frac{|\boldsymbol{\mu}|}{\epsilon' 4\pi}\frac{\rho^2 - 2z^2}{(\rho^2 + z^2)^{5/2}},$$

$$E_\rho = -\frac{\partial \phi_P}{\partial \rho} = \frac{|\boldsymbol{\mu}|}{\epsilon' 4\pi}\frac{3\rho z}{(\rho^2 + z^2)^{5/2}}. \tag{3.20}$$

The field intensity of a dipole varies inversely with the third power of the distance, and its axial component E_z vanishes along the conical surfaces $z = \pm\rho/\sqrt{2}$ (Fig. 3.4.[2]

Placed in an electric field, the two charges of a dipole experience forces **F** in opposite directions. In a homogeneous field these forces are equal in magnitude; hence only a torque **T** results, tending to align the dipole in the field direction. We have already indicated in Fig. 2.5 and Eq. 2.12 that the torque,

$$\mathbf{T} = |\boldsymbol{\mu}|\,|\mathbf{E}|\sin\theta \equiv \boldsymbol{\mu}\times\mathbf{E}, \tag{3.21}$$

can be represented by a vector, the *cross product* $\boldsymbol{\mu}\times\mathbf{E}$. The length of **T** equals the area of the parallelogram formed by the dipole moment and the field strength, whereas its direction coincides with the direction of motion of a right-hand screw when turned from $\boldsymbol{\mu}$ towards **E**.

The potential energy of a dipole in an external field, with the charges $+Q$ and $-Q$ at potentials ϕ_2 and ϕ_1, is

$$U = Q(\phi_2 - \phi_1) \tag{3.22}$$

(see Eq. 3.3). Since the length of a dipole, and therefore the potential drop across it, is, in general, very small,

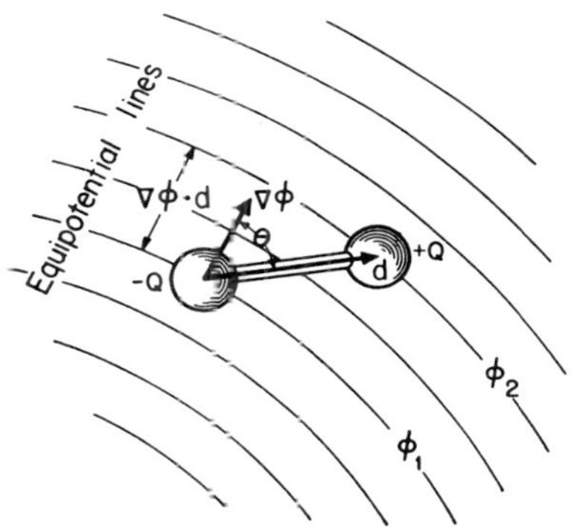

Fig. 3.5. Dipole in external field.

$$\phi_2 \simeq \phi_1 + \nabla\phi\cdot\mathbf{d} \tag{3.23}$$

(Fig. 3.5), this energy expression reduces to

$$U = +\nabla\phi\cdot\boldsymbol{\mu} = -\boldsymbol{\mu}\cdot\mathbf{E} = -|\boldsymbol{\mu}|\,|\mathbf{E}|\cos\theta. \tag{3.24}$$

The dipole will tend to rotate until its potential energy in the external field reaches its minimum at $\theta = 0$, with the moment lined up in the field direction.

In addition to this *orientation by rotation*, a dipole will undergo a translation motion in case the external field is inhomogeneous over its length. The general forces in conservative fields, the sum total of translational

[2] F. Ollendorff, *Potentialfelder der Elektrotechnik*, Springer, Berlin, 1932.

forces **F** and torques **T**, are equal to the negative gradient of the potential energy U. Hence, for our dipole of constant moment, and, since $|\mathbf{E}|$ is a scalar,

$$\mathbf{F} + \mathbf{T} = -\nabla U = -|\boldsymbol{\mu}|\nabla(|\mathbf{E}|\cos\theta)$$
$$= -|\boldsymbol{\mu}|\cos\theta\nabla|\mathbf{E}| + |\boldsymbol{\mu}|\,|\mathbf{E}|\sin\theta. \quad (3.25)$$

The second part of this expression is our previous torque equation (3.21); the first part shows that the *translational force* acting on a dipole of fixed orientation is proportional to the gradient of the field strength.

To arrive at the analogous expressions for a magnetic dipole, we have only to replace the electric moment $\boldsymbol{\mu}$ by the magnetic moment **m** and the electric field strength **E** by the magnetic induction **B** (see Eq. 2.16). The translational force or the torque acting on a dipole in a field can obviously serve for the determination of its dipole moment.

4 · Space-Charge Fields

As in the case of two charges, the potential created at a point P by a *space charge cloud* distributed over a volume V with a charge density ρ can be derived by the superposition of Coulomb potentials as

$$\phi_P = \int_V \frac{\rho_F\, dV}{\epsilon' 4\pi l}. \quad (4.1)$$

Here ρ_F designates the space charge density at a point F of the cloud and l the distance from F to the reference point P. Also the field strength **E** may be found from the charge distribution by integration, because the charge contained in a volume can be measured by counting the total normal flux of lines that emerges from an enclosing surface, or

$$\oint_A \epsilon_0 \mathbf{E}\cdot\mathbf{n}\, dA = \int_V \frac{\rho_F\, dV}{\kappa'}. \quad (4.2)$$

This follows as a logical extension of Eq. 2.4, which establishes a quantitative connection between free charge density and field strength.

Equation 4.2 is a special case of the transformation of a volume into a surface integral known as *Gauss's law*.[1] It states in mathematical language the obvious fact that the difference in the number of deer tracks entering and leaving a parcel of previously empty woodland indicates how many deer are still in the stand.

In general, it is more convenient to derive field distributions by differentiation. We have already made use of this possibility in the previous section, by employing the del operator to obtain the field strength as the negative gradient of the potential or forces and torques as that of the potential energy.

To find the differential relation between field strength and charge density, we apply Gauss's law to an elemental cube $dV = dx\,dy\,dz$ (Fig. 4.1). Here the normal component of the field flux is constant over each side of the cube, and the field strength changes by its first derivative when we pass to the opposite side. The surface integral over the normal flux emanating from the elemental cube may therefore be replaced by the summation

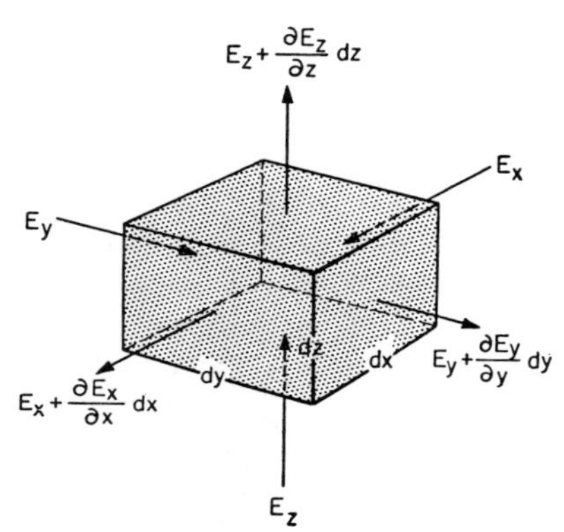

Fig. 4.1. Derivation of divergence in Cartesian co-ordinates.

$$\oint_{\substack{\text{elemental}\\ \text{cube}}} \mathbf{E}\cdot\mathbf{n}\, dA = \left(\frac{\partial E_x}{\partial x} + \frac{\partial E_y}{\partial y} + \frac{\partial E_z}{\partial z}\right) dV$$
$$= \frac{\rho}{\epsilon'}\, dV. \quad (4.3)$$

In Cartesian co-ordinates the field vector **E** and the del operator ∇ (see Eq. 3.18) are given as

$$\mathbf{E} = \mathbf{i}E_x + \mathbf{j}E_y + \mathbf{k}E_z$$

and

$$\nabla = \mathbf{i}\frac{\partial}{\partial x} + \mathbf{j}\frac{\partial}{\partial y} + \mathbf{k}\frac{\partial}{\partial z}, \quad (4.4)$$

where **i**, **j**, and **k** designate the unit vectors in the x,

[1] C. F. Gauss, *Werke*, Göttingen, 1867, Vol. 5, pp. 5–7.

y, and z directions. Hence the sum in Eq. 4.3 corresponds to the vector operation

$$\nabla \cdot \mathbf{E} = \frac{\partial E_x}{\partial x} + \frac{\partial E_y}{\partial y} + \frac{\partial E_z}{\partial z} \equiv \text{div } E, \tag{4.5}$$

called the *divergence of* **E**.

To express the divergence of the vector **E** in the general orthogonal co-ordinates u, v, w (see Sec. 3), we refer to the parallelepiped

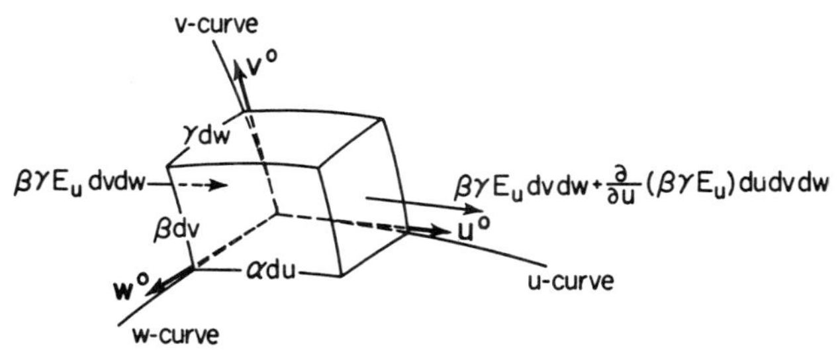

Fig. 4.2. Derivation of divergence in general orthogonal co-ordinates.

(Fig. 4.2) formed by surfaces of constant u, v, and w. The contributions of the field strength

$$\mathbf{E} = \mathbf{u}^0 E_u + \mathbf{v}^0 E_v + \mathbf{w}^0 E_w \tag{4.6}$$

to the normal flux through the elementary surfaces u = constant, and $u + du$ = constant are $-\beta\gamma E_u \, dv \, dw + \frac{\partial}{\partial u}(\beta\gamma E_u)\, du \, dv \, dw$. After computing the corresponding expressions for the other two pairs of surfaces and recalling that the volume element dV in the general co-ordinates corresponds to

$$dV = \alpha\beta\gamma \, du \, dv \, dw, \tag{4.7}$$

we obtain from Gauss's divergence theorem the general expression for the divergence

$$\nabla \cdot \mathbf{E} = \frac{1}{\alpha\beta\gamma}\left[\frac{\partial}{\partial u}(\beta\gamma E_u) + \frac{\partial}{\partial v}(\gamma\alpha E_v) + \frac{\partial}{\partial w}(\alpha\beta E_w)\right]. \tag{4.8}$$

By referring to Tables 3.1 and 3.2 of Sec. 3, we arrive at the formulation of the divergence operator in the three co-ordinate systems

$$\nabla \cdot \mathbf{E} \longrightarrow \begin{cases} \text{Cartesian} & = \frac{\partial E_x}{\partial x} + \frac{\partial E_y}{\partial y} + \frac{\partial E_z}{\partial z}, \\ \text{Cylindrical} & = \frac{1}{\rho}\frac{\partial}{\partial \rho}(\rho E_\rho) + \frac{1}{\rho}\frac{\partial E_\phi}{\partial \phi} + \frac{\partial E_z}{\partial z}, \\ \text{Spherical} & = \frac{1}{r^2}\frac{\partial}{\partial r}(r^2 E_r) + \frac{1}{r \sin\theta}\frac{\partial E_\phi}{\partial \phi} \\ & \quad + \frac{1}{r\sin\theta}\frac{\partial}{\partial\theta}(E_\theta \sin\theta). \end{cases} \tag{4.9}$$

The divergence, by measuring the total flux of field lines per volume element entering and leaving the enclosing surface, measures the density of the free charge per volume element

$$\nabla \cdot \mathbf{E} = \rho/\epsilon'; \tag{4.10}$$

it characterizes the *sources* of the electrostatic field. By replacing in Gauss's law (Eq. 4.2) the charge density by the divergence operator, *Gauss's divergence theorem* is obtained,

$$\oint_A \mathbf{E} \cdot \mathbf{n} \, dA = \int_V \nabla \cdot \mathbf{E} \, dV. \tag{4.11}$$

Since the electric field strength is the negative gradient of the electrostatic potential,† the divergence relation between field strength and free charge density (Eq. 4.10) may be rewritten as a differential equation between the electrostatic potential and the free charge density,

$$\nabla \cdot \nabla \varphi \equiv \nabla^2 \varphi = -\rho/\epsilon'. \tag{4.12}$$

This is *Poisson's equation*,[2] which, for a region free of charge, simplifies to *Laplace's equation* [3]

$$\nabla^2 \varphi = 0. \tag{4.13}$$

The differential operator ∇^2, named *Laplace's operator* or the *Laplacian*, represents in Cartesian co-ordinates the operation

$$\nabla^2 = \frac{\partial^2}{\partial x^2} + \frac{\partial^2}{\partial y^2} + \frac{\partial^2}{\partial z^2}. \tag{4.14}$$

The Laplacian of the potential may be expressed in the general orthogonal co-ordinates by substituting in Eq. 4.8 the components of the field strength by the components of grad ϕ from Eq. 3.17. The result is

$$\nabla^2 \varphi = \frac{1}{\alpha\beta\gamma}\left[\frac{\partial}{\partial u}\frac{\beta\gamma}{\alpha}\frac{\partial \varphi}{\partial u} + \frac{\partial}{\partial v}\frac{\gamma\alpha}{\beta}\frac{\partial \varphi}{\partial v} + \frac{\partial}{\partial w}\frac{\alpha\beta}{\gamma}\frac{\partial \varphi}{\partial w}\right], \tag{4.15}$$

or, by referring to Tables 3.1 and 3.2 of Sec. 3,

$$\nabla^2 \varphi \longrightarrow \begin{cases} \text{Cartesian} & = \frac{\partial^2 \varphi}{\partial x^2} + \frac{\partial^2 \varphi}{\partial y^2} + \frac{\partial^2 \varphi}{\partial z^2}, \\ \text{Cylindrical} & = \frac{\partial^2 \varphi}{\partial \rho^2} + \frac{1}{\rho}\frac{\partial \varphi}{\partial \rho} + \frac{1}{\rho^2}\frac{\partial^2 \varphi}{\partial \phi^2} + \frac{\partial^2 \varphi}{\partial z^2}, \\ \text{Spherical} & = \frac{1}{r^2}\frac{\partial}{\partial r}\left(r^2 \frac{\partial \varphi}{\partial r}\right) + \frac{1}{r^2 \sin^2\theta}\frac{\partial^2 \varphi}{\partial \phi^2} \\ & \quad + \frac{1}{r^2 \sin\theta}\frac{\partial}{\partial \theta}\left(\sin\theta \frac{\partial \varphi}{\partial \theta}\right) \\ & = \frac{1}{r}\frac{\partial^2 (r\varphi)}{\partial r^2} + \frac{1}{r^2 \sin^2\theta}\frac{\partial^2 \varphi}{\partial \phi^2} \\ & \quad + \frac{1}{r^2 \sin\theta}\frac{\partial}{\partial \theta}\left(\sin\theta \frac{\partial \varphi}{\partial \theta}\right). \end{cases} \tag{4.16}$$

To illustrate how we actually handle space charge problems by the integral and the differential approach, we shall calculate the electrostatic field produced by a

† In the present discussion electrostatic potential is represented by φ instead of by ϕ to avoid confusion with the angle ϕ of the cylindrical and spherical co-ordinates.

[2] S. D. Poisson, *Nouveau bull. soc. philomath., Paris, 3*, 388 (1813).

[3] P. S. Laplace, *Ouvres de Laplace*, Gauthier-Villain, Paris, 1782–1786, Vol. 10, p. 312.

spherical space charge cloud of the radius r_0 and of a constant charge density ρ:

Method I (Gauss's law)

The surface integral $\oint \epsilon_0 \mathbf{E}\cdot\mathbf{n}\, dA$, extended over a concentric sphere of radius r, measures the free charge $\oint_V \frac{\rho}{\kappa'}\, dV$ enclosed in the volume of the sphere (see Eq. 4.2). For a sphere smaller than the space-charge cloud ($r \leqslant r_0$) we obtain

$$\epsilon_0 E_r 4\pi r^2 = \frac{\rho}{\kappa'}\frac{4\pi}{3} r^3 \tag{4.17}$$

that is, the field strength *inside* the cloud

$$\mathbf{E}_{r_i} = \frac{\rho}{\epsilon'}\frac{\mathbf{r}}{3} \tag{4.18}$$

rises proportionally to the distance from its center. If, on the other hand, we integrate over the surface of a sphere that is larger than the cloud ($r \geqslant r_0$), the right-hand side of Eq. 4.17 assumes the constant value $\frac{\rho}{\kappa'}\frac{4\pi}{3} r_0^3$; hence the field strength *outside* the space charge sphere,

$$\mathbf{E}_{r_a} = \frac{\rho}{\epsilon'}\frac{r_0^3}{3r^2}\mathbf{r}^0 \tag{4.19}$$

falls as $1/r^2$ (Fig. 4.3).

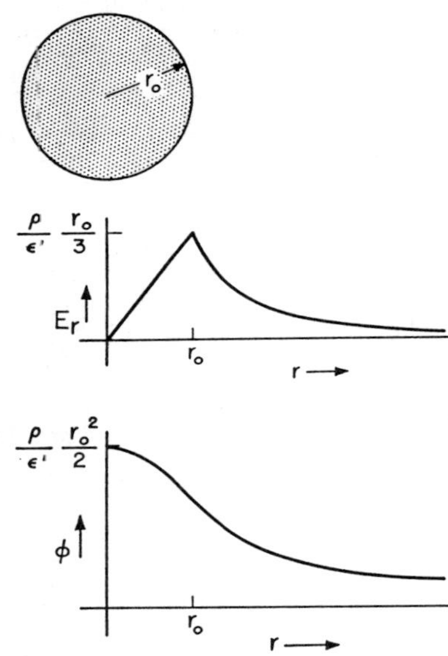

Fig. 4.3. Field strength and potential produced by spherical space-charge cloud of constant density ρ.

The electrostatic potential, derived from the field strength by integration (see Eq. 3.3), is *inside* the cloud

$$\varphi_i = -\int_\infty^r E_r\, dr = -\frac{\rho}{3\epsilon'}\left[\int_\infty^{r_0}\frac{r_0^3}{r^2}\, dr + \int_{r_0}^r r\, dr\right]$$

$$= \frac{\rho}{6\epsilon'}(3r_0^2 - r^2), \tag{4.20}$$

and *outside*

$$\varphi_a = -\frac{\rho}{3\epsilon'}\int_\infty^r \frac{r_0^3}{r^2}\, dr = \frac{\rho}{3\epsilon'}\frac{r_0^3}{r}. \tag{4.21}$$

Method II (superposition of the Coulomb potentials of point charges)

According to Eq. 4.1, the potential produced by the space charge is the sum of the Coulomb potentials of the individual charge elements. Let us designate with r_s the distance of a charge element dV from the cloud center, with l its distance from the reference point P (inside or outside the cloud), and with r the distance of the reference point from the center (Fig. 4.4).

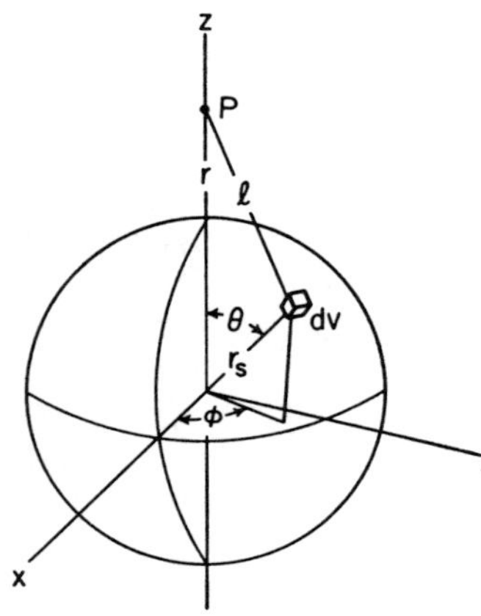

Fig. 4.4. Derivation of space-charge potential at point P by superposition of Coulomb potentials.

Expressing the element dV in spherical co-ordinates

$$dV = (r_s \sin\theta\, d\phi)(r_s\, d\theta)\, dr_s, \tag{4.22}$$

we obtain the potential at the reference point P as a triple integral

$$\varphi_P = \frac{\rho}{\epsilon' 4\pi}\int_0^{r_0}\int_0^{\pi}\int_0^{2\pi}\frac{r_s^2 \sin\theta\, dr_s\, d\theta\, d\phi}{l}. \tag{4.23}$$

The distance l according to the cosine law,

$$l^2 = r^2 + r_s^2 - 2r_p r_s \cos\theta, \tag{4.24}$$

can be used to substitute for the variable θ, since by differentiation

$$l\, dl = r r_s \sin\theta. \tag{4.25}$$

When θ varies from 0 to π, l varies for $r \leqslant r_s$ from $(r_s - r)$ to $(r_s + r)$, and for $r \geqslant r_s$ from $(r - r_s)$ to $(r + r_s)$. Hence the potential inside the cloud ($r < r_0$) becomes

$$\varphi_i = \frac{\rho}{\epsilon' 4\pi}\left[\int_0^r\int_{(r-r_s)}^{(r+r_s)}\int_0^{2\pi}\frac{r_s\, dr_s\, dl\, d\phi}{r} + \int_r^{r_0}\int_{(r_s-r)}^{(r_s+r)}\int_0^{2\pi}\frac{r_s\, dr_s\, dl\, d\phi}{r}\right]$$

$$= \frac{\rho}{6\epsilon'}(3r_0^2 - r^2), \tag{4.26}$$

and outside ($r > r_0$)

$$\varphi_a = \frac{\rho}{\epsilon' 4\pi}\int_0^{r_0}\int_{(r-r_s)}^{(r+r_s)}\int_0^{2\pi}\frac{r_s\, dr_s\, dl\, d\phi}{r} = \frac{\rho}{3\epsilon'}\frac{r_0^3}{r}, \tag{4.27}$$

in agreement with the solution previously obtained.

Method III (Poisson's equation)

Writing the Laplacian in spherical co-ordinates (see Eq. 4.16) and considering that the potential, for reasons of symmetry, can depend only on r, we obtain for the potential inside and outside the space-charge sphere the differential equations (Eqs. 4.12 and 4.13)

$$\frac{d^2\varphi_i}{dr^2} + \frac{2}{r}\frac{d\varphi_i}{dr} = -\frac{\rho}{\epsilon'},$$
$$\frac{d^2\varphi_a}{dr^2} + \frac{2}{r}\frac{d\varphi_a}{dr} = 0. \tag{4.28}$$

The solutions are

$$\varphi_i = -\frac{\rho r^2}{6\epsilon'} + A + \frac{B}{r},$$
$$\varphi_a = \frac{C}{r} + D. \tag{4.29}$$

To determine the constants of integration, we have the following boundary conditions:

(*a*) As $r \to 0$, the potential φ_i has to remain finite, hence $B = 0$.

(*b*) As $r \to \infty$, the potential φ_a has to go to zero, hence $D = 0$.

(*c*) The potential across the surface of the space-charge sphere must be continuous. Hence

$$\varphi_{r0} = -\frac{\rho}{6\epsilon'} r_0^2 + A = \frac{C}{r_0}. \tag{4.30}$$

(d) The field strength across this boundary must also be continuous; hence, since $E_{r_0} = -\nabla\varphi_{r_0} = -r_0 \frac{d\varphi_{r0}}{dr}$ (see Eq. 3.18),

$$E_{r_0} = -\frac{\rho}{3\epsilon'} r_0 = -\frac{C}{r_0^2}. \tag{4.31}$$

By solving the last two equations for A and C and introducing the values for the integration constants in Eqs. 4.29, we end again with the previous result.

Method I obviously allows solving the present problem in the most straightforward manner, but method III proves superior in more complicated cases.

5 · Correlations between Electric and Magnetic Phenomena

The experimental basis of the electrostatic field concept is the existence of positive and negative charges, of the force law (Eq. 2.9), of the phenomenon of polarization leading to the distinction between bound and free charges, and of conductors providing equipotential surfaces. The magnetostatic field concepts are based on the existence of magnetic dipoles, the torque law (Eq. 2.16), and on the phenomenon of magnetization. Free magnetism in the form of magnetic mono poles appears only at the boundaries between materials of different permeability. The H-field arising from free ends of dipole chains is conservative as is the electrostatic E-field (see Eq. 3.4)

$$\oint \mathbf{H}\cdot d\mathbf{l} = 0, \tag{5.1}$$

and concepts such as a scalar potential and a magnetomotive force (cf. Eq. 3.2)

$$\text{mmf} = \int_1^2 \mathbf{H}\cdot d\mathbf{l} \tag{5.2}$$

may be used for its description. The B-field, in contrast, has no sources, but consists of closed induction lines since the magnetostatic H-lines find their continuation in the dipole chains of the materials. The divergence equation (see Eq. 4.10)

$$\nabla\cdot\mathbf{D} = \rho \tag{5.3}$$

characterizes the true sources of the electrostatic field; the statement

$$\nabla\cdot\mathbf{B} = 0 \tag{5.4}$$

expresses the fact that no true magnetic mono poles exist in which the B-field could terminate.

Thus far, electricity and magnetism appear as two new phenomena independent of each other and consequently requiring the introduction of two new fundamental quantities, for example, "electric charge" and "magnetic dipole moment," for their description. Actually, they are interlinked, and only one new, independent quantity may be introduced. One interrelation is given by the fact that an electric current creates a magnetic field according to *Ampère's circuital law,*[1]

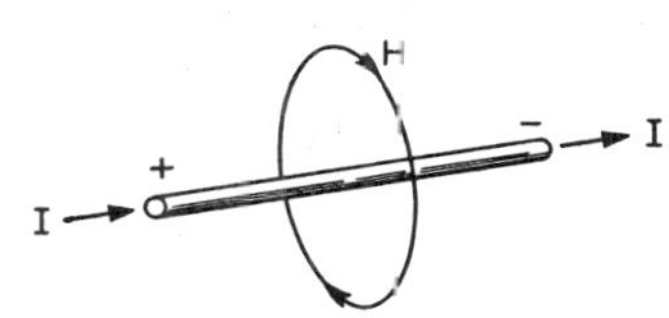

Fig. 5.1. Magnetic field encircling electric current.

$$\oint \mathbf{H}\cdot d\mathbf{l} = I \tag{5.5}$$

(Fig. 5.1). The magnetic field encircling an electric

[1] A. M. Ampère, *Recueil d'observations électrodynamiques* Crochard, Paris, 1820–1833.

current is a whirlpool field of closed lines in contrast to the magnetostatic field that originates in the free ends of dipole chains of a magnetic material. The line integral taken around a closed path is obviously not zero, but depends on the path itself and the direction of passage,

$$\oint \mathbf{H}\cdot d\boldsymbol{l} \neq 0. \tag{5.6}$$

Such *nonconservative* fields consequently may be described quantitatively at any point P in space by the value of this line integral taken along differential paths

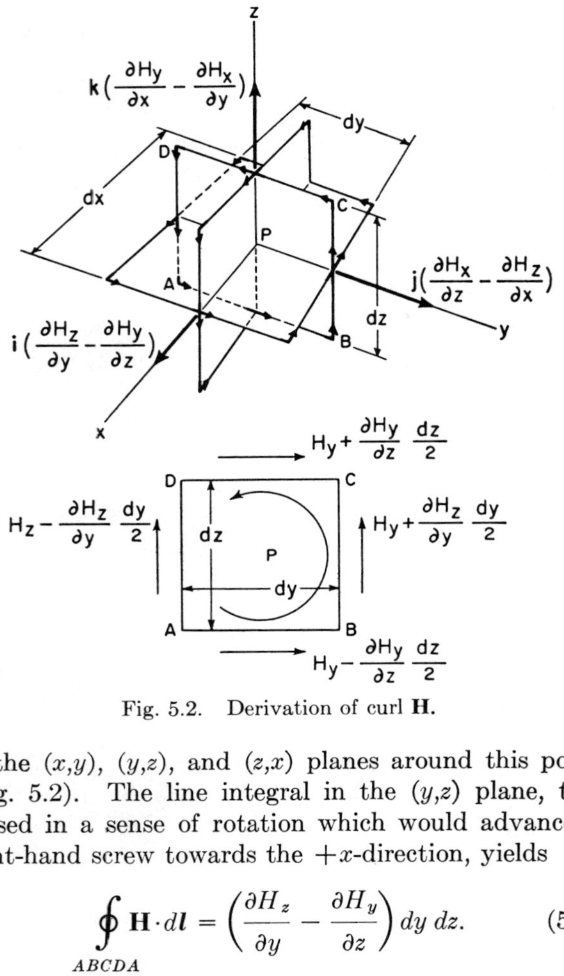

Fig. 5.2. Derivation of curl **H**.

in the (x,y), (y,z), and (z,x) planes around this point (Fig. 5.2). The line integral in the (y,z) plane, traversed in a sense of rotation which would advance a right-hand screw towards the $+x$-direction, yields

$$\oint_{ABCDA} \mathbf{H}\cdot d\boldsymbol{l} = \left(\frac{\partial H_z}{\partial y} - \frac{\partial H_y}{\partial z}\right) dy\, dz. \tag{5.7}$$

Divided by the differential area $dy\,dz$, it corresponds to the x-component of a vector curl **H**. By permutation of the indices, the y and z components, and thus the complete vector curl **H**, is obtained as

$$\text{curl}\,\mathbf{H} = \mathbf{i}\left(\frac{\partial H_z}{\partial y} - \frac{\partial H_y}{\partial z}\right) + \mathbf{j}\left(\frac{\partial H_x}{\partial z} - \frac{\partial H_z}{\partial x}\right) + \mathbf{k}\left(\frac{\partial H_y}{\partial x} - \frac{\partial H_x}{\partial y}\right). \tag{5.8}$$

It describes the turbulence of the H-field around P and may be rewritten alternatively in a determinant form or by use of the *del* operator:

$$\text{curl}\,\mathbf{H} = \begin{vmatrix} i & j & k \\ \frac{\partial}{\partial x} & \frac{\partial}{\partial y} & \frac{\partial}{\partial z} \\ H_x & H_y & H_z \end{vmatrix} = \nabla \times \mathbf{H}. \tag{5.9}$$

To calculate the curl operator in general orthogonal co-ordinates, we have to take line integrals along elementary curvilinear rectangles lying in planes normal to $\mathbf{u}^0$, $\mathbf{v}^0$, and $\mathbf{w}^0$. Except for differentials of higher order, the opposite sides of such a rectangle are equal. Consequently the line integral normal to the u co-ordinate becomes (see Eq. 3.15)

$$\oint_u \mathbf{H}\,dl = (H_v\beta\,dv)_w + (H_w\gamma\,dw)_{v+dv} - (H_v\beta\,dv)_{w+dw} - (H_w\gamma\,dw)_v,$$

or, since

$$(H_w\gamma\,dw)_{v+dv} = (H_w\gamma\,dw)_v + \frac{\partial}{\partial v}(H_w\gamma\,dw)_v\,dv, \text{ etc.,}$$

the u-component of the line integral is

$$\oint_u \mathbf{H}\cdot d\boldsymbol{l} = \frac{\partial}{\partial v}(H_w\gamma\,dw)\,dv - \frac{\partial}{\partial w}(H_v\beta\,dv)\,dw. \tag{5.10}$$

Thus we arrive at the general formulation of the complete vector curl **H** by permutation and dividing by the differential areas $dA_u = \beta\gamma\,dv\,dw$, etc., as

$$\begin{aligned}\text{curl}\,\mathbf{H} = \mathbf{u}^0\frac{1}{\beta\gamma}\left[\frac{\partial}{\partial v}(\gamma H_w) - \frac{\partial}{\partial w}(\beta H_v)\right] \\ + \mathbf{v}^0\frac{1}{\gamma\alpha}\left[\frac{\partial}{\partial w}(\alpha H_u) - \frac{\partial}{\partial u}(\gamma H_w)\right] \\ + \mathbf{w}^0\frac{1}{\alpha\beta}\left[\frac{\partial}{\partial u}(\beta H_v) - \frac{\partial}{\partial v}(\alpha H_u)\right].\end{aligned} \tag{5.11}$$

Returning to the three co-ordinate systems of Tables 3.1 and 3.2 of Sec. 3, we obtain for Cartesian co-ordinates, Eq. 5.8, and for the others:

curl **H**

Cylindrical co-ordinates

$$\begin{cases}\text{curl}_\rho H = \dfrac{1}{\rho}\dfrac{\partial H_z}{\partial \phi} - \dfrac{\partial H_\phi}{\partial z}, \\ \text{curl}_\phi H = \dfrac{\partial H_\rho}{\partial z} - \dfrac{\partial H_z}{\partial \rho}, \\ \text{curl}_z H = \dfrac{1}{\rho}\dfrac{\partial}{\partial \rho}(\rho H_\phi) - \dfrac{1}{\rho}\dfrac{\partial H_\rho}{\partial \phi},\end{cases}$$

Spherical co-ordinates

$$\begin{cases}\text{curl}_r H = \dfrac{1}{r\sin\theta}\dfrac{\partial H_\theta}{\partial \phi} - \dfrac{1}{r\sin\theta}\dfrac{\partial}{\partial \theta}(H_\phi\sin\theta), \\ \text{curl}_\phi H = \dfrac{1}{r}\dfrac{\partial H_r}{\partial \theta} - \dfrac{1}{r}\dfrac{\partial}{\partial r}(rH_\theta), \\ \text{curl}_\theta H = \dfrac{1}{r}\dfrac{\partial}{\partial r}(rH_\phi) - \dfrac{1}{r\sin\theta}\dfrac{\partial H_r}{\partial \phi}.\end{cases} \tag{5.12}$$

The curl operation $\nabla \times \mathbf{H}$ characterizes the sources of the magnetic field (the current density J) just as the divergence operation $\nabla \cdot \mathbf{E}$ describes the sources of the electrostatic field (the free charge density ρ/κ') (see Eq. 4.10). We show this by referring to Fig. 5.3. The line integral over any closed path C bounding a surface A

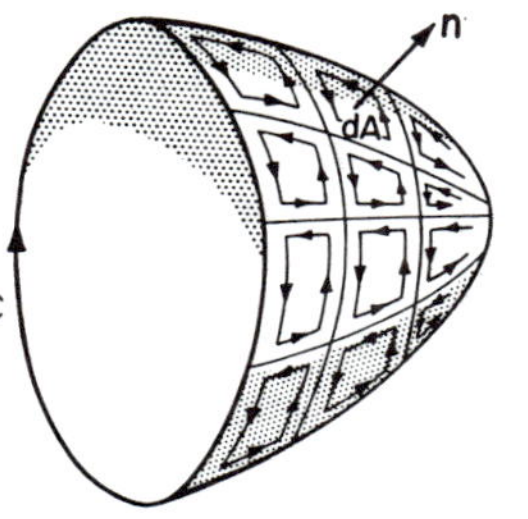

Fig. 5.3. Replacement of line integral by surface integral (Stokes's theorem).

may be found by subdividing A into differential elements dA and summing over the line integral around these elemental areas. The contributions of all interior boundaries cancel since they are traversed twice in opposite directions, and only the contribution of the rim remains. Hence, the left side of Eq. 5.5 may be rewritten as a surface integral by representing the differential line integrals around the elemental areas by their curl components normal to the surface elements

$$\oint_C \mathbf{H} \cdot dl = \int_A (\nabla \times \mathbf{H}) \cdot \mathbf{n}\, dA. \tag{5.13}$$

By introducing the current density $\mathbf{J}$ in Eq. 5.5 and integrating over the cross section A of the conductor, the right side of Eq. 5.5 may also be written as a surface integral,

$$I = \int_A \mathbf{J} \cdot \mathbf{n}\, dA, \tag{5.14}$$

hence Ampère's circuital law assumes the alternative version (Fig. 5.4)

$$\int_A (\nabla \times \mathbf{H}) \cdot \mathbf{n}\, dA = \int_A \mathbf{J} \cdot \mathbf{n}\, dA, \tag{5.15}$$

or, in differential form,

$$\text{curl}\, \mathbf{H} = \nabla \times \mathbf{H} = \mathbf{J}. \tag{5.16}$$

Hence, curl $\mathbf{H}$, the vortex density of the turbulent H-field, is equal to the current density $\mathbf{J}$ creating it.

The mathematical procedure of converting a line integral into a surface integral,

$$\underset{\text{extended over any closed contour}}{\oint \mathbf{H} \cdot dl} = \underset{\text{extended over any surface bounded by that contour}}{\oint (\nabla \times \mathbf{H}) \cdot \mathbf{n}\, dA} \tag{5.17}$$

is known as *Stokes's theorem*.[2] It has the same importance for whirlpool fields that Gauss's law transforming a surface into a volume integral (see Eq. 4.11) has for conservative fields.

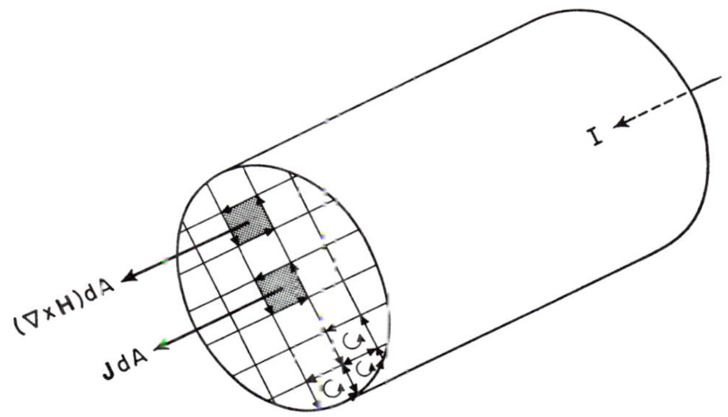

Fig. 5.4. Equivalence between curl **H** and current density.

By applying Ampère's circuital law to a current element, that is, to a differential wire element of the length dl traversed by a current I, we find for the magnetic field at a distance r from the wire

$$d\mathbf{H} = \frac{I}{4\pi r^2}(dl \times \mathbf{r}^0) = \left(\frac{I}{4\pi r^2}\, dl \sin \alpha\right)\mathbf{n} \tag{5.18}$$

(Fig. 5.5). This equation is frequently called the Biot-Savart law.[3]

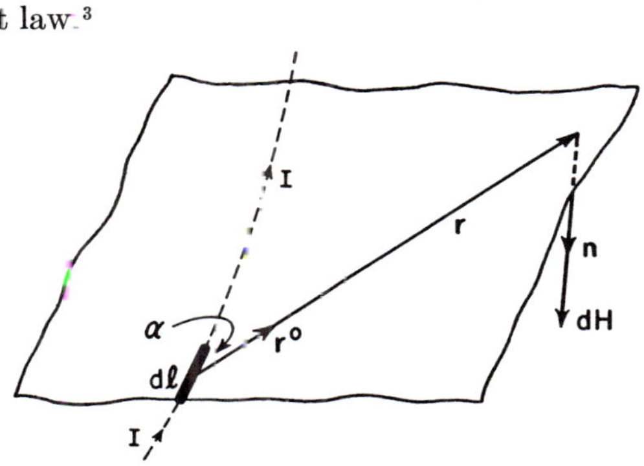

Fig. 5.5. Illustration of the Biot-Savart law.

To derive this result, we introduce conveniently a *vector potential* **O**, from which the magnetic field strength H follows by the curl operation

$$\nabla \times \mathbf{O} = \mathbf{H} \tag{5.19}$$

and the current density consequently (see Eq. 5.16) as

$$\nabla \times (\nabla \times \mathbf{O}) = \mathbf{J}. \tag{5.20}$$

[2] This theorem was first explicitly stated by Lord Kelvin in a letter to Stokes (July 2, 1850) and published in Stokes's *Smith's Prize Examination Paper*, February 1854. See A. A. Stokes, *Collected Mathematical and Physical Papers*, Cambridge University Press, Vol. V, 1905, p. 320.

[3] J. Biot and F. Savart, *Ann. chimie 15*, 220 (1820).

The vector operation on the left-hand side, as we easily confirm, may be rewritten

$$\nabla \times (\nabla \times \mathbf{O}) = \nabla(\nabla \cdot \mathbf{O}) - \nabla \cdot \nabla \mathbf{O}. \tag{5.21}$$

If the vector function $\mathbf{O}$ represents a purely turbulent field without sources, its divergence is zero,

$$\nabla \cdot \mathbf{O} = 0, \tag{5.22}$$

and only the Laplacian of Eq. 5.21 remains. Thus we obtain the analogue of Poisson's equation (see Eq. 4.12) for the turbulent field,

$$\nabla^2 \mathbf{O} = -\mathbf{J}. \tag{5.23}$$

In electrostatics the scalar potential ϕ can be derived from the charge density by integration (Eq. 4.1). By the corresponding integral equation, when ρ/ϵ' is replaced by $\mathbf{J}$, the vector potential at any point P, $\mathbf{O}_P$, is obtained from the space distribution of the current density

$$\mathbf{O}_P = \int \frac{\mathbf{J}_F\, dV}{4\pi r_{P-F}}. \tag{5.24}$$

A conductor element $d\boldsymbol{l}$ carrying a current I gives therefore, at the distance r, a vector potential contribution

$$d\mathbf{O} = \frac{I\, d\boldsymbol{l}}{4\pi r} = \frac{\mathbf{J}\, dV}{4\pi r}, \tag{5.25}$$

or a magnetic field

$$d\mathbf{H} = \nabla \times d\mathbf{O} = \frac{dV}{4\pi}\left(\nabla \times \frac{\mathbf{J}}{r}\right). \tag{5.26}$$

Expansion of the curl operator yields

$$\nabla \times \frac{\mathbf{J}}{r} = \frac{1}{r}\nabla \times \mathbf{J} + \nabla\frac{1}{r} \times \mathbf{J}. \tag{5.27}$$

The first term on the right is zero, since the curl operation refers to the point P and not to the current; thus Eq. 5.18 is obtained.

By integration, we may derive from this law the magnetic field produced by any geometrical configuration

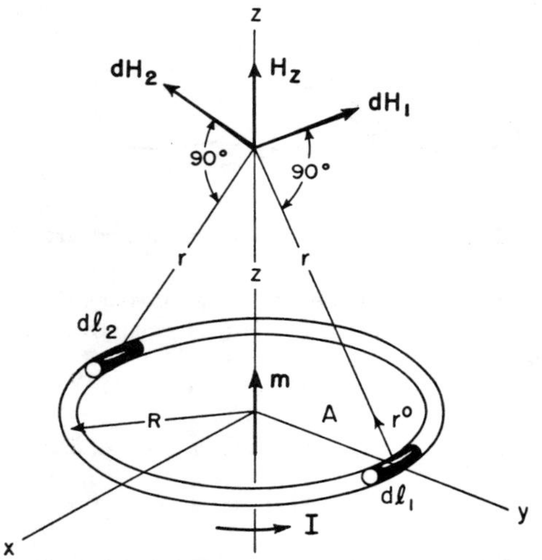

Fig. 5.6. Derivation of magnetic moment of circular current from Biot-Savart's law.

of steady currents. For example, we find that a ring current of the magnitude I encircling an area A (Fig. 5.6) produces at a distance, large in comparison to its radius, a magnetic field which is identical to that of a magnetic dipole of the moment,

$$\mathbf{m} = IA\mathbf{n} \tag{5.28}$$

(Fig. 5.7). This equivalence between the fields of magnetic dipoles and of circular currents allows us to explain the phenomenon "magnetism" and the nonexistence of magnetic monopoles on the basis that the sources of magnetism may be *molecular ring* or *Ampère currents.*

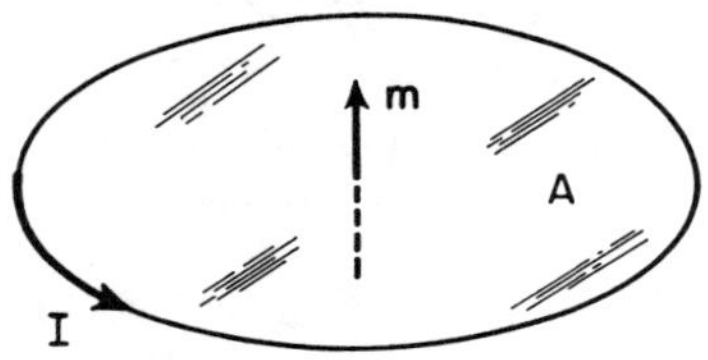

Fig. 5.7. Magnetic moment of circular current.

A second interrelation between electric and magnetic fields is given by *Faraday's induction law.*[4] It states that when the magnetic flux

$$\Phi = \int_A \mathbf{B} \cdot \mathbf{n}\, dA \tag{5.29}$$

traversing a loop of wire changes, whether by a change of the magnetic induction $\mathbf{B}$ or by a change in the position or shape of the loop, an emf is created along the wire, causing an induced voltage $\mathcal{V}_i$ to appear between its ends,

$$\mathcal{V}_i = \int_a^b \mathbf{E} \cdot d\boldsymbol{l} = -\frac{d\Phi}{dt}, \tag{5.30}$$

proportional to the speed of this change. If the loop is closed a current I will flow through the loop resistance R,

$$I = \frac{\mathcal{V}_i}{R} = -\frac{1}{R}\frac{d\Phi}{dt}, \tag{5.31}$$

that causes a magnetic field opposing the change in flux.

Because of the interrelation between magnetic fields and electric currents, the dimensions of the magnetic dipole moment, and with it of all magnetic quantities, can be reduced to electrical and mechanical dimensions.

[4] M. Faraday, *Experimental Researches in Electricity,* Taylor, London, 1839, Vol. I, pp. 1–109.

6 · Maxwell's Field Equations

Ampère's circuital law and Faraday's induction law were discovered and formulated for currents and voltages in wire loops. Maxwell postulated that these laws are valid in space quite independent of the presence of detector loops in which currents and voltages might develop, and thus he arrived at the electromagnetic field equations.[1]

A discussion of the current drawn by a capacitor will make the formulation of Maxwell's first field equation evident. The current of density J streaming into the electrode system is equal to the change of the true charge stored in the condenser, that is,

$$\int_A \mathbf{J}\cdot\mathbf{n}\,dA = \int_A \frac{ds}{dt}\,dA. \tag{6.1}$$

The surface charge density s originates the electric flux density $\mathbf{D}$ (see Eq. 2.3); hence

$$\frac{ds}{dt}\,dA = \frac{d\mathbf{D}}{dt}\cdot\mathbf{n}\,dA. \tag{6.2}$$

The current density may be measured as the time derivative of the electric flux density at the electrode surface,

$$\mathbf{J} = \frac{d\mathbf{D}}{dt}. \tag{6.3}$$

By postulating that this equivalence between the temporal change of the dielectric flux density and an electric current holds also for the interior of a dielectric, that is, that this change in flux produces a magnetic field just like a conduction current, Maxwell arrived at the concept that the conduction current charging a capacitor finds its continuation in a field current traversing the dielectric,

$$\int_A \mathbf{J}\cdot\mathbf{n}\,dA = \int_{A'} \frac{d\mathbf{D}}{dt}\cdot\mathbf{n}\,dA. \tag{6.4}$$

This field current, which extends through the cross section A' of the dielectric as far as the electric field of the capacitor reaches, was named by Maxwell the *displacement current.* By including this displacement current in Ampère's circuital law, *Maxwell's first field equation* results in the integral formulation

$$\oint \mathbf{H}\cdot d\mathbf{l} = \int_A \mathbf{J}\cdot\mathbf{n}\,dA + \int_{A'} \frac{\partial\mathbf{D}}{\partial t}\cdot\mathbf{n}\,dA. \tag{6.5}$$

[1] J. C. Maxwell, *A Treatise on Electricity and Magnetism,* Clarendon Press, Oxford, 1892, Vol. II, pp. 247–262.

With the help of Stokes's theorem the line integral on the left can be transformed into a surface integral, and the differential formulation

$$\nabla\times\mathbf{H} = \mathbf{J} + \frac{\partial\mathbf{D}}{\partial t} \tag{6.6}$$

is obtained (see also Appendix A, I, 6).

The current of the density $\mathbf{J}$ may be a true conduction current obeying Ohm's law,

$$\mathbf{J} = \sigma\mathbf{E}; \tag{6.7}$$

however, from a more general standpoint, the conductivity σ may be interpreted as the dielectric conductivity of Eq. 1.16, representing any energy-consuming process. Thus, by introducing the complex permittivity, the first field equation may be rewritten for sinusoidal fields and isotropic, linear dielectrics † as

$$\nabla\times\mathbf{H} = \epsilon^*\frac{\partial\mathbf{E}}{\partial t}. \tag{6.8}$$

The *second field equation* is a generalization of Faraday's induction law and asserts that the change of a magnetic flux density creates an emf in space quite independently of the presence of a loop for its detection:

$$\oint \mathbf{E}\cdot d\mathbf{l} = -\int_A \frac{\partial\mathbf{B}}{\partial t}\cdot\mathbf{n}\,dA. \tag{6.9}$$

In differential form it becomes

$$\nabla\times\mathbf{E} = -\frac{\partial\mathbf{B}}{\partial t}, \tag{6.10}$$

or, if we recall that also magnetization may lead to energy dissipation and introduce the complex permeability, we arrive at a formulation completely symmetrical to that of the first field equation, except for the negative sign,

$$\nabla\times\mathbf{E} = -\mu^*\frac{\partial\mathbf{H}}{\partial t}. \tag{6.11}$$

Maxwell's field equations thus describe the coupling between the electric and the magnetic field vectors and their interaction with matter in space and time.

† The designation "linear" dielectric signifies that the relation between $\mathbf{D}$ and $\mathbf{E}$ and between $\mathbf{B}$ and $\mathbf{H}$ is a linear one, that is, that the permittivity and permeability are independent of the field strength. The case of anisotropic dielectrics is treated in II, Sec. 26, and that of nonlinear dielectrics in II, Secs. 28 to 30.

7 · Electromagnetic Waves in Unbounded Space

We will, in general, make use of Maxwell's field equations in the differential form for sinusoidal fields just derived:

$$\nabla \times \mathbf{H} = \epsilon^* \frac{\partial \mathbf{E}}{\partial t} \tag{7.1}$$

and

$$\nabla \times \mathbf{E} = -\mu^* \frac{\partial \mathbf{H}}{\partial t}, \tag{7.2}$$

and under the additional assumption that the field contains neither free electric charges

$$\nabla \cdot \mathbf{E} = 0 \tag{7.3}$$

nor free magnetic poles

$$\nabla \cdot \mathbf{H} = 0. \tag{7.4}$$

An interpretation of the field equations requires, as a first step, the separation of the field vectors **E** and **H**. This can be done by differentiating the equations with respect to time and substituting from one equation into the other. Thus we obtain from Eq. 7.1

$$\frac{\partial}{\partial t}(\nabla \times \mathbf{H}) = \nabla \times \frac{\partial \mathbf{H}}{\partial t} = -\frac{1}{\mu^*}\nabla \times (\nabla \times \mathbf{E})$$
$$= \epsilon^* \frac{\partial^2 \mathbf{E}}{\partial t^2}, \tag{7.5}$$

or, after carrying through the vector operation (see Eq. 5.21),

$$\nabla^2 \mathbf{E} = \epsilon^* \mu^* \frac{\partial^2 \mathbf{E}}{\partial t^2}. \tag{7.6}$$

In the same way, by differentiating Eq. 7.2, we have for the magnetic field the completely symmetrical relation

$$\nabla^2 \mathbf{H} = \epsilon^* \mu^* \frac{\partial^2 \mathbf{H}}{\partial t^2}. \tag{7.7}$$

Equations 7.6 and 7.7 are the *wave equations of the electromagnetic field.*

We simplify these equations further by assuming at present that the field vectors **E** and **H** depend only on x and t, that is,

$$\frac{\partial \mathbf{E}}{\partial y} = \frac{\partial \mathbf{E}}{\partial z} = \frac{\partial \mathbf{H}}{\partial y} = \frac{\partial \mathbf{H}}{\partial z} = 0. \tag{7.8}$$

It will prove easy to return later from this special case to the general solution. The wave equations thus become

$$\frac{\partial^2 \mathbf{E}}{\partial x^2} = \epsilon^* \mu^* \frac{\partial^2 \mathbf{E}}{\partial t^2},\qquad \frac{\partial^2 \mathbf{H}}{\partial x^2} = \epsilon^* \mu^* \frac{\partial^2 \mathbf{H}}{\partial t^2}. \tag{7.9}$$

The solution of these differential equations concerning us here is a plane wave,

$$\mathbf{E} = \mathbf{E}_0 e^{j\omega t - \gamma x},\qquad \mathbf{H} = \mathbf{H}_0 e^{j\omega t - \gamma x}, \tag{7.10}$$

varying periodically in time with the frequency

$$\nu = \omega/2\pi \tag{7.11}$$

and advancing in the $+x$ direction through space with a *complex propagation factor*

$$\gamma = j\omega\sqrt{\epsilon^*\mu^*} = \alpha + j\beta; \tag{7.12}$$

α is the *attenuation factor*, and β is the *phase factor* of the wave. Introducing these factors, we may rewrite Eq. 7.10

$$\mathbf{E} = \mathbf{E}_0 e^{-\alpha x} e^{j2\pi(\nu t - \beta x/2\pi)},\qquad \mathbf{H} = \mathbf{H}_0 e^{-\alpha x} e^{j2\pi(\nu t - \beta x/2\pi)}. \tag{7.13}$$

Obviously the wave has a time period

$$T = 1/\nu \tag{7.14}$$

and a space period

$$\lambda = 2\pi/\beta. \tag{7.15}$$

Surfaces of constant phase are given by

$$\nu t - \frac{x}{\lambda} = \text{constant}, \tag{7.16}$$

hence propagate with the *phase velocity*

$$\frac{dx}{dt} = v = \nu\lambda = \frac{\omega}{\beta}. \tag{7.17}$$

For a dielectric without loss ($\epsilon^* = \epsilon'$, $\mu^* = \mu'$), we obtain from Eq. 7.12 the phase factor

$$\beta = \omega\sqrt{\epsilon'\mu'}, \tag{7.18}$$

so that the *phase velocity* in a *loss-free unbounded medium* is

$$v = \frac{1}{\sqrt{\epsilon' \mu'}}. \tag{7.19}$$

To learn about the coupling between the **E** and **H** vectors, we have to return to the field equations and write out the field components. Because of condition Eq. 7.8, only the derivatives with respect to x remain, and the field equations read

$$\begin{array}{rcl|rcl} 0 &=& \epsilon^* \dfrac{\partial E_x}{\partial t} & 0 &=& -\mu^* \dfrac{\partial H_x}{\partial t}, \\ -\dfrac{\partial H_z}{\partial x} &=& \epsilon^* \dfrac{\partial E_y}{\partial t} & -\dfrac{\partial E_z}{\partial x} &=& -\mu^* \dfrac{\partial H_y}{\partial t}, \\ \dfrac{\partial H_y}{\partial x} &=& \epsilon^* \dfrac{\partial E_z}{\partial t} & \dfrac{\partial E_y}{\partial x} &=& -\mu^* \dfrac{\partial H_z}{\partial t}. \end{array} \tag{7.20}$$

In addition, since the divergences are zero (see Eqs. 7.3, 7.4),

$$\frac{\partial E_x}{\partial x} = 0 \qquad \frac{\partial H_x}{\partial x} = 0. \tag{7.21}$$

These component equations contain three statements:

(1) The x components of the field vectors, the longitudinal field components of the electromagnetic wave, are independent of space and time, hence may be assumed to be zero. The plane wave is a *transverse electromagnetic* or TEM wave.

(2) The coupled transversal components of the **E** and **H** waves are *perpendicular* to each other and form, together with the propagation direction, a right-hand coordinate system of the sequence $+x \rightarrow E_y \rightarrow H_z$ (Fig. 7.1). This becomes apparent when we introduce the solution for the **E** and **H** vector (Eq. 7.10) into one of the component equations. Choosing, for example, the relation

$$-\frac{\partial H_z}{\partial x} = \epsilon^* \frac{\partial E_y}{\partial t}, \tag{7.22}$$

we obtain

$$\gamma H_z = j\omega\epsilon^* E_y. \tag{7.23}$$

Hence, the E_y component of the electric field is coupled to the H_z component of the magnetic field, as stated above.

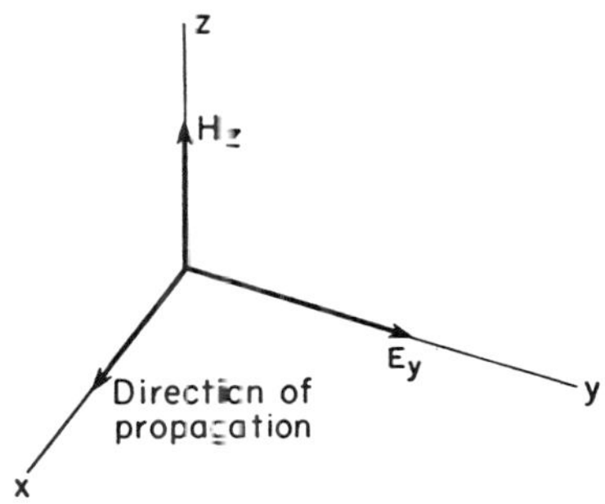

Fig. 7.1. Right-hand co-ordinate system for traveling TEM wave.

(3) The ratio of the coupled electric and magnetic field vectors follows as

$$\frac{|\mathbf{E}|}{|\mathbf{H}|} = \frac{\gamma}{j\omega\epsilon^*} \equiv Z. \tag{7.24}$$

This ratio Z, the *intrinsic impedance of the dielectric*, may be rewritten with the help of Eq. 7.12 in any one of the three versions

$$Z = \frac{\gamma}{j\omega\epsilon^*} = \sqrt{\frac{\mu^*}{\epsilon^*}} = \frac{j\omega\mu^*}{\gamma}. \tag{7.25}$$

8 · Dimensions and Units

With the preceding development of the electric and magnetic field concepts the prerequisites have been established for a reasonable selection of dimensions and units. Length, mass, and time are generally used as the three fundamental dimensions for the description of the mechanical world. Electricity and magnetism represent two new phenomena not contained in the framework of mechanical concepts. They are linked with each other through Ampère's circuital law, Faraday's induction law, or, summarily, through Maxwell's equations; hence it is logical to add one new fundamental quantity to the dimensions of the mechanical system.

The derivations of the preceding sections are independent of this choice, since the dielectric constant ϵ_0 and the permeability μ_0 of free space have been retained in the equations and are available for interpretation.

It was the tendency in earlier times to make the three dimensions of mechanics suffice by prescribing that either the dielectric constant (electrostatic system) or the permeability (electromagnetic system) be a plain number. Such a supposition eliminates a fourth dimension, because ϵ and μ are interlinked through relations, such as the phase velocity (see Eq. 7.19), which contain only mechanical dimensions. However, the result of this artificial reduction is that fractional expo-

Table 8.1. Dimensions and units

Quantity	Symbol	Defining Equation	Dimensions in Mks System		Multiplication Factor for Converting Rationalized Mks Units † to:		Ratio: esu/emu ($c = 3 \times 10^{10}$)
			Primary Units	Derived Units	Esu Units	Emu Units	
Admittance	Y	$Y = I/\mathcal{V}$	sec coul^2/kg m^2	mho	9×10^{11}	1×10^{-9}	c^2
Attenuation constant	α	$\alpha = \frac{1}{x} \ln \frac{E_0}{E}$	m^{-1}	neper			
Capacitance	C	$C = Q/\mathcal{V}$	sec^2 coul^2/kg m^2	farad	9×10^{11}	1×10^{-9}	c^2
Complex dielectric constant	ϵ^*	$\epsilon^* = \epsilon' - j\epsilon''$	sec^2 coul^2/kg m^3	farad/m	$36\pi \times 10^9$	$4\pi \times 10^{-11}$	c^2
Complex index of refraction	n^*	$n^* = n(1 - jk)$					
Complex permeability	μ^*	$\mu^* = \mu' - j\mu''$	kg m/coul^2	henry/m	$\frac{1}{36\pi} \times 10^{-13}$	$\frac{1}{4\pi} \times 10^7$	$1/c^2$
Complex propagation function	γ	$\gamma = \alpha + j\beta$	m^{-1}	m^{-1}	1×10^{-2}	1×10^{-2}	1
Conductance	G	$G = 1/R$	sec coul^2/kg m^2	mho	9×10^{11}	1×10^{-9}	c^2
Conductivity	σ	$\sigma = J/E$	sec coul^2/kg m^3	mho/m (siemens/m)	9×10^9	1×10^{-11}	c^2
Current	**I**	$\lvert \mathbf{I} \rvert = \frac{dQ}{dt}$	coul/sec	amp	3×10^9	1×10^{-1}	c
Current density	**J**	$\lvert \mathbf{J} \rvert = \frac{dI}{dA}$	coul/sec m^2	amp/m^2	3×10^{11}	1×10^1	c
Dielectric conductivity	σ	$\sigma = \omega\epsilon''$	sec coul^2/kg m^3	mho/m	9×10^9	1×10^{-11}	c^2
Dielectric constant (permittivity)	ϵ'	$\epsilon' = \frac{\lvert \mathbf{D} \rvert}{\lvert \mathbf{E} \rvert}$	sec^2 coul^2/kg m^3	farad/m	$36\pi \times 10^9$	$4\pi \times 10^{-11}$	c^2
Dissipation factor (*see* Loss tangent)	$D = \tan \delta$						
Electric charge	Q	Primary unit	coul		3×10^9	1×10^{-1}	c
Electric dipole moment	$\boldsymbol{\mu}$	$\boldsymbol{\mu} = Q\mathbf{d}$	coul m	amp sec m	3×10^{11}	1×10^1	c
Electric field strength	**E**	$\mathbf{E} = \frac{\mathbf{F}}{Q'}$ or $\mathbf{T} = \boldsymbol{\mu} \times \mathbf{E}$	kg m/sec^2 coul	volt/m	$\frac{1}{3} \times 10^{-4}$	1×10^6	$1/c$
Electric flux density (displacement)	**D**	$\oint \mathbf{D} \cdot \mathbf{n}\, dA = \int_V \rho\, dv$	coul/m^2	farad volt/m^2	$12\pi \times 10^5$	$4\pi \times 10^{-5}$	c
Electric loss factor	ϵ''	$\epsilon'' = \frac{J_{\text{loss}}}{\omega E}$	sec^2 coul^2/kg m^3	farad/m	$36\pi \times 10^9$	$4\pi \times 10^{-11}$	c^2
Electric polarization	**P**	$\mathbf{P} = \mathbf{D} - \epsilon_0\mathbf{E}$	coul/m^2	farad volt/m^2	$12\pi \times 10^5$	$4\pi \times 10^{-5}$	c
Electric potential	ϕ	$\phi_1 = -\int_\infty^1 \mathbf{E} \cdot dl$	kg m^2/sec^2 coul	volt	$\frac{1}{300}$	1×10^8	$1/c$
Electric susceptibility	χ	$\chi = \kappa' - 1$					
Electromotive force	emf	$\int_1^2 \mathbf{E} \cdot dl$	kg m^2/sec^2 coul	volt	$\frac{1}{300}$	1×10^8	$1/c$
Electrostatic energy	U	$Q\int_\infty^1 \mathbf{E} \cdot dl$	kg m^2/sec^2	joule	1×10^7	1×10^7	1

Force	$\mathbf{F}$	$\mathbf{F} = d(m\mathbf{v})/dt$	kg m/sec^2	newton	1×10^5	1×10^5	1
Impedance (characteristic)	Z_c	$Z_c = \mathcal{V}/I$	kg m^2/sec coul^2	ohm	$\frac{1}{9} \times 10^{-11}$	1×10^9	$1/c^2$
Impedance (intrinsic)	Z	$Z = E/H$	kg m^2/sec coul^2	ohm	$\frac{1}{9} \times 10^{-11}$	1×10^9	$1/c^2$
Index of absorption	k	$k = \alpha/\beta$					
Index of refraction	n	$n = \lambda_0/\lambda$					
Inductance	L	$L = \dfrac{\mathcal{V}}{\frac{dI_m}{dt}}$	kg m^2/coul^2	henry	$\frac{1}{36\pi} \times 10^{-11}$	$\frac{1}{4\pi} \times 10^9$	$1/c^2$
Loss tangent (dissipation factor)	$\tan\delta$	$\tan\delta = \dfrac{\epsilon''}{\epsilon'}$					
Magnetic dipole moment	$\mathbf{m}$	$\mathbf{m} = IF\mathbf{n}$	coul m^2/sec	amp m^2 = joule/weber/m^2	3×10^{13}	1×10^3	c
Magnetic field strength	$\mathbf{H}$	$\oint \mathbf{H} \cdot dl = \mathbf{I}$	coul/m sec	amp/m	3×10^7	1×10^{-3}	c
Magnetic flux	Φ	$\Phi = \int_A \mathbf{B} \cdot \mathbf{n}\, dA$	kg m^2/sec coul	volt sec = weber	$\frac{1}{3} \times 10^{-2}$	1×10^8	$1/c$
Magnetic flux density (Magnetic induction)	$\mathbf{B}$	$\mathbf{T} = \mathbf{m} \times \mathbf{B}$	kg/sec coul	volt sec/m^2 = weber/m^2	$\frac{1}{3} \times 10^{-4}$	1×10^6	$1/c$
Magnetic loss factor	μ''		kg m/coul^2	henry/m	$\frac{1}{36\pi} \times 10^{-13}$	$\frac{1}{4\pi} \times 10^7$	$1/c^2$
Magnetic permeability	μ'	$\mu' = \mathbf{B}/\mathbf{H}$	kg m/coul^2	henry/m	$\frac{1}{36\pi} \times 10^{-13}$	$\frac{1}{4\pi} \times 10^7$	$1/c^2$
Magnetic susceptibility	χ_m	$\chi_m = \kappa_m' - 1$					
Magnetization	$\mathbf{M}$	$\mathbf{M} = (\mathbf{B}/\mu_0) - \mathbf{H}$	coul/m sec	amp/m	3×10^7	1×10^{-3}	c
Magnetomotive force	mmf	$\text{mmf} = \int_1^2 \mathbf{H} \cdot dl$	coul/sec	amp	3×10^9	1×10^{-1}	c
Permeability (*see* Magnetic permeability)							
Permittivity (*see* Dielectric constant)							
Phase constant	β	$\beta = 2\pi/\lambda$	m^{-1}	m^{-1}	1×10^{-2}	1×10^{-2}	1
Potential (*see* Electric potential)							
Potential difference (*see* Electromotive force)	$\phi_2 - \phi_1$	$\phi_2 - \phi_1 = -\text{emf}$					
Power	P	$P = \dfrac{dW}{dt}$	kg m^2/sec^3	joule/sec = watt	1×10^7	1×10^7	1
Q of a dielectric (quality factor	Q	$Q = \dfrac{1}{\tan\delta}$					
Relative dielectric constant	κ'	$\kappa' = \epsilon'/\epsilon_0$					
Relative magnetic permeability	κ_m'	$\kappa_m' = \mu'/\mu_0$					
Resistance (direct current)	R	$R = \mathcal{V}/I$	kg m^2/sec coul^2	ohm	$\frac{1}{9} \times 10^{-11}$	1×10^9	$1/c^2$
Resistivity	ρ	$\rho = 1/\sigma$	kg m^3/sec coul^2	ohm m	$\frac{1}{9} \times 10^{-9}$	1×10^{11}	$1/c^2$
Torque	$\mathbf{T}$	$\mathbf{T} = \mathbf{F} \times \mathbf{d}$	kg m^2/sec^2	newton m	1×10^7	1×10^7	1
Voltage (*see* Electromotive force)	$\mathcal{V}$	$\mathcal{V} = -\text{emf}$					

† Example for use of conversion factors: given the electric field strength

$E = 31.3$ in. mks system, then $E = 31.3$ kg m/sec^2 coul $= 31.3$ volt/m,
$= 31.3 \times \frac{1}{3} \times 10^{-4} = 10.4 \times 10^{-4}$ statvolt/cm,
$= 31.3 \times 10^6$ abvolt/cm.

Table 8.2. Other units and their relation to the mks units

Quantity	Symbol and Derivation	Mks Unit	Other Units	Conversion Factors
Angle	deg or °	deg	radian (rad)	1 rad = 57.296 deg
Area	A	m^2	cm^2, in^2., ft^2	$1\ cm^2 = 10^{-4}$ $1\ in.^2 = 6.4516 \times 10^{-4}$ $1\ ft^2 = 9.2903 \times 10^{-2}$ } m^2
Attenuation	$\alpha = \frac{1}{x} \ln \frac{E_0}{E} = \frac{1}{2x} \ln \frac{P_0}{P}$	neper (n)	decibel (db) $= 20 \log \frac{E_0}{E}$ $= 10 \log \frac{P_0}{P}$	1 db = 8.686 n
Energy	U	joule = watt sec	erg, kilowatt hour (kwh), cal, Btu	$1\ erg = 10^{-7}$ $1\ kwh = 3.6 \times 10^6$ 1 cal = 4.185 1 Btu = 1055 } joule
Force	F	newton	dyne	$1\ dyne = 10^{-5}$ newton
Length	l	m	cm, mm, micron (μ), mμ, angstrom (A)	$1\ cm = 10^{-2}$ $1\ mm = 10^{-3}$ $1\ \mu = 10^{-6}$ $1\ m\mu = 10^{-9}$ $1\ A = 10^{-10}$ } m
Magnetic field strength	H	amp/m or amp turn/m	oerstedt	1 oerstedt $= \frac{10^3}{4\pi}$ amp/m
Magnetic flux density	B	weber/m^2	gauss	$1\ gauss = 10^{-4}$ weber/m^2
Mass	M	kg	g, pound (lb)	$1\ g = 10^{-3}$ 1 lb = 0.45359 } kg
Power	P	watt	erg sec^{-1}, kw, cal sec^{-1}, Btu hr^{-1}, hp (horsepower)	$1\ erg\ sec^{-1} = 10^{-7}$ $kw = 10^3$ $cal\ sec^{-1} = 4.185$ $Btu\ hr^{-1} = 0.2930$ hp = 746 } watt
Pressure	p	newton/m^2	dyne cm^{-2}, lb $in.^{-2}$, atm, cm Hg, bar	$1\ dyne\ cm^{-2} = 10^{-1}$ $1\ lb\ in.^{-2} = 6.895 \times 10^3$ $1\ atm = 1.013 \times 10^5$ 1 cm Hg = 1333 $1\ bar = 10^5$ } newton/m^2
Volume	V	m^3	cm^3, $in.^3$, liter, gallon	$1\ cm^3 = 10^{-6}$ $1\ in.^3 = 1.6387 \times 10^{-5}$ $1\ liter = 10^{-3}$ $1\ gallon = 3.7854 \times 10^{-3}$ } m^3
Weight	W	newton	dyne	$1\ dyne = 10^{-5}$ newton

nents appear in many of the dimensional equations; fractional dimensions make no physical sense. This difficulty can be avoided and all concepts simplified by referring to four fundamental quantities.[1] Here we adopt *electric charge* as the *fourth dimension.*

As *practical units* we select the meter (m), kilogram (kg), second (sec), and coulomb (coul). In addition to these *primary units, derived units* such as the volt, ampere, and ohm are used to shorten the dimensional equations. Table 8.1 gives a survey of the dimensional relations which are of importance for the macroscopic theory as here presented.

In the formulation of the theory we adopted a *rationalized system* of units by postulating direct equivalence between charge density and electric flux density (Eq. 2.3). This direct correspondence between s and $\mathbf{D}$, assigning one D line to the unit of true charge, instead of the 4π lines of the *unrationalized system,* seems more logical from our approach to the field theory, in which Coulomb's law is not the starting point but only one consequence of the electrostatic concepts. It has the additional advantage of removing the factor 4π from the equations most frequently encountered in dielectric problems.

Rationalization was originally suggested by Heaviside,[2] and a rationalized system based on the meter, kilogram, and second (mks system) was first advocated by Giorgi.[3] The mks system is therefore also called the *Giorgi system.*

To arrive at numerical values requires two more steps: a measurement of the velocity of light

$$c = \frac{1}{\sqrt{\epsilon_0\mu_0}} \tag{8.1}$$

(see Eq. 7.19), which establishes the product $\epsilon_0\mu_0$, and an agreement as to the value of ϵ_0 or μ_0. By international consent the value of μ_0 has been fixed for the rationalized mks system as

$$\mu_0 = 4\pi \times 10^{-7} \simeq 1.257 \times 10^{-6} \quad [\text{henry m}^{-1}]. \tag{8.2}$$

[1] See the discussion of J. A. Stratton, *Electromagnetic Theory,* McGraw-Hill Book Co., New York, 1941, pp. 16 ff.

[2] O. Heaviside, *Electrician 10,* 6 (1882).

[3] G. Giorgi, *Assoc. elettrot. italiana, atti 5,* 402 (1901).

The velocity of light has been measured as

$$c = 2.9979 \times 10^8 \simeq 3 \times 10^8 \quad [\text{m sec}^{-1}]\,^4; \tag{8.3}$$

hence the dielectric constant of free space becomes

$$\epsilon_0 = \frac{1}{36\pi} \times 10^{-9}$$
$$\simeq 8.854 \times 10^{-12} \quad [\text{farad m}^{-1}]. \tag{8.4}$$

The intrinsic impedance Z_0 of free space, determined by the ratio of permeability to permittivity (see Eq. 7.25), follows as

$$Z_0 = \sqrt{\mu_0/\epsilon_0} = 120\pi \simeq 376.6 \quad [\text{ohm}]. \tag{8.5}$$

Table 8.1 contains,[5] in addition to the primary and derived units, their conversion factors to the rationalized electrostatic (esu) and electromagnetic (emu) units. Both these systems use centimeter (cm) and gram (g) as the units of length and mass, and, in addition, the unrationalized esu system postulates $\epsilon_0 = 1$, whereas the unrationalized emu (*Gaussian*) system chooses $\mu_0 = 1$. Consequently, from Eq. 8.1,

$$\frac{\epsilon_0 \text{ esu}}{\epsilon_0 \text{ emu}} = c^2, \qquad \frac{\mu_0 \text{ esu}}{\mu_0 \text{ emu}} = \frac{1}{c^2}. \tag{8.6}$$

Table 8.1 refers only to the primary and derived units of the mks system. Frequently other derived units are customary or of more convenient magnitude; those of importance for this book are listed in Table 8.2, with their conversion factors.

[4] The most accurate measurements of the light velocity are no longer free-space determinations over long distances such as the Michelson-Morley experiment [*Phil. Mag. 13,* 236 (1882); *24,* 449 (1887)] but resonance measurements in wave-guide cavities [see K. D. Froome, *Nature 169,* 107 (1952)].

[5] See also the tabulation of P. Moon and D. E. Spencer, *Am. J. Phys. 16,* 25 (1948).

9 · Description of Dielectrics by Various Sets of Parameters

In Sec. 1 the response of a dielectric material to sinusoidal electric and magnetic fields was expressed by the two complex parameters ϵ^* and μ^* which determine the storage and dissipation of electric and magnetic energy in the medium. These parameters were derived from the amplitude and temporal-phase relations between voltage and current in capacitors and coils. It is now obvious that we are not restricted to using ϵ^* and μ^*, but may refer to alternate parameters that convey the same information.

The power engineer replaces the dielectric constant ϵ' and the loss factor ϵ'' by the combination of ϵ' and *power factor* $\cos\theta$; the radio engineer may choose ϵ' and the *loss tangent* $\tan\delta$, where

$$\tan\delta = \frac{\epsilon''}{\epsilon'} = \frac{\text{loss current}}{\text{charging current}}.\dagger \tag{9.1}$$

Frequently the inverse of the loss tangent, the quality factor Q of the dielectric,

$$\begin{aligned} Q &= \frac{1}{\tan\delta} = \frac{\omega\epsilon' E_0^2}{\omega\epsilon'' E_0^2} \\ &= 2\pi\nu\frac{\frac{1}{2}\epsilon' E_0^2}{\frac{1}{2}\sigma E_0^2} = 2\pi\frac{\text{av. energy stored per half cycle}}{\text{energy dissipated per half cycle}} \\ &= \left[\frac{\text{reactive v-amp}}{\text{watts}}\right], \end{aligned} \tag{9.2}$$

serves as the *figure of merit*, especially in wave-guide problems. An engineer interested in dielectric heating will probably refer to ϵ' and the *dielectric conductivity*

$$\sigma = \omega\epsilon'' \quad [\text{ohm}^{-1}\ \text{m}^{-1}], \tag{9.3}$$

because the power absorbed per unit volume is

$$P = \sigma\frac{E_0^2}{2} \quad [\text{watt m}^{-3}]. \tag{9.4}$$

If, instead of the time relation between current and voltage, the electromagnetic field in space is considered, new substitutes for ϵ^* and μ^* offer themselves. To derive them conveniently, we visualize the spatial electric wave train at some moment t_1,

$$E_y = E_1 e^{-\gamma x} = E_1 e^{-\alpha x} e^{-j2\pi\frac{x}{\lambda}} \tag{9.5}$$

(cf. Eq. 7.13). The wave amplitude oscillates in space with a periodicity λ; it is enclosed between exponential envelopes determined by the attenuation constant α (Fig. 9.1*a*). Alternatively, in polar co-ordinates, the wave amplitude may be depicted as a radius vector which, rotating clockwise as the distance increases, describes a logarithmic spiral (Fig. 9.1*b*). The parameter x is replaced in the latter representation by the phase angle ϕ according to the relation

$$\frac{x}{\lambda} = \frac{\phi}{2\pi}, \tag{9.6}$$

and the electric field strength is rewritten as

$$E_y = E_1 e^{-\phi\left(\frac{\alpha\lambda}{2\pi}+j\right)}. \tag{9.7}$$

In vacuum the wavelength is λ_0, and the wave travels with the velocity of light (cf. Eq. 7.19),

$$c = \lambda_0\nu = \frac{1}{\sqrt{\epsilon_0\mu_0}}. \tag{9.8}$$

In other media the wavelength normally shortens and the phase velocity slows down. The ratio of the wavelength or phase velocity in vacuum to that in the dielectric designates the *index of refraction* of the dielectric medium,

$$n \equiv \frac{\lambda_0}{\lambda} = \frac{c}{v} = \frac{\lambda_0}{2\pi}\beta. \tag{9.9}$$

For a loss-free medium this equation simplifies to

$$n = \sqrt{\epsilon'\mu'/\epsilon_0\mu_0} \equiv \sqrt{\kappa'\kappa_m'}. \tag{9.10}$$

If, in addition, the magnetization can be neglected ($\mu' = \mu_0$), the well-known *Maxwell relation* † results,

$$n^2 = \frac{\epsilon'}{\epsilon_0} = \kappa'. \tag{9.11}$$

† It should be noted that, since $\cos\theta = \sin\delta$, the power factor and loss tangent (dissipation factor) may be considered equal only for sufficiently small *loss angles* δ, where $\sin\delta \simeq \tan\delta$ because $\cos\delta \simeq 1$.

† This relation has been abused frequently in predicting static dielectric constants from optical refraction data. Actually, it states only that the square of the index of refraction of a nonabsorbing, nonmagnetic material is equal to the relative permittivity at that frequency.

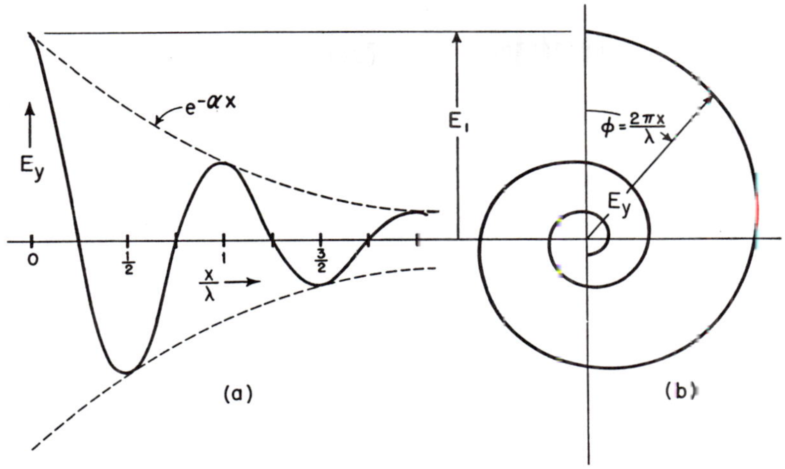

Fig. 9.1. Electric wave train in space.

The physicist normally uses the index of refraction as one of his parameters and pairs with it, by making use of the polar representation of the wave, the attenuation per radian called *index of absorption,*

$$k = \frac{\alpha\lambda}{2\pi} = \frac{\alpha}{\beta}. \tag{9.12}$$

By substituting these indices of refraction and absorption for the attenuation factor α and the phase factor β of the propagation factor in Eq. 7.12 we obtain

$$\gamma = j\frac{2\pi}{\lambda_0}n(1 - jk) = j\frac{2\pi}{\lambda_0}n^*. \tag{9.13}$$

The propagation factor γ used by the communication engineer may thus be replaced by the *complex index of refraction*

$$n^* = n(1 - jk) \tag{9.14}$$

employed in the calculations of physical optics.

The propagation factor γ is proportional to the product $\sqrt{\epsilon^*\mu^*}$, whereas the intrinsic impedance Z is equal to the ratio $\sqrt{\mu^*/\epsilon^*}$. Both complex quantities have to be determined to obtain ϵ^* and μ^* individually.

From the intrinsic impedance

$$Z = \frac{E}{H} = \sqrt{\frac{\mu^*}{\epsilon^*}} \tag{9.15}$$

in polar form,

$$Z = |Z|e^{j\zeta} = \left[\frac{(\epsilon'\mu' + \epsilon''\mu'')^2 + (\epsilon''\mu' - \epsilon'\mu'')^2}{(\epsilon'^2 + \epsilon''^2)^2}\right]^{1/4} e^{j\zeta} \tag{9.16}$$

with

$$\tan 2\zeta = \frac{\epsilon''\mu' - \epsilon'\mu''}{\epsilon'\mu' + \epsilon''\mu''}, \tag{9.17}$$

we can derive the phase relation between the electric and the magnetic wave. It is evident that the electric field vector is advanced or retarded with respect to the magnetic vector in temporal phase, depending on the preponderance of the term pertaining to the electric or the magnetic loss. For negligible magnetic loss ($\mu'' = 0$),

$$\tan 2\zeta = \tan\delta = \frac{2k}{1 - k^2} \tag{9.18}$$

or

$$\tan\zeta = k; \tag{9.19}$$

the phase advance of the electric wave is equal to the arc tangent of the index of absorption. In a loss-free medium in unbounded space the electric and magnetic field vectors of an electromagnetic wave are exactly in phase (Fig. 9.2).

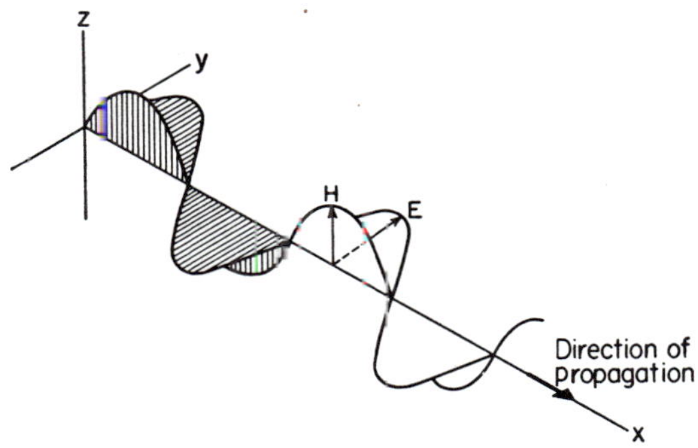

Fig. 9.2. Traveling TEM wave in loss-free dielectric.

The general characterization of a dielectric as the carrier of an electromagnetic field requires two independent complex parameters which have to be determined by four independent measurements; however, the situation fortunately simplifies in practice. Ferromagnetics excepted, the magnetic polarization is, in general, so weak that μ^* may be replaced by the permeability μ_0 of free space for all practical purposes. Thus, two measurements normally suffice to determine the dielectric response of homogeneous isotropic materials at a given frequency. Consequently, in most cases, our dielectric characteristics show only the specific dielectric constant,

$$\kappa' = \epsilon'/\epsilon_0, \tag{9.20}$$

and the loss tangent $\tan\delta$.

To allow a convenient change-over from these to other parameters, we equate the real and imaginary parts of Eq. 7.12. It follows for the attenuation factor of a transversal electromagnetic wave (TEM wave)

$$\alpha = \frac{\lambda\omega^2}{4\pi}(\epsilon'\mu'' + \epsilon''\mu'), \tag{9.21}$$

and for the phase factor

$$\beta = \frac{2\pi}{\lambda} = \omega\left[\frac{(\epsilon'\mu' - \epsilon''\mu'')}{2} \times\left\{1 + \sqrt{1 + \left(\frac{\epsilon'\mu'' + \epsilon''\mu'}{\epsilon'\mu' + \epsilon''\mu''}\right)^2}\right\}\right]^{1/2}. \tag{9.22}$$

Thus we arrive at the conversion formulas:

For materials with negligible magnetic loss ($\mu'' = 0$) we obtain from Eq. 9.22 for the wavelength the simplified expression

$$\lambda = \frac{1}{\nu}\frac{1}{[\frac{1}{2}\epsilon'\mu'\{1 + \sqrt{1 + \tan^2\delta}\}]^{1/2}}. \tag{9.23}$$

If, in addition, the permeability is that of vacuum ($\mu' = \mu_0$), we may write for the index of refraction

$$n = \frac{\lambda_0}{\lambda} = \left[\frac{1}{2}\kappa'\{\sqrt{1 + \tan^2\delta} + 1\}\right]^{1/2}. \tag{9.24}$$

Similarly the attenuation factor becomes

$$\alpha = \frac{2\pi}{\lambda_0}\left[\frac{1}{2}\kappa'\{\sqrt{1 + \tan^2\delta} - 1\}\right]^{1/2}, \tag{9.25}$$

and the index of absorption

$$k = \frac{\alpha}{\beta} = \left[\frac{\sqrt{1 + \tan^2\delta} - 1}{\sqrt{1 + \tan^2\delta} + 1}\right]^{1/2}. \tag{9.26}$$

These equations show that it is convenient to discuss the effect of the dielectric loss on other parameters by studying the three boundary cases: $\tan^2\delta \ll 1$, $\tan^2\delta \simeq 1$, and $\tan^2\delta \gg 1$.

The attenuation produced by a dielectric is frequently expressed as the *attenuation distance* $1/\alpha$ through which the field strength decays to $1/e = 0.368$ of its original value,

$$\frac{1}{\alpha} = \frac{\lambda_0}{2\pi}\left[\frac{2}{\kappa'(\sqrt{1 + \tan^2\delta} - 1)}\right]^{1/2} \quad [\text{m}], \tag{9.27}$$

or as the attenuation in *decibels per meter* produced by the material. If the field strength falls from $E(0)$ to $E(x)$, or the power from $P(0)$ to $P(x)$, over a length x of the dielectric, this decibel loss is defined as

$$20\log\frac{E(0)}{E(x)} = 10\log\frac{P(0)}{P(x)} = 8.686\alpha x \quad [\text{db}]; \tag{9.28}$$

that is, the decibel loss per meter is given as

$$8.686\alpha = 8.686\frac{2\pi}{\lambda_0}\left[\frac{1}{2}\kappa'\{\sqrt{1 + \tan^2\delta} - 1\}\right]^{1/2} \left[\frac{\text{db}}{\text{m}}\right]. \tag{9.29}$$

For low-loss materials ($\tan\delta \ll 1$) this loss becomes simply

$$8.686\frac{\pi}{\lambda_0}\sqrt{\kappa'}\tan\delta = 1637\frac{\sigma}{\sqrt{\kappa'}} \quad \left[\frac{\text{db}}{\text{m}}\right]. \tag{9.30}$$

For the convenience of the reader *nomographic charts* have been provided which allow a quick evaluation of $1/\alpha$ (Charts 9.1–9.3), of the decibel loss per meter (Charts 9.4–9.6), of β (Chart 9.7), and of n and k (Chart 9.8), when the specific dielectric constant and the loss tangent are given. Chart 9.9, in addition, gives the dielectric conductivity in terms of the same parameters calculated from the equation

$$\sigma = \omega\epsilon'' = 5.56 \times 10^{-11}\nu\kappa'\tan\delta \quad [\text{ohm}^{-1}\,\text{m}^{-1}]. \tag{9.31}$$

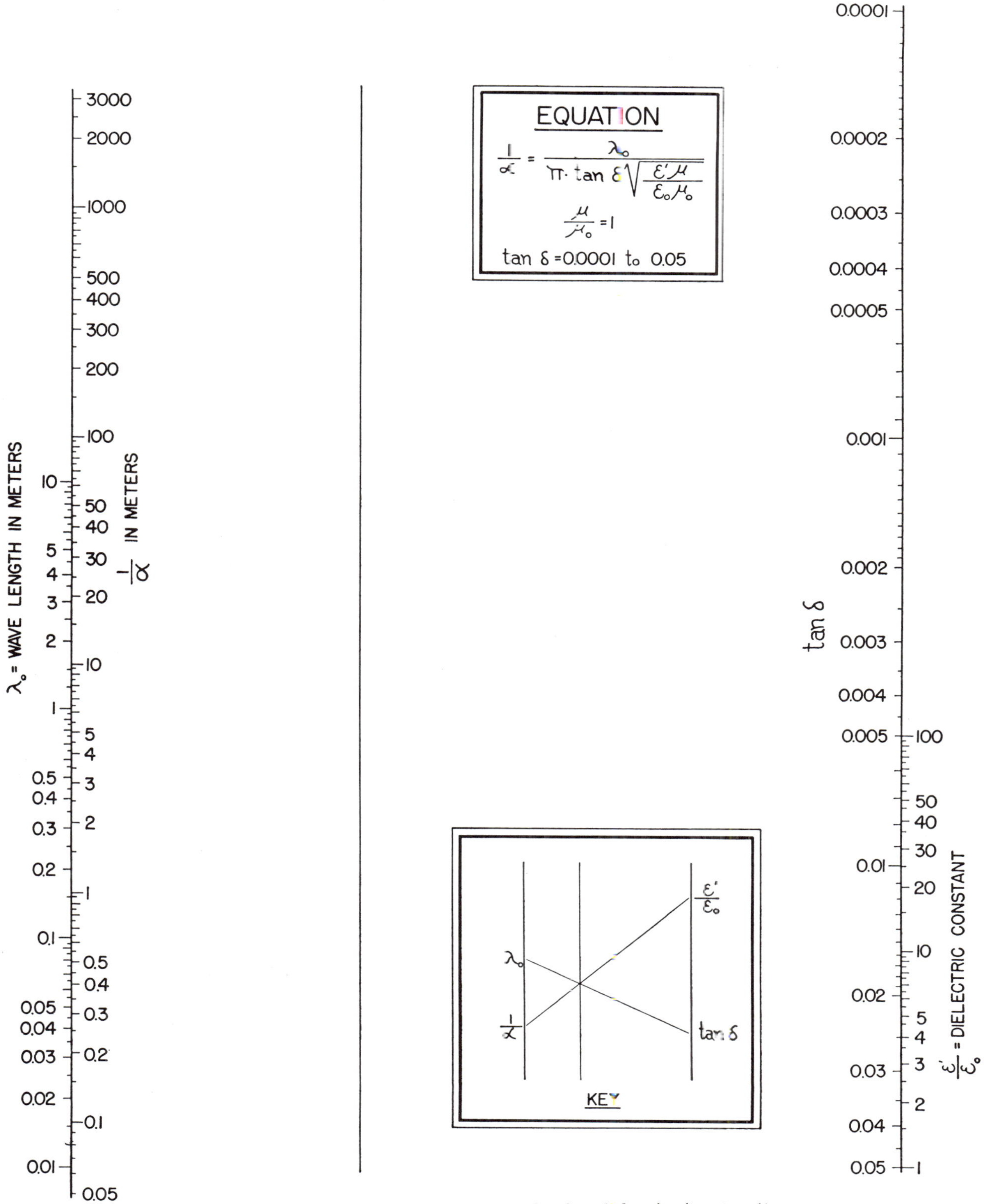

Chart 9.1. Attenuation distance $1/\alpha$ for low-loss dielectrics ($\tan\delta \ll 1$).

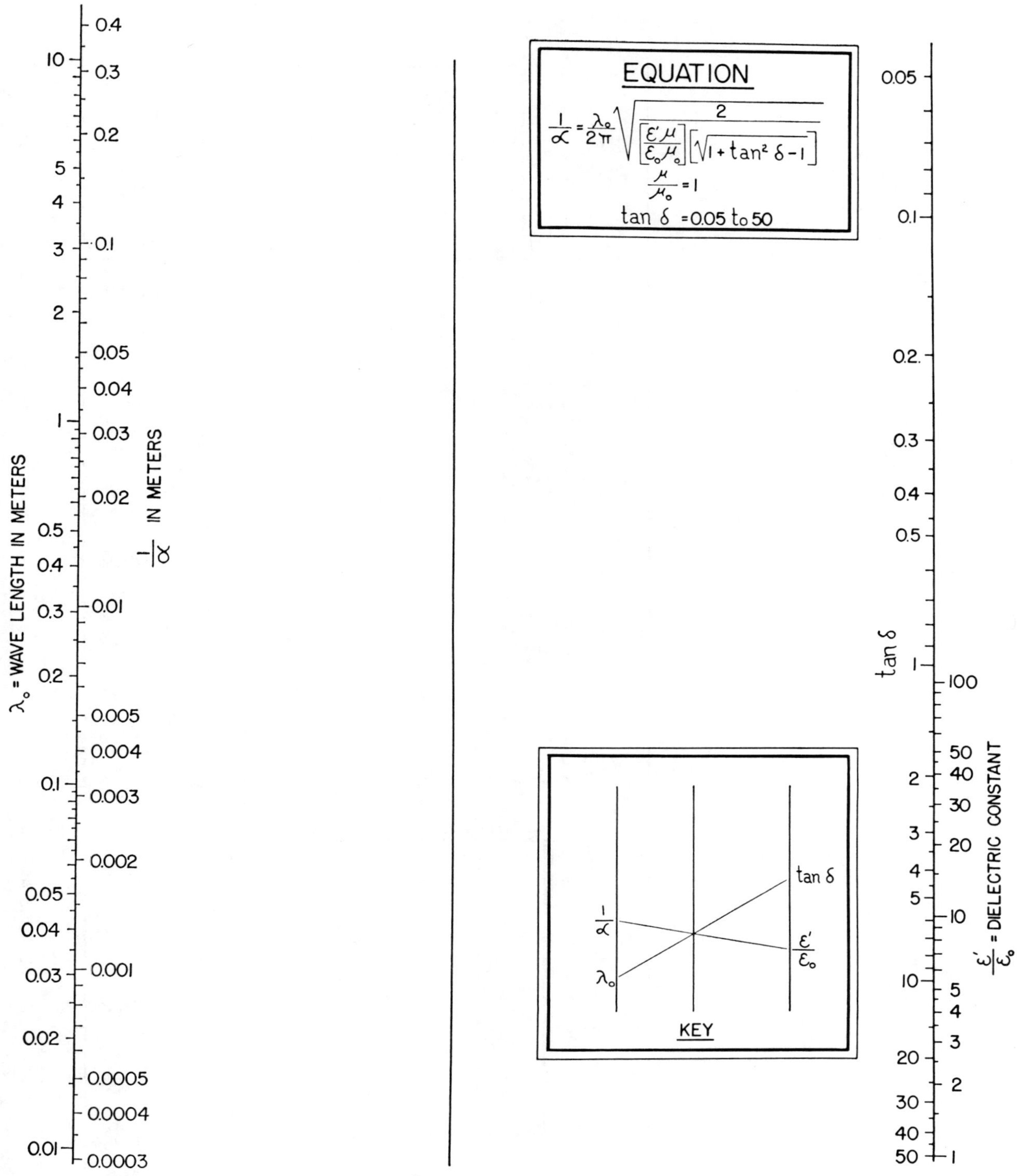

Chart 9.2. Attenuation distance 1/α for medium-loss dielectrics (tan δ comparable to 1).

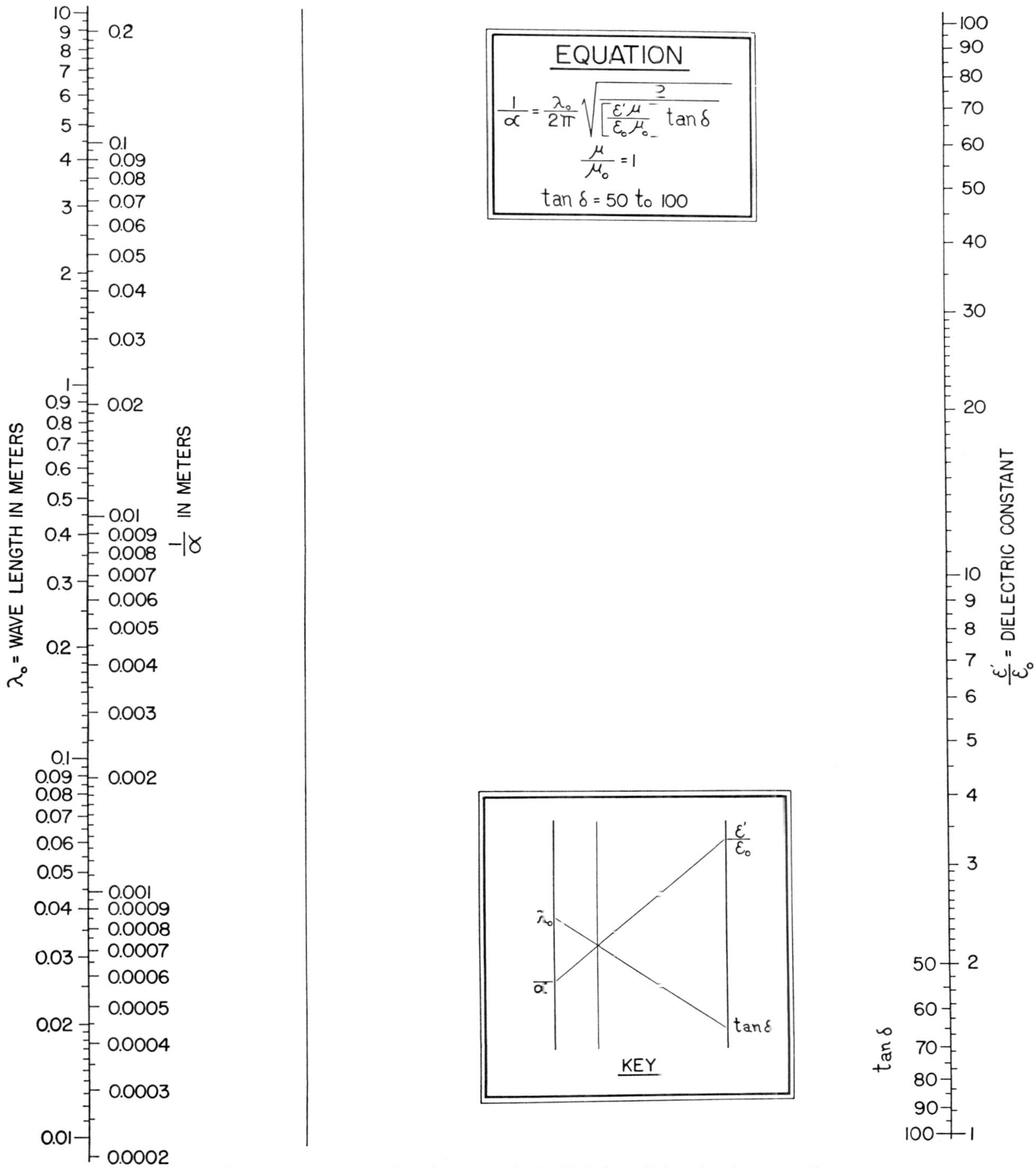

Chart 9.3. Attenuation distance $1/\alpha$ for high-loss dielectrics ($\tan\delta \gg 1$).

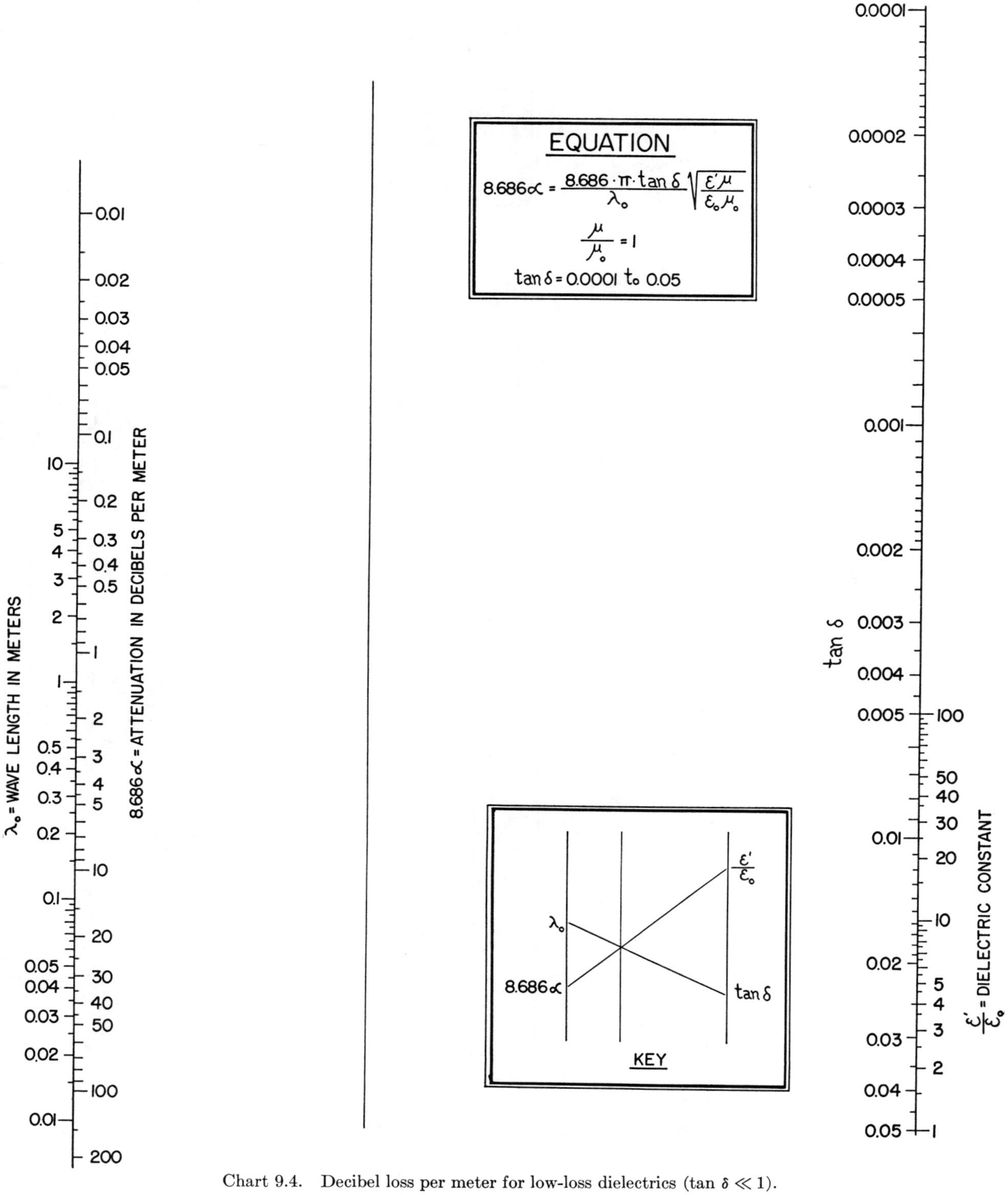

Chart 9.4. Decibel loss per meter for low-loss dielectrics ($\tan \delta \ll 1$).

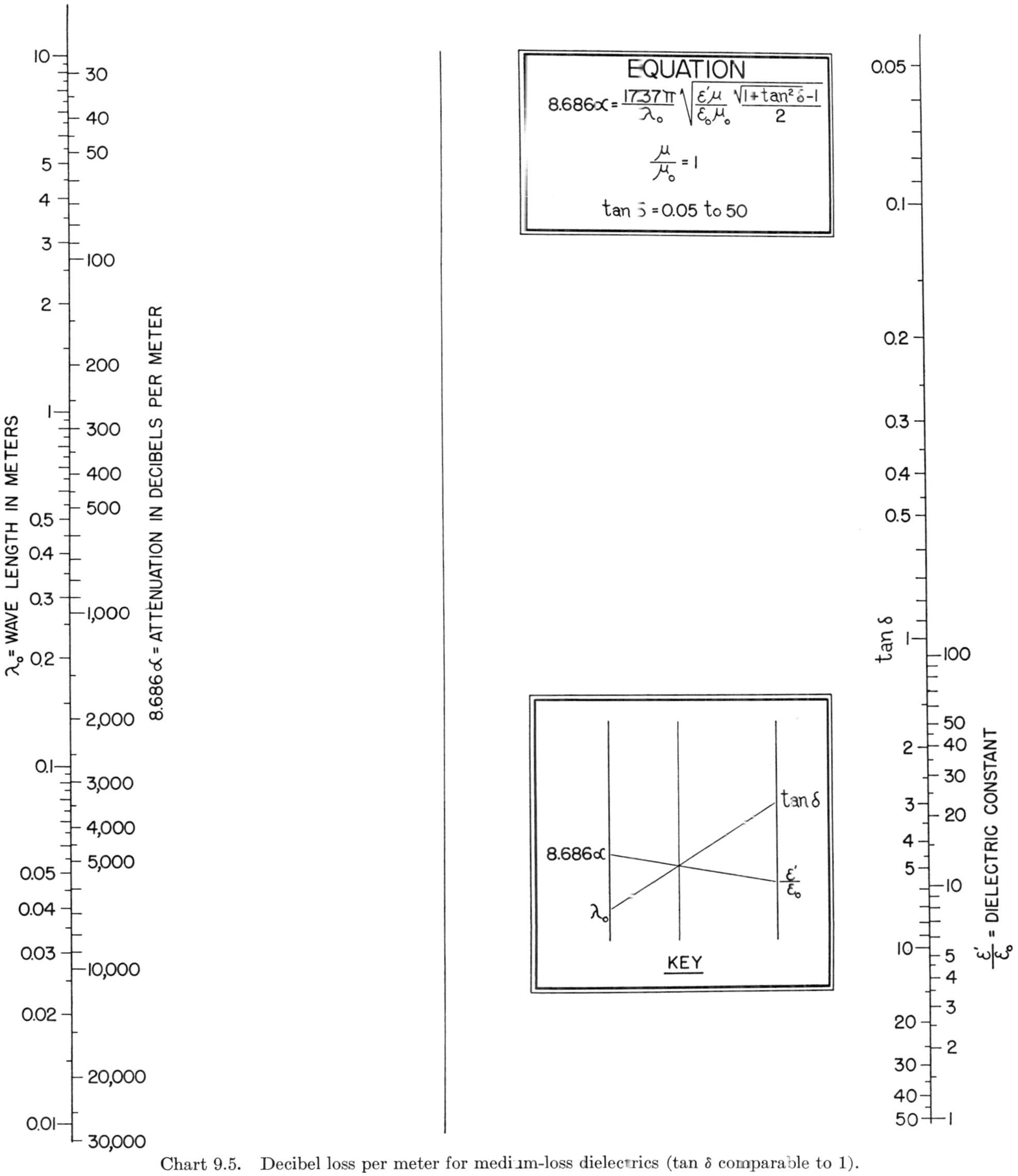

Chart 9.5. Decibel loss per meter for medium-loss dielectrics (tan δ comparable to 1).

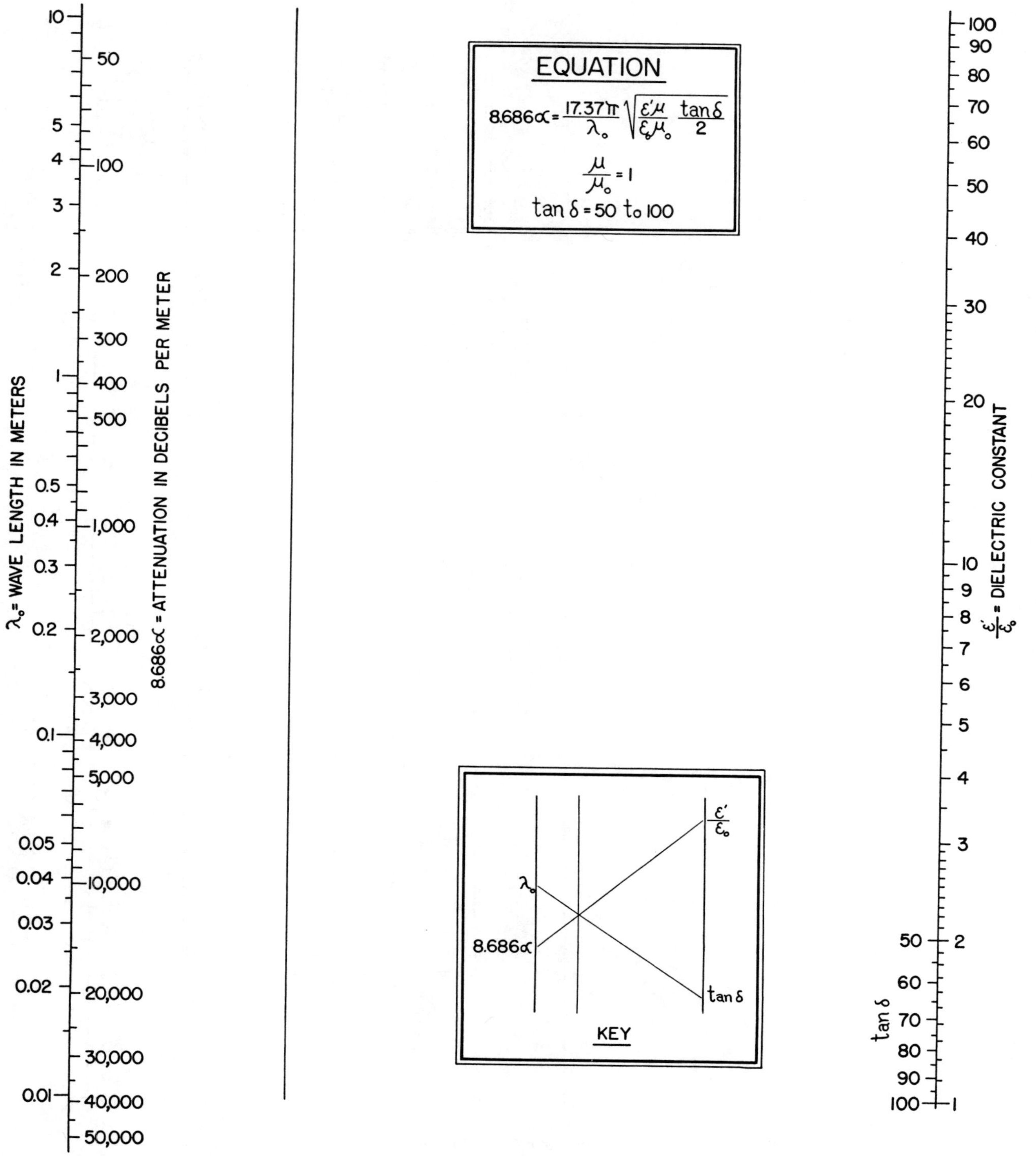

Chart 9.6. Decibel loss per meter for high-loss dielectrics ($\tan\delta \gg 1$).

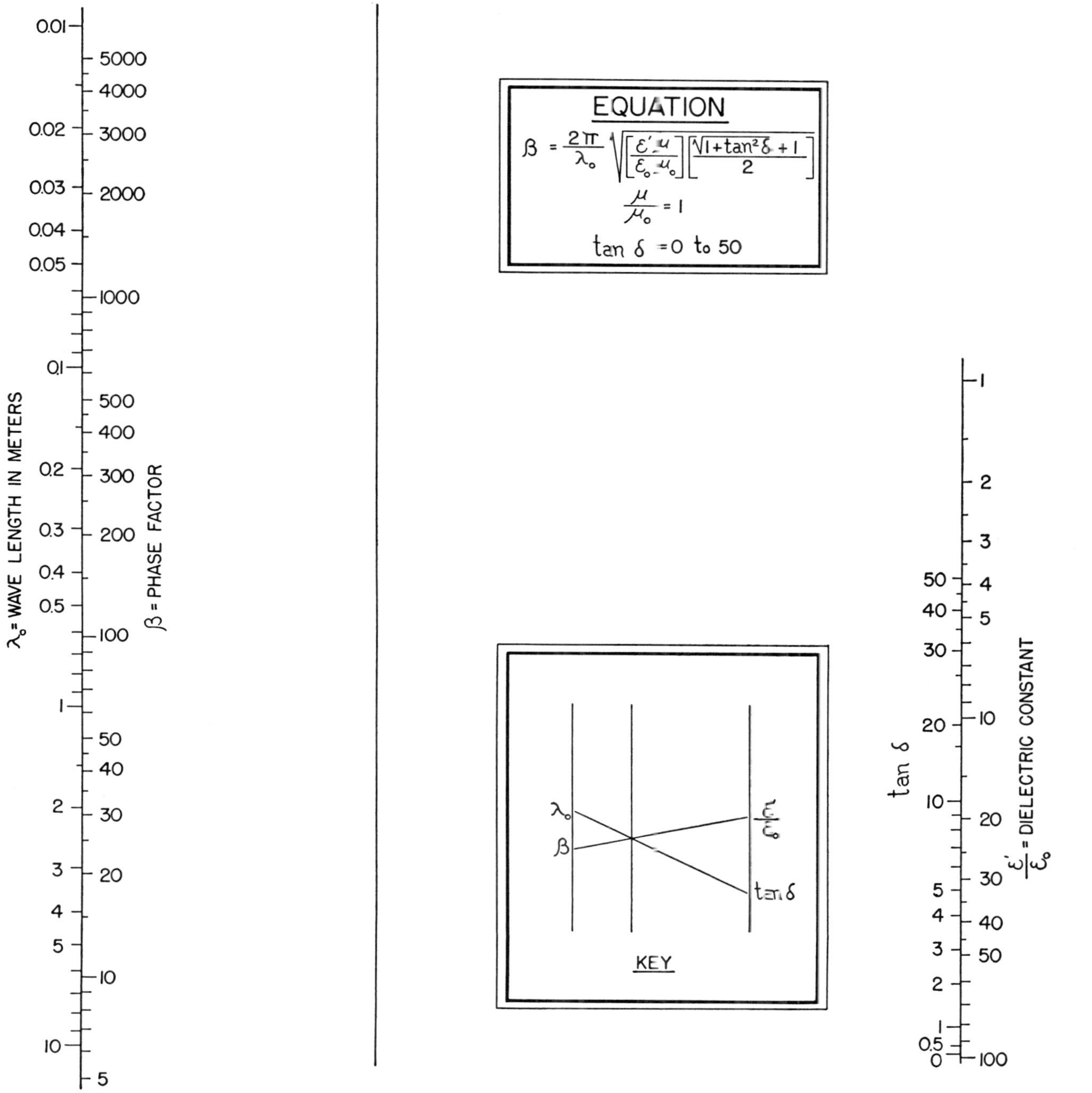

Chart 9.7. Phase factor β as function of dielectric constant and loss.

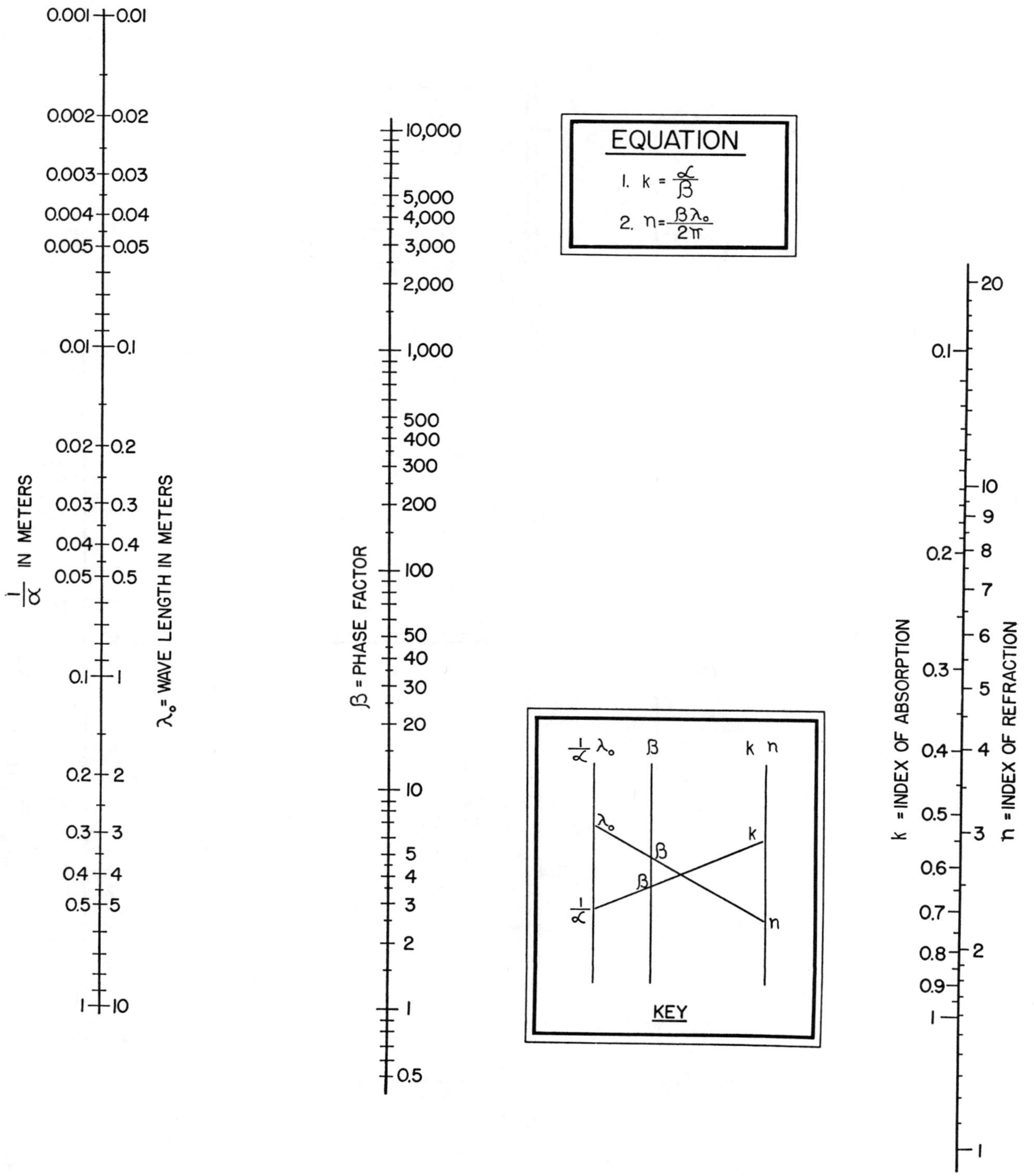

Chart 9.8. Complex index of refraction as function of attenuation distance $1/\alpha$ and phase factor β.

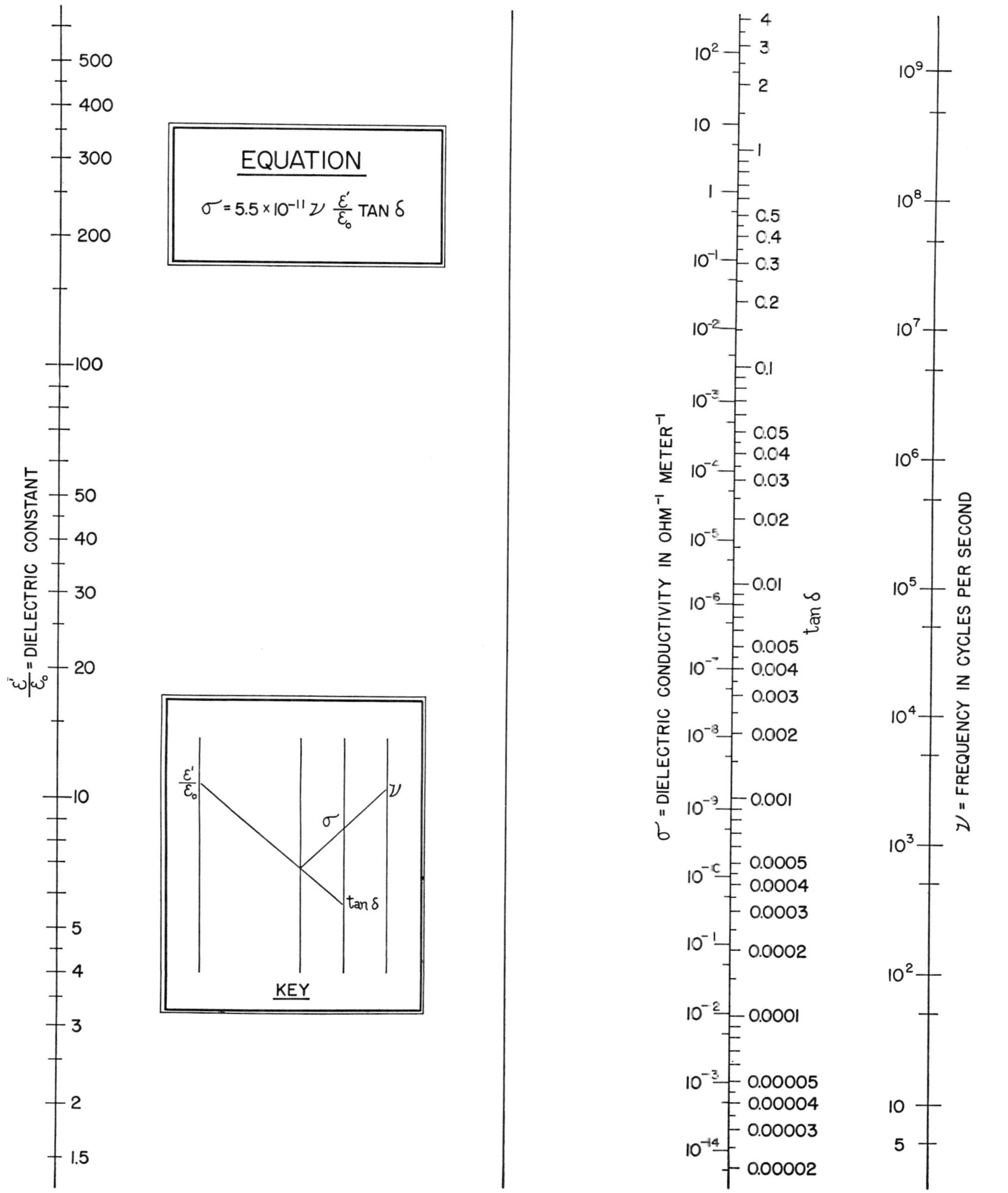

Chart 9.9. Dielectric conductivity as function of frequency, dielectric constant, and loss.

10 · Forces

The force of an electrostatic field of the strength **E** acting on a probe charge Q' was postulated to be (see Eq. 2.9)

$$\mathbf{F}_e = Q'\mathbf{E}. \tag{10.1}$$

This formulation rests on the tacit assumption that the detector charge Q' is so small that the field generated by it may be neglected in comparison to **E**. Acceding first to this limitation, we may derive a variety of force expressions for specific field constellations.

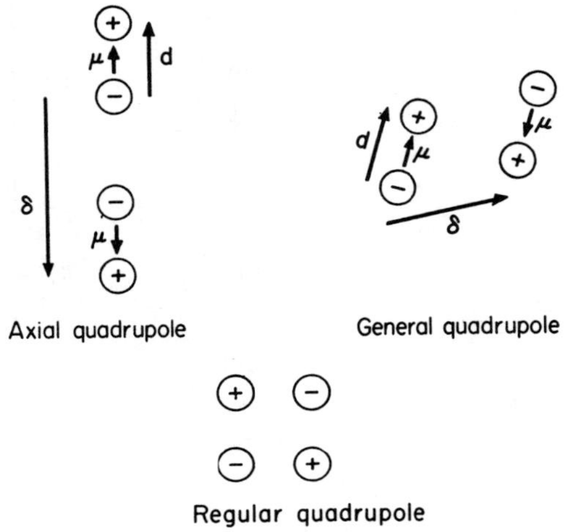

Fig. 10.1. Quadrupoles.

If the field **E** originates from a point charge Q, Coulomb's law results (see Eq. 3.8),

$$\mathbf{F}_e = \frac{QQ'}{\epsilon' 4\pi r^2}\mathbf{r}^0; \tag{10.2}$$

because of the spherical symmetry of the field, its strength decreases inversely proportional to the second power of the distance from the generating charge Q. If the field stems from an electric dipole, a force decreasing inversely proportional to the third power of the distance is obtained (see Eq. 3.19). A combination of two dipoles, a *quadrupole*, produces a force varying inversely proportional to the fourth power of the distance; an *octupole*, formed by two quadrupoles, acts on Q' with a force $\sim 1/r^5$; and any higher multipole arrangements lead to forces represented by the corresponding higher members of a power series.[1]

[1] See, for example, J. C. Slater and N. H. Frank, *Electromagnetism*, McGraw-Hill Book Company, New York, 1947, pp. 227 ff.

This type of dependence can be perceived easily, if we remember that the electrostatic potentials of the generating charges superpose arithmetically (see Eq. 4.1). The dipole arrangement of Fig. 3.2 thus produces, at a distance $r \gg d$, a potential ϕ_d proportional

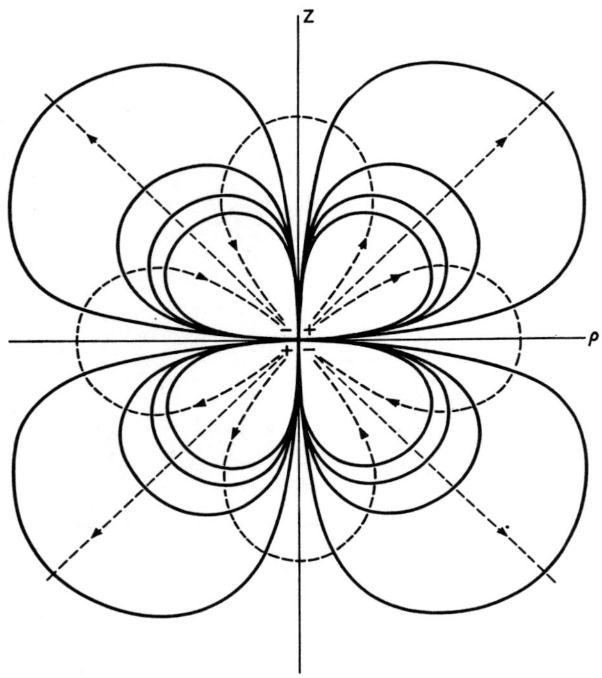

Fig. 10.2. Field of regular quadrupole (—— equipotential lines, ----- field lines).

to the gradient of the Coulomb potential (see Eq. 3.10),

$$\phi_d = \frac{\boldsymbol{\mu}\cdot\mathbf{r}^0}{\epsilon' 4\pi r^2} = -\mathbf{d}\cdot\nabla\phi. \tag{10.3}$$

Since the Coulomb potential varies as $1/r$, the dipole potential has to vary with $1/r^2$ in accordance with the gradient operation (Eq. 3.18). Quadrupoles (Fig. 10.1) create correspondingly, at a distance $r \gg \delta$, a potential ϕ_{qu} proportional to the gradient of the dipole potential

$$\phi_{qu} = -\boldsymbol{\delta}\cdot\nabla\phi_d, \tag{10.4}$$

that is, ϕ_{qu} varies as $1/r^3$, and so on (Fig. 10.2).

Although arrangements of alternating positive and negative charges thus reduce the range of the electrostatic forces below that of a Coulomb field, a favorable arrangement of like charges may have the opposite effect. Well-known examples are the field strength in a cylindrical capacitor which varies proportional to $1/r$ only, and the field strength in the homogeneous part of a plate condenser which is even independent of distance.

Frequently we are not concerned with the action of a field **E** on a simple point charge detector Q', but on some more complicated charge constellation. In Sec. 3 we derived the fact that a dipole in a homogeneous field is subject to a torque, whereas in an inhomogeneous field, in addition, a translational force acts proportional to the gradient of the field strength (see Eq. 3.25). Since dielectric bodies in electrostatic fields polarize, the forces exerted on them have to be calculated as a field action on more or less complicated dipole constellations. The same is true for magnetized materials in magnetic fields (see Sec. 5).

A cylindrical rod of length d and cross section $q = \rho^2\pi$, polarized homogeneously parallel to its axis, can be treated for $d \gg \rho$ simply as a dipole of the pole strength $\mathbf{P}q$ and of the moment

$$\boldsymbol{\mu} = \mathbf{P}qd = \mathbf{P}V. \tag{10.5}$$

The vector **P**, the polarization, represents, as previously, the dipole moment per unit volume (see Fig. 5.2).

In a few more complicated cases an "equivalent" dipole moment $\boldsymbol{\mu}$ of the body may be calculated. A typical example is the case of an uncharged dielectric sphere of radius R placed in a homogeneous field $\mathbf{E}_0$ pointing in the $+z$-direction (Fig. 10.3).

The field distribution, produced by the polarization of the sphere, must be obviously symmetrical around the z-axis, that is, independent of the angle ϕ. The Laplace equation for the spherical case (see Eqs. 4.13 and 4.16) therefore simplifies to

$$\nabla^2\phi = \frac{\partial^2\phi}{\partial r^2} + \frac{2}{r}\frac{\partial\phi}{\partial r} + \frac{1}{r^2}\frac{\partial^2\phi}{\partial\theta^2} + \frac{1}{r^2}\frac{\cos\theta}{\sin\theta}\frac{\partial\phi}{\partial\theta} = 0. \tag{10.6}$$

If ϕ_a and ϕ_i designate the potential outside and inside the sphere, respectively, we obtain the solution

$$\phi_a = \left(\frac{A}{r^2} + Br\right)\cos\theta, \qquad \phi_i = \left(\frac{C}{r^2} + Dr\right)\cos\theta. \tag{10.7}$$

Since the field intensity at the surface of the sphere remains finite, ϕ_a must become identical with ϕ_i at this boundary. According to Eq. 2.3, furthermore, the normal component of D must be continuous if no true surface charge exists. Hence

$$\left.\begin{aligned}\phi_a &= \phi_i\\ \epsilon_1'\frac{\partial\phi_a}{\partial r} &= \epsilon_2'\frac{\partial\phi_i}{\partial r}\end{aligned}\right\}\quad \text{at } r = R. \tag{10.8}$$

At a very great distance from the sphere the external field will remain undisturbed,

$$\phi_a = -E_0 r\cos\theta = -E_0 z\cos\theta \quad \text{for } r = \infty, \tag{10.9}$$

and in the center of the sphere the potential must stay finite,

$$\phi_i = \text{finite} \qquad \text{for } r = 0. \tag{10.10}$$

These boundary conditions (Eqs. 10.8 to 10.10) determine the coefficients of Eqs. 10.7 as

$$A = \frac{\epsilon_2' - \epsilon_1'}{\epsilon_2' + 2\epsilon_1'}R^3E_0,\quad B = -E_0,\quad C = 0,\quad D = -\frac{3\epsilon_1'}{\epsilon_2' + 2\epsilon_1'}E_0. \tag{10.11}$$

The potential outside and inside the dielectric sphere thus becomes

$$\phi_a = \left(\frac{\epsilon_2' - \epsilon_1'}{\epsilon_2' + 2\epsilon_1'}\frac{R^3}{r^3} - 1\right)E_0z, \qquad \phi_i = -\frac{3\epsilon_1'}{\epsilon_2' + 2\epsilon_1'}E_0z. \tag{10.12}$$

Consequently the field strength in the interior of the sphere,

$$\mathbf{E}_{iz} = -\nabla\phi_i = \frac{3\epsilon_1'}{\epsilon_2' + 2\epsilon_1'}\mathbf{E}_0, \tag{10.13}$$

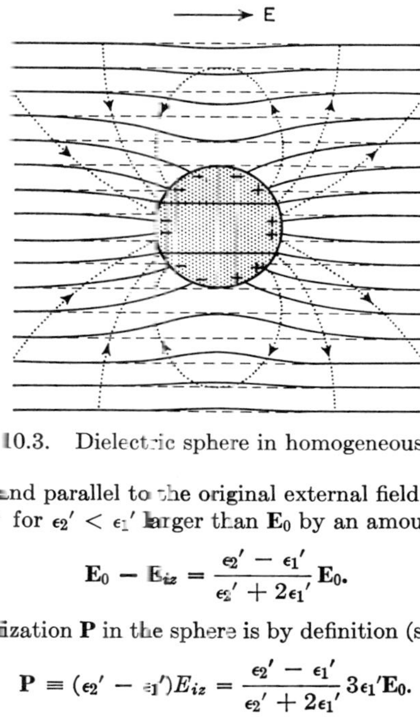

Fig. 10.3. Dielectric sphere in homogeneous field.

is constant and parallel to the original external field; for $\epsilon_2' > \epsilon_1'$ it is smaller for $\epsilon_2' < \epsilon_1'$ larger than $\mathbf{E}_0$ by an amount

$$\mathbf{E}_0 - \mathbf{E}_{iz} = \frac{\epsilon_2' - \epsilon_1'}{\epsilon_2' + 2\epsilon_1'}\mathbf{E}_0. \tag{10.14}$$

The polarization **P** in the sphere is by definition (see Eq. 2.7)

$$\mathbf{P} \equiv (\epsilon_2' - \epsilon_1')E_{iz} = \frac{\epsilon_2' - \epsilon_1'}{\epsilon_2' + 2\epsilon_1'}3\epsilon_1'\mathbf{E}_0. \tag{10.15}$$

Since this is the dipole moment per unit volume, the sphere acts to the outside with a dipole moment

$$\boldsymbol{\mu} = V\mathbf{P} = \frac{4\pi}{3}R^3\mathbf{P} = 4\pi R^3\epsilon_1'\frac{\epsilon_2' - \epsilon_1'}{\epsilon_2' + 2\epsilon_1'}\mathbf{E}_0 \tag{10.16}$$

placed in the center of the sphere. Figure 10.3 illustrates the actual field distribution resulting from the superposition of the original field E_0 and this dipole field of the polarized sphere (see also Appendix B, 7).

In an inhomogeneous field $\mathbf{E}_0$ a translational force would act on the sphere (see Eq. 3.25)

$$\mathbf{F} = \boldsymbol{\mu}\cdot\nabla\mathbf{E}_0 = 2\pi R^3\epsilon_1'\frac{\epsilon_2' - \epsilon_1'}{\epsilon_2' + 2\epsilon_1'}\nabla(\mathbf{E}_0^2), \tag{10.17}$$

driving it, for $\epsilon_2 > \epsilon_1$, into the region of highest and for $\epsilon_2 < \epsilon_1$ into the region of lowest field strength.

The basic force equation (Eq. 10.1) remains valid in electromagnetic fields; it describes the action of the electric field strength **E** on a charge Q', whether or not **E** can be derived from a scalar potential. If the charge is in motion, an additional force arises due to the action of the magnetic field. That such a force must exist can

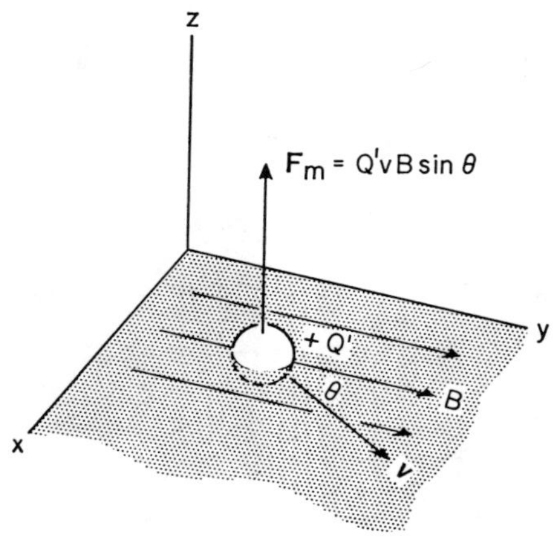

Fig. 10.4. Magnetic force on moving charge Q'.

be foreseen from the fact that a charge Q' moving with the velocity $\mathbf{v}$ corresponds to a current

$$\mathbf{I} = Q'\mathbf{v}. \tag{10.18}$$

A current, according to Ampère's circuital law (Eq. 5.5), is surrounded by a magnetic field. For a ring current this field is equivalent to that of a magnetic dipole (Eq. 5.28), and a magnetic dipole of the moment $\mathbf{m}$ suffers a torque in an external magnetic field (Eq. 2.16)

$$\mathbf{T} = \mathbf{m} \times \mathbf{B}. \tag{10.19}$$

Thus a linear current element $I\,d\mathbf{l}$ will be subject to a force. Its value is found to be

$$d\mathbf{F}_m = I\,d\mathbf{l} \times \mathbf{B}. \tag{10.20}$$

In contrast to the electric force, which is directed along the vector **E**, the magnetic force operates normal to the plane defined by the current element and the vector **B** (Fig. 10.4).

The total force for a charge Q' moving with the velocity v in an electromagnetic field becomes thus simply (see Eqs. 10.18 and 10.20)

$$\mathbf{F} = \mathbf{F}_e + \mathbf{F}_m = Q'(\mathbf{E} + \mathbf{v} \times \mathbf{B}) \quad \text{[newton]}, \tag{10.21}$$

if Q' is expressed in coulombs, **E** in volts per meter, **v** in meters per second, and **B** in webers per square meter (see Table 8.1). In the more general case of a volume V filled with true charge of the density ρ and currents of the density **J**, an electromagnetic field produces the net mechanical force on this volume

$$\mathbf{F} = \int_V \rho\mathbf{E}\,dV + \int_V (\mathbf{J} \times \mathbf{B})\,dV. \tag{10.22}$$

In closed systems it frequently proves convenient to derive the forces acting on an object in electric and magnetic fields from the maximum decrease in free energy caused by a differential motion of the body. Equation 3.25 illustrates this procedure.

11 · Field Energy and Radiation

The energy content of the electromagnetic field is stored in the electric and in the magnetic field. In addition, energy may be dissipated as heat during the electric and magnetic polarization cycle. Let us start, as in the preceding section, with the electrostatic case and then proceed by analogy.

That an electrostatic field represents stored energy follows directly from the fact that work is required for its creation. Let us visualize a charge Q on a sphere of radius r_1, assembled successively from infinity in n charge elements dQ. At each step of the transfer, work has to be done against the Coulomb force of the charge already present (see Eqs. 3.7 and 3.8). Hence the total work required to concentrate the charge on the sphere is

$$-W = 0 + dQ\frac{dQ}{\epsilon'4\pi r_1} + dQ\frac{2dQ}{\epsilon'4\pi r_1} + \cdots + dQ\frac{(n-1)\,dQ}{\epsilon'4\pi r_1}$$

$$= \frac{n(n-1)(dQ)^2}{2\epsilon'4\pi r_1} = \frac{1}{2}Q\phi\left(1 - \frac{1}{n}\right). \tag{11.1}$$

In the limiting case of an infinite number of steps ($n \to \infty$, $dQ \to 0$, with $n\,dQ = Q$) there will be stored the potential energy

$$U = -W = \tfrac{1}{2}Q\phi. \tag{11.2}$$

In Eq. 11.2 the charge is visualized as the seat of the electric energy. Alternatively, the field in space, connected to the charge by the defining equations 2.3 to 2.5, may be considered as the carrier of this energy. Each volume element of space stores

$$dU = \frac{\mathbf{E}\cdot\mathbf{D}}{2}\,dV, \tag{11.3}$$

hence the total volume V filled by the field contains the electric energy

$$U = \tfrac{1}{2}\int_V \mathbf{E}\cdot\mathbf{D}\,dV = \tfrac{1}{2}\int_V \epsilon' E^2\,dV \quad \text{[joule]}. \tag{11.4}$$

That Eqs. 11.2 and 11.4 represent alternative expressions may be verified readily in the special case of a charged sphere (Fig. 11.1). The field strength $\mathbf{E}$ is

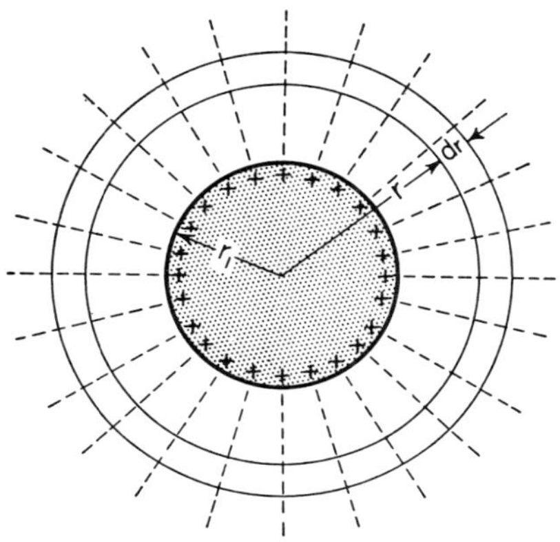

Fig. 11.1. Energy content of Coulomb field.

essentially constant within concentric volume elements dV and is identical with the Coulomb field of Eq. 3.6; consequently

$$U = \frac{1}{2}\int_V \epsilon' E^2\,dV$$

$$= \frac{1}{2}\int_{r_1}^{\infty} \frac{Q^2}{\epsilon'(4\pi)^2 r^4}\,4\pi r^2\,dr = \frac{1}{2}\,Q\phi, \tag{11.5}$$

as asserted. The general equivalence of Eqs. 11.2 and 11.4 can be based on the fact that any electrostatic field can be built up from the Coulomb fields of point charges.

In the example of a single charged sphere the neutralizing countercharge $-Q$ resides at infinity. If, instead, it is located at a finite distance, the potential difference $\mathcal{V}$ between the two charges replaces the potential ϕ in Eq. 11.2, that is,

$$U = \tfrac{1}{2}Q\mathcal{V} = \tfrac{1}{2}C\mathcal{V}^2, \tag{11.6}$$

with C the capacitance of the condenser system.

According to the same type of argument, either the current I traversing a coil of the inductance L may be visualized as carrier of the *magnetic energy*

$$U = \tfrac{1}{2}LI^2, \tag{11.7}$$

or the magnetic field in space may be considered as its seat,

$$U = \tfrac{1}{2}\int_V \mathbf{H}\cdot\mathbf{B}\,dV = \tfrac{1}{2}\int_V \mu' H^2\,dV \quad \text{[joule]}. \tag{11.8}$$

By summing over the electric and magnetic energy, stored and dissipated, we obtain the *electromagnetic energy* of sinusoidal fields in isotropic dielectrics as

$$U = \tfrac{1}{2}\int_V \epsilon^* E^2\,dV + \tfrac{1}{2}\int_V \mu^* H^2\,dV \quad \text{[joule]}. \tag{11.9}$$

The electric and magnetic field strengths of a TEM wave in an unbounded medium are interrelated by the intrinsic impedance (see Eqs. 7.24 and 7.25) as

$$\frac{E^2}{H^2} = \frac{\mu^*}{\epsilon^*}. \tag{11.10}$$

Equation 11.9, rewritten as

$$U = \int_V \epsilon^* E^2\,dV = \int_V \mu^* H^2\,dV, \tag{11.11}$$

shows that in traveling waves the electric and magnetic energy densities in each volume element are equal at any moment.

The rate of decrease of the electromagnetic energy in a volume V, due to other causes than heat loss, measures the outward flow of the energy through the enclosing surface. The intensity of this electromagnetic radiation, the energy per second crossing a unit area normal to the direction of propagation, can be described by the *Poynting vector* [1]

$$\mathbf{S} = \mathbf{E}\times\mathbf{H} \quad [\text{watt m}^{-2}], \tag{11.12}$$

and the total radiation emanating from a volume V becomes

$$-\int_V \left(\epsilon^* E\,\frac{\partial E}{\partial t} + \mu^* H\,\frac{\partial H}{\partial t}\right) dV$$

$$= \oint_A (\mathbf{E}\times\mathbf{H})\cdot\mathbf{n}\,dA \quad \text{[watt]}. \tag{11.13}$$

The energy flow in an electromagnetic field cannot be observed for a *monochromatic wave*,

$$E = ae^{(j\omega t - \beta x)}, \tag{11.14}$$

because the wave train would have infinite extent and duration. A wave train of finite length, and therefore

[1] See the discussion of the Poynting theorem in J. A. Stratton, *Electromagnetic Theory*, McGraw-Hill Book Co., New York, 1941, p. 131.

of finite energy content, cannot be monochromatic but consists, according to Fourier analysis,[2] of a wave group formed by the superposition of waves. In the simplest case a *wave group* or *signal* can be composed of two waves, E_1 and E_2, of equal amplitude deviating from each other by a differential value $d\omega$ in frequency and $d\beta$ in phase as

$$E_1 = ae^{j\left[\left(\omega-\frac{d\omega}{2}\right)t-\left(\beta-\frac{d\beta}{2}\right)x\right]},$$
$$E_2 = ae^{j\left[\left(\omega+\frac{d\omega}{2}\right)t-\left(\beta+\frac{d\beta}{2}\right)x\right]}. \tag{11.15}$$

The resultant field strength

$$E = E_1 + E_2 = ae^{j(\omega t-\beta x)} 2\cos\left(\frac{d\beta}{2}x - \frac{d\omega}{2}t\right) \tag{11.16}$$

is *modulated*, with maxima (antinodes) appearing for $\frac{d\beta}{2}x - \frac{d\omega}{2}t = m\pi$ (where $m = 0, 1, 2, \cdots$) and minima (nodes) at $\frac{d\beta}{2}x - \frac{d\omega}{2}t = m\frac{\pi}{2}$ (where $m = 1, 3, 5, \cdots$) (Fig. 11.2). This is the familiar phenomenon of *beat frequencies*. The propagation velocity of a specific beat,

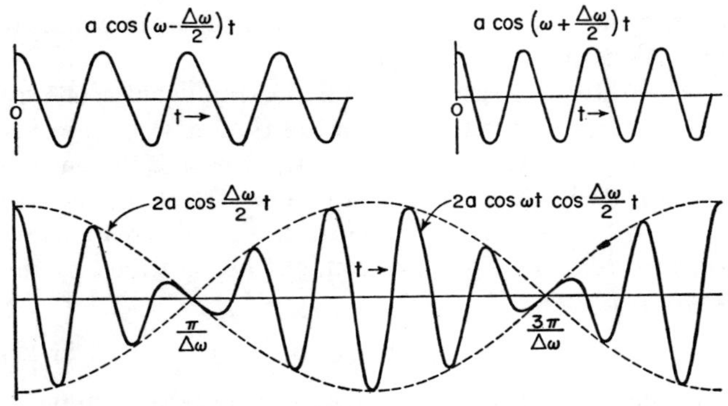

Fig. 11.2. Beat frequencies produced by two waves of slightly different wavelengths.

the *signal velocity* or *group velocity* of the modulated wave, is given by

$$\frac{dx}{dt} \equiv v_g = \frac{d\omega}{d\beta} = dv\, d\lambda, \tag{11.17}$$

[2] For example, see E. A. Guillemin, *The Mathematics of Circuit Analysis*, John Wiley and Sons, New York, 1949.

since a plane of constant *modulated amplitude* is described by

$$\frac{d\beta}{2}x - \frac{d\omega}{2}t = \text{constant}. \tag{11.18}$$

In comparing the group velocity v_g of a modulated wave with the phase velocity v of a monochromatic wave (cf. Eq. 7.19), we obtain

$$v_g = \frac{1}{\frac{d\beta}{d\omega}} = v + \beta\frac{dv}{d\beta} = v - \lambda\frac{dv}{d\lambda}. \tag{11.19}$$

Group and phase velocity are identical only in media in which the propagation velocity is frequency independent, that is, in *nondispersive dielectrics*.

The superposition of two waves leads to a periodic sequence of beats. By the superposition of many plane waves, covering a frequency interval with prescribed amplitudes, we may construct pulses or *wave packets* of any desired shape (Fig. 11.3). The fact that any wave

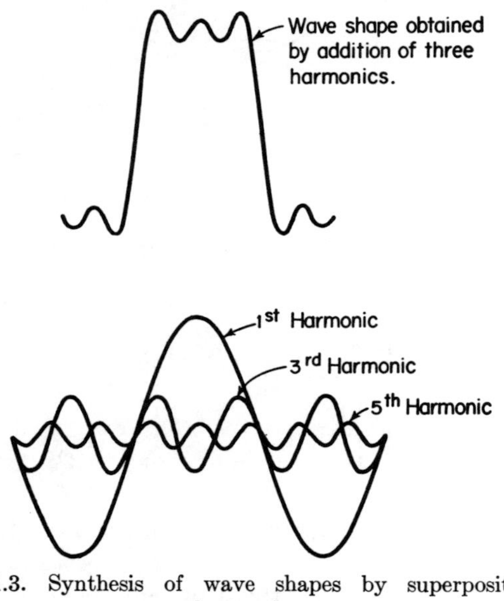

Fig. 11.3. Synthesis of wave shapes by superposition of harmonics.

phenomenon may be compounded from plane waves justifies simplifying the discussion of the field equations by the condition of Eq. 7.8 without loss in generality.

12 · Polarized Radiation

The theory outlined thus far leads to the following arrangement for measuring ϵ^* and μ^* in unbounded space (Fig. 12.1): An incoming modulated TEM wave

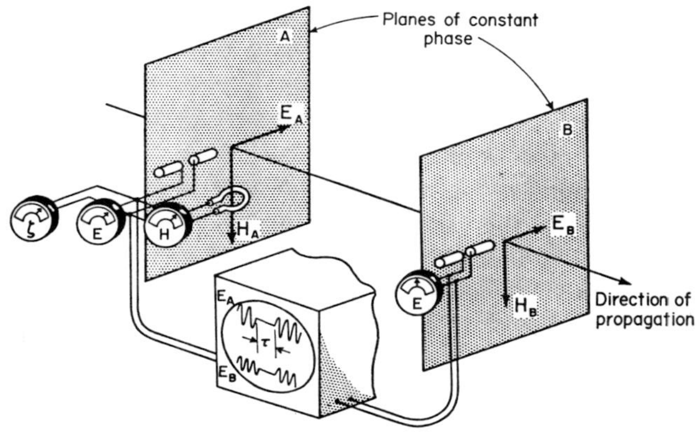

Fig. 12.1. Arrangement for measuring ϵ^* and μ^* in unbounded space.

passes in succession the planes A and B, oriented perpendicularly to the direction of propagation. At A, an electric and a magnetic receiver, arranged normally to each other, measure the ratio E/H; at B, a second E or H detector compares the signal with that in A as to magnitude and time difference τ of arrival. The first measurement gives the intrinsic impedance or $\sqrt{\mu^*/\epsilon^*}$; the second, the propagation factor or $\sqrt{\epsilon^*\mu^*}$ for a known signal frequency.

This measurement presupposes that the dielectric is homogeneous and isotropic, that is, that the *state of polarization* of the wave does not change from A to B. In other words, the field vectors **E** and **H**, when projected on a screen normal to the direction of propagation, must traverse the same geometrical pattern in the same space orientation at any distance x.

To derive a more quantitative description of this wave polarization, let the electric field be composed of the components

$$E_y = a\cos\omega t$$

and (12.1)

$$E_z = b\cos(\omega t + \delta) = b(\cos\omega t\cos\delta - \sin\omega t\sin\delta).$$

These components oscillate on the screen in the y- and z-directions of a Cartesian co-ordinate system, with amplitudes a and b, and follow each other with a relative shift in temporal phase, δ (Fig. 12.2). If we rewrite Eq. 12.1,

$$\frac{E_y}{a}\sin\delta = \cos\omega t\sin\delta,$$

$$\frac{E_y}{a}\cos\delta - \frac{E_z}{b} = \sin\omega t\sin\delta, \qquad (12.2)$$

we obtain, by squaring and adding, the orbital equation

$$\left(\frac{E_y}{a}\right)^2 - 2\frac{E_y}{a}\frac{E_z}{b}\cos\delta + \left(\frac{E_z}{b}\right)^2 = \sin^2\delta. \qquad (12.3)$$

For a phase difference of 90° between the components $\left(\delta = \pm\frac{\pi}{2}\right)$, this reduces to

$$\left(\frac{E_y}{a}\right)^2 + \left(\frac{E_z}{b}\right)^2 = 1, \qquad (12.4)$$

the well-known normal form for an ellipse, oriented with its axes $2a$ and $2b$ in the y- and z-directions, respectively.

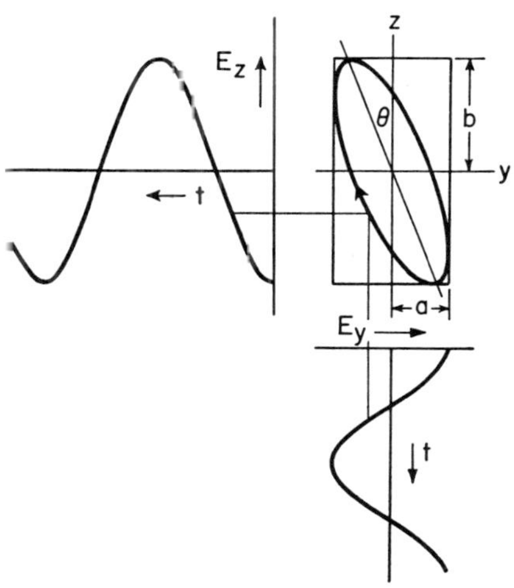

Fig. 12.2. Elliptically polarized radiation.

The general equation (12.3) depicts an ellipse inscribed within a rectangle of the sides $2a$ and $2b$; the principal axes of the ellipse are turned by an angle θ into the y'- and z'-directions (Fig. 12.3). This can be shown as follows. The ellipse is represented as

$$\left(\frac{E_{y'}}{u}\right)^2 + \left(\frac{E_{z'}}{w}\right)^2 = 1, \tag{12.5}$$

where

$$E_{y'} = u \cos(\omega t - \theta) = E_y \cos\theta + E_z \sin\theta,$$
$$E_{z'} = w \sin(\omega t - \theta) = -E_y \sin\theta + E_z \cos\theta. \tag{12.6}$$

By introducing these expressions for the components into Eq. 12.5, the form of the general equation (12.3) is regained as

$$E_y^2\left(\frac{\cos^2\theta}{u^2} + \frac{\sin^2\theta}{w^2}\right) - 2E_yE_z \sin\theta\cos\theta\left(\frac{1}{w^2} - \frac{1}{u^2}\right) + E_z^2\left(\frac{\cos^2\theta}{w^2} + \frac{\sin^2\theta}{u^2}\right) = 1. \tag{12.7}$$

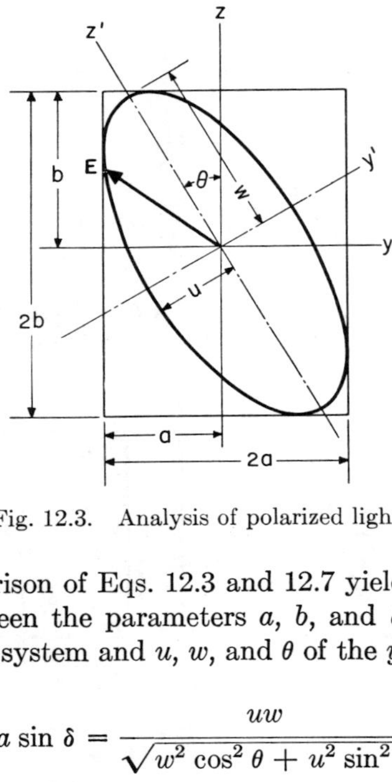

Fig. 12.3. Analysis of polarized light.

A comparison of Eqs. 12.3 and 12.7 yields the correlation between the parameters a, b, and δ of the y, z co-ordinate system and u, w, and θ of the y', z' system:

$$a \sin\delta = \frac{uw}{\sqrt{w^2\cos^2\theta + u^2\sin^2\theta}},$$
$$b \sin\delta = \frac{uw}{\sqrt{u^2\cos^2\theta + w^2\sin^2\theta}}, \tag{12.8}$$
$$\frac{ab\sin\delta}{\cos\delta} = \frac{u^2w^2}{\sin\theta\cos\theta\{u^2 - w^2\}}.$$

In addition,

$$a^2 + b^2 = u^2 + w^2. \tag{12.9}$$

If the parameters a, b, and δ are given, u, w, and θ may be found as

$$\tan 2\theta = \frac{2ab}{a^2 - b^2}\cos\delta = \tan 2\beta\cos\delta,$$
$$\tan\beta \equiv \frac{b}{a},$$
$$\sin 2\alpha = \pm\sin 2\beta\sin\delta, \tag{12.10}$$
$$\tan\alpha \equiv \frac{w}{u}.$$

A temporal phase shift δ between the two field components creates, therefore, in general, *elliptically polarized radiation* (Fig. 12.4). If the two components have equal amplitudes and a phase shift of an odd multiple of $\pi/2$, *circularly polarized radiation* results. If the observer, facing the oncoming wave, sees a clockwise rotation of the radius vector, the polarization is called *right-hand circular*, whereas a counterclockwise rotation corresponds to *left-hand circular polarization*. For a phase

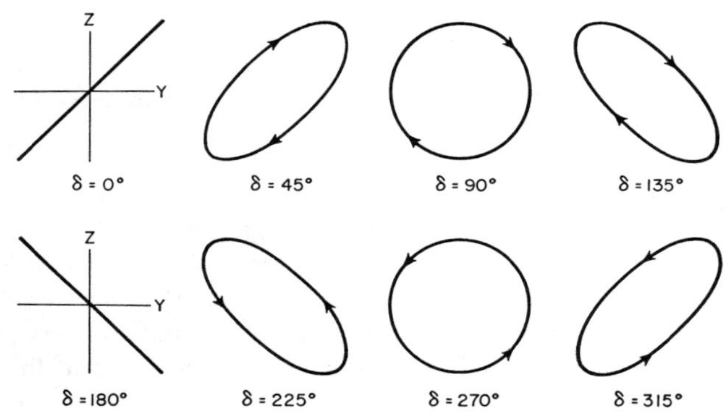

Fig. 12.4. Various states of polarization ($a = b$).

shift of a multiple of π, the ellipse degenerates into a straight line ($\delta = \pm m\pi$, $m = 1, 2, 3, \ldots$), the wave is *linearly polarized*, and, according to Eq. 12.10,

$$\tan\theta = \frac{b}{a}(-1)^m. \tag{12.11}$$

The *state of polarization of a wave*, that is, a/b and δ, may be found by determining first the position of the principal axes y', z' in space and then delaying one of the two field components E_y or E_z, by an additional quarter wavelength in order to produce linearly polarized radiation. By measuring the angle α of this linear radiation with respect to the y'-axis, the ratio w/u is established (see Eq. 12.10); by the position of the y'-axis, in relation to the y-axis, the angle θ is found; hence a/b and δ are obtained.

13 · Dipole Radiation

The theory of polarization and magnetization is based on the dipole concept. In Sec. 3 we derived the electric field surrounding a static dipole of the permanent moment $\boldsymbol{\mu}$. Frequently, however, we do not deal with such permanent moments, but with moments induced by the external field and varying with it as a periodic function of time,

$$\boldsymbol{\mu}(t) = Q\mathbf{d}e^{j\omega t} = \boldsymbol{\mu}_0 e^{j\omega t}. \qquad (13.1)$$

An *induced dipole moment* $\mu(t)$ can be visualized as a dipole of the constant length d, in which the charge alternates periodically between a $(+-)$ and a $(-+)$ constellation. Such a dipole is equivalent to a linear dipole antenna of the length d traversed by an alternating current (Fig. 13.1),

$$I = \frac{dQ}{dt} = \frac{1}{d}\frac{|d\boldsymbol{\mu}(t)|}{dt}. \qquad (13.2)$$

Dipole antennas are the simplest source of polarized radiation. In fact, Maxwell's electromagnetic theory

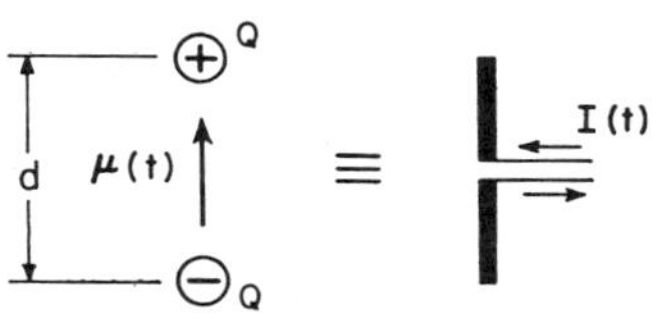

Fig. 13.1. Equivalence of alternating dipole and dipole antenna.

was confirmed when Hertz[1] succeeded in calculating this radiation and in verifying its existence by experiment. The electromagnetic field of an alternating dipole is thus of great importance for the macroscopic as well as the molecular theory. We will analyze it briefly.

The field vectors **E** and **H** have to satisfy Maxwell's equations, and therefore the wave equations (7.6 and 7.7). Assuming that the dipole is surrounded by an isotropic medium without loss, we may write these equations

$$\nabla^2\mathbf{E} = \frac{1}{v^2}\frac{\partial^2\mathbf{E}}{\partial t^2},$$
$$\nabla^2\mathbf{H} = \frac{1}{v^2}\frac{\partial^2\mathbf{H}}{\partial t^2}, \qquad (13.3)$$

where v is the phase velocity of the electromagnetic waves in the dielectric (see Eq. 7.19).

[1] H. Hertz, *Ann. Physik 36*, 1 (1888).

The wave equations hold when the divergence conditions, Eqs. 7.3 and 7.4, are fulfilled. Hence they are valid everywhere in space except at the location of the charges and the current elements of the dipole proper. However, we know already the magnetic field in the immediate surroundings of a current element. It is described by Biot-Savart's law (Eq. 5.18), which, for our dipole antenna of the length d, may be rewritten according to Eq. 13.2:

$$\mathbf{H} = \frac{I}{4\pi r^2}(\mathbf{d}\times\mathbf{r}^0) = \frac{1}{4\pi r^2}\frac{d|\boldsymbol{\mu}(t)|}{dt}\frac{(\mathbf{d}\times\mathbf{r}^0)}{d}. \qquad (13.4)$$

Our first problem is therefore to find an expression for the magnetic field strength **H** that obeys the wave equation in space and reduces to the Biot-Savart equation (13.4) at the dipole.

We can guess at the proper solution by the following consideration. In Sec. 5 we derived the Biot-Savart law by the introduction of a vector potential **O** from which the magnetic field strength was obtained by the curl operation (see Eq. 5.19). This vector potential is spherically symmetric and can be written for the dipole antenna according to Eqs. 5.25, 13.1, and 13.2 as

$$\mathbf{O} = \frac{\mathbf{I}d}{4\pi r} = \frac{1}{4\pi r}\frac{d\boldsymbol{\mu}(t)}{dt} = \frac{j\omega}{4\pi r}\boldsymbol{\mu}_0 e^{j\omega t}. \qquad (13.5)$$

The electromagnetic field created by the dipole antenna will travel outward with a phase velocity v, hence must appear at a distance r from the wire delayed by a temporal phase

$$e^{-j\omega\frac{r}{v}} = e^{-j\frac{2\pi}{\lambda}r}. \qquad (13.6)$$

The simplest assumption is that the vector potential **O** of Biot-Savart's law has only to be modified, by the introduction of this *retardation factor*, as

$$\mathbf{O} = \frac{j\omega}{4\pi r}\boldsymbol{\mu}_0 e^{j\omega\left(t-\frac{r}{v}\right)} \qquad (13.7)$$

to give the magnetic field of the alternating dipole.

Carrying through the curl operation (see Eqs. 5.12 and 5.27), we obtain the field expression

$$\mathbf{H} = \nabla\times\mathbf{O}$$
$$= \frac{|\boldsymbol{\mu}_0|v}{4\pi}e^{j\omega\left(t-\frac{r}{v}\right)}\left\{-\frac{\omega^2}{rv^2} + j\frac{\omega}{r^2 v}\right\}\frac{(\mathbf{d}\times\mathbf{r}^0)}{d}; \qquad (13.8)$$

or, written in spherical co-ordinates:

$$H_r = 0,$$
$$H_\theta = 0, \qquad (13.9)$$
$$H_\phi = \frac{|\,\boldsymbol{\mu}(t)\,|\,v}{4\pi} e^{-j\omega\frac{r}{v}}\left(-\frac{\omega^2}{rv^2} + j\frac{\omega}{r^2 v}\right)\sin\theta.$$

In the *near zone* of the dipole ($r \ll \lambda$) the term in $1/r^2$ dominates, and the magnetic field becomes identical with the Biot-Savart field of Eq. 13.4 save for the retarding phase factor. In the *far zone*, at great distance from the dipole ($r \gg \lambda$), only the term varying with $1/r$ remains.

To calculate the accompanying electric field, we have to introduce the expression for **H** of Eq. 13.8 into Maxwell's first field equation (Eq. 7.1), written for a dielectric medium without loss

$$\nabla \times \mathbf{H} = \epsilon' \frac{\partial \mathbf{E}}{\partial t}. \qquad (13.10)$$

The curl operation, carried through as above, leads to the electric field components

$$E_r = \frac{|\,\boldsymbol{\mu}(t)\,|}{\epsilon' 4\pi} e^{-j\omega\frac{r}{v}}\left(\frac{2}{r^3} + j\frac{2\omega}{r^2 v}\right)\cos\theta,$$
$$E_\theta = \frac{|\,\boldsymbol{\mu}(t)\,|}{\epsilon' 4\pi} e^{-j\omega\frac{r}{v}}\left(\frac{1}{r^3} + j\frac{\omega}{r^2 v} - \frac{\omega^2}{rv^2}\right)\sin\theta, \qquad (13.11)$$
$$E_\phi = 0.$$

In the near zone of the dipole ($r \ll \lambda$) the term in $1/r^3$ dominates, and the electric field intensity becomes

$$E_r = \frac{|\,\boldsymbol{\mu}(t)\,|\cos\theta}{\epsilon' 2\pi r^3} e^{-j\omega\frac{r}{v}},$$
$$E_\theta = \frac{|\,\boldsymbol{\mu}(t)\,|\sin\theta}{\epsilon' 4\pi r^3} e^{-j\omega\frac{r}{v}}. \qquad (13.12)$$

It is identical, as a comparison with Eq. 3.19 shows, with the field of an electrostatic dipole, save that now the dipole moment is a periodic function of time and that the retardation factor $e^{-j\omega\frac{r}{v}}$ appears again.

Thus in the near zone the electromagnetic field of an alternating electric dipole or a linear dipole antenna consists of the components

$$E_r = \frac{|\,\boldsymbol{\mu}(t)\,|\cos\theta}{\epsilon' 2\pi r^3} e^{-j\omega\frac{r}{v}},$$
$$E_\theta = \frac{|\,\boldsymbol{\mu}(t)\,|\sin\theta}{\epsilon' 4\pi r^3} e^{-j\omega\frac{r}{v}}, \qquad (13.13)$$
$$H_\phi = \frac{d|\,\boldsymbol{\mu}(t)\,|}{dt}\,\frac{\sin\theta}{4\pi r^2} e^{-j\omega\frac{r}{v}}.$$

It is built up by the electric field of a dipole that consists of a radial component E_r and of a component E_θ tangential to the circles of longitude of a polar sphere. Superposed on it is the magnetic field of the dipole antenna current H_ϕ oriented tangentially in the direction circles of latitude of the polar sphere.

For the far zone ($r \gg \lambda$) only the terms varying with $1/r$ remain of importance, that is, the electromagnetic field consists of the two tangential components (Fig. 13.2),

$$E_\theta = \frac{d^2|\,\boldsymbol{\mu}(t)\,|}{dt^2}\,\frac{\sin\theta}{\epsilon' 4\pi r v^2} e^{-j\omega\frac{r}{v}},$$
$$H_\phi = \frac{d^2|\,\boldsymbol{\mu}(t)\,|}{dt^2}\,\frac{\sin\theta}{4\pi r v} e^{-j\omega\frac{r}{v}}. \qquad (13.14)$$

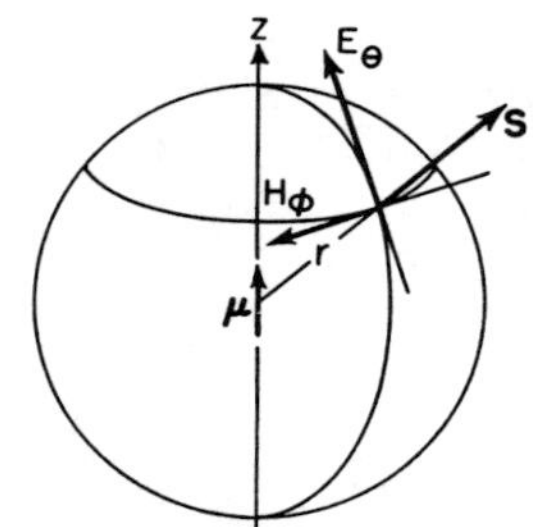

Fig. 13.2. Tangential components (far zone) and Poynting vector.

This field at great distance from the dipole antenna consists of TEM waves propagating radially outwards. The field amplitudes diminish proportional to $1/r$, that is, the field energy with $1/r^2$, as we would expect for any point source of radiation. They are furthermore proportional to the second derivative of the dipole moment, that is, the first derivative of the antenna current. Finally, the radiation pattern is directive; the intensity is a maximum perpendicular to the alternating dipole and zero in the direction of the dipole moment.

Equations 13.9 and 13.11 represent Hertz's famous solution of Maxwell's equations for spherical waves, which has the unusual feature of describing simultaneously the *light source* (alternating dipole of near field) and its *radiation field in space* (antenna field of far zone) (Fig. 13.3).[2]

The energy density of the radiation field in a volume element dV at a distance r from the source is (see Eq. 11.9)

$$dU = \rho\, dV = \tfrac{1}{2}(\epsilon' E^2 + \mu' H^2)\, dV \qquad (13.15)$$

or, by introducing the field intensities from Eq. 13.14,

$$dU = \frac{\mu'\omega^4\boldsymbol{\mu}^2(t)}{(4\pi)^2 r^2 v^2}\sin^2\theta\, dV. \qquad (13.16)$$

[2] G. Joos, *Lehrbuch der theoretischen Physik*, Becker und Erler Kom.-Ges., 6th ed., 1945.

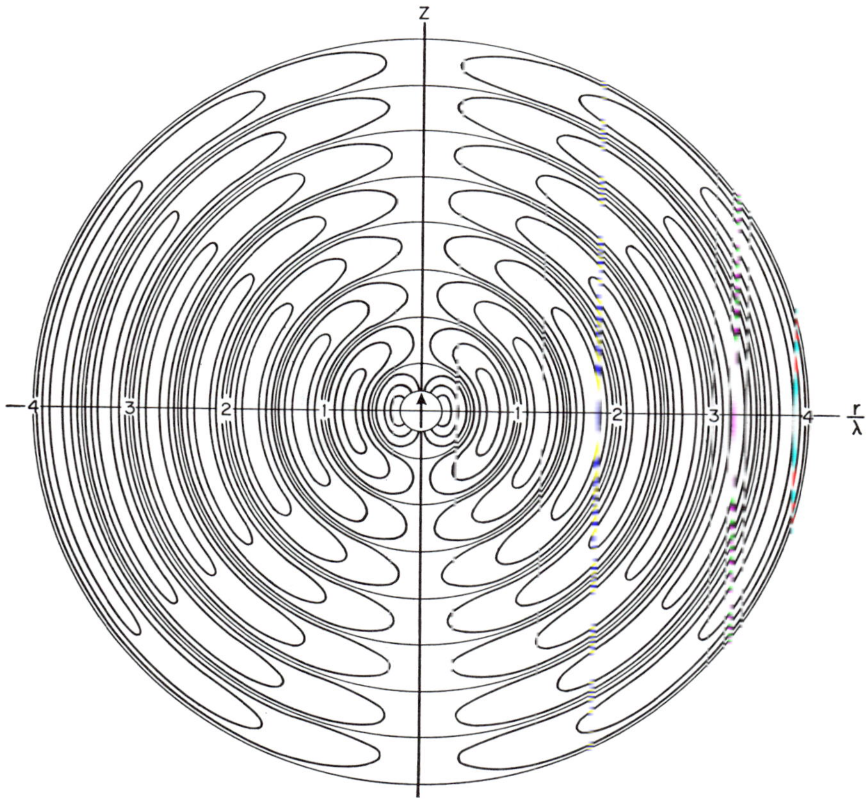

Fig. 13.3. Radiation pattern of **E**-field of dipole antenna. (After Joos.[2])

It streams outwards with the velocity v; the energy flow per second through a surface element dA of the polar sphere is therefore (see Eq. 11.12)

$$\bar{\mathbf{S}}\,dA = \mathbf{r}^0 v\rho\,dA, \tag{13.17}$$

and the total energy $\bar{P}$ radiated away per second

$$\bar{P} = \oint \bar{\mathbf{S}}\,dA = \frac{\mu'\omega^4}{(4\pi)^2 r^2 v}\frac{\boldsymbol{\mu}_0{}^2}{2}2r^2\pi\int_0^\pi \sin^2\theta\sin\theta\,d\theta$$
$$= \frac{\mu'\omega^4}{12\pi v}\boldsymbol{\mu}_0{}^2 \quad \text{[watt].}\dagger \tag{13.18}$$

This power loss of a linear dipole antenna due to radiation can be expressed as being equivalent to that of a nonradiating antenna of the resistance R_s, in which the same power

$$P = \frac{I_0{}^2}{2}R_s \tag{13.19}$$

is annihilated in Joule heat. By equating the two expressions for P and recalling that $\boldsymbol{\mu}_0$ stands for the maximum dipole moment of the antenna and I_0 for the maximum current traversing it,

$$\boldsymbol{\mu}_0 = Q_0\mathbf{d},$$
$$|I_0| = \left|\frac{dQ_0}{dt}\right| = \omega|Q_0|, \tag{13.20}$$

we obtain for the *equivalent radiation resistance of the dipole antenna* in vacuum [$v = c$; $\mu'c = Z_0 = 120\pi$ (see Eq. 8.5)],

$$R_s = 80\pi^2\left(\frac{d}{\lambda}\right)^2 \quad \text{[ohm].} \tag{13.21}$$

It will prove useful for the later molecular discussion to represent the power loss of a dipole antenna alternatively as caused by a *radiation friction force* $\mathbf{F}_s$ attenuating the dipole oscillation. In this equivalent picture the dipole moment changes with time, not because its charge alternates but because its length d varies with time as $\mathbf{d} = \mathbf{d}_0 e^{j\omega t}$. The factor $\omega^4\boldsymbol{\mu}^2(t)$ of Eq. 13.16 has therefore to be interpreted as representing

$$\omega^4\boldsymbol{\mu}^2(t) = Q^2\left[\frac{d^2(\mathbf{d})}{dt^2}\right]^2 = Q^2\omega^4\mathbf{d}^2. \tag{13.22}$$

The power emitted in a time interval $0 \rightarrow t$ becomes equal to the work done in changing the length of the

† $\bar{S}$ is the mean value of the energy crossing a unit area of the polar sphere in the direction of the normal r^0; it can also be calculated as a real part of the complex Poynting vector, Re (S^*). (See footnote, Sec. 1, p. 3, and J. A. Stratton, *Electromagnetic Theory*, p. 436.)

dipole against the action of an opposing force F_s,

$$\int_0^t P\,dt = \frac{\mu'}{6\pi v}\int_0^t Q^2\left[\frac{d^2(\mathbf{d})}{dt^2}\right]^2 dt$$
$$\equiv \int_0^t \mathbf{F}_s\cdot\frac{d(\mathbf{d})}{dt}\,dt. \qquad (13.23)$$

By partial integration we change the left-hand side of this integral equation into the form

$$\frac{\mu'}{6\pi v}\int_0^t Q^2\left[\frac{d^2(\mathbf{d})}{dt^2}\right]^2 dt = -\frac{\mu'}{6\pi v}\int_0^t Q^2\frac{d^3(\mathbf{d})}{dt^3}\frac{d(\mathbf{d})}{dt}\,dt + \left[\frac{\mu'}{6\pi v}Q^2\frac{d^2(\mathbf{d})}{dt^2}\frac{d(\mathbf{d})}{dt}\right]_0^t. \qquad (13.24)$$

The second term vanishes, when the attenuation is small, and the radiation friction force becomes

$$\mathbf{F}_s = -\frac{\mu' Q^2}{6\pi v}\frac{d^3(\mathbf{d})}{dt^3}. \qquad (13.25)$$

In contrast to ordinary friction forces which change proportionally to the first derivative of the position vector, that is, proportional to the velocity of motion, the friction force simulating radiation damping changes proportional to the third derivative of the dipole length. This becomes obvious when we consider that the second derivative of the dipole moment represents the change in current that originates the electromagnetic field.

14 · Boundary Conditions

Electromagnetic waves are, in general, not observed in unbounded space but confined by various types of boundaries. This is no disadvantage; on the contrary, the reflection and refraction of fields by matter provides the essential means for directing and modifying such fields and for measuring the dielectric properties of materials.

The interaction of static fields with boundaries is simple. The surfaces of conductors in electrostatic fields are, by definition, equipotential surfaces; hence the field lines must be perpendicular to them. At the interface between two nonconducting media the normal component of the electric flux density **D** has to be continuous when there is no surface charge (see Eq. 2.3),

$$D_{n_1} = D_{n_2}. \qquad (14.1)$$

Consequently, there must be a jump in the amplitude of the normal component of the field strength **E** at the interface in the inverse ratio of the dielectric constants of the two media (see Eq. 2.6)

$$\frac{E_{n_1}}{E_{n_2}} = \frac{\epsilon_2'}{\epsilon_1'} \qquad (14.2)$$

because the free ends of dipole chains produce free charges. The tangential component of **E** on the other hand must be continuous,

$$E_{t_1} = E_{t_2}, \qquad (14.3)$$

by reason of the fact that the field is conservative (see Eq. 3.4). A closed path may be chosen which hugs the boundary and consists for all practical purposes of only the tangential field components in the two media (Fig. 14.1). From Eqs. 14.2 and 14.3 it follows that field **E**,

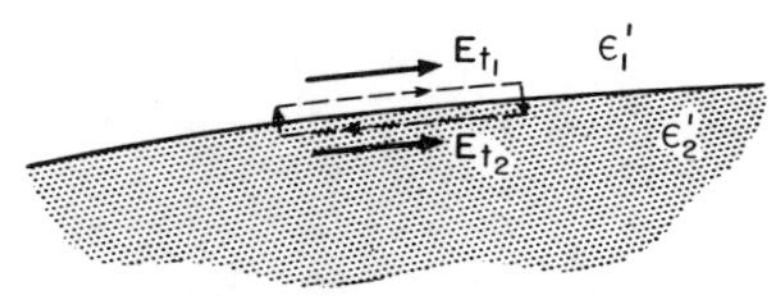

Fig. 14.1. Continuity of the tangential field components.

inclined in medium 1 at an angle α_1 with respect to the normal, will continue in medium 2 at an angle α_2 (Fig. 14.2) in accordance with the *refraction law for electrostatic field lines*

$$\frac{\tan\alpha_1}{\tan\alpha_2} = \frac{\epsilon_1'}{\epsilon_2'}. \qquad (14.4)$$

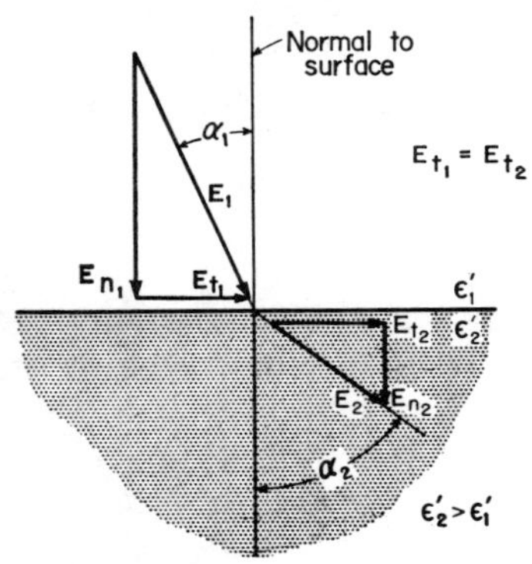

Fig. 14.2. Refraction law for static field.

Analogous reasoning leads for magnetostatic fields to the *magnetostatic refraction law*

$$\frac{\tan \alpha_1}{\tan \alpha_2} = \frac{\mu_1'}{\mu_2'}. \tag{14.5}$$

For electromagnetic waves striking the boundary between two media a much more complicated situation arises because the waves are partly reflected and partly refracted, thus changing their direction of propagation as well as their amplitudes and phases. The basic facts will become clear by reference to Fig. 14.3.[1]

Let O be a fixed origin located on the interface plane S, and $\mathbf{r}$ be a position vector drawn from O to any point P in medium 1 or 2. Furthermore let $\mathbf{k}^0$ be a unit vector pointing in the direction of propagation of a TEM wave, and let the indices i, r, t signify the incident, reflected, and transmitted waves, respectively. By specifying a unit vector $\mathbf{n}$ normal to the boundary and pointing from medium 2 into medium 1, we may define the interface S as

$$\mathbf{n}\cdot\mathbf{r} = 0, \tag{14.6}$$

whereas

$$\mathbf{k}^0\cdot\mathbf{r} = \text{constant} \tag{14.7}$$

designates a plane of constant phase of the TEM wave. The electric and magnetic field strengths of the incident, reflected, and refracted waves may now be written (cf. Eqs. 7.10 and 7.24)

$$\begin{aligned}
\mathbf{E}_i &= \mathbf{E}_0 e^{j\omega t - \gamma_1 \mathbf{k}_i^0\cdot\mathbf{r}},\\
\mathbf{H}_i &= \mathbf{H}_0 e^{j\omega t - \gamma_1 \mathbf{k}_i^0\cdot\mathbf{r}} = \frac{1}{Z_1}(\mathbf{k}_i^0 \times \mathbf{E}_i);\\
\mathbf{E}_r &= \mathbf{E}_1 e^{j\omega t - \gamma_1 \mathbf{k}_r^0\cdot\mathbf{r}},\\
\mathbf{H}_r &= \mathbf{H}_1 e^{j\omega t - \gamma_1 \mathbf{k}_r^0\cdot\mathbf{r}} = \frac{1}{Z_1}(\mathbf{k}_r^0 \times \mathbf{E}_r);\\
\mathbf{E}_t &= \mathbf{E}_2 e^{j\omega t - \gamma_2 \mathbf{k}_t^0\cdot\mathbf{r}},\\
\mathbf{H}_t &= \mathbf{H}_2 e^{j\omega t - \gamma_2 \mathbf{k}_t^0\cdot\mathbf{r}} = \frac{1}{Z_2}(\mathbf{k}_t^0 \times \mathbf{E}_t).
\end{aligned} \tag{14.8}$$

By introducing the general distance parameter $\mathbf{k}^0\cdot\mathbf{r}$ of a phase plane from the origin in place of a definite space co-ordinate, we have gained the freedom of treating the direction of propagation as a variable of our problem. The alternative expressions for the **H** vectors in Eqs. 14.8 relating them to the **E** vectors result from the definition of the intrinsic impedance (see Eq. 7.25) and from the condition that $\mathbf{k}^0$, **E**, and **H** must form a right-hand co-ordinate system (see Fig. 7.1).

[1] See also J. A. Stratton, *Electromagnetic Theory*, McGraw-Hill Book Co., New York, 1941, pp. 490 ff.

The incident, reflected, and refracted waves are interconnected by the *boundary condition:* the tangential components of **E** and **H** must be continuous in traversing the interface S, or

$$\begin{aligned}
\mathbf{n}\times(\mathbf{E}_i + \mathbf{E}_r) &= \mathbf{n}\times\mathbf{E}_t,\\
\mathbf{n}\times(\mathbf{H}_i + \mathbf{H}_r) &= \mathbf{n}\times\mathbf{H}_t.
\end{aligned} \tag{14.9}$$

This condition is a consequence of Ampère's circuital law and of Faraday's induction law where applied to a

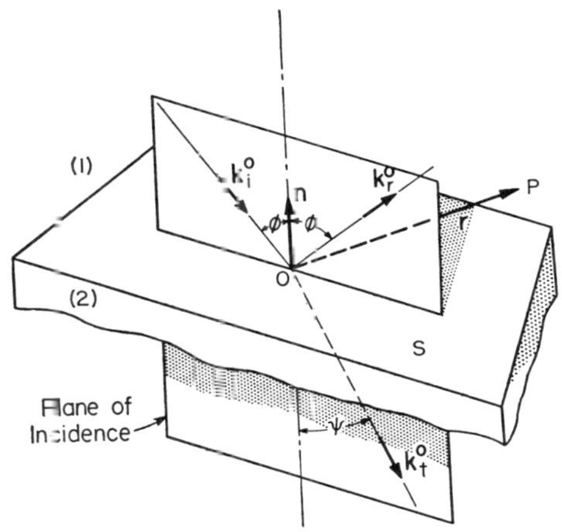

Fig. 14.3. Reference system for reflection and refraction of TEM wave.

path hugging the boundary as in the static case (see Fig. 14.1); again only the tangential components contribute and have to equal each other (see also Appendix A, I, 13).

The boundary condition can be fulfilled only if the amplitudes as well as the exponential functions of the tangential components obey it on the interface, that is, Eq. 14.9 splits into the two sets of conditions:

$$\begin{aligned}
\mathbf{n}\times(\mathbf{E}_0 + \mathbf{E}_1) &= \mathbf{n}\times\mathbf{E}_2,\\
\mathbf{n}\times(\mathbf{H}_0 + \mathbf{H}_1) &= \mathbf{n}\times\mathbf{H}_2,
\end{aligned} \tag{14.10}$$

and

$$\begin{aligned}
\gamma_1\mathbf{k}_i^0\cdot\mathbf{r} &= \gamma_1\mathbf{k}_r^0\cdot\mathbf{r},\\
\gamma_1\mathbf{k}_i^0\cdot\mathbf{r} &= \gamma_2\mathbf{k}_t^0\cdot\mathbf{r}.
\end{aligned} \tag{14.11}$$

The first condition, prescribing that the sum of the amplitudes for the tangential components of the incident and reflected waves at the interface must be equal to the amplitude of the tangential component of the transmitted wave, leads to *Fresnel's equations.*[2] The second condition, demanding equality of the phases at

[2] A. Fresnel, *Mém. acad. sciences 11*, 393 (1832). The complete Fresnel equations were first presented to the Académie des Sciences in 1823, but the manuscript was lost. Nine years later it was found among Fourier's belongings and posthumously published.

the boundary, contains *Snell's laws of reflection and refraction.*[3]

To analyze the phase condition, Eq. 14.11, we split the position vector **r** into its components parallel and perpendicular to the interface (Fig. 14.4),

$$\mathbf{r} = (\mathbf{n}\cdot\mathbf{r})\mathbf{n} - \mathbf{n}\times(\mathbf{n}\times\mathbf{r}). \qquad (14.12)$$

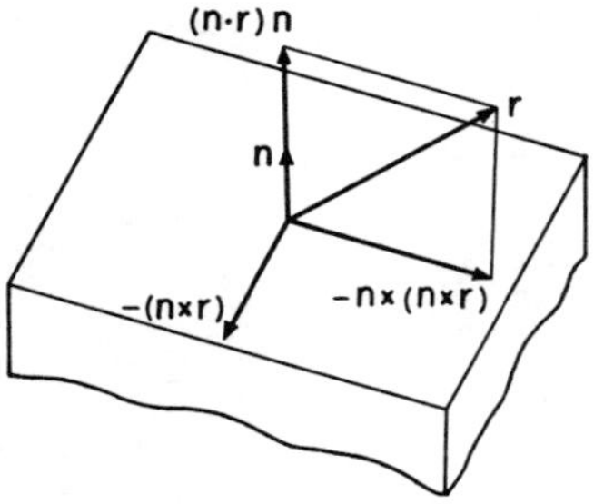

Fig. 14.4. Components of position vector.

Since $\mathbf{n}\cdot\mathbf{r}$ at the interface is zero (see Eq. 14.6) and by making use of the vector rule

$$\mathbf{k}^0\cdot\mathbf{n}\times(\mathbf{n}\times\mathbf{r}) = (\mathbf{k}^0\times\mathbf{n})\cdot(\mathbf{n}\times\mathbf{r}), \qquad (14.13)$$

we can rewrite the phase condition (Eq. 14.11)

$$(\mathbf{k}_i^0\times\mathbf{n} - \mathbf{k}_r^0\times\mathbf{n})\cdot(\mathbf{n}\times\mathbf{r}) = 0,$$

$$(\mathbf{k}_i^0\times\mathbf{n} - \frac{\gamma_2}{\gamma_1}\mathbf{k}_t^0\times\mathbf{n})\cdot(\mathbf{n}\times\mathbf{r}) = 0. \qquad (14.14)$$

[3] W. Snell established this law experimentally, probably in 1621. A theoretical derivation was given by Descartes in his *Dioptrique* (Leyden, 1638), but no credit was given Snell. Huygens states in one of his letters that Snell's results were known to Descartes prior to the publication of the *Dioptrique* (E. T. Whittaker, *A History of the Theories of Aether and Electricity*, Thomas Nelson and Sons, London, 1951, p. 10).

The cross products of the three propagation vectors $\mathbf{k}^0$ and the normal **n** are equal to the sines of the *angles of incidence* (ϕ), *reflection* (ϕ'), and *refraction* (ψ) (see Fig. 14.3), or

$$\begin{aligned}\mathbf{k}_i^0\times\mathbf{n} &= \sin\phi,\\ \mathbf{k}_r^0\times\mathbf{n} &= \sin\phi',\\ \mathbf{k}_t^0\times\mathbf{n} &= \sin\psi.\end{aligned} \qquad (14.15)$$

Hence, the first line of Eq. 14.14 states that $\mathbf{k}_i^0$, $\mathbf{k}_r^0$, and **n** lie in one plane and that

$$\sin\phi = \sin\phi'; \qquad (14.16)$$

the angle of incidence is equal to the angle of reflection (*Snell's reflection law*). According to the second line of Eq. 14.14 the unit vectors $\mathbf{k}_i^0$, $\mathbf{k}_t^0$, and **n** are also coplanar and

$$\sin\phi = \frac{\gamma_2}{\gamma_1}\sin\psi. \qquad (14.17)$$

This is *Snell's refraction law*, which may be written in terms of the alternative parameters ϵ^*, μ^*, or Z (see Eqs. 7.12 and 7.25) as

$$\frac{\sin\phi}{\sin\psi} = \sqrt{\frac{\epsilon_2^*\mu_2^*}{\epsilon_1^*\mu_1^*}} = \frac{Z_1\mu_2^*}{Z_2\mu_1^*}. \qquad (14.18)$$

For a medium without magnetization and loss ($\mu^* = \mu_0$, $\epsilon^* = \epsilon'$) it simplifies to

$$\frac{\sin\phi}{\sin\psi} = \sqrt{\frac{\epsilon_2'}{\epsilon_1'}} = \frac{\lambda_1}{\lambda_2} = \frac{n_2}{n_1} \equiv n_{21}. \qquad (14.19)$$

The common plane of the unit vectors, containing the direction of incidence and the normal to the interface, is called the *plane of incidence.*

15 · Fresnel's Equations

Returning to the amplitude conditions (Eq. 14.10), we have to prescribe the state of polarization of the incident TEM wave. The field vectors are perpendicular to the direction of incidence and normal to each other,

$$\begin{aligned}\mathbf{k}_i^0\cdot\mathbf{E}_0 &= 0,\\ \mathbf{k}_i^0\cdot\mathbf{H}_0 &= 0,\\ \mathbf{E}_0\cdot\mathbf{H}_0 &= 0,\end{aligned} \qquad (15.1)$$

but their orientation relative to the plane of incidence may still be arbitrarily chosen and will influence reflection and refraction. We therefore resolve each field vector into two components, an n component pointing normal to the plane of incidence (that is, tangent to the interface S) and a p component lying parallel to this plane (Fig. 15.1), and investigate the two components separately.

(a) The electric field normal to the plane of incidence

Since these $\mathbf{E}_n$ components are normal to **n**,

$$\mathbf{n}\cdot\mathbf{E}_{0,n} = \mathbf{n}\cdot\mathbf{E}_{1,n} = \mathbf{n}\cdot\mathbf{E}_{2,n} = 0, \qquad (15.2)$$

the first of Eqs. 14.10 reads simply

$$\mathbf{E}_{0,\,n} + \mathbf{E}_{1,\,n} = \mathbf{E}_{2,\,n}. \tag{15.3}$$

The magnetic components of Eq. 14.10, rewritten in terms of the electric field

$$\mathbf{n} \times (\mathbf{k}_i^{\,0} \times \mathbf{E}_0 + \mathbf{k}_r^{\,0} \times \mathbf{E}_1)\,\frac{1}{Z_1} = \mathbf{n} \times (\mathbf{k}_t^{\,0} \times \mathbf{E}_2)\,\frac{1}{Z_2}, \tag{15.4}$$

and expanded, lead to the additional interrelation

$$\cos\phi\,(\mathbf{E}_{0,\,n} - \mathbf{E}_{1,\,n}) = \frac{Z_1}{Z_2}\cos\psi\,\mathbf{E}_{2,\,n}. \tag{15.5}$$

Equations 15.3 and 15.5 together allow us to express the amplitudes of the reflected and refracted beams in

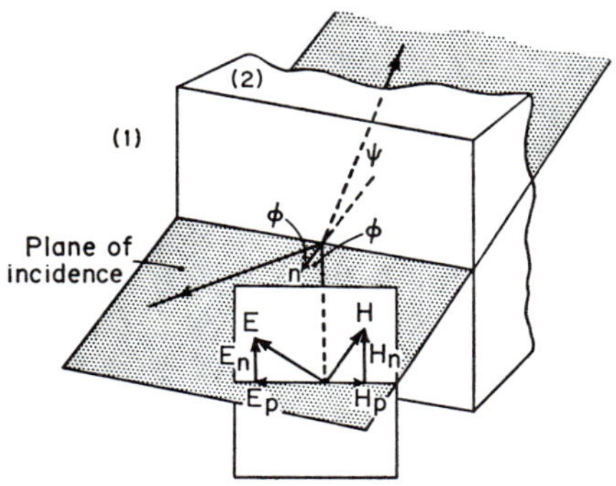

Fig. 15.1. Plane of incidence and field components.

terms of the incident amplitude, the angles ϕ and ψ, and the impedances of the two media,

$$\mathbf{E}_{1,\,n} = \frac{Z_2\cos\phi - Z_1\cos\psi}{Z_2\cos\phi + Z_1\cos\psi}\,\mathbf{E}_{0,\,n},$$
$$\mathbf{E}_{2,\,n} = \frac{2Z_2\cos\phi}{Z_2\cos\phi + Z_1\cos\psi}\,\mathbf{E}_{0,\,n}. \tag{15.6}$$

(b) The electric field parallel to the plane of incidence

Since in this case the magnetic components are perpendicular to the plane of incidence, we can repeat the calculation just performed for the $\mathbf{H}_n$ vectors and obtain in place of Eqs. 15.3 and 15.5

$$\mathbf{H}_{0,\,n} + \mathbf{H}_{1,\,n} = \mathbf{H}_{2,\,n},$$
$$\cos\phi(\mathbf{H}_{0,\,n} - \mathbf{H}_{1,\,n}) = \frac{Z_2}{Z_1}\cos\psi\mathbf{H}_{2,\,n}. \tag{15.7}$$

Thus, in analogy to Eq. 15.6, the reflected and transmitted normal magnetic field components become

$$\mathbf{H}_{1,\,n} = \frac{Z_1\cos\phi - Z_2\cos\psi}{Z_1\cos\phi + Z_2\cos\psi}\,\mathbf{H}_{0,\,n},$$
$$\mathbf{H}_{2,\,n} = \frac{2Z_1\cos\phi}{Z_1\cos\phi + Z_2\cos\psi}\,\mathbf{H}_{0,\,n}. \tag{15.8}$$

By making use of the impedance relations

$$\mathbf{E}_0 = -Z_1(\mathbf{k}_i^{\,0} \times \mathbf{H}_0), \text{ etc.}, \tag{15.9}$$

these magnetic components may be transformed into the accompanying electric field components parallel to the plane of incidence

$$\mathbf{E}_{1,\,p} = \frac{Z_2\cos\psi - Z_1\cos\phi}{Z_1\cos\phi + Z_2\cos\psi}\,\mathbf{E}_{0,\,p},$$
$$\mathbf{E}_{2,\,p} = \frac{2Z_2\cos\phi}{Z_1\cos\phi + Z_2\cos\psi}\,\mathbf{E}_{0,\,p}. \tag{15.10}$$

In the same way the still missing relation for the magnetic field components parallel to the plane of incidence may be derived from Eq. 15.6, as

$$\mathbf{H}_{1,\,p} = \frac{Z_1\cos\psi - Z_2\cos\phi}{Z_1\cos\psi + Z_2\cos\phi}\,\mathbf{H}_{0,\,p},$$
$$\mathbf{H}_{2,\,p} = \frac{2Z_1\cos\phi}{Z_1\cos\psi + Z_2\cos\phi}\,\mathbf{H}_{0,\,p}. \tag{15.11}$$

Equations 15.6, 15.8, 15.10, and 15.11 together represent a *general formulation of Fresnel's equations;* they prescribe the amplitudes and phases of the reflected and refracted waves in their dependence on the amplitude and angle of incidence of the incident wave and on the dielectric parameters (intrinsic impedance and propagation function) of media 1 and 2. The cosine of the angle of refraction in these equations can be replaced by the angle of incidence, since, according to the refraction law Eq. 14.17,

$$\cos\psi = \sqrt{1 - \frac{\gamma_1^{\,2}}{\gamma_2^{\,2}}\sin^2\phi}. \tag{15.12}$$

In the discussion of reflection and refraction phenomena, the relative amplitudes and intensities of the partial beams are normally of importance, but not their absolute values. Defining the ratio of the reflected to the incident amplitude at the boundary as the *reflection coefficient*

$$r_E = \mathbf{E}_1/\mathbf{E}_0, \quad r_H = \mathbf{H}_1/\mathbf{H}_0, \tag{15.13}$$

and the corresponding ratio of transmitted to incident amplitude as the transmission coefficients,

$$t_E = \mathbf{E}_2/\mathbf{E}_0, \quad t_H = \mathbf{H}_2/\mathbf{H}_0, \tag{15.14}$$

we may reformulate Fresnel's equations for the components normal and parallel to the plane of incidence as

$$r_{E_n} = \left(\frac{E_1}{E_0}\right)_n = \frac{Z_2 \cos\phi - Z_1 \cos\psi}{Z_2 \cos\phi + Z_1 \cos\psi} = -r_{H_p},$$
$$r_{E_p} = \left(\frac{E_1}{E_0}\right)_p = \frac{Z_2 \cos\psi - Z_1 \cos\phi}{Z_2 \cos\psi + Z_1 \cos\phi} = -r_{H_n}, \qquad (15.15)$$

and

$$t_{E_n} = \left(\frac{E_2}{E_0}\right)_n = \frac{2Z_2 \cos\phi}{Z_2 \cos\phi + Z_1 \cos\psi} = \frac{Z_2}{Z_1} t_{H_p},$$
$$t_{E_p} = \left(\frac{E_2}{E_0}\right)_p = \frac{2Z_2 \cos\phi}{Z_2 \cos\psi + Z_1 \cos\phi} = \frac{Z_2}{Z_1} t_{H_n}. \qquad (15.16)$$

For *perpendicular incidence* ($\phi = \psi = 0$) the distinction between the n- and p-components disappears, and Fresnel's equations simplify to

$$r_E = \frac{Z_2 - Z_1}{Z_2 + Z_1} = -r_H,$$
$$t_E = \frac{2Z_2}{Z_2 + Z_1} = \frac{Z_2}{Z_1} t_H. \qquad (15.17)$$

It is important to note that only for perpendicular incidence the amplitude of the incident wave equals the sum of the amplitudes of the reflected and transmitted waves. At oblique incidence the refracted beam changes cross section and the relation holds only for the components normal to the plane of incidence. These components are parallel to each other and tangent to the interface and therefore have to be continuous according to the boundary condition, Eq. 14.9. This statement reads, expressed with reflection and transmission coefficients,

$$t_{E_n} - r_{E_n} = 1,$$
$$t_{H_n} - r_{H_n} = 1. \qquad (15.18)$$

The minus sign enters, because the direction of propagation is reversed for the reflected beam.

The intensity of the average energy flow normal to the interface, on the other hand, must be continuous according to the energy principle, that is, if $\overline{\mathbf{S}}$ represents the time average of the Poynting vector † (see Eq. 11.12),

$$\mathbf{n}\cdot(\overline{\mathbf{S}}_i + \overline{\mathbf{S}}_r) = \mathbf{n}\cdot\overline{\mathbf{S}}_t. \qquad (15.19)$$

Equation 15.19 may be rewritten by changing over from the $\mathbf{H}$ to the $\mathbf{E}$ vector with help of the impedance concept,

$$\frac{1}{2Z_1}(E_0^2 - E_1^2)\cos\phi = \frac{1}{2Z_2} E_2^2 \cos\psi. \qquad (15.20)$$

By introducing the ratio of reflected to incident normal energy flow as the *energy coefficient of reflection* or the *reflectivity*

$$R \equiv \frac{\mathbf{n}\cdot\overline{\mathbf{S}}_r}{\mathbf{n}\cdot\overline{\mathbf{S}}_i} = \frac{E_1^2}{E_0^2}, \qquad (15.21)$$

and the corresponding ratio of transmitted to incident normal energy flow as the *energy coefficient of transmission* or the *transmissibility*

$$T = \frac{\mathbf{n}\cdot\overline{\mathbf{S}}_t}{\mathbf{n}\cdot\overline{\mathbf{S}}_i} = \frac{E_2^2 Z_1 \cos\psi}{E_0^2 Z_2 \cos\phi}, \qquad (15.22)$$

it follows that

$$R + T = 1 \qquad (15.23)$$

as required.

A comparison of the amplitude coefficients of reflection and transmission with the energy coefficients leads to the conversion formulas

$$R = |r_E^2|,$$
$$T = \left| t_E^2 \frac{Z_1 \cos\psi}{Z_2 \cos\phi} \right|. \qquad (15.24)$$

† Note that the mean intensity of the energy flow is $\mathbf{S} = \mathrm{Re}(\mathbf{E}) \times \mathrm{Re}(\mathbf{H}) = \frac{1}{2}\mathrm{Re}(\mathbf{E} \times \tilde{\mathbf{H}})$, with $\tilde{\mathbf{H}}$ indicating the conjugate of $\mathbf{H}$ in accordance with the algebra of complex quantities (see J. A. Stratton, *Electromagnetic Theory*, 1941, p. 136).

16 · Reflection and Refraction of Plane Waves by Loss-Free Dielectrics

Snell's laws and Fresnel's equations together determine unambiguously the direction, intensity, and polarization of the reflected and refracted beams as a function of the properties of the incident TEM wave and of the dielectric characteristics of the two abutting media.

In the usual case of nonmagnetics ($\mu_1{}^* = \mu_2{}^* = \mu_0$) Snell's refraction law (Eq. 14.18) simplifies to

$$\frac{\sin\phi}{\sin\psi} = \frac{Z_1}{Z_2} = \sqrt{\frac{\epsilon_2{}^*}{\epsilon_1{}^*}}, \tag{16.1}$$

and Fresnel's equations (Eqs. 15.15 and 15.16) assume the convenient form

$$r_{E_n} = -r_{H_p} = -\frac{\sin(\phi-\psi)}{\sin(\phi+\psi)},$$
$$r_{E_p} = -r_{H_n} = -\frac{\tan(\phi-\psi)}{\tan(\phi+\psi)}; \tag{16.2}$$

$$t_{E_n} = t_{H_p}\frac{Z_2}{Z_1} = \frac{2\sin\psi\cos\phi}{\sin(\phi+\psi)},$$
$$t_{E_p} = t_{H_n}\frac{Z_2}{Z_1} = \frac{2\sin\psi\cos\phi}{\sin(\phi+\psi)\cos(\phi-\psi)}. \tag{16.3}$$

Even this case represents a very complex situation. We shall therefore begin with a further simplification and first treat *dielectrics without loss* ($\epsilon'' = 0$), where

$$\frac{\sin\phi}{\sin\psi} = \sqrt{\frac{\epsilon_2{}'}{\epsilon_1{}'}} \equiv n_{21}. \tag{16.4}$$

(a) Waves incident on a medium of higher dielectric constant ($\epsilon_2{}' > \epsilon_1{}'$)

Because the angle of refraction is always smaller than the angle of incidence ($\psi < \phi$), a refracted beam appears at a real angle for any angle of incidence. The coefficients of transmission are positive (see Eq. 16.3), showing that the transmitted and incident waves are in phase at the boundary. Of the reflected field components, E_n always undergoes a phase jump of π with respect to the incident wave as the minus signs of Eqs. 16.2 indicate, whereas H_p stays in phase. For the other pair, E_p jumps by π whereas H_n stays in phase as long as the sum of the angles $(\phi + \psi) < 90°$; at angles $(\phi+\psi) > 90°$ the signs reverse. At $(\phi+\psi) = 90°$, the denominator $\tan(\phi+\psi)$ becomes infinite, and the component pair E_p, H_n of the reflected beam vanishes.

Figure 16.1 illustrates the situation for optical waves and for microwaves striking the surface of water at various angles of incidence.

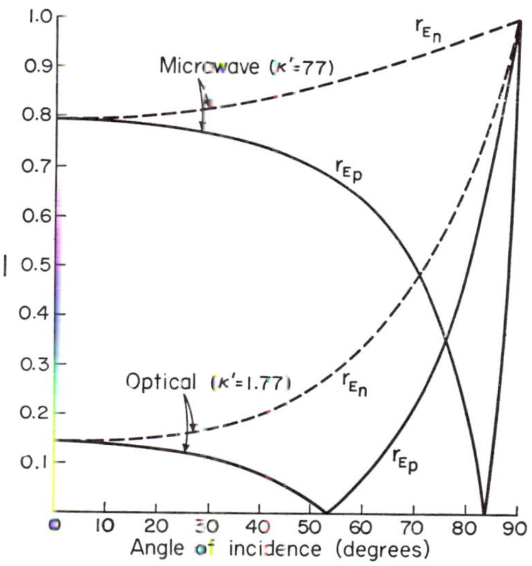

Fig. 16.1. Reflection of optical and microwaves from water surface.

The condition for the disappearance of the reflected pair of components

$$\phi + \psi = 90° \quad \text{or} \quad \cos\phi = \sin\psi \tag{16.5}$$

may be written, in view of Eq. 16.4, as

$$\tan\phi = \sqrt{\frac{\epsilon_2{}'}{\epsilon_1{}'}} \equiv n_{21}. \tag{16.6}$$

This angle ϕ, at which circularly polarized incident light becomes linearly polarized by reflection, is called the *polarizing* or *Brewster angle*, after D. Brewster,[1] who first derived the equation.

The phase relations between the **E** and **H** vectors are due to the fact that the direction of propagation reverses for the reflected beam, and that consequently

[1] D. Brewster, *Phil. Trans.*, 1815, p. 125.

either the **E** or the **H** vector must reverse phase in order that the right-hand co-ordinate system (Fig. 7.1) and the correct direction of power flow be maintained.

(b) Waves incident on a medium of lower dielectric constant ($\epsilon_2' < \epsilon_1'$)

Since the angle of refraction is now larger than the angle of incidence, the situation becomes more involved. For $\sin \psi < 1$ the preceding discussion still holds, the only difference being that the phases of the reflected components are inverted. The Brewster angle still occurs at $\tan \phi = n_{21}$; but at a larger angle of incidence, where

$$\sin \psi = \frac{\sin \phi_t}{n_{21}} = 1,$$

that is,

$$\sin \phi_t = n_{21}, \tag{16.7}$$

total reflection sets in.

This statement can be confirmed as follows: for $\sin \psi > 1$, $\cos \psi$ becomes imaginary,

$$\begin{aligned} \cos \psi &= -\sqrt{1 - \sin^2 \psi} \\ &= -\frac{j}{n_{21}} \sqrt{\sin^2 \phi - n_{21}{}^2} \equiv \frac{j}{n_{21}} u, \end{aligned} \tag{16.8}$$

the angle of refraction becomes complex, and Fresnel's equations 16.2 and 16.3 may be rewritten

$$\begin{aligned} r_{E_n} &= -r_{H_p} = \frac{\cos \phi - ju}{\cos \phi + ju}, \\ r_{E_p} &= -r_{H_n} = \frac{-n_{21}{}^2 \cos \phi + ju}{n_{21}{}^2 \cos \phi + ju}; \end{aligned} \tag{16.9}$$

$$\begin{aligned} t_{E_n} &= t_{H_p} \frac{Z_2}{Z_1} = \frac{2 \cos \phi}{\cos \phi + ju}, \\ t_{E_p} &= t_{H_n} \frac{Z_2}{Z_1} = \frac{2 n_{21} \cos \phi}{n_{21}{}^2 \cos \phi + ju}. \end{aligned} \tag{16.10}$$

Changing over from amplitude to energy coefficients of reflection and transmission (see Eqs. 15.24), we find

$$\begin{aligned} R_n &= R_p = r_E \tilde{r}_E = 1 \\ T &= 1 - R = 0, \end{aligned} \tag{16.11}$$

that is, the average power carried by the reflected wave is equal to that carried by the incident wave; this means total reflection.

The refracted wave, however, has not disappeared for $\sin \psi > 1$; it has only assumed some new and surprising aspects. To discuss these we refer to a definite co-ordinate system with the y-z plane representing the boundary S, the x-z plane that of incidence and the $+x$-direction pointing normal to the boundary and directed into medium 2 (Fig. 16.2). The transmitted electric wave becomes (see Eq. 14.8)

$$E_t = E_2 e^{j\omega t - \gamma_2 (x \cos \psi + z \sin \psi)}. \tag{16.12}$$

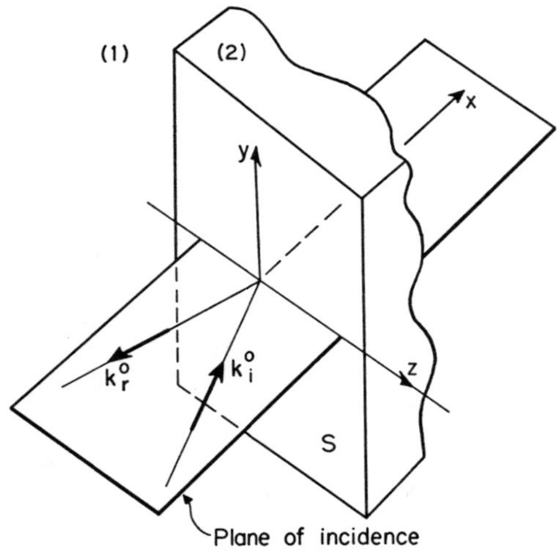

Fig. 16.2. Reference system for discussion of total reflection.

Since we specified nonmagnetic, loss-free media, it follows from Snell's refraction law that

$$\gamma_2 = \gamma_1 n_{21} = j \frac{2\pi}{\lambda_1} n_{21}, \tag{16.13}$$

and Eq. 16.12 may be rewritten

$$E_t = E_2 e^{-\frac{2\pi x}{\lambda_1} \sqrt{\sin^2 \phi - n_{21}{}^2}} e^{j2\pi \left(\nu t - \frac{z \sin \phi}{\lambda_1} \right)}. \tag{16.14}$$

A refracted wave results which is exponentially attenuated in depth as it penetrates into medium 2; it propagates parallel to the surface of the dielectric with a phase velocity

$$v_2 = \frac{dz}{dt} = \frac{v_1}{\sin \phi}, \tag{16.15}$$

which depends on the angle of incidence and is larger than that in medium 1.

The planes of constant amplitude are parallel to the interface and the planes of constant phase normal to it; since they do not coincide, the transmitted surface wave is a *nontransverse wave.* How the energy penetrates in the region of total reflection into the medium 2 of lower index of refraction, and then returns again into medium 1, has been demonstrated recently by Goos and Hänchen [2] in a remarkable experiment.

These authors measured, by multireflection, how an optical beam PQ, striking the interface at Q under an angle of total reflection ($\phi \geq \phi_t$), is not directly reflected as a beam QR but penetrates for some distance

[2] F. Goos and H. Hänchen, *Ann. Phys.* [6] *1,* 333 (1947).

into medium 2 and emerges as a beam ST, displaced parallel by a distance D (Fig. 16.3). They found for this displacement distance the expression

$$D = kn_2 \frac{\lambda_1}{\sqrt{\sin^2 \phi - n_{21}{}^2}}, \tag{16.16}$$

where $k = 0.52$ and $n_2 = \lambda_0/\lambda_2$. The distance is dependent on the depth of penetration, since Eq. 16.14 may be reformulated

$$E_t = E_2 e^{-2\pi k n_2 \frac{x}{D}} e^{j2\pi\left(\nu t - \frac{z \sin \phi}{\lambda_1}\right)}. \tag{16.17}$$

As we decrease the angle of incidence towards the limiting angle of total reflection (ϕ_t), the displacement distance and penetration depth increase. At ϕ_t itself D becomes infinite, the attenuation in medium 2 disappears, and the transmitted wave travels as a homogeneous surface wave parallel to the boundary. The boundary acts as a *wave guide* (see Sec. 21). The existence of a nontransverse surface wave in the region of

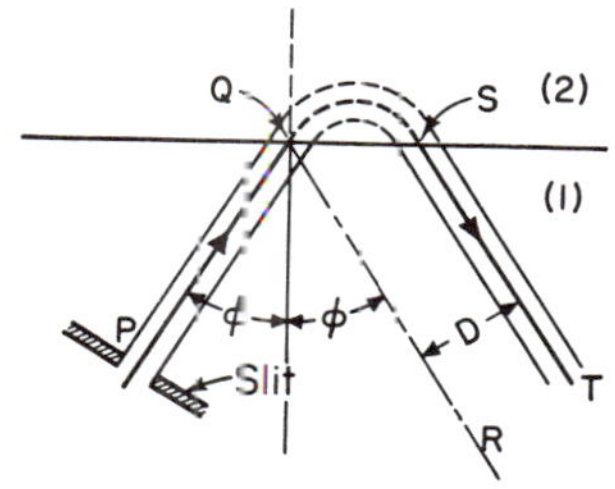

Fig. 16.3. Experiment on total reflection. (Goos and Hänchen.[2])

total reflection ($\phi > \phi_t$) causes the reflected wave to be *elliptically polarized*, as the complex character of the reflection coefficients (Eq. 16.9) indicates (see Sec. 20).

17 · Standing Waves

At normal incidence the incoming and reflected waves superpose and form, by interference, a standing-wave pattern. This standing-wave pattern in space can serve for a determination of the dielectric properties of the abutting media.

Let the incident wave $\mathbf{E}_y$, $\mathbf{H}_z$ propagate in the $+x$-direction through medium 1 and strike the boundary of medium 2 (Fig. 17.1) at $x = 0$. The reflected wave,

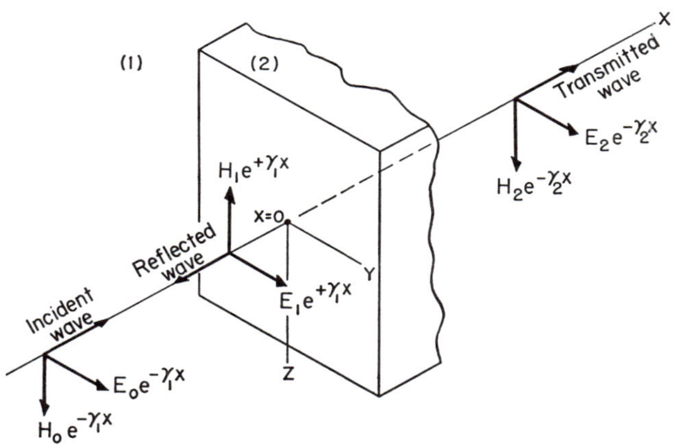

Fig. 17.1. Formation of standing waves.

returning in the $-x$-direction, combines with it to yield the resulting field strength

$$\begin{aligned} E_{y_1} &= E_0 e^{j\omega t - \gamma_1 x} + E_1 e^{j\omega t + \gamma_1 x}, \\ H_{z_1} &= H_0 e^{j\omega t - \gamma_1 x} + H_1 e^{j\omega t + \gamma_1 x}. \end{aligned} \tag{17.1}$$

By introducing the reflection coefficients at the boundary $x = 0$, r_0 and $-r_0$ for the electric and the magnetic waves, and by expressing the magnetic field in terms of the electric field and impedance, we may rewrite this equation as

$$\begin{aligned} E_{y_1} &= E_0 e^{j\omega t}(e^{-\gamma_1 x} + r_0 e^{\gamma_1 x}), \\ H_{z_1} &= \frac{E_0}{Z_1} e^{j\omega t}(e^{-\gamma_1 x} - r_0 e^{\gamma_1 x}). \end{aligned} \tag{17.2}$$

Let us assume for the present that the transmitted wave continues in medium 2 without further reflection,

$$\begin{aligned} E_{y_2} &= E_2 e^{j\omega t - \gamma_2 x}, \\ H_{z_2} &= \frac{E_2}{Z_2} e^{j\omega t - \gamma_2 x}. \end{aligned} \tag{17.3}$$

The reflection coefficient for normal incidence is, in this case (see also Eq. 15.17).

$$r_0 = \frac{Z_2 - Z_1}{Z_2 + Z_1}. \tag{17.4}$$

There will be no reflection ($r_0 = 0$) when the intrinsic impedances of the two media are matched, that is, if

$$\frac{\epsilon_1{}^*}{\epsilon_2{}^*} = \frac{\mu_1{}^*}{\mu_2{}^*}, \tag{17.5}$$

or, for loss-free media

$$\frac{D_1}{D_2} = \frac{B_1}{B_2}. \tag{17.6}$$

This *invisibility condition* for a boundary at normal incidence postulates that the electric and magnetic flux densities must change by equal ratios in traversing the interface if reflection is to be avoided.

The opposite extreme, *total reflection*, requires a complete mismatch of the two flux densities, whether by adjoining media with very low and very high permittivity or permeability, or by a mismatch in losses, as, for instance, by choosing as medium 2 a metal. In metals, up to the optical frequency range, the conductivity dominates over the real part of the dielectric constant. Consequently, if magnetic loss can be neglected ($\mu'' = 0$), the *intrinsic impedance of a metal* may be written (see Eq. 7.25)

$$Z_2 = \sqrt{\mu_2'/-j\epsilon_2''} = \sqrt{j\omega\mu_2'/\sigma_2}. \qquad (17.7)$$

Because of the high conductivity σ_2, $|Z_2| \ll |Z_1|$; the reflection coefficient at the metal boundary thus becomes

$$r_0 = \frac{Z_2 - Z_1}{Z_2 + Z_1} \simeq -1. \qquad (17.8)$$

Under this *condition of total reflection* and for a loss-free medium 1 $\left(\gamma_1 = j\dfrac{2\pi}{\lambda_1}\right)$, Eq. 17.2 simplifies to

$$E_{y_1} = -j2E_0 e^{j\omega t}\sin\frac{2\pi x}{\lambda_1},$$
$$H_{z_1} = \frac{2E_0}{Z_1} e^{j\omega t}\cos\frac{2\pi x}{\lambda_1}, \qquad (17.9)$$

or, by changing over to the real field components,

$$\text{Re}\,(E_{y_1}) = 2E_0 \sin\omega t \sin\frac{2\pi x}{\lambda_1},$$
$$\text{Re}\,(H_{z_1}) = \frac{2E_0}{Z_1}\cos\omega t\cos\frac{2\pi x}{\lambda_1}. \qquad (17.10)$$

This equation represents a standing wave for which the electric-field strength has its minima or *nodes* at the metal surface and at distances $-x = n\dfrac{\lambda_1}{2}$ ($n = 1, 2, 3, \ldots$), whereas the *antinodes* (maxima) of **E** are found at $-x = m\dfrac{\lambda_1}{4}$ ($m = 1, 3, 5, \ldots$) (Fig. 17.2). The nodes and antinodes of the magnetic wave appear in the reverse order.

In comparing Figs. 9.2 and 17.2 and the corresponding equations of a progressing and a standing TEM wave in a loss-free medium, we find an essential difference between the two wave types. Although no phase difference exists between the electric and the magnetic fields in the traveling wave, in the standing wave the two fields are displaced by 90° with respect to each other in spatial and temporal phase. In the traveling wave the energy is equally divided between the electric and the magnetic fields; in the standing wave it alternates between total electric and total magnetic energy storage.

To visualize the standing-wave pattern in the general case of partial reflection, we rewrite the electric wave of Eq. 17.2

$$E_y = E_0 e^{j\omega t - \gamma_1 x}(1 + r_0 e^{2\gamma_1 x}), \qquad (17.11)$$

that is, we express it as the product of the incident wave and of a term in brackets which gives rise to the standing-wave pattern.

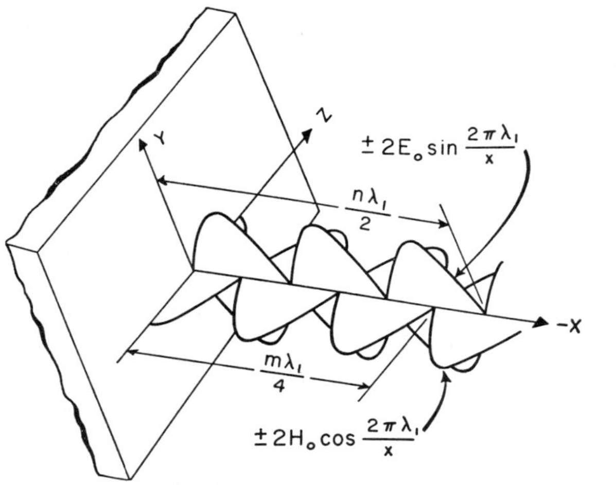

Fig. 17.2. Standing wave in loss-free dielectric in front of total reflector.

By writing the distance and the reflection coefficient in polar form,

$$x = \frac{\lambda_1}{2\pi}\phi,$$
$$r_0 = |r_0|e^{-j2\psi}, \qquad (17.12)$$

and by introducing the index of absorption k (see Eq. 9.12), this term may be reformulated as

$$1 + r_0 e^{2\gamma_1 x} = 1 + |r_0|e^{2k\phi}e^{j2(\phi-\psi)}. \qquad (17.13)$$

Its second part, $r_0 e^{2\gamma_1 x}$, now represents a radius vector in the complex plane which, at the interface $\phi = 0$, has the length $|r_0|$, and points at an angle -2ψ (Fig. 17.3). As we move from the boundary into medium 1 in the $-x$-direction, the radius vector spins clockwise one full turn for each half wavelength ($\phi = \pi, 2\pi$, etc.) and shrinks simultaneously, thus describing a logarithmic spiral. The vector $1 + r_0 e^{2\gamma_1 x}$ originates at the point -1 and connects to the tip of this radius vector. Similarly, the vector $1 - r_0 e^{2\gamma_1 x}$ of the magnetic wave connects from the point −1 to the tip of the radius vector which starts its spiral 180° displaced.

Reverting from polar to linear scale (Fig. 17.4), we obtain the relative amplitude of the standing-wave pattern and may read off from both diagrams immediately the dielectric information which the standing-wave pattern contains.

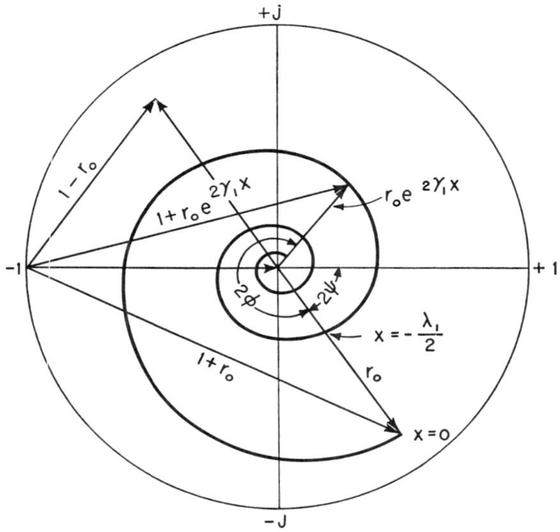

Fig. 17.3. Polar diagram of standing wave.

The ratio of the two field vectors at the boundary ($x = 0$),

$$Z(0) = \frac{E_1(0)}{H_1(0)} = Z_1 \frac{1 + r_0}{1 - r_0}, \qquad (17.14)$$

known as the *terminating impedance* of medium 1, can be found by measuring the amplitude $|r_0|$ and the phase angle 2ψ of the reflection coefficient r_0. Obviously, the first minimum of the E wave is reached when the radius vector has rotated through an angle

$$2\phi_0 = \pi - 2\psi, \qquad (17.15)$$

or, in linear scale,

$$2\phi_0 = \frac{4\pi x_0}{\lambda_1}. \qquad (17.16)$$

Hence, by measuring the distance x_0 of the first minimum from the boundary of the dielectric, the phase angle is obtained as

$$2\psi = 4\pi\left(\frac{1}{4} - \frac{x_0}{\lambda_1}\right). \qquad (17.17)$$

The ratio of the electric to the magnetic field strength at this minimum is

$$\frac{E_{\min}}{H_{\max}} = Z_1 \frac{1 - |r_0| e^{-2\alpha_1 x_0}}{1 + |r_0| e^{-2\alpha_1 x_0}}. \qquad (17.18)$$

If the attenuation in medium 1 can be neglected ($\alpha_1 \simeq 0$), $H_{\max} = \dfrac{E_{\max}}{Z_1}$ and Eq. 17.18 simplifies to

$$\frac{E_{\max}}{E_{\min}} = \frac{1 + |r_0|}{1 - |r_0|}. \qquad (17.19)$$

This ratio of maximum to minimum electric field intensity is called the *voltage standing wave ratio* (VSWR).

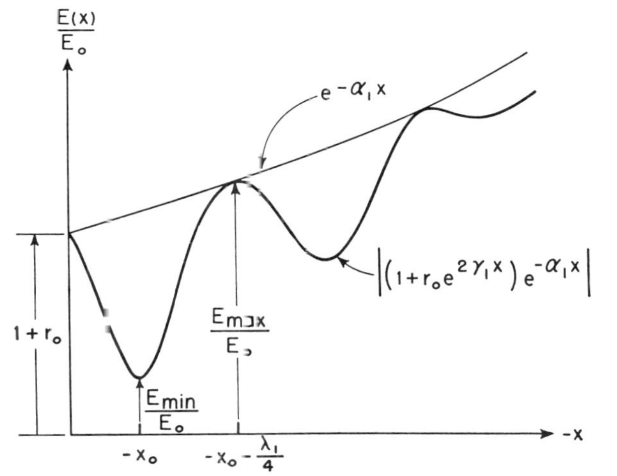

Fig. 17.4. Linear diagram of standing wave (attenuation in medium 1 overemphasized).

Thus, by measuring the ratio of maximum to minimum field strength and the distance of the first minimum from the dielectric boundary, the magnitude and phase of the reflection coefficient r_0 are determined. With it the terminating impedance $Z(0)$ is found if the intrinsic impedance Z_1 of medium 1 is known.

Since the tangential field components are continuous through the boundary, the terminating impedance for normal incidence is the same whether seen from medium 1 or 2:

$$Z(0) = \frac{E_1(0)}{H_1(0)} = \frac{E_2(0)}{H_2(0)}. \qquad (17.20)$$

Consequently, if no further reflection takes place in medium 2 (see Eq. 17.3), the terminating impedance measured by the standing-wave pattern in medium 1 is directly equal to the intrinsic impedance of medium 2, $Z(0) = Z_2$.

In case the media 1 and 2 are loss-free, and medium 2 is of infinite extent, the reflection coefficient at the boundary is real (see Eq. 17.4); the phase jump 2ψ is either zero or π, and the VSWR simplifies for $\epsilon_2' > \epsilon_1'$ ($r_0 = -|r_0|$) to

$$\frac{E_{\max}}{E_{\min}} = \frac{Z_1}{Z_2} = \sqrt{\frac{\epsilon_2'}{\epsilon_1'}} = n_{21}; \qquad (17.21)$$

it is equal to the relative index of refraction of the two media.

18 · Measurement of Dielectrics by Standing Waves;[1] Interference Optics

The dielectric medium 2 of the preceding section will, in general, not approximate infinite length, but will consist of a layer of thickness d, followed by a medium 3, etc. (Fig. 18.1). Additional reflections will therefore occur at the boundary 2–3 and possibly subsequent boundaries, and a part of this reflected energy will return into medium 1. The reflected wave in medium 1, thus composed of several partial beams, will itself be the resultant of an interference phenomenon.

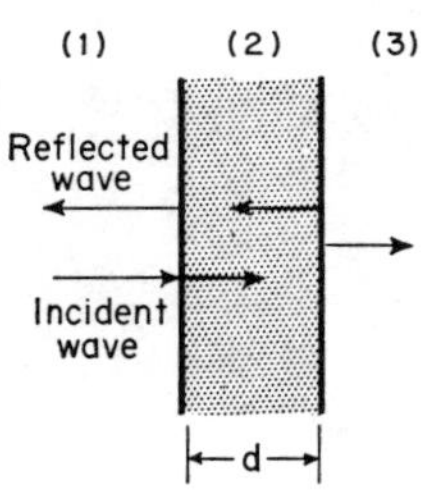

Fig. 18.1. Wave reflected on dielectric layer.

This more complex situation does not affect the measurement of the standing-wave pattern in medium 1, but it changes the interpretation of the reflection coefficient r_0 and terminating impedance $Z(0)$. Before proceeding with this discussion, we shall reformulate $Z(0)$ in terms of the directly measurable parameters E_{min}/E_{max}, x_0, and λ_1, to arrive at an expression useful in actual calculations.

Setting

$$r_0 \equiv e^{-2u} \qquad (\text{where } u = \rho + j\psi), \tag{18.1}$$

we may write the terminating impedance (Eq. 17.14)

$$Z(0) = Z_1 \frac{1 + r_0}{1 - r_0} = Z_1 \coth u, \tag{18.2}$$

and the inverse standing-wave ratio (Eq. 17.19)

$$\frac{E_{min}}{E_{max}} = \frac{1 - e^{-2\rho}}{1 + e^{-2\rho}} = \tanh \rho. \tag{18.3}$$

By expanding coth u and recalling the relation between ψ and x_0 (Eq. 17.17), we obtain the desired expression

$$Z(0) = Z_1 \frac{\tanh \rho - j \cot \psi}{1 - j \tanh \rho \cot \psi} = Z_1 \frac{\dfrac{E_{min}}{E_{max}} - j \tan \dfrac{2\pi x_0}{\lambda_1}}{1 - j \dfrac{E_{min}}{E_{max}} \tan \dfrac{2\pi x_0}{\lambda_1}}. \tag{18.4}$$

[1] S. Roberts and A. von Hippel, Publication of the Massachusetts Institute of Technology, March 1941; *J. Appl. Phys. 17*, 610 (1946).

The terminating impedance is thus determined experimentally by measurements on the standing-wave pattern in medium 1, as discussed. To obtain from it the properties of medium 2, we consider the situation at the boundary 2–3 at $x = d$.

At this boundary a reflection coefficient r_{23} produces a reflected wave which, together with the incident wave, sets up a standing wave (cf. Eq. 17.2) in medium 2,

$$\begin{aligned} E_{y_2} &= E_2 e^{j\omega t} \{e^{-\gamma_2(x-d)} + r_{23} e^{\gamma_2(x-d)}\}, \\ H_{z_2} &= \frac{E_2}{Z_2} e^{j\omega t} \{e^{-\gamma_2(x-d)} - r_{23} e^{\gamma_2(x-d)}\}. \end{aligned} \tag{18.5}$$

The terminating impedance at $x = 0$, seen from medium 2, is

$$Z(0) = \frac{E_2(0)}{H_2(0)} = Z_2 \frac{e^{\gamma_2 d} + r_{23} e^{-\gamma_2 d}}{e^{\gamma_2 d} - r_{23} e^{-\gamma_2 d}}. \tag{18.6}$$

In case the transmitted wave continues in medium 3 without further reflection, the reflection coefficient at the boundary 2, 3 becomes (cf. Eq. 17.4)

$$r_{23} = \frac{Z_3 - Z_2}{Z_3 + Z_2}. \tag{18.7}$$

If medium 3 is a metal, approximately total reflection takes place at d,

$$r_{23} \simeq -1 \tag{18.8}$$

(cf. Eq. 17.8), and the terminating impedance for this *short-circuit measurement* of medium 2 simplifies to

$$Z(0) = Z_2 \tanh \gamma_2 d. \tag{18.9}$$

In case the shorting metal plate is placed a distance Δ behind d (Fig. 18.2), a standing wave forms in medium 3,

$$\begin{aligned} E_{y_3} &= E_3 e^{j\omega t}(e^{-\gamma_3(x-d-\Delta)} - e^{\gamma_3(x-d-\Delta)}), \\ H_{z_3} &= \frac{E_3}{Z_3} e^{j\omega t}(e^{-\gamma_3(x-d-\Delta)} + e^{\gamma_3(x-d-\Delta)}). \end{aligned} \tag{18.10}$$

Since the impedance at $x = d$ is the same, whether seen from medium 2 or 3

$$\left(\frac{E_{y_2}}{H_{z_2}}\right)_{x=d} = \left(\frac{E_{y_3}}{H_{z_3}}\right)_{x=d}, \tag{18.11}$$

the reflection coefficient r_{23} can be expressed in its dependence on the layer thickness Δ as

$$r_{23} = \frac{Z_3(1 - e^{-2\gamma_3\Delta}) - Z_2(1 + e^{-2\gamma_3\Delta})}{Z_3(1 - e^{-2\gamma_3\Delta}) + Z_2(1 + e^{-2\gamma_3\Delta})}. \quad (18.12)$$

For $\Delta = \lambda_3/4$ and a loss-free medium 3 $\left(\gamma_3 = j\frac{2\pi}{\lambda_3}\right)$, we obtain

$$r_{23} = +1; \quad (18.13)$$

the combination, a quarter-wavelength section termi-

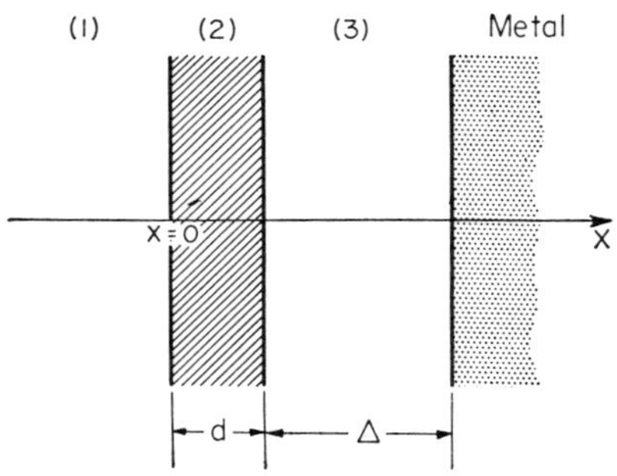

Fig. 18.2. Arrangement for measuring a dielectric 2 by standing wave in medium 1.

nated by a short circuit, corresponds to an *open circuit*, and the terminating impedance becomes

$$Z(0) = Z_2 \coth \gamma_2 d. \quad (18.14)$$

Thus, by performing an open-circuit and a short-circuit measurement of the terminating impedance, we obtain two equations for Z_2 and γ_2 and can calculate the complex dielectric constant and the complex permeability of medium 2. For a nonmagnetic dielectric ($\mu_2^* = \mu_0$), one of the measurements suffices.

The problem just discussed is a typical problem of interference optics which the physicist customarily treats by superposing partial beams and introducing reflection coefficients only. We can change over to this language by referring to the termination reflection coefficient

$$r_0 = \frac{Z(0) - Z_1}{Z(0) + Z_1}, \quad (18.15)$$

to the reflection coefficient of the partial beam at the interface 1–2

$$r_{12} = \frac{Z_2 - Z_1}{Z_2 + Z_1}, \quad (18.16)$$

and to the reflection coefficient which measures the reflection of the boundaries 2–3, etc., as it appears at $x = 0$ (see Eq. 18.6)

$$r_{23}e^{-2\gamma_2 d} = \frac{Z(0) - Z_2}{Z(0) + Z_2}. \quad (18.17)$$

In terms of these reflection coefficients Eq. 18.6 can be rewritten

$$r_0 = \frac{r_{12} + r_{23}e^{-2\gamma_2 d}}{1 + r_{12}r_{23}e^{-2\gamma_2 d}}. \quad (18.18)$$

As the layer thickness d increases, the radius vector r_0 winds eccentrically to the asymptotic value r_{12} because the partial beams returning from the inner boundaries 2–3, and so on, become attenuated.

If medium 3 is a metal ($r_{23} \simeq -1$), the expression simplifies to

$$r_0 = \frac{r_{12} - e^{-2\gamma_2 d}}{1 - r_{12}e^{-2\gamma_2 d}}. \quad (18.19)$$

A comparison of Eqs. 18.9 and 18.19 makes it obvious that the impedance concept leads to the more convenient formulation.

Two additional cases of interference may be discussed briefly since they are of great practical importance.

(1) Plane-parallel dielectric between two identical media

The operation of most interferometers and the design of radomes is based on this situation; here $r_{12} = -r_{23}$, hence

$$r_0 = \frac{r_{12}(1 - e^{-2\gamma_2 d})}{1 - r_{12}^2 e^{-2\gamma_2 d}}. \quad (18.20)$$

Figure 18.3 shows the reflection on such a dielectric layer in free space at normal incidence as a function

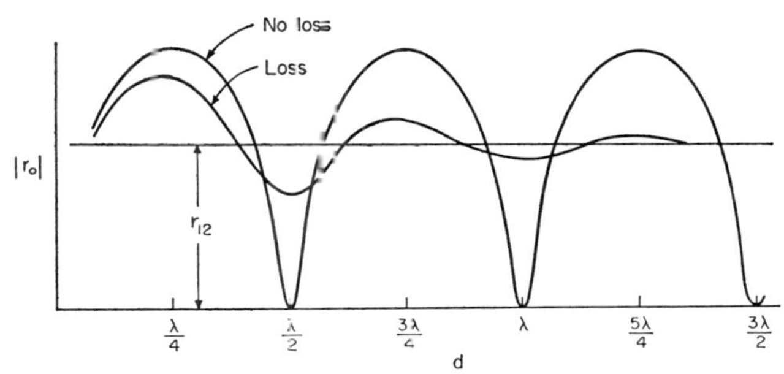

Fig. 18.3. Reflection on dielectric at normal incidence as function of the layer thickness *d*.

of the layer thickness. Without attenuation, a layer of the thickness $d = n\frac{\lambda_2}{2}$ (where $n = 1, 2, 3, \ldots$) is invisible.

That the interference maxima occur at odd multiples of $\lambda_2/4$ and the minima at multiples of $\lambda_2/2$, becomes obvious from the phase relations of the partial beams. The electric wave, when reflected from the front surface of the interference plate (for $\epsilon_2' > \epsilon_1'$ and $\tan\delta \ll 1$), suffers a phase change of π, whereas the beam reflected at the back surface stays in phase with the incident wave; for $\epsilon_2' < \epsilon_1'$, the role of the partial beams is reversed (see Sec. 16). The total phase difference of the two partial beams is therefore

$$\Delta = \frac{2\pi}{\lambda_2} 2d \pm \pi. \tag{18.21}$$

They are in phase for

$$2d = 2m\frac{\lambda_2}{4} \quad (m = 1, 3. 5 \ldots) \tag{18.22}$$

and in opposition for

$$2d = 2m\frac{\lambda_2}{2} \quad (m = 0, 1, 2 \ldots), \tag{18.23}$$

as indicated.

For oblique incidence, the geometrical path difference between the two partial beams increases from $2d$ to the distance $A \rightarrow B \rightarrow C$ in medium 2 minus the

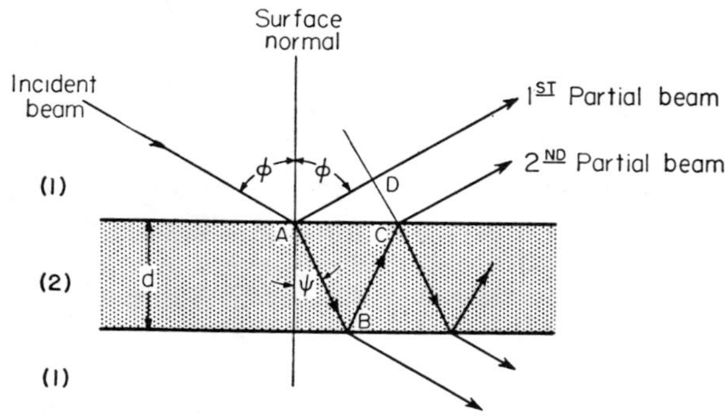

Fig. 18.4. Reflection at oblique incidence.

distance $A \rightarrow D$ in medium 1 (Fig. 18.4). Thus the total optical phase difference becomes

$$\Delta = \frac{2\pi}{\lambda_2}\frac{2d}{\cos\psi} - \frac{2\pi}{\lambda_1} 2d \tan\psi \sin\phi \pm \pi. \tag{18.24}$$

By substituting for $\sin\phi$ the expression $\lambda_1/\lambda_2 \sin\psi$ according to Snell's refraction law, we obtain the well-known *phase shift equation of interference optics*,

$$\Delta = \frac{4\pi d}{\lambda_2}\cos\psi \pm \pi, \tag{18.25}$$

which applies, for example, in the discussion of the color of thin layers and the evaluation of interference patterns [2] (see also Appendix A, I, 14).

(2) Nonreflecting surfaces

It has become a technique of great practical importance (nonreflecting lenses,[3] camouflage, etc.) to make the interface between two media 1 and 2 invisible by the insertion of a thin dielectric layer 2. At normal incidence we obtain from Eq. 18.18 the *condition for invisibility* ($r_0 = 0$),

$$r_{12} + r_{23}e^{-2\gamma_2 d} = 0. \tag{18.26}$$

It simplifies for nonmagnetic media of small absorption $\left(\gamma_2 \simeq j\frac{2\pi}{\lambda_2}\right)$ to the two conditions: (1) The thickness of the inserted layer must correspond to an odd multiple of $\lambda_2/4$ to fulfill the phase relation

$$e^{-j\frac{4\pi}{\lambda_2}d} = -1. \tag{18.27}$$

(2) The ratio of the dielectric constants of the three media must fulfill the condition

$$r_{12} = r_{23}. \tag{18.28}$$

Equation 18.28 may be rewritten in the impedance terminology as

$$\frac{Z_2}{Z_1} = \frac{Z_3}{Z_2} \quad \text{(see Eqs. 18.7 and 18.16)}; \tag{18.29}$$

by referring to the relative indices of refraction it becomes

$$n_{21} = n_{32} \quad \text{(see Eq. 14.19)}, \tag{18.30}$$

or, in terms of the dielectric constants,

$$\frac{\epsilon_2'}{\epsilon_1'} = \frac{\epsilon_3'}{\epsilon_2'}. \tag{18.31}$$

Hence the boundary becomes invisible when the thickness of the inserted layer is properly chosen and its dielectric constant is selected as the geometric means

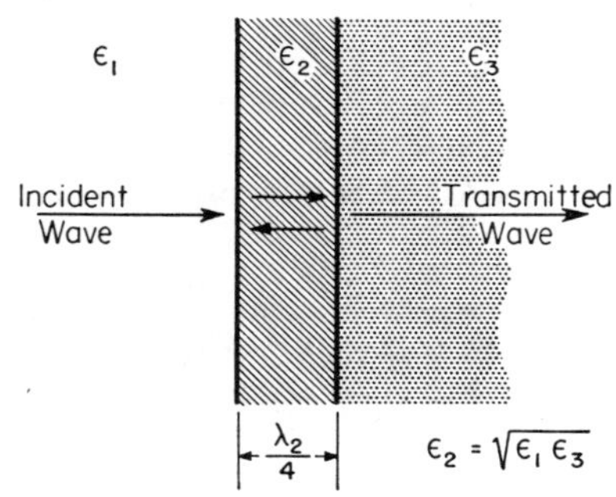

Fig. 18.5. Nonreflecting surface produced by insertion of dielectric layer.

of those of the adjacent media. In this case the partial beams reflected from the two surfaces of the interposed layer cancel each other in amplitude and phase (Fig. 18.5). The electric engineer calls this procedure *impedance matching*.

[2] See, for example, M. Born, *Optik*, Springer, Berlin, 1933, pp. 118 ff.

[3] A. Smakula, DRP 685767 (1935), *Z. Instrumentenkunde 60*, 33 (1940); C. H. Cartwright and A. F. Turner, *Phys. Rev. 55*, 595 (1939); K. B. Blodgett, *Phys. Rev. 55*, 391 (1939).

19 · Skin Effect

Metals have been treated as total reflectors in the preceding discussion; actually, a highly attenuated wave penetrates into their interior.

If medium 2 represents a metal of negligible magnetic loss, its propagation factor is, according to Eq. 17.7,

$$\gamma_2 = \alpha_2 + j\beta_2 = \frac{j\omega\mu_2'}{Z_2} = \sqrt{j\omega\mu_2'\sigma_2}. \tag{19.1}$$

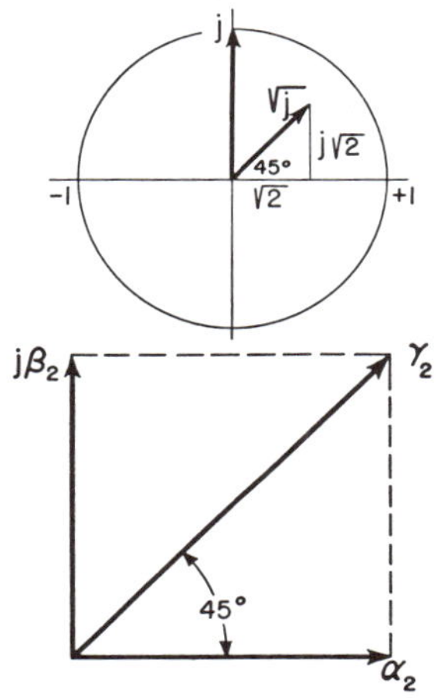

Fig. 19.1. Propagation factor of metal.

Since γ_2 is oriented at 45° to the real axis of the complex plane,

$$|\alpha_2| = |\beta_2| = \sqrt{\pi\nu\mu_2'\sigma_2} \tag{19.2}$$

(Fig. 19.1). The wavelength in the metal

$$\lambda_2 = \frac{2\pi}{\beta_2} = \sqrt{\frac{4\pi}{\nu\mu_2'\sigma_2}}, \tag{19.3}$$

and the field strength falls to $1/e = 0.368$ of its surface value at a depth

$$d = \frac{1}{\alpha_2} = \frac{1}{\sqrt{\pi\nu\mu_2'\sigma_2}} = \frac{\lambda_2}{2\pi}. \tag{19.4}$$

In angular measure this depth of penetration d corresponds to a distance of one radian.

The impedance

$$Z_2 = \frac{E_2}{H_2} = \sqrt{\frac{j\omega\mu_2'}{\sigma_2}} \tag{19.5}$$

indicates that the **E** vector leads the **H** vector in temporal phase by 45°. As the frequency is increased, the magnitude of the electric field in the metal increases relative to that of the magnetic field, but for reasonably low frequencies the intensity of the electric field in good conductors is much smaller than that of the magnetic field. The wave propagates in the metal with a phase velocity

$$v_2 = \nu\lambda_2 = \sqrt{2\omega/\mu_2'\sigma_2}, \tag{19.6}$$

that is, v_2 is zero for direct current or for infinite conductivity or permeability, and increases with the square root of the frequency. This peculiar behavior of waves in metals is caused by the conduction current set up by the electric field.

The field at any depth x in the interior may be expressed in terms of the field at the surface ($x = 0$) as

$$E_2 = E_0 e^{-\frac{x}{d}} e^{j\left(\omega t - \frac{2\pi x}{\lambda_2}\right)}. \tag{19.7}$$

It produces, according to Ohm's law, a current density

$$J = \sigma_2 E_0 e^{-\frac{x}{d}} e^{j\left(\omega t - \frac{2\pi x}{\lambda_2}\right)} \tag{19.8}$$

which is large at the surface and attenuates rapidly with depth (Fig. 19.2). An equivalent current of the uniform density $\sigma_2 E_0 e^{j(\omega t - 2\pi x/\lambda_2)}$, confined to a layer of the depth d, would cause a power loss identical with that actually observed. Since the current is limited to small depth, the conduction phenomenon is called *skin effect*, and the distance d is known as the *skin depth* of the metal.

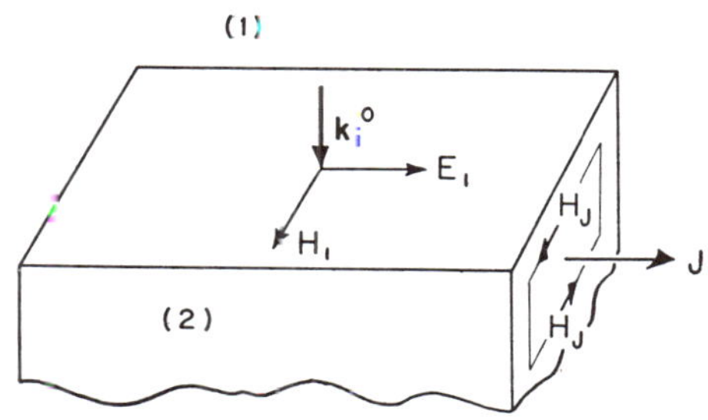

Fig. 19.2. Current and magnetic field caused by TEM wave penetrating into metal.

The nomographic chart (Chart 19.1) allows a quick evaluation of the wavelength λ_2 and phase velocity v_2 for TEM waves in nonmagnetic metals as a function of frequency and d-c conductivity. Chart 19.2 gives the skin depth as a function of resistivity and free-space wavelength.

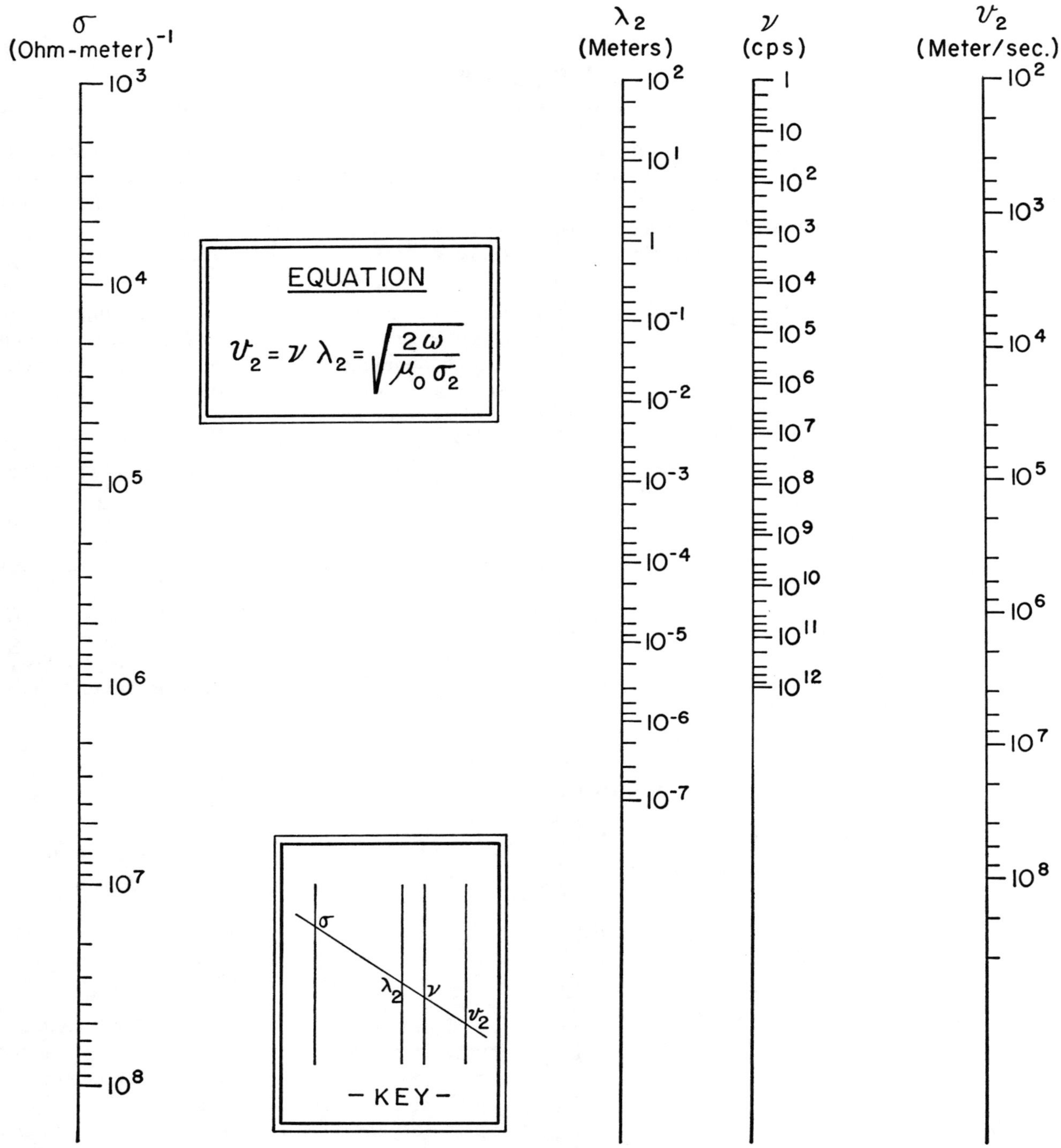

Chart 19.1. Phase velocity and wavelength in metal as function of conductivity and frequency.

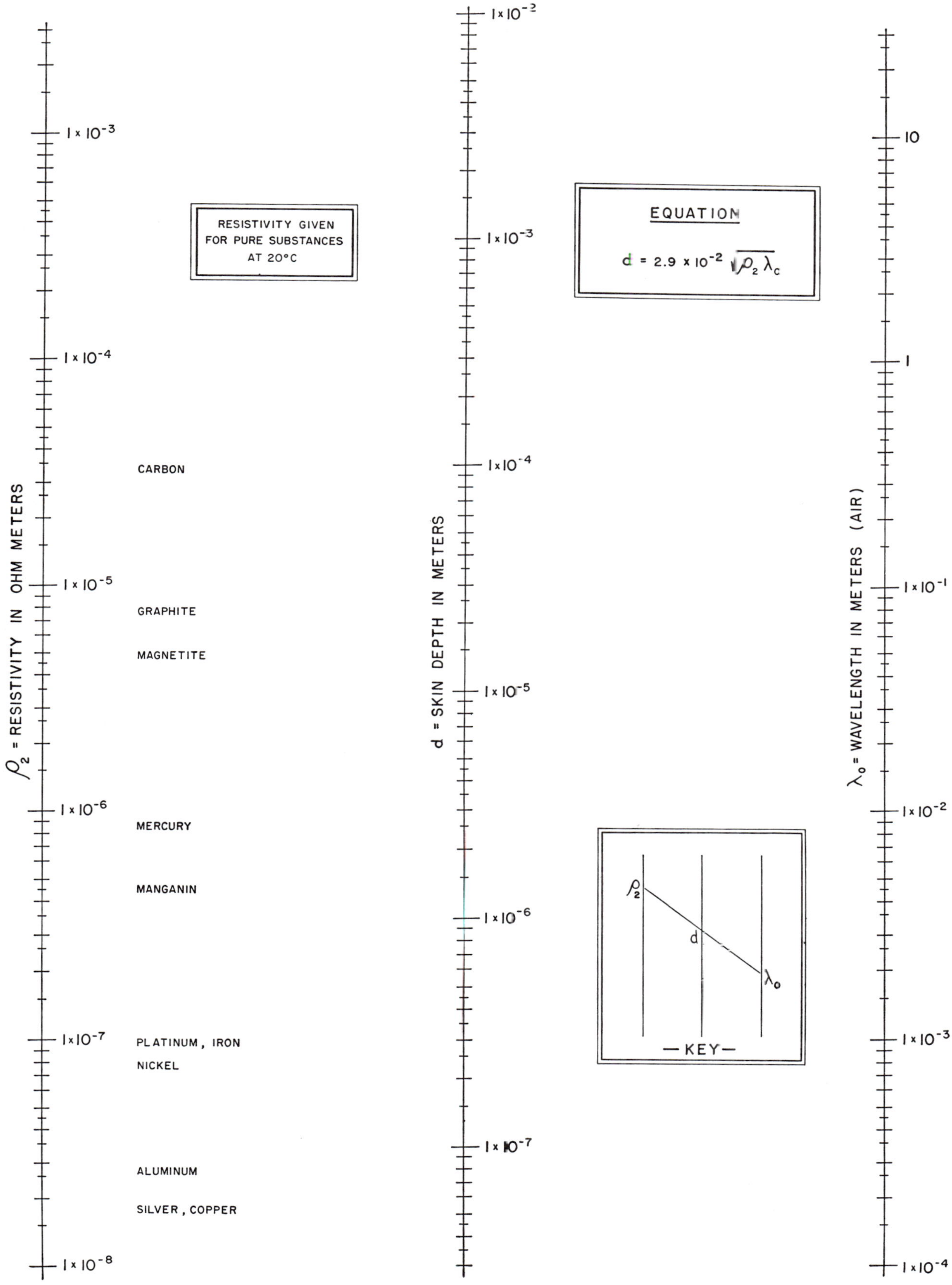

Chart 19.2. Skin depth of metal as function of resistivity and of wavelength in air.

20 · Reflection and Refraction by Media with Loss

In the preceding sections we have treated special cases of the general problem as to how electromagnetic plane waves are reflected and refracted by dielectrics: loss-free dielectrics (Sec. 16) and metals (Sec. 19) represent the two extremes in materials; for media of any arbitrary absorption we prescribed normal incidence (Secs. 17 and 18).

The general case of arbitrary dielectrics and oblique incidence is of great complexity. We approach it here with the restriction that medium 1 be loss-free, whereas medium 2 may be chosen at will, that is,

$$\gamma_1 = j\beta_1 = j\frac{2\pi}{\lambda_1}, \qquad \gamma_2 = \alpha_2 + j\beta_2. \tag{20.1}$$

The ratio of the propagation function is now complex and may be expressed conveniently by the relative complex index of refraction (see Eq. 9.14),

$$\frac{\gamma_2}{\gamma_1} \equiv n_{21}^*(1 - jk_2), \tag{20.2}$$

with

$$n_{21} = \beta_2/\beta_1, \qquad k_2 = \alpha_2/\beta_2. \tag{20.3}$$

According to Snell's refraction law (see Eq. 14.17),

$$\gamma_2 \sin\psi = \gamma_1 \sin\phi = j\beta_1 \sin\phi, \tag{20.4}$$

whereas

$$\gamma_2 \cos\psi = \gamma_2\sqrt{1 - \sin^2\psi} \equiv p + jq. \tag{20.5}$$

The transmitted electric wave may be written in the co-ordinate system of Fig. 16.2 as

$$E_t = E_2 e^{j\omega t - \gamma_2(x\cos\psi + z\sin\psi)}. \tag{20.6}$$

By substituting the expressions of Eqs. 20.4 and 20.5, we reformulate this equation

$$E_t = E_2 e^{-px} e^{j(\omega t - qx - \beta_1 z \sin\phi)}. \tag{20.7}$$

The surfaces of constant amplitude (px = constant) of this transmitted wave are parallel to the interface, whereas its planes of constant phase

$$qx + \beta_1 z \sin\phi \equiv x\cos\psi' + z\sin\psi' = \text{constant} \tag{20.8}$$

propagate at a true angle of refraction ψ' (Fig. 20.1).

Since

$$\sin\psi' = \frac{\beta_1 \sin\phi}{\sqrt{q^2 + \beta_1^2\sin^2\phi}}, \tag{20.9}$$

a geometrical index of refraction $n(\phi)$ can be defined which relates this true angle ψ' to the angle of incidence according to a modified Snell's refraction law

$$\frac{\sin\phi}{\sin\psi'} = \frac{\sqrt{q^2 + \beta_1 \sin^2\phi}}{\beta_1} \equiv n(\phi). \tag{20.10}$$

This index of refraction for the planes of constant phase is a function of the angle of incidence; it can be measured by the customary optical methods as long as medium 2 is of sufficient transparency.

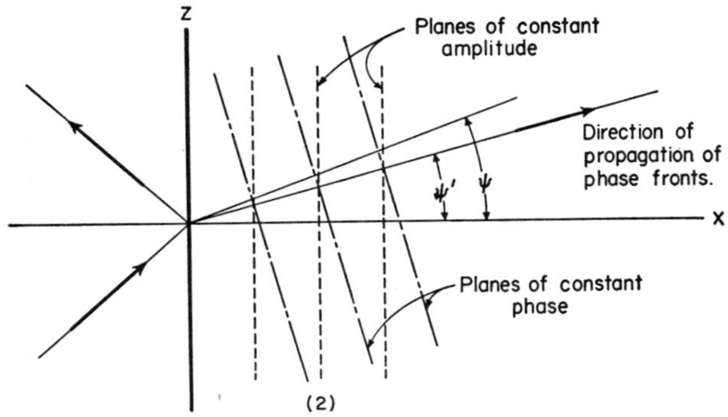

Fig. 20.1. Reflection and refraction on medium with loss.

A detailed calculation shows that the propagation factor β_2 and loss factor α_2 of the material are related to $n(\phi)$, ψ', p, q, and β_1 as

$$\beta_1^2 n(\phi) - p^2 = \beta_2^2 - \alpha_2^2, \qquad pq = \alpha_2\beta_2, \qquad \beta_1 n(\phi) p = \frac{\alpha_2\beta_2}{\cos\psi'}. \tag{20.11}$$

Again we treat first the extreme cases of the general refraction phenomenon: total reflection from a loss-free medium and reflection from a metal (Fig. 20.2). In the former instance, as a comparison of Eqs. 16.8 and 16.13 with Eq. 20.5 shows, $q = 0$, that is, the planes of constant amplitude and of constant phase are normal to each other, and the refracted wave propagates parallel to the surface as previously deduced.

In the case of metals, we define the wavelength in medium 2 in terms of the geometrical index of refraction as

$$\lambda_2 \equiv \frac{\lambda_1}{n(\phi)} = \frac{2\pi}{\sqrt{q^2 + \beta_1{}^2 \sin^2 \phi}}. \tag{20.12}$$

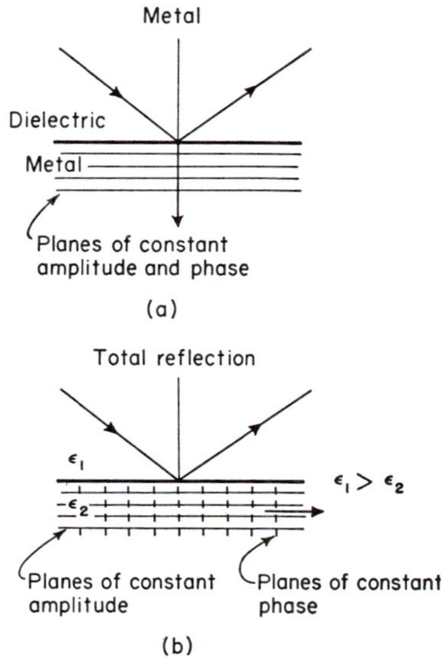

Fig. 20.2. Two extremes in refraction: refraction accompanying (*a*) reflection from metal, (*b*) total reflection from loss-free medium.

For normal incidence ($\sin \phi = 0$), the wavelength was given by Eq. 19.3, hence

$$q = \beta_2. \tag{20.13}$$

Since for metals in general

$$\frac{\lambda_2}{\lambda_1} = \frac{\beta_1}{\beta_2} \ll 1, \tag{20.14}$$

it follows from Eq. 20.9 that $\sin \psi' \ll 1$ or ψ' approaches zero. Thus in metals up to the optical range (as long as the conductivity dominates) the planes of constant phase are essentially parallel to the planes of constant amplitude and propagate normal to the conductor surface.

It will not be possible, in general, to observe the refracted wave in a highly absorbing medium, but the dielectric properties can be derived from the state of polarization of the reflected wave. Returning to Fresnel's equations for nonmagnetic dielectrics (see Eq. 16.2), we obtain for the ratio of the reflection coefficients parallel and normal to the plane of incidence

$$\frac{r_{Ep}}{r_{En}} = \frac{\cos(\phi + \psi)}{\cos(\phi - \psi)} \equiv \rho e^{-j\delta}. \tag{20.15}$$

Here ρ stands for the ratio of the amplitudes and

$$\delta = \delta_p - \delta_n \tag{20.16}$$

for the temporal phase difference of the two components of the reflected polarized radiation (see Eq. 12.1).

Let the incident light be linearly polarized at 45° to the plane of incidence ($E_{0,p} = E_{0,n}$). Since the sine and cosine of the angle of refraction are complex (see Eqs. 20.4 and 20.5), the phase shift between the reflected components will usually be $\delta \neq 0$. Linearly polarized incident light becomes, in general, elliptically polarized by reflection from an absorbing medium. Exceptions are normal incidence ($\phi = 0$, $\rho = 1$, $\delta = 0$) and grazing incidence ($\phi = \pi/2$, $\rho = 1$, $\delta = \pi$), in which cases the linear polarization is preserved.

Obviously there must exist an intermediate angle of incidence at which the relative phase shift between the two reflected components becomes $\pi/2$,

$$\delta = \pi/2, \quad e^{-j\delta} = -j \quad \text{at } \phi = \phi_0. \tag{20.17}$$

This *principal angle of incidence*, ϕ_0, at which an incident TEM wave, linearly polarized at 45° to the plane of incidence, becomes circularly polarized by reflection, is identical with the Brewster angle for loss-free dielectrics (see Sec. 16). When there is no loss the phase difference jumps abruptly from zero to π at this angle; with loss in medium 2 the phase change becomes more and more gradual as the absorption increases (Fig. 20.3).

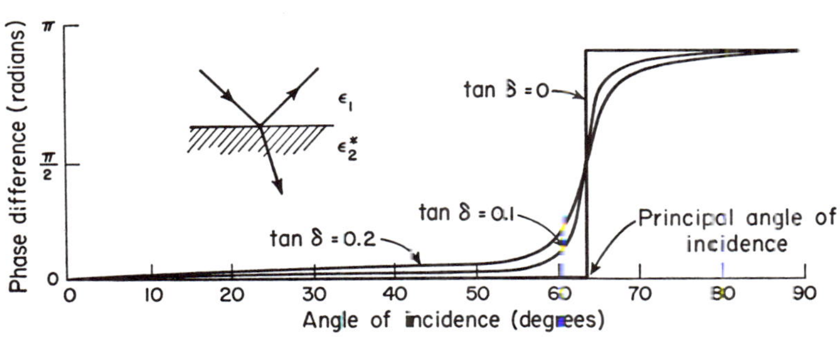

Fig. 20.3. Relative phase shift between E_p and E_n for various dissipation factors of the reflecting medium 2.

To discuss the dependence of this principal angle on the dielectric properties of the two media, we form, with the help of Eqs. 20.4, 20.5, and 20.15, the expression

$$\frac{1+\rho e^{-j\delta}}{1-\rho e^{-j\delta}} = \frac{\cos\phi\cos\psi}{\sin\phi\sin\psi} = \frac{\sqrt{(\gamma_1/\gamma_2)^2 - \sin^2\phi}}{\sin\phi\tan\phi}. \qquad (20.18)$$

By replacing the ratio of the propagation factors with the complex index of refraction (see Eq. 20.2), and designating the amplitude ratio of the reflected components at the principal angle ϕ_0 by ρ_0, we rewrite Eq. 20.18 for this angle

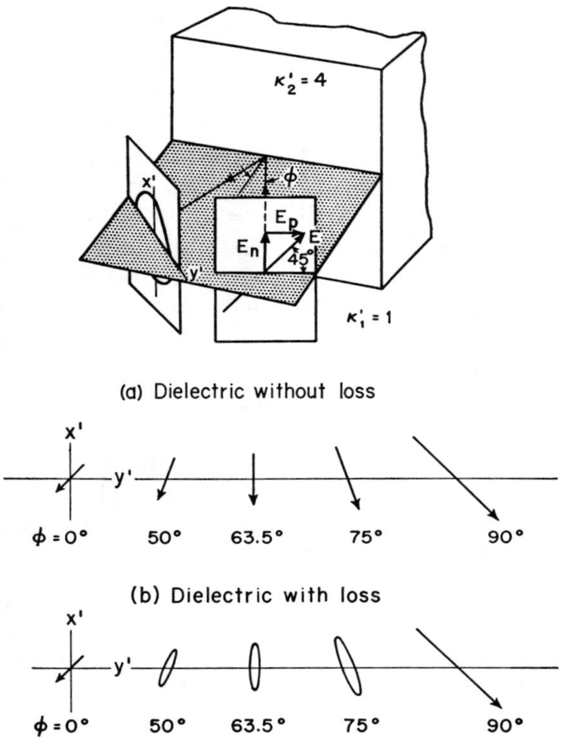

Fig. 20.4. Polarization by reflection on dielectric.

$$\frac{1-j\rho_0}{1+j\rho_0} = \frac{\sqrt{(n_{21}{}^*)^2 - \sin^2\phi_0}}{\sin\phi_0\tan\phi_0}. \qquad (20.19)$$

Finally, if we square this equation, multiply each term by its conjugate and expand, we obtain ϕ_0 in terms of the relative index of refraction and the index of absorption of medium 2 as

$$\sin^4\phi_0\tan^4\phi_0 = n_{21}{}^4(1+k_2{}^2)^2 + 2n_{21}{}^2(1-k_2{}^2)\sin^2\phi_0 + \sin^4\phi_0.^1 \qquad (20.20)$$

[1] See M. Born, *Optik*, Springer, Berlin, 1933, pp. 261 ff.

If medium 2 is a metal, this expression can be simplified appreciably. Below the optical frequency range, the equations derived in Sec. 19 are valid; hence

$$n_{21} \gg 1, \quad k_2 = 1, \qquad (20.21)$$

and the first member on the right-hand side of Eq. 20.20 dominates. Thus

$$\sin\phi_0\tan\phi_0 = n_{21}\sqrt{2} = \frac{\lambda_1}{\lambda_2}\sqrt{2}. \qquad (20.22)$$

For the very high frequencies of the optical spectrum ($\nu > 3 \times 10^{13}\ \text{sec}^{-1}$) the simple skin-effect equations begin to lose their validity, but still only the first member on the right side of Eq. 20.20 need be considered and

$$\sin\phi_0\tan\phi_0 = n_{21}\sqrt{1+k_2{}^2}. \qquad (20.23)$$

A general treatment of the reflection as a function of the angle of incidence requires discussion of Eq. 20.18 itself.[2] By multiplying numerator and denominator by $(1-\rho e^{j\delta})$ and defining

$$\rho \equiv \tan\beta, \qquad (20.24)$$

the left-hand side of this equation reads

$$\frac{(1+\rho e^{-j\delta})(1-\rho e^{+j\delta})}{(1-\rho e^{-j\delta})(1-\rho e^{+j\delta})} = \frac{1-\rho^2-2j\rho\sin\delta}{1+\rho^2+2\rho\cos\delta} = \frac{\cos 2\beta - j\sin 2\beta\sin\delta}{1+\sin 2\beta\cos\delta}. \qquad (20.25)$$

The right-hand side of Eq. 20.18 can be simplified, as long as

$$n_{21}{}^2(1-k_2{}^2) \gg \sin^2\phi, \qquad (20.26)$$

by the approximation

$$\frac{\sqrt{n_{21}{}^{2*} - \sin^2\phi}}{\sin\phi\tan\phi} \simeq \frac{n_{21}(1-jk_2)}{\sin\phi\tan\phi}. \qquad (20.27)$$

Hence

$$\frac{\cos 2\beta - j\sin 2\beta\sin\delta}{1+2\sin 2\beta\cos\delta} = \frac{n_{21}(1-jk_2)}{\sin\phi\tan\phi}, \qquad (20.28)$$

or, split up into the real and the imaginary part,

$$\frac{\cos 2\beta}{1+2\sin 2\beta\cos\delta} = \frac{n_{21}}{\sin\phi\tan\phi}, \qquad \frac{\sin 2\beta\sin\delta}{1+2\sin 2\beta\cos\delta} = \frac{n_{21}k_2}{\sin\phi\tan\phi}. \qquad (20.29)$$

From these equations we obtain the relative index of refraction and the index of absorption of medium 2 as

[2] See also J. A. Stratton, *Electromagnetic Theory*, McGraw-Hill Book Co., New York, 1941, pp. 505 ff.

a function of the angle of incidence ϕ, of the amplitude ratio of the reflected components ($\rho \equiv \tan \beta$), and of the temporal phase shift δ between these components as

$$n_{21} = \frac{\sin \phi \tan \phi \cos 2\beta}{1 + 2 \sin 2\beta \cos \delta}, \qquad k_2 = \sin \delta \tan 2\beta. \tag{20.30}$$

At the principal angle of circular polarization ($\sin \delta = 1$)

$$k_2 = \tan 2\beta_0 \qquad (\text{where } \rho_0 \equiv \tan \beta_0). \tag{20.31}$$

Figure 20.4 illustrates the state of polarization (relative amplitudes in the x',y' plane) of the reflected beam as a function of the angle of incidence for (*a*) a loss-free and (*b*) a lossy material. The incident wave is linearly polarized at 45°, the principal angle observed at $\phi = 63.5°$.

Summarizing: (1) Refraction of TEM waves by absorbing media produces, in general, a transmitted beam which has longitudinal field components since the planes of constant amplitude and of constant phase do not coincide; the planes of constant amplitude are always parallel to the interface. (2) In metals, the planes of constant phase are practically parallel to those of constant amplitude, and the transmitted beam propagates therefore about normal to the metal boundary for all angles of incidence as long as the skin-effect equations are valid. (3) Linearly polarized incident light becomes, in general, elliptically polarized by reflection from an absorbing medium. (4) In place of Brewster's angle for loss-free media, a principal angle of incidence may be observed at which the reflected wave is circularly polarized. (5) The complex permittivity of an absorbing medium may be calculated from the geometrical index of refraction, the principal angle of incidence or, quite generally, from the state of polarization of the reflected wave as a function of the angle of incidence.

21 · Guided Waves

At normal incidence, as discussed in Sec. 17, incoming and reflected waves superpose to form a standing-wave pattern. At oblique incidence, an interference pattern forms in the crossover region of the incident and the reflected beam (Fig. 21.1). In this case, however, the interference phenomenon is not standing in space but glides along the interface.

To see this in mathematical formulation we refer to the co-ordinate system of Fig. 21.1, for which the normal distance of the planes of constant phase of the two waves from the origin becomes

$$\begin{aligned} k_i{}^0 \cdot r &= x \cos \phi + z \sin \phi, \\ k_r{}^0 \cdot r &= -x \cos \phi + z \sin \phi. \end{aligned} \tag{21.1}$$

By superposing both waves, the field strengths in the interference region may be written in analogy to Eq. 17.2

$$\begin{aligned} E_{y1} &= E_0 e^{j\omega t - \gamma_1 z \sin \phi}(e^{-\gamma_1 x \cos \phi} + r_0 e^{\gamma_1 x \cos \phi}), \\ H_{z1} &= \frac{E_0}{Z_1} e^{j\omega t - \gamma_1 z \sin \phi}(e^{-\gamma_1 x \cos \phi} - r_0 e^{\gamma_1 x \cos \phi}). \end{aligned} \tag{21.2}$$

If medium 1 is loss-free $\left(\gamma_1 = j\dfrac{2\pi}{\lambda_1}\right)$ and medium 2 a metal ($r_0 \simeq -1$), the real field components of the interference pattern are simply

$$\begin{aligned} \mathrm{Re}\,(E_{y1}) &= 2E_0 \sin\left(\omega t - \frac{2\pi z \sin \phi}{\lambda_1}\right) \times \sin\left(\frac{2\pi x}{\lambda_1} \cos \phi\right), \\ \mathrm{Re}\,(H_{z1}) &= 2\frac{E_0}{Z_1} \cos\left(\omega t - \frac{2\pi z \sin \phi}{\lambda_1}\right) \times \cos\left(\frac{2\pi x}{\lambda_1} \cos \phi\right). \end{aligned} \tag{21.3}$$

As in the case of the standing waves (see Eq. 17.10), a system of interference fringes has formed parallel to the interface, with the nodes of the electric field (or antinodes of the magnetic field) at the distances

$$x = n\frac{\lambda_1}{2}\frac{1}{\cos \phi} \qquad (\text{where } n = 0, 1, 2, 3 \ldots) \tag{21.4}$$

in front of the boundary. However, the entire pattern glides along the surface in the $+z$-direction with a phase velocity

$$\frac{dz}{dt} = \frac{\omega\lambda_1}{2\pi \sin \phi} = \frac{v_1}{\sin \phi} \equiv v_i, \tag{21.5}$$

which is larger than the phase velocity of the incident wave and dependent on the angle of incidence.

The existence of phase velocities which may exceed the velocity of light was encountered previously for the surface waves forming at total reflection (see Eq.

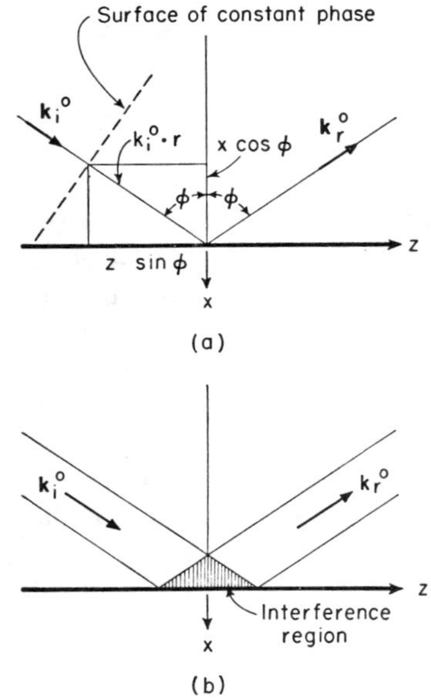

Fig. 21.1. Interference at oblique incidence: (*a*) distance of plane of constant phase from origin, (*b*) cross-over region of incident and reflected beam.

16.15). The phenomenon is easily explained (Fig. 21.2). In the direction of propagation the planes of identical phase follow each other at distances equal to the wavelength λ_1; parallel to the interface, however. they are separated by the larger distance $\frac{\lambda_1}{\sin\phi}$. This distance

$$\lambda_i = \frac{\lambda_1}{\sin\phi} \tag{21.6}$$

is the wavelength of the interference wave. A point P of constant phase, traced parallel to the surface in the

Fig. 21.2. Derivation of wavelength and phase velocity of interference wave.

interference pattern, has, therefore, to travel with the velocity

$$v_i = \lambda_i\nu = \frac{v_1}{\sin\phi}, \tag{21.7}$$

as just derived.

The energy flow along the surface, on the other hand, proceeds with a group velocity $v_g < v_1$. Traveling in the direction of propagation of the incident and reflected wave, the energy traverses per second only the distance

$$v_g = v_1 \sin\phi \tag{21.8}$$

parallel to the boundary. The product of phase and group velocity of this interference wave is thus equal

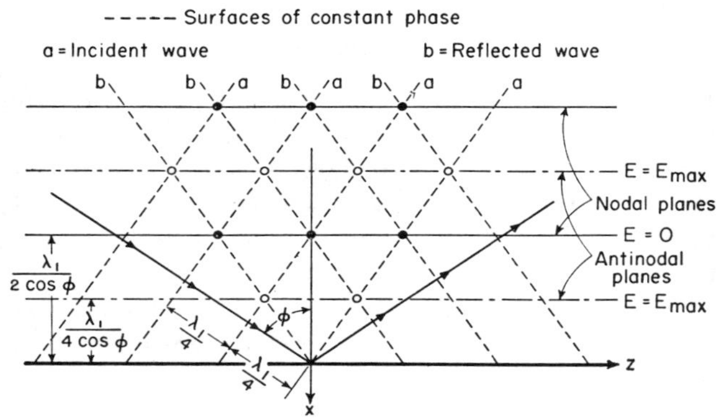

Fig. 21.3. Interference pattern in cross-over region.

to the square of the phase velocity v_1 in unbounded space,

$$v_i v_g = v_1^2. \tag{21.9}$$

Since in the nodes of the electric wave the state of the electromagnetic field is the same as at the boundary of a perfect metal (Fig. 21.3), we may place a second metal plate parallel to the first one at a distance (see Eq. 21.4)

$$b = n\frac{\lambda_1}{2}\frac{1}{\cos\phi} \tag{21.10}$$

at one of the interference minima without causing a disturbance of the field pattern between the plates (Fig. 21.4). Although the reflected radiation previously escaped into space, it is now guided between the two plates by multireflection; the two parallel metal mirrors represent the *prototype of a wave guide.*

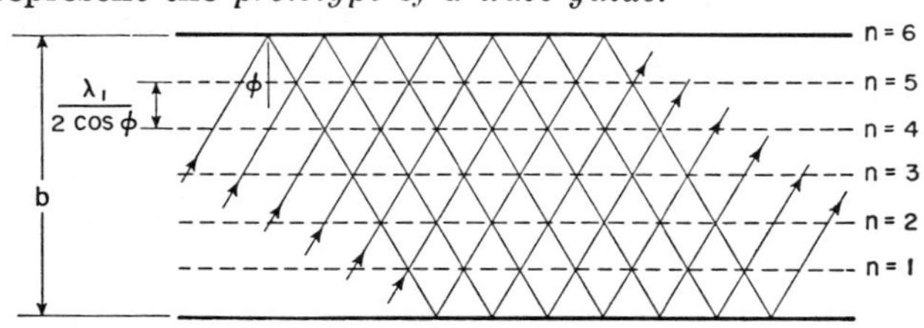

Fig. 21.4. Plane-parallel wave guide.

The properties of such a wave guide can be discussed without explicit reference to the angle of incidence by combining Eqs. 21.7 and 21.10 as $\sin^2\phi + \cos^2\phi = 1$ to form the expression

$$\left(\frac{v_1}{v_i}\right)^2 + \left(\frac{n\lambda_1}{2b}\right)^2 = 1. \tag{21.11}$$

The ratio of the phase velocities in unbounded space and in the guide, plotted as function of the ratio wavelength in free space to plate separation, is represented by the elliptical characteristic of Fig. 21.5. As the

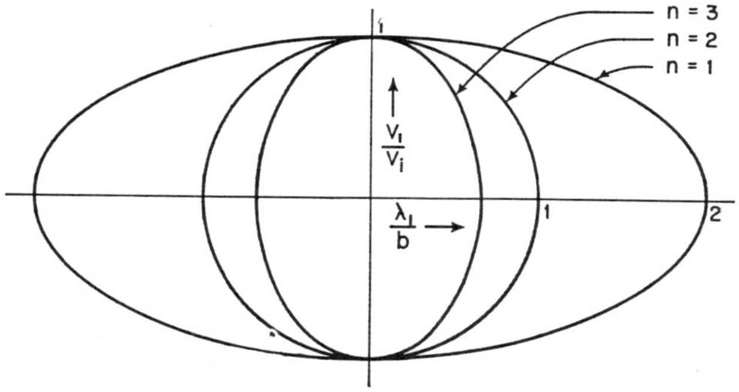

Fig. 21.5. Relative phase velocity as function of wavelength to plane separation.

spacing b of the parallel-plane guide shrinks, we approach a critical *cut-off wavelength*

$$\lambda_{1c} = 2b/n \tag{21.12}$$

or cut-off frequency

$$\nu_c = \frac{v_1}{\lambda_{1c}} = \frac{nv_1}{2b}, \tag{21.13}$$

at which

$$v_i = \infty, \text{ that is, } v_g = 0. \tag{21.14}$$

This situation, according to Eq. 21.7, corresponds to the case of perpendicular incidence ($\phi = 0$). The interference pattern appears here simultaneously parallel to the whole surface of the mirror, that is, with infinite phase velocity. The interference phenomenon, however, has come to a standstill; hence no energy is transferred parallel to the boundaries ($v_g = 0$).

For longer wavelengths, that is, lower frequencies, the wave guide cannot accommodate n interference fringes; it blocks the propagation of the corresponding *electromagnetic mode*. The longest mode possible in the parallel plane guide contains one fringe ($n = 1$); $\lambda_1/2$ in this case is equal to the guide separation.

Although Rayleigh [1] as early as 1897 discussed the propagation of waves through tubes, open transmission lines or coaxial cables were used exclusively until 1936, because the cut-off properties of hollow wave guides made them impractical at the frequencies then available. The development of magnetrons and other short-wave oscillators overcame this obstacle; and in 1936 the properties of hollow and of dielectric guides were rediscovered independently by Barrow [2] and Southworth.[3] Since that time, the theory and applications of wave guides have developed rapidly under the impact of radar,[4] and provide today a well-founded basis for modern microwave techniques.

[1] Lord Rayleigh, *Phil. Mag. 43,* 125 (1897).
[2] W. L. Barrow, *Proc. Inst. Rad. Eng. 24,* 1298 (1936).
[3] G. C. Southworth, *Bell. Syst. Tech. J. 15,* 284 (1936).
[4] See Massachusetts Institute of Technology Radiation Laboratory Series, McGraw-Hill Book Co., New York, Vols. 8–13.

22 · Electromagnetic Fields in Wave Guides

In the preceding treatment of the interference phenomenon between parallel plates we considered only the exponential phase functions of the incident and reflected waves without regard to the state of polarization of the wave pattern. If we now include the orientation of the electric and magnetic field vectors in our discussion, obviously two principal interference pattern types may be distinguished (Fig. 22.1). If the direct wave is polarized with the electric vector normal to the plane of incidence, the interference pattern gliding along the plates will have transverse electric components, but its magnetic components will be transverse and longitudinal ones. In the language of the wave-guide theory, these are the *TE modes* of propagation. The opposite case of polarization leads to the transverse magnetic or *TM modes.* It is convenient to break down the wave patterns developing in any type of guide into these two classes of TE modes and TM modes.

A general discussion of wave propagation in guides of arbitrary cross section requires that we turn from

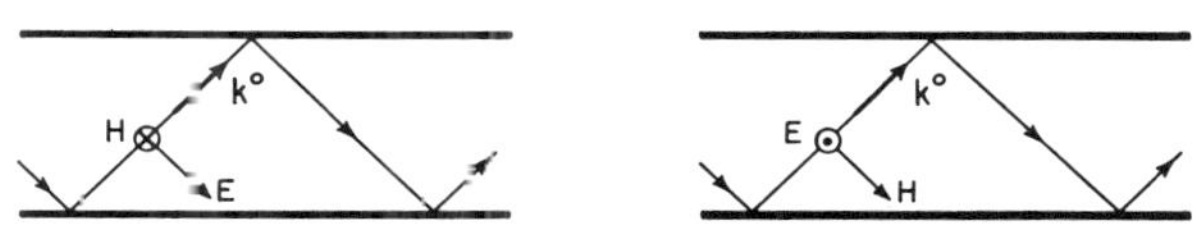

Fig. 22.1. TM waves and TE waves.

the instructive graphical approach of superposing incident and reflected waves to a mathematical treatment of the wave equations of the electromagnetic field (Eqs. 7.6 and 7.7) with prescribed boundary conditions. We will carry the analysis through for the rectangular guide (Fig. 22.2) and then draw some general conclusions.

The wave pattern progressing through such a guide in the $+x$-direction has to satisfy the wave equations

$$\nabla^2 \mathbf{E} = \epsilon_1{}^* \mu_1{}^* \frac{\partial^2 \mathbf{E}}{\partial t^2}, \tag{22.1}$$

$$\nabla^2 \mathbf{H} = \epsilon_1{}^* \mu_1{}^* \frac{\partial^2 \mathbf{H}}{\partial t^2}. \tag{22.2}$$

We assume the solution

$$\begin{aligned} \mathbf{E} &= \mathbf{E}_0(y, z) e^{j\omega t - \gamma_1' x}, \\ \mathbf{H} &= \mathbf{H}_0(y, z) e^{j\omega t - \gamma_1' x}. \end{aligned} \tag{22.3}$$

In addition to the fact that the electric and magnetic field amplitudes are now functions of y and z, this solution differs from the one previously assumed for TEM

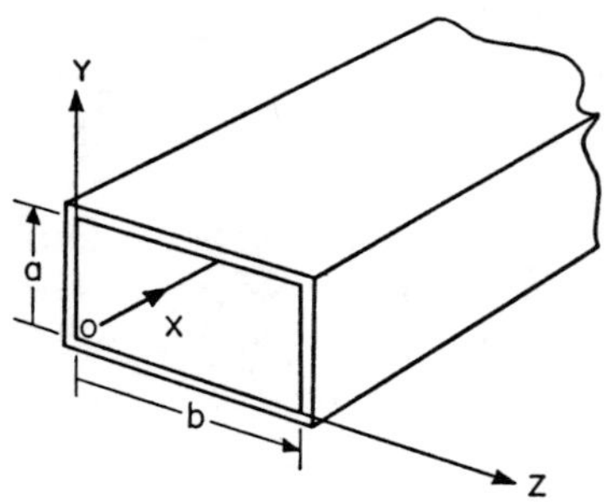

Fig. 22.2. Rectangular wave guide.

waves in unbounded space (see Eq. 7.10) as follows. The complex permittivity and permeability, $\epsilon_1{}^*$ and $\mu_1{}^*$ of the wave equations refer to the dielectric properties of an isotropic and homogeneous medium 1 in unbounded space, and determine the *intrinsic complex propagation factor* (Eq. 7.12)

$$\gamma_1 = j\omega\sqrt{\epsilon_1{}^* \mu_1{}^*} \tag{22.4}$$

of the dielectric material. In Eq. 22.3 there appears in place of this intrinsic propagation factor of the free medium a *characteristic propagation factor* γ_1' which allows for the fact that the field is confined. The distributed capacitance, inductance, and resistance of the bounding wave guide have changed the energy storage, dissipation, and propagation of the electromagnetic field.

Introducing the solution (Eq. 22.3) into the wave equations, we obtain for the x-components of the electric and magnetic field strengths the differential equations

$$\begin{aligned} \frac{\partial^2 E_x}{\partial y^2} + \frac{\partial^2 E_x}{\partial z^2} + (\gamma_1'^2 - \gamma_1^2) E_x &= 0, \\ \frac{\partial^2 H_x}{\partial y^2} + \frac{\partial^2 H_x}{\partial z^2} + (\gamma_1'^2 - \gamma_1^2) H_x &= 0. \end{aligned} \tag{22.5}$$

Hence *longitudinal field components* E_x and H_x have become possible owing to the stabilizing influence of the confining walls, expressed by the difference between the characteristic and the intrinsic propagation factors.

As stated previously, two types of vibrations may coexist independently of each other, the TE and the TM modes. In the reference system of Fig. 22.2 the field components of these modes are

$$\begin{aligned} &\text{TE} \begin{cases} E_x = 0, \\ E_y, E_z, H_x, H_y, H_z; \end{cases} \\ &\text{TM} \begin{cases} H_x = 0, \\ H_y, H_z, E_x, E_y, E_z. \end{cases} \end{aligned} \tag{22.6}$$

For both types we have to introduce the boundary condition (assuming infinite conductivity of the metal walls) that the tangential electric field components at the metal surfaces are short-circuited, that is, must be zero.

Starting with the TM modes we have to fulfill, in consequence, the conditions at the boundaries parallel to the x-y plane:

$$\left.\begin{aligned} E_x &= 0 \\ E_y &= 0 \end{aligned}\right\} \quad \text{at } z = 0 \text{ and } z = b, \tag{22.7}$$

and parallel to the x-z plane:

$$\left.\begin{aligned} E_x &= 0 \\ E_z &= 0 \end{aligned}\right\} \quad \text{at } y = 0 \text{ and } y = a. \tag{22.8}$$

The differential equation for the E_x component (Eq. 22.5) is a partial differential equation of the second order in two variables. It may be solved by assuming that the solution can be expressed as the product of two functions,

$$E_x = YZ, \tag{22.9}$$

where Y is a function of y and Z a function of z only. Substituting this expression into Eq. 22.5 and dividing by YZ, we obtain, by this *method of separation of variables*, the total differential equation

$$\frac{1}{Y}\frac{d^2Y}{dy^2} + \frac{1}{Z}\frac{d^2Z}{dz^2} = -(\gamma_1'^2 - \gamma_1^2). \tag{22.10}$$

Since the quantity on the right side of this equation is independent of y and z, each term on the left side must correspond to a constant, or

$$\frac{1}{Y}\frac{d^2Y}{dy^2} = -A_1, \quad \frac{1}{Z}\frac{d^2Z}{dz^2} = -A_2. \tag{22.11}$$

Hence Y and Z can be represented by the exponential functions

$$\begin{aligned} Y &= C_1 e^{j\sqrt{A_1}\,y} + C_2 e^{-j\sqrt{A_1}\,y}, \\ Z &= C_3 e^{j\sqrt{A_2}\,z} + C_4 e^{-j\sqrt{A_2}\,z}. \end{aligned} \tag{22.12}$$

The boundary condition for E_x (Eq. 22.8) demands $Y = 0$ for $y = 0$; thus $C_1 = -C_2$. Furthermore $Y = 0$ for $y = a$; hence $\sqrt{A_1} = m\pi/a$ and

$$Y = C_1(e^{j\frac{m\pi}{a}y} - e^{-j\frac{m\pi}{a}y}) = 2jC_1 \sin\left(\frac{m\pi}{a}\right)y. \qquad (22.13)$$

In analogy we obtain from the boundary condition, Eq. 22.7,

$$Z = C_3(e^{j\frac{n\pi}{b}z} - e^{-j\frac{n\pi}{b}z}) = 2jC_3 \sin\left(\frac{n\pi}{b}\right)z; \qquad (22.14)$$

consequently E_x, the product of the two functions, becomes

$$E_x = -C \sin\left(\frac{m\pi}{a}\right)y \sin\left(\frac{n\pi}{b}\right)z. \qquad (22.15)$$

From Eqs. 22.10 and 22.11, it follows, in addition, that

$$A_1 + A_2 = \left(\frac{m\pi}{a}\right)^2 + \left(\frac{n\pi}{b}\right)^2 = \gamma_1'^2 - \gamma_1^2. \qquad (22.16)$$

The integers $m = 0, 1, 2, 3 \ldots$ and $n = 0, 1, 2, 3 \ldots$ designate the number of half wavelengths occurring across the height a and the width b of the wave guide.

To find the other field components we have to return to Maxwell's equations (7.1 and 7.2). In view of the solution, Eq. 22.3, the curl components can be written

$$\frac{\partial H_z}{\partial y} - \frac{\partial H_y}{\partial z} = j\omega\epsilon^* E_x, \qquad \frac{\partial E_z}{\partial y} - \frac{\partial E_y}{\partial z} = -j\omega\mu^* H_x,$$

$$\frac{\partial H_x}{\partial z} + \gamma_1' H_z = j\omega\epsilon^* E_y, \qquad \frac{\partial E_x}{\partial z} + \gamma_1' E_z = -j\omega\mu^* H_y,$$

$$-\gamma_1' H_y - \frac{\partial H_x}{\partial y} = j\omega\epsilon^* E_z, \qquad -\gamma_1' E_y - \frac{\partial E_x}{\partial y} = -j\omega\mu^* H_z. \qquad (22.17)$$

Since for the TM modes $H_x = 0$ and E_x is given by Eq. 22.15, we obtain for the remaining four field components

$$E_y = -\frac{\gamma_1'}{\gamma_1'^2 - \gamma_1^2}\frac{\partial E_x}{\partial y}$$

$$= \frac{\gamma_1'}{\gamma_1'^2 - \gamma_1^2}\left(\frac{m\pi}{a}\right) C \cos\left(\frac{m\pi}{a}\right)y \sin\left(\frac{n\pi}{b}\right)z$$

$$E_z = -\frac{\gamma_1'}{\gamma_1'^2 - \gamma_1^2}\frac{\partial E_x}{\partial z}$$

$$= \frac{\gamma_1'}{\gamma_1'^2 - \gamma_1^2}\left(\frac{n\pi}{b}\right) C \sin\left(\frac{m\pi}{a}\right)y \cos\left(\frac{n\pi}{b}\right)z \qquad (22.18)$$

$$H_y = -\frac{j\omega\epsilon^*}{\gamma_1'} E_z$$

$$H_z = \frac{j\omega\epsilon^*}{\gamma_1'} E_y.$$

Equation 22.16 allows us to express the characteristic propagation factor γ_1' by the intrinsic propagation factor γ_1, the mode integers n and m and the guide dimensions a and b as

$$\gamma_1' = \sqrt{\gamma_1^2 + \left(\frac{m\pi}{a}\right)^2 + \left(\frac{n\pi}{b}\right)^2}$$

$$= j\sqrt{\omega^2\epsilon^*\mu^* - \left[\left(\frac{m\pi}{a}\right)^2 + \left(\frac{n\pi}{b}\right)^2\right]}. \qquad (22.19)$$

This factor must be complex or imaginary if the wave is to be propagated, hence the angular frequency must satisfy the condition

$$\omega \geq \sqrt{\frac{\left(\frac{m\pi}{a}\right)^2 + \left(\frac{n\pi}{b}\right)^2}{\epsilon^*\mu^*}}. \qquad (22.20)$$

The low-frequency limit designates the *cut-off frequency of the rectangular wave guide* for the mode characterized by the integers m and n. Its value for a loss-free medium 1 is therefore

$$\nu_c = \sqrt{\frac{\left(\frac{m}{2a}\right)^2 + \left(\frac{n}{2b}\right)^2}{\epsilon'\mu'}}$$

$$= v_1\sqrt{\left(\frac{m}{2a}\right)^2 + \left(\frac{n}{2b}\right)^2}. \qquad (22.21)$$

By choosing $a = \infty$, the rectangular guide is transformed into the parallel-plane guide, and Eq. 22.21 becomes identical with Eq. 21.13.

As long as the loss in the dielectric and in the walls of the guide can be neglected, Eq. 22.19 may be simplified to

$$\gamma_1' = j\beta_1' = j\frac{\omega}{v_1}\sqrt{1 - \left(\frac{\nu_c}{\nu}\right)^2}$$

$$= \gamma_1\sqrt{1 - \left(\frac{\nu_c}{\nu}\right)^2}. \qquad (22.22)$$

The phase velocity of the wave pattern in the guide becomes correspondingly

$$v_i = \frac{\omega}{\beta_1'} = \frac{v_1}{\sqrt{1 - \left(\frac{\nu_c}{\nu}\right)^2}}, \qquad (22.23)$$

its wavelength

$$\lambda_i = \frac{2\pi}{\beta_1'} = \frac{v_1}{\sqrt{\nu^2 - \nu_c^2}}, \qquad (22.24)$$

and group velocity

$$v_g = \frac{1}{\dfrac{d\beta_1'}{d\omega}} = v_1 \sqrt{1 - \left(\frac{\nu}{\nu_c}\right)^2} \quad \text{(cf. Eq. 11.19).} \qquad (22.25)$$

The product $v_i \cdot v_g$ is again equal to the square of the free space velocity as established in Eq. 21.9. Figure 22.3 illustrates the frequency dependence of phase velocity, group velocity, and wavelength.

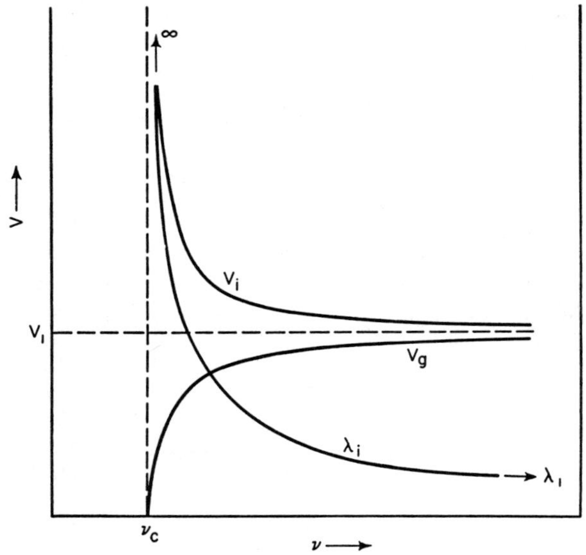

Fig. 22.3. Frequency dependence of phase velocity, group velocity and wavelength of TM wave in rectangular guide.

The wave guide, as this figure shows, behaves like a highly dispersive dielectric; the product of permittivity and permeability falls below that of free space. This can be seen by specifying a *permittivity* ϵ_{g1}' *and permeability* μ_{g1}' of *medium 1 in the guide* and a phase velocity of the interference pattern as

$$v_i = \frac{1}{\sqrt{\epsilon_{g1}'\mu_{g1}'}}, \qquad (22.26)$$

in analogy to the expression for the phase velocity in the unbounded medium

$$v_1 = \frac{1}{\sqrt{\epsilon_1'\mu_1'}}. \qquad (7.20)$$

The ratio of the guide value to the free space values of the dielectric constants and permeabilities becomes (Eq. 22.23)

$$\left(\frac{v_1}{v_i}\right)^2 = \frac{\epsilon_{g1}'\mu_{g1}'}{\epsilon_1'\mu_1'} = 1 - \left(\frac{\nu_c}{\nu}\right)^2. \qquad (22.27)$$

For a loss-free medium 1, it follows from Eqs. 22.19 to 22.22 that

$$\frac{\gamma_1'}{\gamma_1'^2 - \gamma_1^2} = \sqrt{\left(\frac{\nu}{\nu_c}\right)^2 - 1} = \frac{\nu}{\nu_c}\frac{v_1}{v_i},$$

$$\frac{j\omega\epsilon_1'}{\gamma_1'^2 - \gamma_1^2} = \frac{\nu}{\nu_c}\sqrt{\frac{\epsilon_1'}{\mu_1'}} = \frac{\nu}{\nu_c}\frac{1}{Z_1}. \qquad (22.28)$$

By substituting these expressions into Eq. 22.18, we arrive at the final formulation for the field components of the *TM modes* in the rectangular loss-free wave guide:

$$\begin{aligned}
E_x &= -C \sin\left(\frac{m\pi}{a}\right) y \sin\left(\frac{n\pi}{b}\right) z,\\
E_y &= C \frac{\nu}{\nu_c}\frac{v_1}{v_i}\left(\frac{m\pi}{a}\right) \cos\left(\frac{m\pi}{a}\right) y \sin\left(\frac{n\pi}{b}\right) z,\\
E_z &= C \frac{\nu}{\nu_c}\frac{v_1}{v_i}\left(\frac{n\pi}{b}\right) \sin\left(\frac{m\pi}{a}\right) y \cos\left(\frac{n\pi}{b}\right) z,\\
H_x &= 0,\\
H_y &= -\frac{C}{Z_1}\frac{\nu}{\nu_c}\left(\frac{n\pi}{b}\right) \sin\left(\frac{m\pi}{a}\right) y \cos\left(\frac{n\pi}{b}\right) z,\\
H_z &= \frac{C}{Z_1}\frac{\nu}{\nu_c}\left(\frac{m\pi}{a}\right) \cos\left(\frac{m\pi}{a}\right) y \sin\left(\frac{n\pi}{b}\right) z.
\end{aligned} \qquad (22.29)$$

It is apparent that near cut-off the longitudinal E_x component dominates the electric field, but with increasing frequency the transversal components become decisive and the wave approaches the character of a normal TEM wave, as already evident from Fig. 22.3.

Returning to Eq. 22.17, we can derive, with the condition $E_x = 0$, the *TE modes* of the rectangular wave guide and find for a loss-free medium 1, in analogy to Eqs. 22.29, the field components

$$\begin{aligned}
E_x &= 0,\\
E_y &= BZ_1 \frac{\nu}{\nu_c}\left(\frac{n\pi}{b}\right) \cos\left(\frac{m\pi}{a}\right) y \sin\left(\frac{n\pi}{b}\right) z,\\
E_z &= BZ_1 \frac{\nu}{\nu_c}\left(\frac{m\pi}{a}\right) \sin\left(\frac{m\pi}{a}\right) y \cos\left(\frac{n\pi}{b}\right) z,\\
H_x &= -B \cos\left(\frac{m\pi}{a}\right) y \cos\left(\frac{n\pi}{b}\right) z,\\
H_y &= -B \frac{\nu}{\nu_c}\frac{v_1}{v_i}\left(\frac{m\pi}{a}\right) \sin\left(\frac{m\pi}{a}\right) y \cos\left(\frac{n\pi}{b}\right) z,\\
H_z &= B \frac{\nu}{\nu_c}\frac{v_1}{v_i}\left(\frac{n\pi}{b}\right) \cos\left(\frac{m\pi}{a}\right) y \sin\left(\frac{n\pi}{b}\right) z.
\end{aligned} \qquad (22.30)$$

The integers m and n indicate the number of half wavelength or nodal lines which the interference wave possesses in the y–z cross section of the rectangular pipe. Thus the checkerboard pattern of Fig. 22.4 results, in which each individual mode of vibration can be identified by the double integers m and n. These

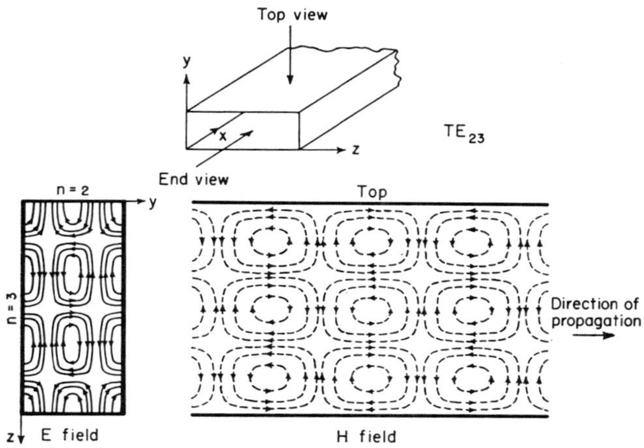

Fig. 22.4. Identification of wave modes in rectangular guide by integers m and n.

integers are used as subscripts in the mode designation, that is, we refer to TE_{mn} and TM_{mn} modes. For the index values $m = n = 0$, the TM field components vanish, and the TE_{00} mode degenerates into a constant longitudinal magnetic component; for $m = 0$ or $n = 0$, no TM modes are possible. The lowest-order modes that may be supported by a rectangular guide are the TE_{01} and the TE_{10} modes (Fig. 22.5). The lowest-order mode of a wave-guide system is called its *dominant mode.*

If the rectangular guide is opened up into a parallel-plane arrangement ($a = \infty$), the periodic functions containing a in Eqs. 22.29 and 22.30 are removed, the components with $1/a$ in the denominator become zero, and

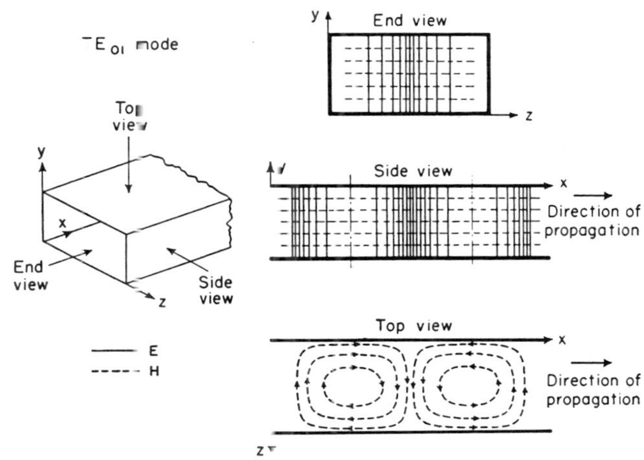

Fig. 22.5. Lowest order modes in rectangular guide.

the factors $(n\pi/b)$ are removed. The result is that for $n = 0$ no TE wave exists between parallel-plate conductors, and that the TM_0 mode degenerates into a TEM wave with only transversal field components. This wave is generated by connecting the two conductor plates to opposite poles of an oscillator. Such a TEM or *principal* wave can obviously exist also in a coaxial line, but not in any uniconductor wave guide.

23 · Measurement of Dielectrics in Shorted Wave Guides

In Secs. 16 and 20 it was shown how the dielectric properties of a medium can be derived from the angles of incidence and refraction and from the state of polarization of the reflected and refracted beams. This is the normal optical procedure employing traveling waves; it is based on Snell's laws and Fresnel's equations. Carried over into the microwave range these methods become definitely inferior. Boundary effects arise because the wavelength is comparable to the sample dimensions, and standing waves are formed between transmitter and receiver, producing intensity fluctuations.

The macroscopic size of the wavelength, on the other hand, proves of great advantage if the standing-wave methods discussed in Sec. 18 are applied, because a detector may travel directly through the profile of the wave pattern. Drude's two classical methods [1] already utilize this possibility and have since been employed in many variations.[2] The use of open transmission lines (*Lecher systems* [3]) of the type shown in Fig. 23.1 has been a handicap. An empirical calibration of the condenser system was required, and extreme care had to be

[1] P. Drude, *Wied. Ann.* *55*, 633 (1895); *61*, 466 (1897).

[2] See, for example, G. Potapenko, *Z. Physik* *20*, 21 (1923); M. Seeberger, *Ann. Physik* *16*, 77 (1933); R. King, *Rev. Sci. Instr.* *8*, 201 (1937); K. E. Slevogt, *Ann. Physik* *36*, 141 (1939); H. Slätis, *Ann. Physik* *36*, 397 (1939).

[3] E. Lecher, *Wied. Ann.* *41*, 850 (1890).

taken to avoid perturbation of the waves by the detector system and disturbance of the detector by stray fields. These limitations have been overcome by enclosing the electromagnetic field in wave guides.[4] All boundary and stray effects thus disappear automatically, and small amounts of a dielectric can be measured with precision on the basis of the theory given in Sec. 18.

The type of wave-guide instrument, developed for these measurements in the Laboratory for Insulation Research during World War II,[5] is shown in Fig. 23.2. A stabilized klystron oscillator radiates monochromatic waves of a prescribed frequency into one end of a coaxial line or hollow wave guide; they are reflected by a shorting metallic boundary at the other end. Standing

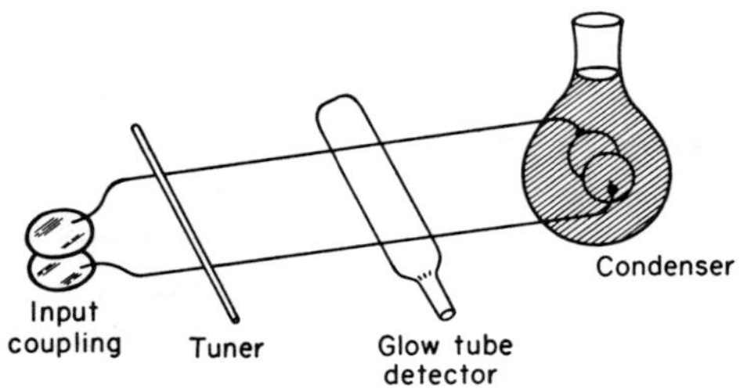

Fig. 23.1. Drude's arrangement for measuring the dielectric constant of liquids in the microwave range.

waves are set up and can be measured by a probe detector traveling along a narrow slot cut in the guide parallel to its axis. The dielectric is inserted in the closed end of the guide opposite the transmitter. For the short-circuit measurement, the sample is placed in direct contact with the metal short; for the open-circuit measurement, it is located a quarter wavelength ahead of the termination (see Sec. 18). It is essential that the sample intimately fit the walls of the guide and the shorting plate, and that its faces be perpendicular to the guide axis.

The standing wave is measured in air (medium 1) above the dielectric sample of thickness d (medium 2); the terminating impedance $Z(0)$ of medium 1 is found by determining the ratio of minimum to maximum electric field strength and the distance x_0 of the first minimum from the dielectric boundary. Since the detector would be overloaded and the field distribution disturbed if $E_{\max}$ were measured directly, the ratio $E_{\min}/E_{\max}$ is usually found indirectly by measuring the distance Δx between the two positions where the current in the detector doubles its minimum value

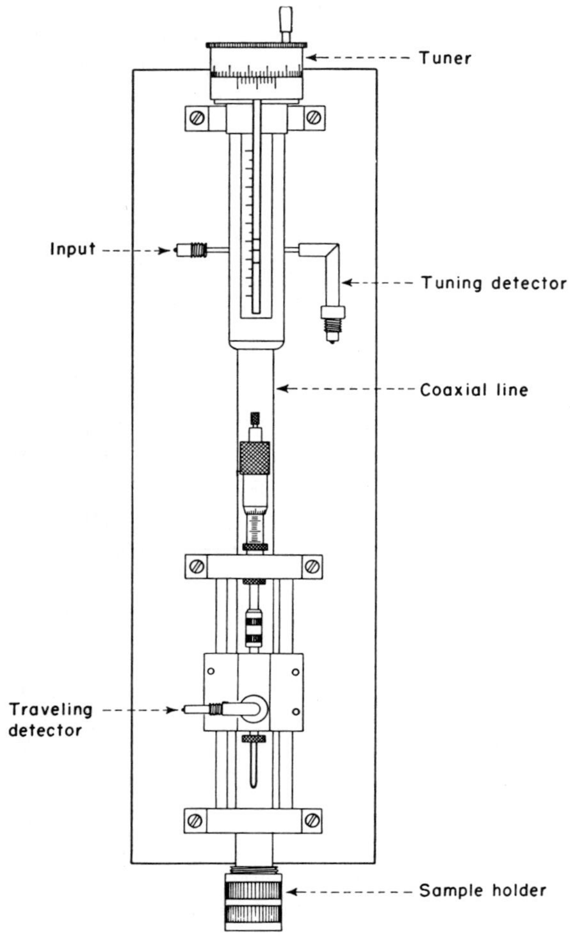

Fig. 23.2. M.I.T. coax instrument.

(Fig. 23.3). For a detector of "square-law" response ($I \sim E^2$), for example, a germanium diode, it can be easily shown [4,5] that

$$\frac{E_{\min}}{E_{\max}} = \frac{\pi\,\Delta x}{\lambda_1'} \qquad \left(\text{for } \frac{E_{\min}}{E_{\max}} \leq 0.1\right). \tag{23.1}$$

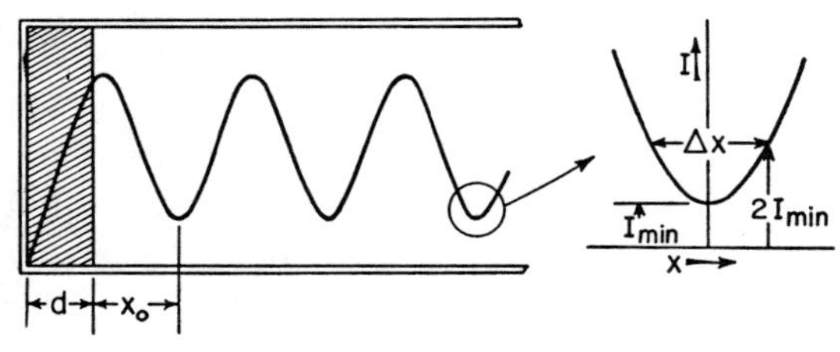

Fig. 23.3. Measurement of standing-wave ratio with square law detector.

The half wavelength $\lambda_1'/2$ in the air space of the guide may be obtained by directly measuring either the distance between two minima of the standing wave with

[4] S. Roberts and A. von Hippel, *Phys. Rev.* *57*, 1056 (1940); *J. Appl. Phys.* *17*, 610 (1946).

[5] A. von Hippel, D. G. Jelatis, and W. B. Westphal "The Measurement of Dielectric Constant and Loss with Standing Waves in Coaxial Wave Guides," NDRC Contract OEMsr-191, Laboratory for Insulation Research, Massachusetts Institute of Technology, April 1943.

the traveling detector or by setting the tuning plunger in the head of the guide to two consecutive resonance positions.

As shown in the preceding chapter, the impedance and the propagation function of a wave in a hollow wave guide differ from their values in unbounded space. We denote the quantities pertaining to the guide by a prime and write the terminating impedance expression of Sec. 18 (Eq. 18.4) as

$$Z'(0) = Z_1' \frac{\dfrac{E_{\min}}{E_{\max}} - j \tan \dfrac{2\pi x_0}{\lambda_1'}}{1 - j \dfrac{E_{\min}}{E_{\max}} \tan \dfrac{2\pi x_0}{\lambda_1'}}. \tag{23.2}$$

Using the short-circuit method we have the additional equation for the terminating impedance, as seen from medium 2 (Eq. 18.9)

$$Z'(0) = Z_2' \tanh \gamma_2' d. \tag{23.3}$$

The parameters Z_1' and Z_2' may be called the *characteristic wave impedances* of media 1 and 2 as distinguished from the intrinsic impedances. For the γ' parameters we introduced in Sec. 22 the name *characteristic propagation factor*. All field impedances (E/H) used here relate only to the transverse field components because only these obey the continuity condition (Eq. 14.9).

To measure the transverse components of the electric field, we operate conveniently with TE waves and a detector probe coupling with this electric field. Figure 23.4 refers in particular to the dominant TE_{01} mode in a rectangular wave guide and shows the orientation of slot and detector in relation to the wave pattern.

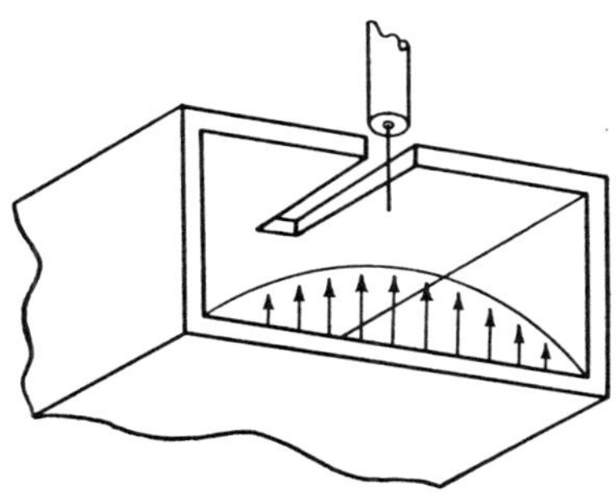

Fig. 23.4. TE_{01} mode and standing-wave detector.

The characteristic wave impedance for TE waves, according to Eq. 22.30, is given as

$$Z' = \frac{E_y}{H_z} = -\frac{E_z}{H} = Z\frac{v_i}{v}, \tag{23.4}$$

whereas the characteristic propagation factor follows from Eqs. 22.22 and 22.27 as

$$\gamma' = \gamma \frac{v}{v_i}; \tag{23.5}$$

hence

$$Z'\gamma' = Z\gamma. \tag{23.6}$$

Since for nonmagnetic dielectrics $Z_1\gamma_1 = Z_2\gamma_2$ (see Eq. 7.25), it follows that

$$Z_2' = Z_1' \frac{\gamma_1'}{\gamma_2'}. \tag{23.7}$$

Hence Eq. 23.3 may be rewritten for nonmagnetic dielectrics

$$Z'(0) = Z_1'\gamma_1' d \frac{\tanh \gamma_2' d}{\gamma_2' d}. \tag{23.8}$$

Equating Eqs. 23.2 and 23.8 and noting that medium 1 is loss-free $\left(\gamma_1' = j\dfrac{2\pi}{\lambda_1'}\right)$, we can derive γ_2' from the measured quantities by the expression

$$\frac{\tanh \gamma_2' d}{\gamma_2' d} = \frac{-j\lambda_1'}{2\pi d} \frac{\dfrac{E_{\min}}{E_{\max}} - j \tan \dfrac{2\pi x_0}{\lambda_1'}}{1 - j \dfrac{E_{\min}}{E_{\max}} \tan \dfrac{2\pi x_0}{\lambda_1'}} \equiv Ce^{j\zeta}. \tag{23.9}$$

The function $Ce^{j\zeta}$ is found by measuring the thickness d of the sample, the wavelength λ_1' in the air-filled pipe, the inverse VSWR $E_{\min}/E_{\max}$, and the distance x_0 of the first minimum from the surface of medium 2. Next the function

$$\gamma_2' d \equiv Te^{j\tau} \tag{23.10}$$

has to be determined from charts or from a series approximation of the function

$$\frac{\tanh Te^{j\tau}}{Te^{j\tau}} = Ce^{j\zeta}. \tag{23.11}$$

In Fig. 23.5 a survey map drawn by S. Roberts [4] is given, with the argument τ the ordinate, the absolute value of T the abscissa, whereas C and ζ are parameters of intersecting curves; τ and ζ are expressed in degrees.[6] The hyperbolic functions are multivalued so that measurements with a single thickness d of the dielectric may leave the value γ_2' in doubt. But if two different thicknesses are used, only one set of values T/d and τ will satisfy both experimental results.

[6] Detailed charts may be found, for example, in the companion volume, *Dielectric Materials and Applications*, Technology Press of M.I.T. and John Wiley and Sons, New York, 1954, Sec. II A, 2.

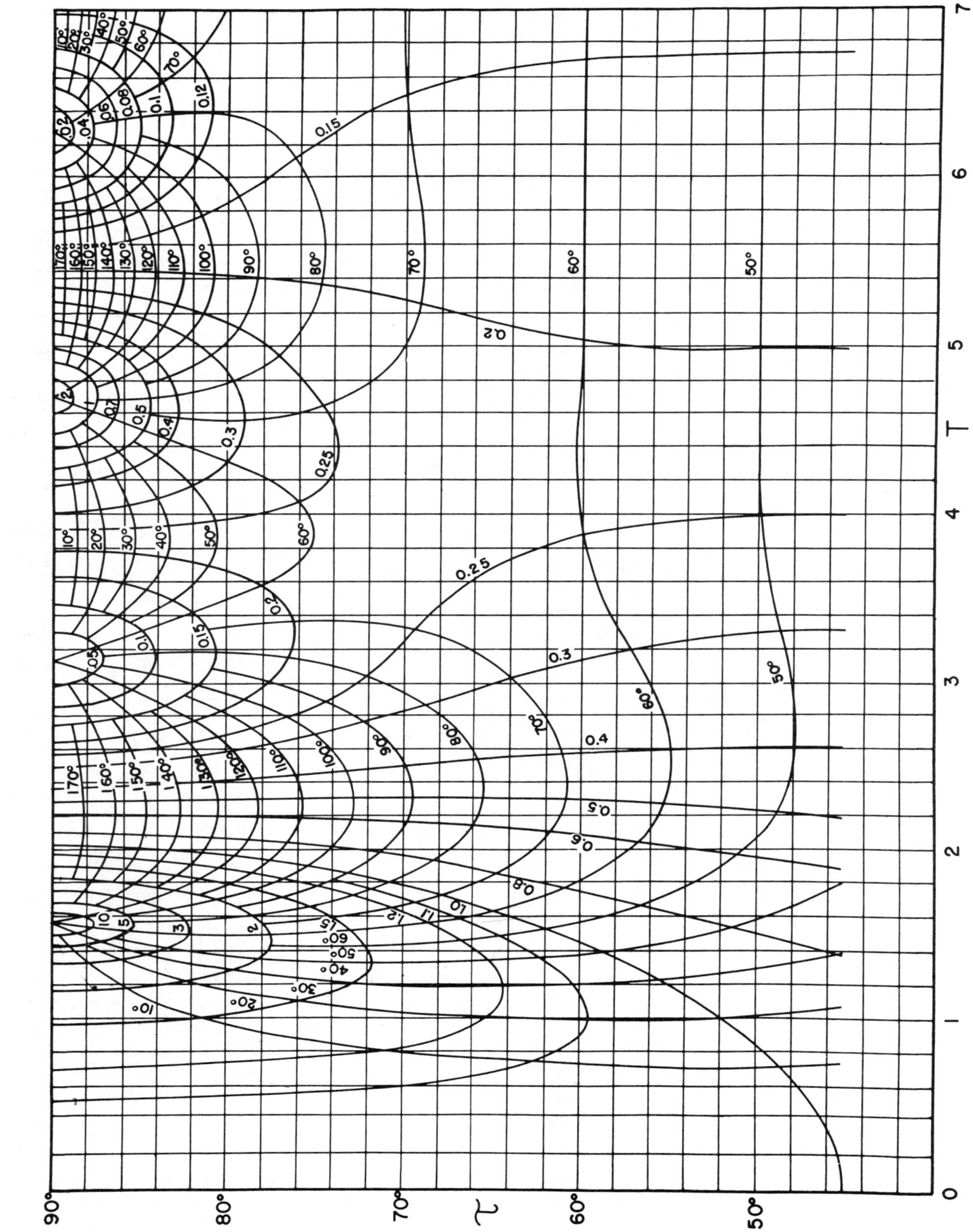

Fig. 23.5. Survey map of $Ce^{i\xi}$ function as $f(Te^{i\tau})$.

After the characteristic propagation factor γ_2' is obtained, the intrinsic propagation factor γ_2 of the dielectric in free space can be calculated from Eq. 22.19 as

$$\gamma_2'^2 = \gamma_2^2 + \left(\frac{m\pi}{a}\right)^2 + \left(\frac{n\pi}{b}\right)^2. \qquad (23.12)$$

We may rewrite this expression by referring to the cut-off wavelength λ_c of the guide. From Eq. 22.21 it follows that

$$\lambda_c = \frac{v_1}{\nu_c} = \frac{1}{\sqrt{\left(\frac{m}{2a}\right)^2 + \left(\frac{n}{2b}\right)^2}}, \qquad (23.13)$$

hence, according to Eq. 23.12,

$$\gamma_2'^2 = \gamma_2^2 + \left(\frac{2\pi}{\lambda_c}\right)^2. \qquad (23.14)$$

Since for air as the loss-free medium 1 the corresponding equation may be written as

$$\gamma_1'^2 = -\left(\frac{2\pi}{\lambda_1'}\right)^2 = -\omega^2\epsilon_0\mu_0 + \left(\frac{2\pi}{\lambda_c}\right)^2, \qquad (23.15)$$

we may express the complex dielectric constant of a nonmagnetic medium 2 finally in the convenient form

$$\epsilon_2^* = \frac{-\gamma_2^2}{\omega^2\mu_0} = \epsilon_0 \frac{\left(\frac{1}{\lambda_c}\right)^2 - \left(\frac{\gamma_2'}{2\pi}\right)^2}{\left(\frac{1}{\lambda_c}\right)^2 + \left(\frac{1}{\lambda_1'}\right)^2}. \qquad (23.16)$$

The cut-off wavelength for the lowest order waves in the customary types of wave guide are

rectangular pipe $\lambda_c = 2 \times$ width (see Fig. 22.5),

round pipe $\lambda_c = 1.71$ diameter,

coaxial line $\lambda_c = \infty$.

For the TEM wave of the coaxial line Eq. 23.16 reduces therefore to

$$\epsilon_2^* = \epsilon_0 \left(\frac{\gamma_2}{\gamma_1}\right)^2. \qquad (23.17)$$

Summarizing, the procedure of determining ϵ_2^* begins with measurements of $E_{\min}/E_{\max}$, x_0, d, and λ_1'. From these measured values, C and ζ are calculated with the help of Eq. 23.9. From C and ζ the values of T and τ are found from charts or by a series approximation. By dividing $Te^{-\tau}$ by d, γ_2' is obtained; finally ϵ_2^* is determined from γ_2' by Eq. 23.16 or 23.17.

In complete analogy to this short-circuit method based on the equation

$$Z'(0)_s = Z_2' \tanh \gamma_2' d, \qquad (23.3)$$

we can use the open-circuit method, in which a quarter wavelength of dielectric is inserted between medium 2 and the shorting plate. This arrangement (see Eq. 18.14) leads to the expression

$$Z'(0)_0 = Z_2' \coth \gamma_2' d. \qquad (23.18)$$

Since the hyperbolic tangent and cotangent are reciprocals of each other, the product of the two types of impedances is

$$Z'(0)_s Z'(0)_0 = Z_2'^2. \qquad (23.19)$$

The characteristic wave impedance of a dielectric is the geometrical mean of its short-circuited and open-circuited impedances. Thus Z_2' can be found without reference to hyperbolic functions if both input impedances for the shorted and for the open dielectric are measured. This suffices for nonmagnetic materials, since Z_2' determines γ_2' (see Eq. 23.7). Magnetic materials, on the other hand, require a separate calculation of Z_2' and γ_2' because the impedance contains the ratio, and the propagation factor the product of μ_2^* and ϵ_2^*. In this case we have to exploit the short-circuit method as well as the open-circuit method to their full extent.[7]

[7] For permissible simplifications of these methods, see W. B. Westphal in Ref. 6.

24 · Short-Circuited Guides and Cavity Resonators

The standing-wave method of Sec. 23 has the advantage of depending only on the wave-guide section between detector probe and shorting plate. The input part of the guide between the detector and tuning plunger (cf. Fig. 23.2) does not enter into the final result; the setting of the plunger affects only the intensity of the standing-wave pattern. Thus we are freed of the errors inherent in the imperfect contact between tuning plunger and guide walls, but we have to put up with a traveling detector requiring a slotted line and appropriate shielding. The shielding problem has been solved satisfactorily, and most of the precision instruments used today employ traveling detectors and slotted wave guides. However, if we want to avoid this approach, we have the alternative of measuring the input impedance of the whole guide as a function of its length or of the exciting frequency. The guide in this case serves as a cavity resonator.

The impedance of a shorted line

We have already encountered resonance phenomena in our discussion of interference optics (see Sec. 18) without referring to them explicitly. Equation 18.9 describes the input impedance of a dielectric resonator as a function of its length, and Fig. 18.3 illustrates how the reflection from a dielectric layer varies from resonance to antiresonance as the layer thickness increases. Let us return to this equation and examine in more detail the input impedance of a wave-guide section extending from $x = 0$ to $x = l$ and terminated first at l only, later at both ends, by a metallic short circuit.

Seen from $x = 0$, the input impedance of the shorted line filled with a medium 1 is

$$Z'(0) = Z_1' \tanh \gamma_1' l; \tag{24.1}$$

the primed quantities refer, as previously, to the field in the guide. Without attenuation the equation simplifies to

$$Z'(0) = Z_1' j \tan \frac{2\pi}{\lambda_1'} l; \tag{24.2}$$

the field in the guide consists of the real components (see Eq. 17.10)

$$\begin{aligned} \mathrm{Re}\,(E_{y1}) &= 2E_0 \sin \omega t \sin \frac{2\pi}{\lambda_1'}(x - l) \\ \mathrm{Re}\,(H_{z1}) &= \frac{2E_0}{Z_1'} \cos \omega t \cos \frac{2\pi}{\lambda_1'}(x - l). \end{aligned} \tag{24.3}$$

The standing electric and magnetic waves in front of the short circuit (Fig. 24.1) are displaced with respect

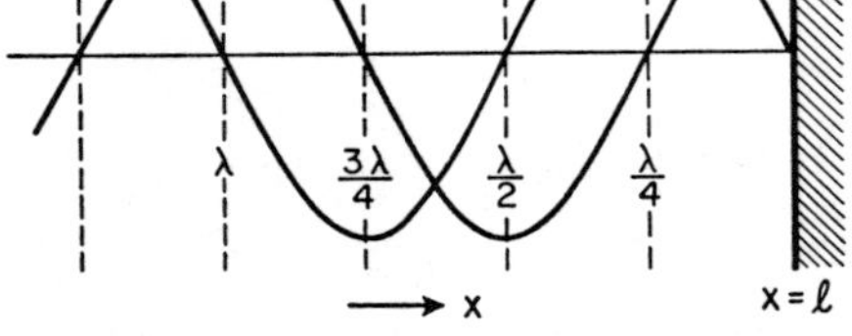

Fig. 24.1. Standing-wave in front of short-circuited guide.

to each other by a quarter wavelength in space and by a quarter period in time.

The impedance $Z'(0)$ is obtained by measuring at $x = 0$ the amplitude ratio of the E and H field and the temporal phase angle between the field vectors. **E** and **H** of Eq. 24.3 are in time quadrature, and the impedance is purely reactive. It is zero at the end of the line (at l), varies from 0 to $+\infty$ as the reference point is moved backwards a quarter of a wavelength, at this point jumps to $-\infty$, and returns to 0 again at $l - \dfrac{\lambda_1'}{2}$.

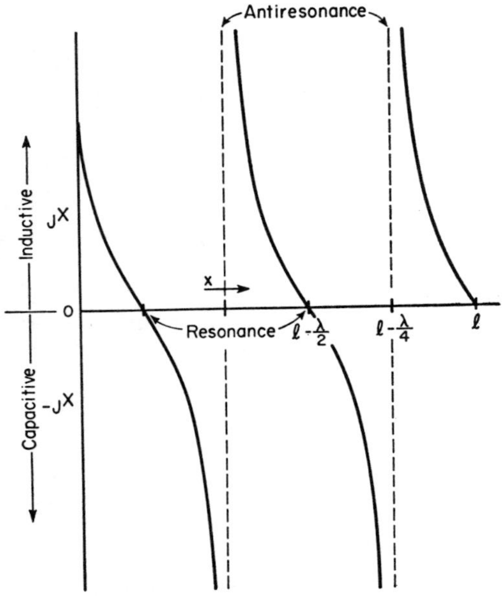

Fig. 24.2. Reactance of loss-free line, shorted at $x = l$.

At an even number of quarter wavelengths the line is in *resonance*, at an odd number of quarter wavelengths in *antiresonance*. While lengthening the line from resonance to antiresonance, we note a *positive* or *inductive* reactance; further extending it from antiresonance to resonance, we observe a *negative* or *capacitive* reactance (Fig. 24.2).

For the general case of a shorted line with attenuation ($\gamma_1' = \alpha + j\beta$), Eq. 24.1 may be rewritten

$$Z'(0) = Z_1' \tanh (\alpha + j\beta)l = R + jX. \tag{24.4}$$

Since

$$\begin{aligned} \sinh (\alpha + j\beta) &= \sinh \alpha \cos \beta + j \cosh \alpha \sin \beta, \\ \cosh (\alpha + j\beta) &= \cosh \alpha \cos \beta + j \sinh \alpha \sin \beta, \end{aligned} \tag{24.5}$$

the *resistance* R and the *reactance* X of the guide section become

$$\begin{aligned} R &= Z_1' \frac{\sinh \alpha l \cosh \alpha l}{\cosh^2 \alpha l \cos^2 \beta l + \sinh^2 \alpha l \sin^2 \beta l}, \\ X &= Z_1' \frac{\sin \beta l \cos \beta l}{\cosh^2 \alpha l \cos^2 \beta l + \sinh^2 \alpha l \sin^2 \beta l}. \end{aligned} \tag{24.6}$$

These relations may be mapped conveniently in the complex plane by plotting the normalized resistance and reactance,

$$r = R/Z_1', \qquad ⓧ = X/Z_1', \tag{24.7}$$

as the abscissa and ordinate respectively, and by drawing, in addition, lines of constant attenuation, αl, and of constant phase, βl. These lines prove to be two families of circles (Fig. 24.3).[1] The circles of constant

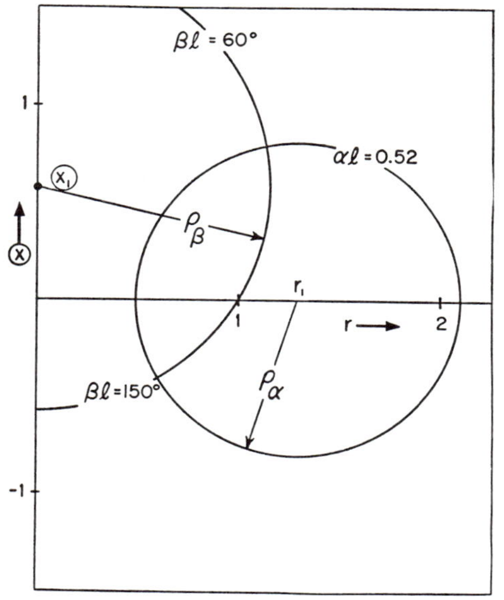

Fig. 24.3. Circles of constant attenuation (αl) and constant phase (βl).

αl have their centers on the r-axis at an abscissa r_1; their radius is ρ_α, where

$$r_1 = \coth 2\alpha l$$
$$(r - r_1)^2 + ⓧ^2 = \rho_\alpha{}^2 \tag{24.8}$$
$$\rho_\alpha = \csc 2\alpha l = \frac{2}{e^{2\alpha l} - e^{-2\alpha l}}.$$

The circles of constant βl have their centers on the ⓧ axis at an ordinate $ⓧ_1$ and have a radius ρ_β, where

$$ⓧ_1 = -\cot 2\beta l,$$
$$r^2 + (ⓧ - ⓧ_1)^2 = \rho_\beta{}^2, \tag{24.9}$$
$$\rho_\beta = \csc 2\beta l.$$

Thus the *impedance circle diagram* of Fig. 24.4 results, in which all the circles of constant attenuation surround the point $r = 1$, $ⓧ = 0$, whereas all the circles of constant phase, βl, pass through this point. We read this diagram as follows. If the attenuation of the wave guide is constant, we trace out a circle αl = constant as the length of the line increases. Since the phase function varies with $2\beta l = \frac{4\pi}{\lambda'} l$, this circle is described

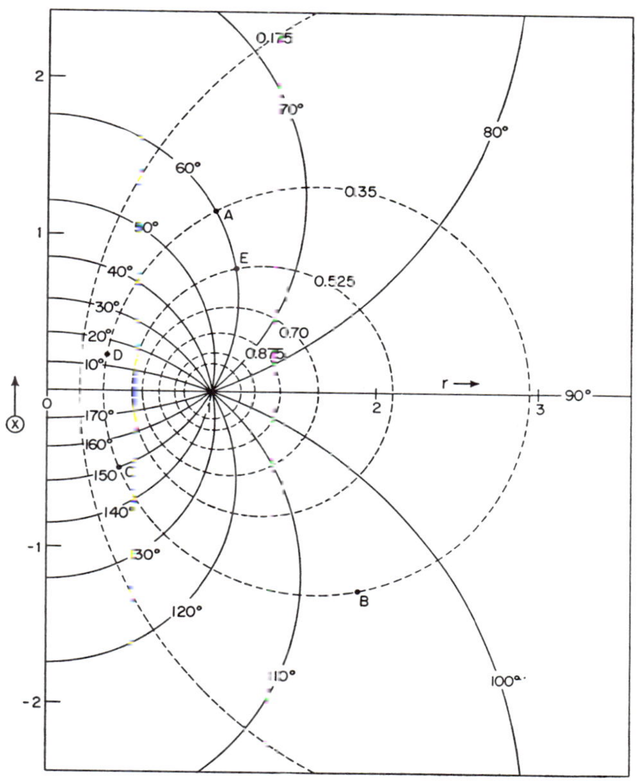

Fig. 24.4. Impedance circle diagram of transmission line.

once for each increase of the guide length by $\lambda_1'/2$. Starting, for example, from A we arrive at B, C, D, and A again by lengthening the line in successive steps of one eighth of a wavelength. Actually, the total attenuation increases with the length of the guide; hence we shall not arrive at A again after one full turn but at some point E of higher attenuation. We do not travel on a circle αl = constant, but on a spiral, which gradually winds up to the point $r = 1$, $ⓧ = 0$ (Fig. 24.5). This situation is identical with that discussed previously for the reflection coefficient r_0 as $f(d)$ (see Eq. 18.18).

The radius vector from the origin of the circle diagram to any point of the spiral represents in magnitude and phase the normalized impedance $Z'(0)/Z_1'$. It is small and purely resistive when the length of the wave guide corresponds to an even number of quarter wavelengths, and large and purely resistive when the length

[1] See, for example, J. C. Slater, *Microwave Transmission*, McGraw-Hill Book Co., New York, 1942, pp. 30 ff.; W. Jackson, *High Frequency Transmission Lines*, Methuen and Co., London, 1945.

corresponds to an odd number. These are the cases of *resonance* and *antiresonance;* at intermediate lengths the reactance is inductive or capacitive in accordance with Fig. 24.2.

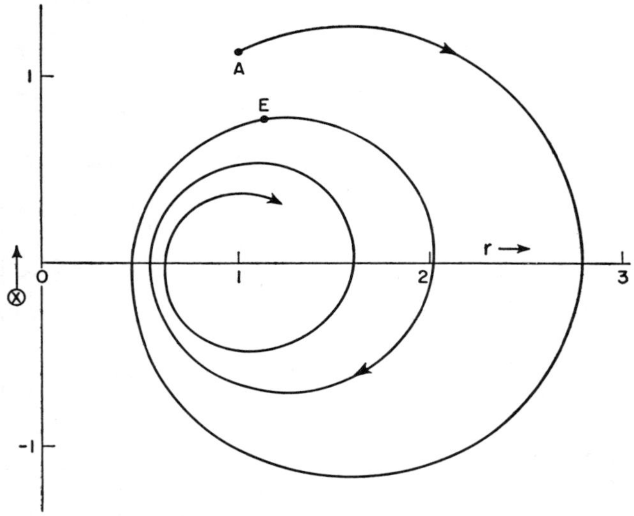

Fig. 24.5. Impedance spiral as function of line length.

The Q of cavity resonators

The preceding discussion of the shorted line prescribed only a short circuit at $x = l$ while leaving the boundary condition at the input end ($x = 0$) unspecified. The term *resonance* really implies the existence of some reflector at $x = 0$ and the build-up of high field amplitudes by *multireflection* between the two terminations. We have to ask whether or not the introduction of a real boundary at $x = 0$ changes the impedance expression (Eq. 24.4).

Obviously, if we tune the guide to a multiple of half wavelengths,

$$l = n\lambda_1'/2 \qquad (\text{with } n = 1, 2, 3 \ldots), \tag{24.10}$$

and transform it into a *cavity resonator* by shorting at $x = 0$, nothing will change (Fig. 24.6). The wave pat-

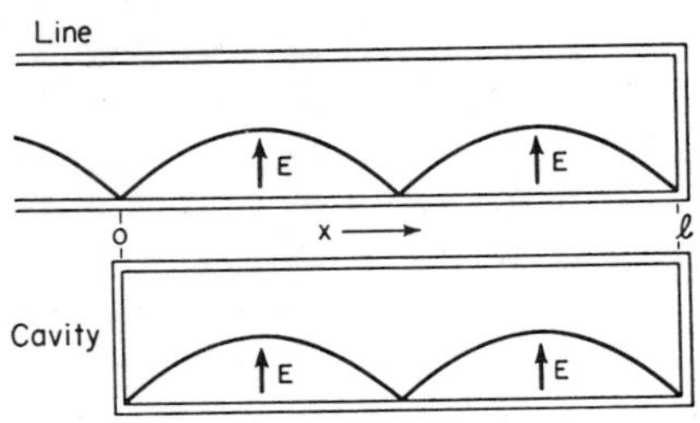

Fig. 24.6. Transformation of line into cavity resonator.

tern in this special case can fulfill the short-circuit condition at both ends. For any arbitrary length l, however, a phase mismatch results between the standing wave building up in front of the shorting plate at $x = l$ and that required by the boundary at $x = 0$. Destructive interference ensues, and decreases the standing wave amplitude. But the impedance expression (Eq. 24.4) remains valid for $x > 0$ since its only prerequisite is that a wave entering from the left is totally reflected at $x = l$ and superposes with the returning wave to a standing-wave pattern.

The field energy of the standing wave alternates between electric and magnetic energy storage. For a resonator filled with a nonmagnetic medium, this energy is (see Eq. 11.11)

$$U = \int_V \mu_0 H^2\, dV = \int_V \mu_0 Y'^2 E^2\, dV; \tag{24.11}$$

the field strength and the *characteristic wave admittance* $Y' = 1/Z'$ in the cavity varies from point to point, depending on the geometry of the resonator and the type of mode excited.

The energy oscillating in the cavity can be characterized by the energy input sustaining it. If the voltage across the input probe is kept constant, the field strength E remains constant and the sustaining energy is proportional to the square of the input admittance $Y'(0)$. By plotting $Y'(0)^2$ as $f(l)$ or $f(\nu)$, a *resonance characteristic* of the cavity is obtained.

The input admittance of the resonator follows from Eq. 24.4 as

$$Y'(0) = \frac{1}{Z'(0)} = \frac{1}{Z_1'} \coth{(\alpha + j\beta)l}. \tag{24.12}$$

The absolute value of the coth function is the square root of the sum of its squared real and imaginary parts or

$$\left|\coth{(\alpha + j\beta)l}\right| = \left|\frac{\cosh{(\alpha + j\beta)l}}{\sinh{(\alpha + j\beta)l}}\right| = \sqrt{\frac{\sinh^2 \alpha l \cosh^2 \alpha l + \sin^2 \beta l \cos^2 \beta l}{(\sinh^2 \alpha l \cos^2 \beta l + \cosh^2 \alpha l \sin^2 \beta l)^2}}. \tag{24.13}$$

If we are interested only in the behavior of cavities of low attenuation and in small phase deviations Δ from resonance, the approximations

$$\begin{aligned} \sinh^2 \alpha l &\to (\alpha l)^2, \\ \cosh^2 \alpha l &\to 1, \\ \sin^2 \beta l &\to \Delta^2, \\ \cos^2 \beta l &\to 1, \end{aligned} \tag{24.14}$$

hold. Hence Eq. 24.13 simplifies to

$$\left|\coth{(\alpha + j\beta)l}\right| = \sqrt{\frac{1}{\Delta^2 + (\alpha l)^2}}. \tag{24.15}$$

When the guide is approximately tuned to resonance, l equals about a multiple of half wavelengths (see Eq. 24.10), and αl may be replaced by

$$\alpha l \simeq \frac{\alpha n \lambda_1'}{2} = \frac{\alpha}{\beta} n\pi. \qquad (24.16)$$

Thus the square of the input admittance may finally be written for small variations from resonance

$$Y'(0)^2 = \frac{1}{Z_1'^2} \frac{1}{\Delta^2 + \left(\frac{\alpha}{\beta} n\pi\right)^2}. \qquad (24.17)$$

The energy delivered to the guide varies with $Y'(0)^2$ and falls to its half value, when the line is detuned from resonance by a change in phase

$$\Delta = \frac{\Delta_h}{2} = \pm \frac{\alpha}{\beta} n\pi. \qquad (24.18)$$

This detuning can be accomplished by changing the cavity from its resonance length l to $l \pm \frac{\Delta x_h}{2}$. The change in length producing the required phase change is (see Eqs. 24.18 and 24.16)

$$\frac{\Delta x_h}{2} = \frac{\lambda_1}{2\pi} \frac{\Delta_h}{2} = \pm \frac{\alpha}{\beta} l. \qquad (24.19)$$

Alternatively, the phase shift can be produced by changing the input frequency from its resonance value ν_0 to $\nu_0 \pm \frac{\Delta \nu_h}{2}$. Since

$$\frac{\Delta x_h}{l} = \frac{\Delta \nu_h}{\nu_0}, \qquad (24.20)$$

the energy input falls to its half value, when

$$\frac{\Delta \nu_h}{2} = \pm \frac{\alpha}{\beta} \nu_0. \qquad (24.21)$$

Δx_h and $\Delta \nu_h$, the *half-width* of the resonance characteristic, represent the distance between the two half-power points in the line-length scale and in the frequency scale, respectively (Fig. 24.7). The *relative half-width* or relative phase change between the half-power points,

$$\frac{\Delta x_h}{l} = \frac{\Delta \nu_h}{\nu_0} = 2\frac{\alpha}{\beta} = \frac{\Delta_h}{n\pi}, \qquad (24.22)$$

is identical in both scales.

To interpret this characteristic property of a resonator, we recall that, according to Eq. 9.26, for TEM waves and low loss ($\tan^2 \delta \ll 1$)

$$2\frac{\alpha}{\beta} \simeq \tan \delta. \qquad (24.23)$$

Hence the relative half-width is equal to the loss tangent, or its inverse, according to Eq. 9.2, is equal to the Q of the resonator,

$$Q = \frac{\beta}{2\alpha} = \frac{n\pi}{\Delta_h} \left[\frac{\text{reactive v-amp}}{\text{watts}} \right]. \qquad (24.24)$$

This quality factor of a resonator can be visualized in a still different way by referring to the transient behavior of the cavity. The energy content of a free

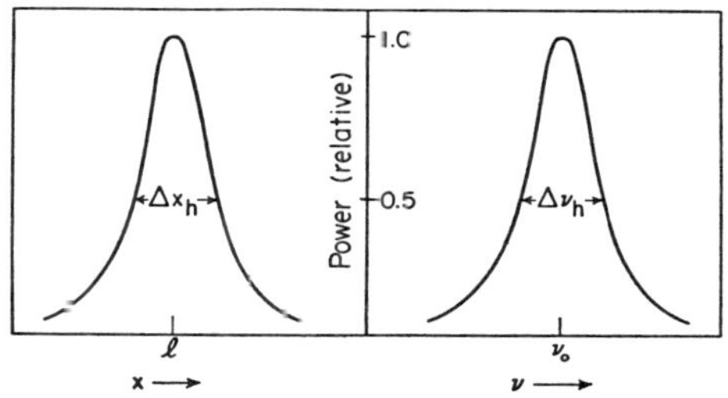

Fig. 24.7. Half width of resonance characteristic.

oscillator, if not replenished, is attenuated exponentially and falls to $1/e$ of its original value after a time (see II, Eq. 4.18),

$$\tau = \frac{1}{2\pi \Delta \nu_h} \text{ [sec]}. \qquad (24.25)$$

Consequently,

$$\tau = \frac{Q}{\omega_0} \text{ [sec]} \qquad (24.26)$$

is the *ringing time* of the resonator; or, if $T_0 = 1/\nu_0$ designates the time required for one oscillation, the ratio

$$\frac{\tau}{T_0} = \frac{Q}{2\pi} \qquad (24.27)$$

represents the number of free oscillations required to reduce the energy content of the resonator to 0.368 of its starting value.

The great advantage of cavity resonators is their excellent "ringing" quality in comparison with that of open-line resonators. Since no energy is lost by radiation, Q values of 10^4 can be obtained without much difficulty, and even 10^6 may be realized by the excitation of favorable modes in large cavities, whereas values not higher than 10^2 may be expected from an open transmission line in the microwave region. It is this high Q which enables us to measure low-loss dielectrics in closed-wave guides with high precision.

Measurement of dielectrics in tuned guides

In Sec. 23 we have shown how the dielectric properties of a material can be evaluated in a shorted line from the ratio of minimum to maximum field strength and the distance of the first minima from the dielectric boundary. In place of this "inside" method, which determines the standing-wave pattern in the line with a traveling detector, we now have an alternative "outside" method which obtains the Q of the cavity and its resonant length or resonant frequency from the input impedance. Obviously, the two methods are closely interrelated, and, for low loss, the choice between them is essentially one of experimental convenience. For high loss, the *traveling detector method* is superior because the resonance characteristic becomes asymmetrical, and the distance between the half-power points loses its simple meaning.

To establish the connection between the two techniques for low loss, we return once more to the standing

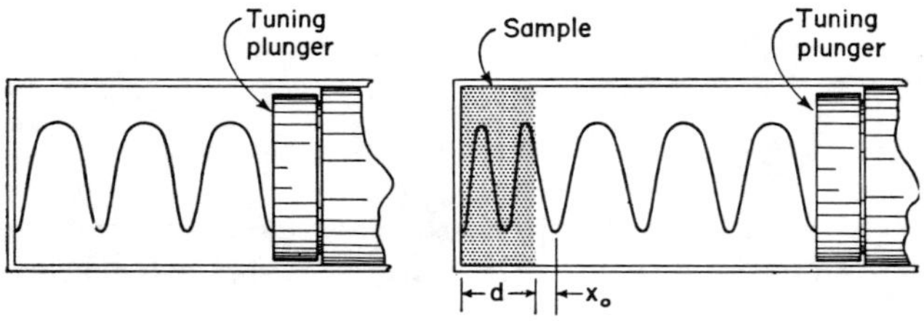

Fig. 24.8. Effect of dielectric on resonator length.

wave in the shorted line. Through superposition of incident and reflected wave, the standing electric wave results,

$$E(x) = E_0 e^{j\omega t}\{e^{-\gamma_1'(x-l)} - e^{\gamma_1'(x-l)}\}$$
$$= -2E_0 e^{j\omega t} \sinh\{(\alpha + j\beta)(x - l)\}. \quad (24.28)$$

By expanding this equation and using the approximations of Eq. 24.14, we obtain the simplified expression

$$E(x) = -2E_0 e^{j\omega t}\{\alpha(x - l)\cos\beta(x - l) + j\sin\beta(x - l)\}. \quad (24.29)$$

For the resonating wave guide $\left(l = \dfrac{n\lambda_1'}{2}\right)$ we observe at $x = 0$ a field strength minimum

$$E_{\min} = \alpha l 2E_0 e^{j\omega t} = \alpha l E_{\max}, \quad (24.30)$$

hence (cf. Eq. 24.24)

$$\frac{E_{\min}}{E_{\max}} = \alpha l = \frac{2\alpha}{\beta}\frac{\beta l}{2} = \frac{1}{Q}\frac{\pi l}{\lambda_1'}. \quad (24.31)$$

By inserting a dielectric of the thickness d into a tuned guide, we alter its original resonance length $x_1 = m_1\lambda_1'/2$ to a new resonance length (Fig. 24.8)

$$x_2 = \frac{m_2\lambda_1'}{2} + x_0 + d \quad \text{(where } m = 0, 1, 2, 3, \ldots). \quad (24.32)$$

Thus the retuning distance, $x_2 - x_1$, is a measure of x_0.

The Q has to be measured for the empty and for the filled guide in order to separate the losses of the cavity proper from those of the dielectric. Since the *wall losses* of the cavity produce an attenuation factor α_w which adds to the α of the dielectric material, the observed loss tangent is the sum of the loss tangent of dielectric and guide.

$$\tan\delta = \tan\delta_d + \tan\delta_w$$

or

$$\frac{1}{Q} = \frac{1}{Q_d} + \frac{1}{Q_w}. \quad (24.33)$$

After the Q_d of the dielectric has been obtained, its propagation factor γ_2' can be evaluated [2] as in the previous method (see Eq. 23.9), except that the ratio $E_{\min}/E_{\max}$ is replaced by the expression of Eq. 24.31. Thus

$$\frac{\tanh\gamma_2'd}{\gamma_2'd} = -\frac{j\lambda_1'}{2\pi d}\,\frac{\dfrac{\pi l}{Q_d\lambda_1'} - j\tan\dfrac{2\pi x_0}{\lambda_1'}}{1 - j\dfrac{\pi l}{Q_d\lambda_1'}\tan\dfrac{2\pi x_0}{\lambda_1'}}. \quad (24.34)$$

[2] See, for details of this resonance method, W. B. Westphal in Ref. 6, Sec. II A, 2.

25 · Treatment of Field Phenomena by Equivalent Circuits

Thus far, we have used the language of the field theory in discussing the propagation of electromagnetic waves and their reflection and refraction on dielectric and metallic boundaries. Frequently, however, it may prove convenient to handle field phenomena by an equivalence approach in which the role of the electric and magnetic fields in space corresponds to the interplay of voltages and currents in electric circuits composed of the proper capacitance, inductance, and resistance elements.

This equivalence, in the case of guided waves, may be a real one: the electromagnetic field patterns in space are one aspect of the physical phenomenon, the charge and current distribution in the guide structure its complement. Knowing the one we can derive the other unambiguously. Let us illustrate this complementarity for the TEM mode of a coaxial wave guide (Fig. 25.1).

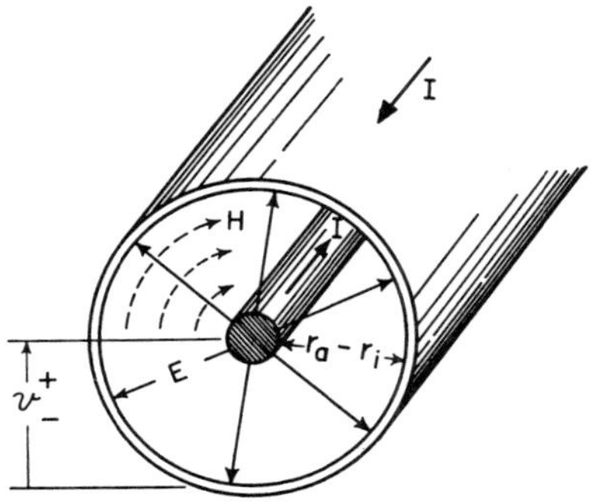

Fig. 25.1. Relation between *E-H* field and voltage-current distribution in coaxial line.

The electric flux density, and with it the electric field strength, in a system of cylindric symmetry, falls with $1/r$; hence the field intensity at any distance r from the center and the total voltage $\mathcal{V}$ between the conductors can be expressed as

$$E_r = \frac{E_i r_i}{r}$$

$$\mathcal{V} = \int_{r_i}^{r_a} E_r\,dr = E_i r_i \ln\frac{r_a}{r_i}. \tag{25.1}$$

The field strength E_i at the center conductor is prescribed by the free-charge density on this conductor surface (see Eq. 2.4). If dQ designates the true charge density on a section element dl of the coaxial cable, this field intensity is

$$E_i = \frac{dQ}{\epsilon' 2 r_i \pi\, dl}. \tag{25.2}$$

Introduced into Eq. 25.1, it determines the capacitance of the cable element as

$$dC \equiv \frac{dQ}{\mathcal{V}} = \frac{\epsilon' 2\pi\, dl}{\ln (r_a/r_i)}. \tag{25.3}$$

Thus the voltage $\mathcal{V}$ between the conductors and the *capacitance per unit length of coaxial cable*

$$C = \frac{2\pi\epsilon'}{\ln (r_a/r_i)} \quad [\text{farad m}^{-1}] \tag{25.4}$$

are given when the geometry of the electric field, its magnitude at one point, and the dielectric constant of the filling medium are prescribed.

The current flowing in the cable may be found from the line integral of the magnetic field encircling the conductors (Ampère's circuital law, Eq. 5.5). For any distance r from the center we obtain for the current enclosed

$$I = \oint \mathbf{H}_r \cdot d\mathbf{l} = 2r\pi H_r. \tag{25.5}$$

In the field space between the conductors this line integral is constant, since (see Eq. 25.1)

$$rH_r = r\frac{E_r}{Z} = \frac{\mathcal{V}}{Z \ln (r_a/r_i)}; \tag{25.6}$$

the enclosed current is that traversing the inner conductor. For $r > r_a$ or $r < r_i$, that is, in the metal walls of the conductors, the electric field E_r, and with it H_r and the line integral of the magnetic field, falls rapidly to zero, indicating that no net current is enclosed. The current streaming along the outer conductor is equal and opposite to that passing along the center conductor, and both currents are limited to surface layers as already discussed in Sec. 19.

From Eqs. 25.5 and 25.6 we obtain the ratio of voltage to current in a differential length of the coaxial line,

$$\frac{\mathcal{V}}{I} = Z\frac{\ln (r_a/r_i)}{2\pi}. \tag{25.7}$$

The ratio of transversal voltage to longitudinal current is called the *characteristic impedance* of a wave guide,

$$Z_c \equiv \mathcal{V}/I \tag{25.8}$$

It plays the same role in the discussion of transmission line *circuits* that is assigned in the treatment of electromagnetic *fields* to the ratio of transversal electric to transversal magnetic field intensity, that is, to the characteristic impedance Z'. (For TEM waves, Z' reduces to the intrinsic impedance Z of the dielectric.)

Knowing current, voltage, and capacitance of a line section we can derive its inductance from the fact that the electromagnetic field carries equal amounts of electric and magnetic energy (see Eqs. 11.6, 11.7, and (11.11). Thus

$$LI^2 = C\mathcal{V}^2, \tag{25.9}$$

and the *inductance per unit length of coaxial cable*

$$L = CZ_c^{\,2} = \frac{\mu'}{2\pi}\ln\frac{r_a}{r_i} \quad [\text{henry m}^{-1}]. \tag{25.10}$$

A circuit description by capacitance and inductance alone obviously neglects the existence of dielectric and conductor losses in the cable. To include such losses properly in the equivalent picture requires a separate consideration of the electric and of the magnetic field action.

The electric field causes a polarization loss in the cable segment dl,

$$\left(\tfrac{1}{2}\int_{r_i}^{r_a} \sigma E_r^2\,dr\right) dl = \left(\tfrac{1}{2}\epsilon''\omega \int_{r_i}^{r_a} E_r^2\,dr\right) dl$$

$$\equiv \left(\tfrac{1}{2}G\mathcal{V}^2\right) dl, \tag{25.11}$$

due to the dielectric conductivity σ of the filling medium (see Eq. 1.16). Thus an equivalent *conductance per unit length of coaxial cable* can be defined,

$$G = \epsilon''\omega \frac{\dfrac{1}{r_i} - \dfrac{1}{r_a}}{\left(\ln \dfrac{r_a}{r_i}\right)^2} \quad [\text{mho m}^{-1}], \tag{25.12}$$

which shunts the capacitance of the wave guide. The magnetic field, if $\mu'' \neq 0$, produces an analogous power loss in the dielectric during the magnetization cycle

$$\left(\tfrac{1}{2}\int_{r_i}^{r_a} \mu''\omega H_r^2\,dr\right) dl = \left(\tfrac{1}{2}R_m I^2\right) dl. \tag{25.13}$$

In addition, by inducing the current I, it causes a power loss in the wall resistance of the guide section,

$$dP_w = \left(\tfrac{1}{2}R_w I^2\right) dl. \tag{25.14}$$

Both these losses weaken the induced current; they act like an equivalent resistor

$$R = R_m + R_w \tag{25.15}$$

placed in series with the inductance L in the current path. Consequently the complete circuit analogue of the cable section dl is the equivalent circuit of Fig. 25.2.

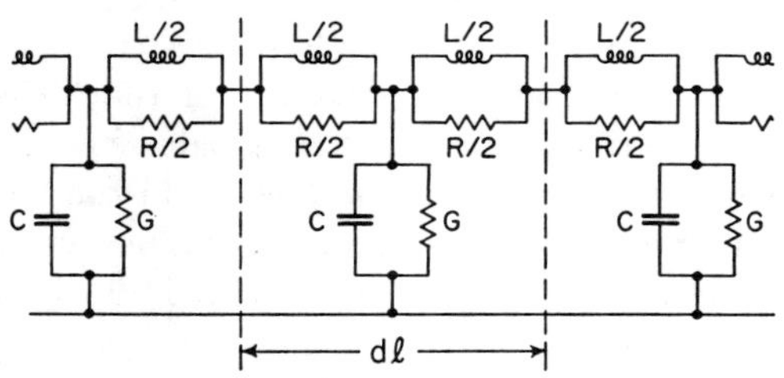

Fig. 25.2. Equivalent circuit of coaxial cable section.

The power loss due to wall resistance may be calculated from the magnetic and electric field intensity at the conductor surfaces. The relation between these fields is given by the intrinsic impedance Z_2 of the metal (see Eq. 19.5), and shows that the electric field

$$E_w = H_w Z_2 = H_w\sqrt{j\omega\mu_2'/\sigma_2} \tag{25.16}$$

leads the magnetic field in temporal phase by 45°; thus the magnitude of the in-phase electric component is $E_w/\sqrt{2j}$. Recalling that $I = 2r\pi H_r$, we may express the wall loss in the section dl of the cable by the product of the magnetic and the in-phase electric field at the surface of the inner and outer conductor as

$$dP_w = \left(\frac{1}{2}H_a 2r_a\pi \frac{H_a Z_2}{\sqrt{2j}} + \frac{1}{2}H_i 2r_i\pi \frac{H_i Z_2}{\sqrt{2j}}\right) dl$$

$$= \left[\frac{1}{2}\frac{I^2}{2\pi}\left(\frac{1}{r_a} + \frac{1}{r_i}\right)\frac{Z_2}{\sqrt{2j}}\right] dl. \tag{25.17}$$

Equating the expressions 25.14 and 25.17 yields the *equivalent wall resistance per unit length of coaxial cable*

$$R_w = \frac{1}{2\pi}\sqrt{\frac{\omega\mu_2'}{2\sigma_2}}\left(\frac{1}{r_a} + \frac{1}{r_i}\right) \quad [\text{ohm m}^{-1}]. \tag{25.18}$$

The power dissipated in a wave guide section is, in general, only a small fraction of the transmitted power P_t, that is,

$$dP_w = P_t(1 - e^{-2\alpha_w})\,dl \simeq P_t 2\alpha_w\,dl. \tag{25.19}$$

Since the transmitted power is determined by current and characteristic impedance as

$$P_t\,dl = \tfrac{1}{2}I^2 Z_c\,dl, \tag{25.20}$$

the *attenuation constant of the coaxial cable* due to wall losses follows as

$$\alpha_w = \frac{\tfrac{1}{2}I^2 R_w}{I^2 Z_c}$$

$$= \frac{1}{2Z}\sqrt{\frac{\omega\mu_2'}{2\sigma_2}}\left(\frac{1}{r_a} + \frac{1}{r_i}\right)\frac{1}{\ln(r_a/r_i)} \quad [\text{neper m}^{-1}]. \tag{25.21}$$

For TEM waves there exists thus a true complementarity between the electromagnetic field in space and the current-voltage distribution in the guide system. The equivalent circuit of Fig. 25.2 corresponds to an actual alternative picture. Extended to all types of guided waves the equivalence loses this physical meaning and becomes a formal representation based on the requirement that the power flow and the energy storage in the equivalent circuit shall be the same as in the wave guide.

The parameters of the equivalent circuit (Fig. 25.2) in this general case have to be determined by equating the field energies of the electric and magnetic field, stored and dissipated, to the energies stored and dissipated in the corresponding circuit elements, that is,

$$\begin{aligned} \tfrac{1}{2}CV^2\,dl &= \tfrac{1}{2}\int \epsilon' E^2\,dV, \\ \tfrac{1}{2}LI^2\,dl &= \tfrac{1}{2}\int \mu' H^2\,dV, \\ \tfrac{1}{2}RI^2\,dl &= \tfrac{1}{2}\int \mu''\omega H^2\,dV + P_w\,dl, \\ \tfrac{1}{2}GV^2\,dl &= \tfrac{1}{2}\int \epsilon''\omega E^2\,dV. \end{aligned} \qquad (25.22)$$

Current and voltage are assumed to be constant over the length dl of the infinitesimal guide section whereas the electric and magnetic fields may vary over the cross section of the guide in accordance with its geometry and the type of mode excited.

For TEM modes it follows from these equivalence expressions that we can change over from the equations of the field theory to those of the circuit theory, and vice versa, by replacing

$$\begin{aligned} E &\leftrightarrow \mathcal{V}, \\ H &\leftrightarrow I, \\ \epsilon' &\leftrightarrow C, \\ \mu' &\leftrightarrow L, \\ \epsilon''\omega &\leftrightarrow G, \\ \mu''\omega &\leftrightarrow R. \end{aligned} \qquad (25.23)$$

The complex propagation factor of the electromagnetic waves

$$\gamma = j\omega\sqrt{\epsilon^*\mu^*} = \sqrt{(\mu''\omega + j\omega\mu')(\epsilon''\omega + j\omega\epsilon')} \qquad (25.24)$$

is thus transformed into the propagation factor of a transmission line

$$\gamma = \alpha + j\beta = \sqrt{(R + j\omega L)(G + j\omega C)}, \qquad (25.25)$$

and the intrinsic impedance of a dielectric

$$Z = \sqrt{\frac{\mu^*}{\epsilon^*}} = \sqrt{\frac{\mu''\omega + j\omega\mu'}{\epsilon''\omega + j\omega\epsilon'}} \qquad (25.26)$$

into the characteristic line impedance

$$Z_c = \sqrt{\frac{R + j\omega L}{G + j\omega C}}. \qquad (25.27)$$

The wave equations of the electromagnetic field for E (and H),

$$\nabla^2\mathbf{E} = \epsilon^*\mu^*\frac{\partial^2\mathbf{E}}{\partial t^2} = \gamma^2\mathbf{E}, \qquad (25.28)$$

change over into the so-called *telegrapher's equations* for $\mathcal{V}$ (and I),

$$\frac{\partial^2\mathcal{V}}{\partial l^2} = RG\mathcal{V} + (LG + RC)\frac{\partial\mathcal{V}}{\partial t} + LC\frac{\partial^2\mathcal{V}}{\partial t^2}. \qquad (25.29)$$

The theory of propagation, reflection, and refraction of TEM waves in space becomes the current-voltage theory of transmission lines, where each line with its characteristic network elements (Fig. 25.3) is the

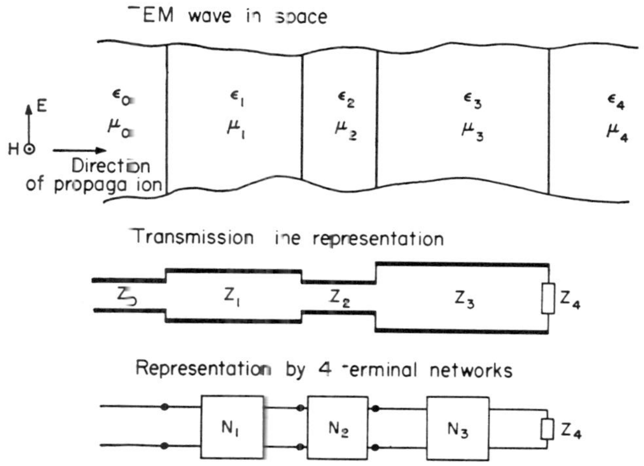

Fig. 25.3. Equivalent representations of sequence of dielectric layers traversed by TEM wave.

equivalent of a dielectric bounded laterally by some conducting structure, and where the lengths, terminations, and combinations of such lines are the variables of the problem.

The resonance discussion of the preceding section was based on the special case of low loss ($\alpha l \ll 1$, see Eq. 24.14). In the language of the line theory this means that the power transmitted through a line section is much larger than the power dissipated in it, or $G \ll \omega C$ and $R \ll \omega L$. Under these conditions the attenuation factor α and phase factor β of Eq. 25.25 simplify to

$$\begin{aligned} \alpha &= \frac{R}{2}\sqrt{\frac{C}{L}} + \frac{G}{2}\sqrt{\frac{L}{C}} \quad [\mathrm{m}^{-1}] \\ \beta &= \omega\sqrt{LC} \quad [\mathrm{m}^{-1}] \end{aligned} \qquad (25.30)$$

The current and voltage waves are propagated along the line with a phase velocity (see Eq. 7.17).

$$v = \frac{\omega}{\beta} = \frac{1}{\sqrt{LC}}. \qquad (25.31)$$

If the line is shorted at l and tuned to resonance (see Eq. 24.10),

$$\beta l = n\pi \qquad (n = 1, 2, 3 \ldots); \tag{25.32}$$

the resonance frequency becomes

$$\omega_0 = \frac{\beta}{\sqrt{LC}} = \frac{n\pi}{l\sqrt{LC}}. \tag{25.33}$$

The Q of the resonator (see Eq. 24.24) for dominating wall losses ($R \gg G$) is accordingly

$$Q = \frac{\beta}{2\alpha} = \frac{\omega_0 L}{R} = 2\pi \frac{\frac{1}{2}LI^2}{\left(\frac{1}{2\nu}\right) RI^2}$$

$$= 2\pi \times \frac{\text{average energy stored per half cycle}}{\text{energy dissipated per half cycle}}, \tag{25.34}$$

in agreement with our earlier definition (see Eq. 9.2). In the opposite case of dominant conductance loss in the dielectric ($G \gg R$), the Q expression becomes

$$Q = \frac{\beta}{2\alpha} = \frac{\omega_0 C}{G}$$

$$= 2\pi \frac{\frac{1}{2}C\mathcal{V}^2}{\left(\frac{1}{2\nu}\right) G\mathcal{V}^2} \left[\frac{\text{volt} \times \text{ampere}}{\text{watt dissipated}}\right]. \tag{25.35}$$

Equations 25.34 and 25.35 represent the Q of a *resonating coil* and of a *resonating condenser*, respectively.

It is this complementarity between electromagnetic fields in bounded space and current-voltage distributions in the guide system that confers on the microwave techniques their special flexibility and fascination. In normal optics the propagation of waves can, in general, be influenced by dielectric means only. In microwave optics it may be altered by changes in the dielectric medium or in the geometry of the guiding walls (see also Appendix A, I, 15).

The correlation between fields and current-voltage distributions ceases to be of importance when the wavelength becomes either very much smaller or very much larger than the guide dimensions. In the former case the waves lose their guided character, and we approach the laws of propagation in free space as formulated in the language of the field theory in previous sections. In the latter case the space dependence of the wave phenomena can be neglected. The line, previously composed of *distributed parameters*, that is, of a sequence of network elements where voltage and current vary from element to element, reduces to one single circuit of *lumped parameters* in which all parts of a condenser electrode are at the same voltage and all sections of an inductance or resistance are traversed by the same current at any given moment. Only the time dependence of current and voltage is essential in lumped circuit analysis; space considerations have become unimportant except for an actual determination of the magnitude of the circuit parameters by analytical calculation or geometrical field mapping.

In spite of this disappearance of a physical complementarity, it may still prove useful to compare the time and frequency dependence of some phenomenon with the response of an electric equivalent circuit composed of the proper lumped network elements. Once the rules of translation are established, an electric network analysis may prove a powerful method of elucidating phenomena which are not directly accessible. We shall illustrate this "approach by analogy" in the next section by giving a circuit interpretation of some typical dielectric characteristics.

26 · Representation of Dielectrics by Lumped Circuit Equivalents

In Sec. 1 of Part I of the book we introduced the concepts of a complex permittivity and complex permeability by considering the current-voltage characteristics of a capacitor and of a coil. We then incorporated these concepts into the field theory and studied the interaction between fields and matter in its macroscopic ramifications. Thus it became clear how static fields as well as electromagnetic waves are influenced by the polarization and magnetization of dielectrics, and how this interplay between fields and matter leads to measuring techniques for determining ϵ^* and μ^*. What really happens in dielectrics the macroscopic theory cannot reveal; it can only give some phenomenological description in pictures taken from its own range of experience. Customarily it takes the standpoint of the electrical engineer and visualizes the material as equivalent to a two-terminal network of lumped circuit components. With a consideration of this interpretation our macroscopic discussion completes its full cycle and returns to the capacitor of Sec. 1.

The current-voltage diagram of Fig. 1.2, measured for a capacitor dielectric at one frequency only, allows any number of circuit interpretations. The simplest equivalent picture would be that the capacitor housing contains an ideal condenser C_i and a resistor R in series or in parallel combination (Fig. 26.1). How well either

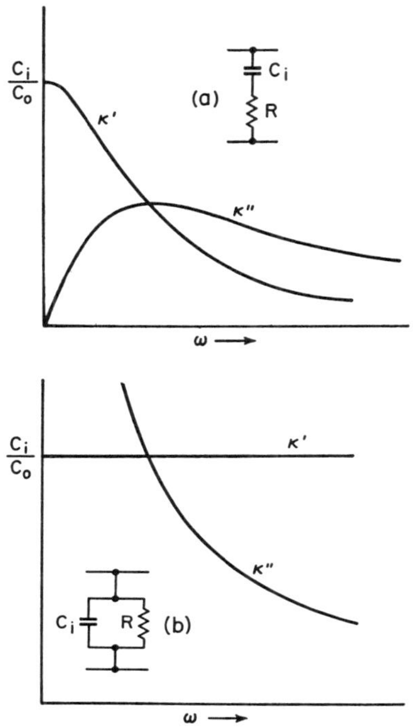

Fig. 26.1. Simplest equivalent circuits for capacitor dielectric.

one of these circuits simulates the behavior of the actual dielectric can be established only by calculating the frequency response of these networks and then comparing them with the dielectric response characteristic actually observed.

In the series arrangement the applied sinusoidal voltage equals the sum of the voltage drops across the resistor and the capacitor; in the parallel arrangement the total sinusoidal current is equal to the sum of the currents passing the two circuit elements. Hence we obtain:

Series Arrangement	*Parallel Arrangement*	
$\mathcal{V} = IR + \dfrac{\int I\,dt}{C_i}$,	$I = \dfrac{\mathcal{V}}{R} + C_i \dfrac{d\mathcal{V}}{dt}$,	
$= I\left(R + \dfrac{1}{j\omega C_i}\right)$,	$= \mathcal{V}\left(\dfrac{1}{R} + j\omega C_i\right)$,	(26.1)
$Z = \dfrac{\mathcal{V}}{I} = R + \dfrac{1}{j\omega C_i}$,	$Y = \dfrac{I}{\mathcal{V}} = \dfrac{1}{R} + j\omega C_i$.	

In terms of the complex permittivity, a capacitor of the geometrical capacitance C_0 has the admittance (cf. Eq. 1.10)

$$Y = \frac{1}{Z} = (\epsilon'' + j\epsilon')\,\frac{\omega C_0}{\epsilon_0}. \tag{26.2}$$

By equating the admittances of the two circuit arrangements to this expression, we arrive at the equivalent values for the relative dielectric constant, loss factor, and loss tangent:

Series Arrangement	*Parallel Arrangement*	
$\kappa' \equiv \dfrac{\epsilon'}{\epsilon_0} = \dfrac{C_i}{C_0[1 + (\omega R C_i)^2]}$,	$\kappa' = \dfrac{C_i}{C_0}$,	
$\kappa'' \equiv \dfrac{\epsilon''}{\epsilon_0} = \dfrac{\omega R C_i^2}{C_0[1 + (\omega R C_i)^2]}$,	$\kappa'' = \dfrac{1}{\omega R C_0}$,	(26.3)
$\tan\delta = \omega R C_i$,	$\tan\delta = \dfrac{1}{\omega R C_i}$.	

Obviously, as Fig. 26.1 shows, the frequency dependence of these two elementary circuits has completely different trends. Proper combinations of both circuits may therefore produce a variety of response curves with maxima or minima of the loss tangent which may fit an actually measured frequency characteristic when neither of the elementary circuits will. Figures 26.2 and 26.3 show two typical examples: the frequency characteristics of water and of the capacitor

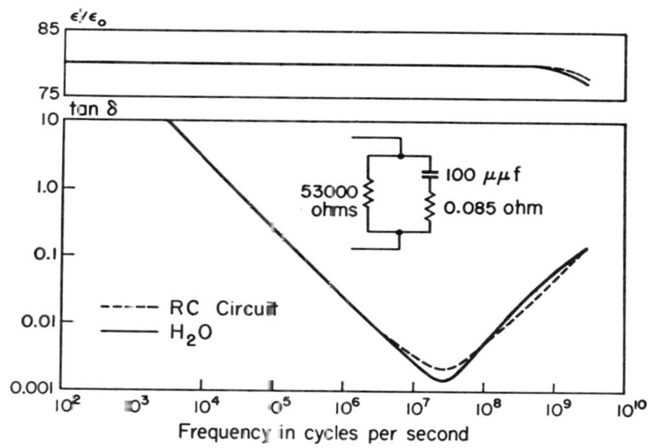

Fig. 26.2. Frequency response of water and its representation by equivalent circuit.

fluid, *Pyranol*,† at room temperature. Obviously, even a simple three-element equivalent circuit may provide a relatively good fit over an appreciable frequency range. That the equivalence between the lumped cir-

† Pyranol 1476 (a mixture of isomeric pentachlorodiphenyls), General Electric Co., Pittsfield, Mass.

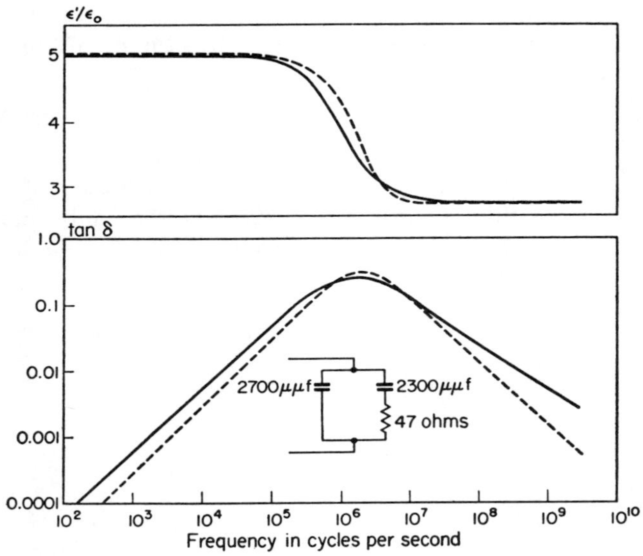

Fig. 26.3. Dielectric characteristic of Pyranol and approximation by equivalent circuit.

cuit and the dielectric material is a purely formal one is demonstrated in Fig. 26.4 by the alternative placing of the series capacitor shown in Fig. 26.3. The two

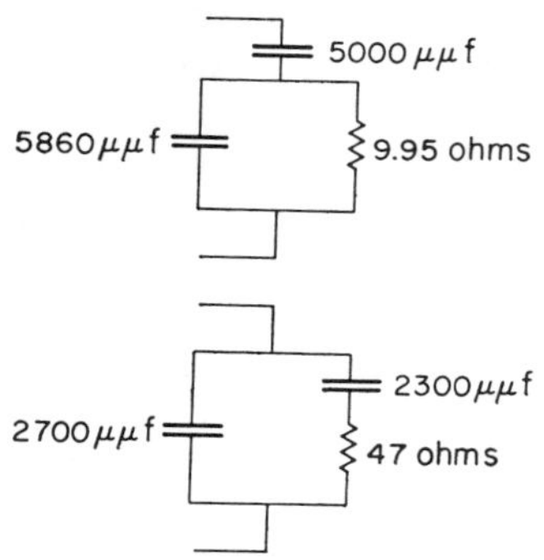

Fig. 26.4. Two equivalent networks.

circuits are undistinguishable by two-terminal measurements.

Dielectric spectra that can be represented by combinations of *RC* circuits are called *relaxation spectra.*

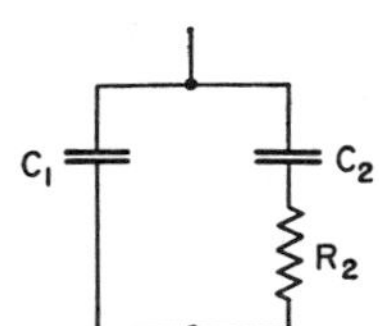

Fig. 26.5. Network representing the simplest type of relaxation spectrum of a polar material.

A common criterion of such spectra is that, whereas the loss tangent as function of the frequency may rise and fall, the dielectric constant can only stay constant or fall, as the frequency increases.

The network corresponding to the Pyranol curve represents the simplest type of relaxation spectrum encountered in polar dielectrics and may therefore be discussed in some detail for future reference (Fig. 26.5). Its admittance is

$$Y = \left(j\omega C_1 + \frac{1}{Z_2}\right),$$

where

$$Z_2 = R_2 + \frac{1}{j\omega C_2}. \tag{26.4}$$

If the capacitor C_2 is charged to a voltage $\mathcal{V}_0$ and then the circuit shorted across the R_2C_2 combination, the voltage $\mathcal{V}_2$ across the condenser will decrease as function of time and cause a current flow $C_2 \dfrac{d\mathcal{V}_2}{dt}$ through the resistor. Hence, for the partial circuit R_2C_2, the transient equation holds

$$R_2C_2 \frac{d\mathcal{V}_2}{dt} + \mathcal{V}_2 = 0 \tag{26.5}$$

with the solution

$$\mathcal{V}_2 = \mathcal{V}_0 e^{-\frac{t}{R_2C_2}}. \tag{26.6}$$

The voltage drops exponentially with a *relaxation time*

$$\tau_2 \equiv R_2C_2. \tag{26.7}$$

Introducing τ_2 into Eq. 26.4 and separating the real from the imaginary part, we obtain for the network admittance the expression

$$Y = \frac{\omega^2 C_2 \tau_2}{1 + \omega^2\tau_2^2} + j\omega C_1 + \frac{j\omega C_2}{1 + \omega^2\tau_2^2}. \tag{26.8}$$

By equating Eq. 26.8 with Eq. 26.2, the relative dielectric constant and loss of the equivalent circuit become

$$\kappa' = \frac{C_1}{C_0} + \frac{C_2}{C_0}\frac{1}{1 + \omega^2\tau_2^2},$$

$$\kappa'' = \frac{C_2}{C_0}\frac{\omega\tau_2}{1 + \omega^2\tau_2^2}; \tag{26.9}$$

hence its complex relative permittivity is

$$\kappa^* = \frac{C_1}{C_0} + \frac{C_2}{C_0}\frac{1}{1 + j\omega\tau_2}. \tag{26.10}$$

For the series branch R_2C_2 alone, the loss tangent is

$$\tan \delta_2 = \omega\tau_2, \tag{26.11}$$

but the by-pass condenser C_1 reduces the overall loss

tangent by carrying more current as the frequency increases. Therefore,

$$\tan\delta = \frac{\tan\delta_2}{1 + \dfrac{C_1}{C_2}(1 + \omega^2\tau_2{}^2)}. \qquad (26.12)$$

It proves convenient to introduce the specific permittivities for zero frequency, κ_s', and for infinite frequency ($\omega \gg 2\pi/\tau_2$), κ_∞', as well as their differences into these equations by defining

$$\kappa_s' \equiv \frac{C_1}{C_0} + \frac{C_2}{C_0},$$
$$\kappa_\infty' \equiv \frac{C_1}{C_0}, \qquad (26.13)$$
$$\kappa_s' - \kappa_\infty' = \frac{C_2}{C_0} \equiv S.$$

Equation 26.10 thus becomes

$$\kappa^* = \kappa_\infty' + \frac{S}{1 + j\omega\tau_2},$$

with

$$\kappa' = \kappa_\infty' + \frac{S}{1 + \omega^2\tau_2{}^2},$$
$$\kappa'' = \frac{S\omega\tau_2}{1 + \omega^2\tau_2{}^2}. \qquad (26.14)$$

The loss factor reaches its maximum $S/2$ at the frequency

$$\omega_s = 1/\tau_2, \qquad (26.15)$$

at which $\kappa' - \kappa_\infty'$ has fallen to its half value $S/2$ (Fig. 26.6). Plotted on a logarithmic frequency scale the

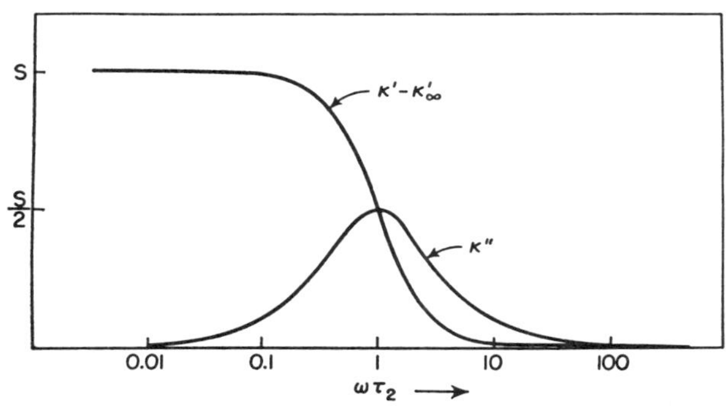

Fig. 26.6. Frequency response of equivalent circuit of Fig. 26.5.

loss factor describes a symmetrical, bell-shaped absorption band centered at ω_s and covering about two decades above and below this value, whereas κ' traverses a falling, S-shaped characteristic limited essentially to the range of one decade on both sides of the center frequency. This characteristic corresponds to the relaxation spectrum of an electric dipole, as will become evident in II, Sec. 22.

Relaxation spectra corresponding to RL circuits may arise in magnetic materials, and finally *resonance spectra* may occur. Resonance spectra can be easily distinguished from relaxation spectra by the anomalous dispersion of the dielectric constant (see II, Sec. 4).

Let us consider the frequency dependence of the series RLC circuit of Fig. 26.7. The applied sinusoidal

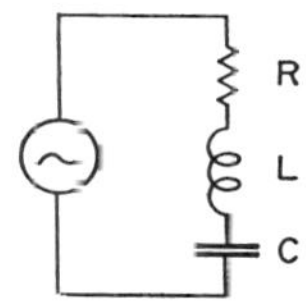

Fig. 26.7. Series resonance circuit.

voltage $\mathcal{V}$ is balanced by the voltages across the three circuit elements, that is the differential equation holds,

$$L\frac{dI}{dt} + RI + \frac{1}{C}\int I\,dt = \mathcal{V}. \qquad (26.16)$$

Its steady state solution for sinusoidal currents is consequently

$$I = \frac{\mathcal{V}}{R + j\omega L + \dfrac{1}{j\omega C}} = \mathcal{V}Y. \qquad (26.17)$$

The current reaches a maximum when the two reactance components compensate each other, that is, at the *resonance frequency*

$$\omega_0 = 1/\sqrt{LC} \qquad (26.18)$$

(Fig. 26.8). For $\mathcal{V} = 0$, the differential equation (26.16)

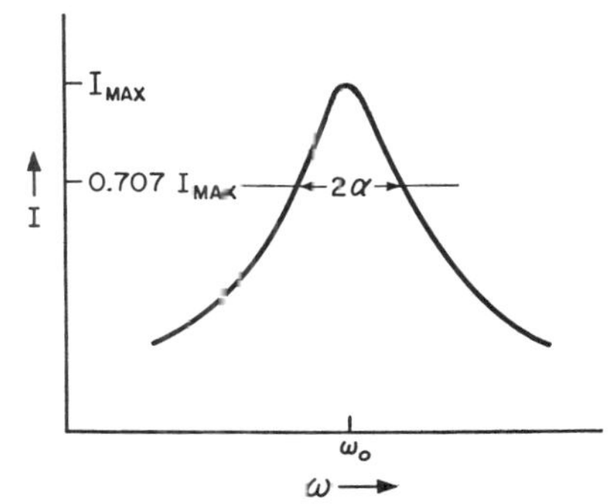

Fig. 26.8. Resonance curve of series circuit.

transforms into the transient equation of the series resonance circuit. By replacing the current I with the charge Q of the capacitor, it may be written

$$L\frac{d^2Q}{dt^2} + R\frac{dQ}{dt} + \frac{1}{C}Q = 0. \qquad (26.19)$$

Assuming the solution

$$Q = Ae^{pt}, \tag{26.20}$$

and substituting it into Eq. 26.19, we find that p must fulfill the condition

$$Lp^2 + Rp + \frac{1}{C} = 0. \tag{26.21}$$

Hence

$$p = -\frac{R}{2L} \pm \sqrt{\left(\frac{R}{2L}\right)^2 - \frac{1}{LC}}$$
$$\equiv -\alpha \pm j\omega_d. \tag{26.22}$$

The freely oscillating RLC circuit has an attenuation factor

$$\alpha = \frac{R}{2L} \tag{26.23}$$

and a frequency

$$\omega_d = \sqrt{\frac{1}{LC} - \left(\frac{R}{2L}\right)^2} = \sqrt{\omega_0^2 - \alpha^2}, \tag{26.24}$$

which for the undamped circuit is identical with the resonance frequency of the steady state (Eq. 26.18).

By introducing α and ω_0 into the steady state solution (Eq. 26.17), we obtain the admittance of the resonance circuit as

$$Y = \frac{j\dfrac{\omega}{L}}{\omega_0^2 - \omega^2 + j\omega 2\alpha}. \tag{26.25}$$

In case of low attenuation ($2\alpha \ll \omega_0$), it is convenient to refer, not to the frequency itself, but to the frequency deviation from resonance by approximating:

$$\Delta\omega \equiv \omega_0 - \omega,$$
$$\omega \simeq \omega_0, \tag{26.26}$$

$$Y \simeq \frac{j}{2L}\frac{1}{\Delta\omega + j\alpha}. \tag{26.27}$$

Equating Eqs. 26.2 and 26.27, we arrive at the relative permittivity, loss factor, and loss tangent of a *resonating* dielectric:

$$\kappa' = \frac{B\Delta\omega}{(\Delta\omega)^2 + \alpha^2}$$
$$\kappa'' = \frac{B\alpha}{(\Delta\omega)^2 + \alpha^2} \qquad \left(\text{where } B = \frac{1}{\omega C_0 2L}\right), \tag{26.28}$$
$$\tan\delta = \frac{\alpha}{\Delta\omega}$$

(Fig. 26.9). The dielectric constant rises with frequency, reaches its maximum $B/2\alpha$ at $\Delta\omega = +\alpha$, falls through zero at the resonance frequency to its minimum $-B/2\alpha$ at $\Delta\omega = -\alpha$, and approaches asymptotically zero again. This anomalous dispersion characteristic of the dielectric constant is typical for resonance spectra. It is accompanied by a bell-shaped absorption characteristic which passes through its maximum B/α

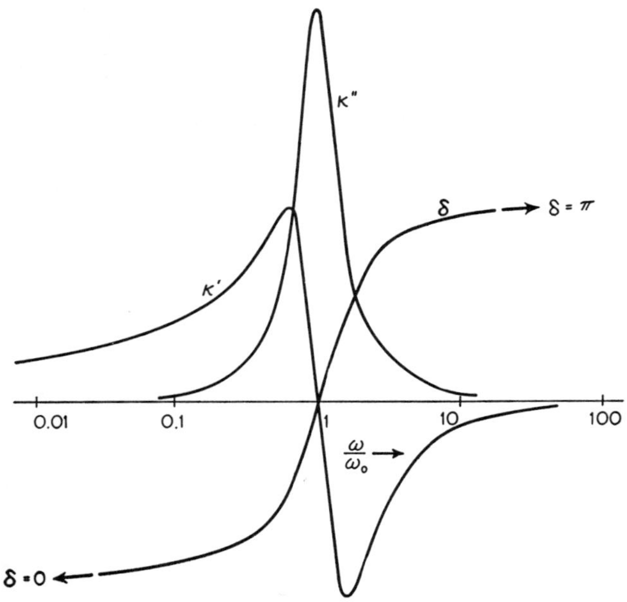

Fig. 26.9. Resonance response of dielectric plotted on a normalized frequency scale.

at ω_0. The phase angle δ swings from zero at low frequency over $\pi/2$ at resonance to π at high frequencies, passing the values $\pi/4$ and $3\pi/4$ at the points $\Delta\omega = \pm\alpha$.

The power delivered to the resonance circuit by a sinusoidal voltage source of constant amplitude varies proportionally to I^2, that is, proportionally to

$$Y^2 = \frac{1}{4L^2(\Delta\omega^2 + \alpha^2)}. \tag{26.29}$$

The points $\Delta\omega = \pm\alpha$ are the half-power points of the series resonance circuit for low attenuation (see Fig. 26.8), or

$$\Delta\omega_h = 2\alpha \tag{26.30}$$

is the *half-width* of the resonance absorption. The *relative half-width*

$$\frac{\Delta\omega_h}{\omega_0} = \frac{2\alpha}{\omega_0} = \frac{R}{\omega_0 L} \equiv \frac{1}{Q} \tag{26.31}$$

designates the Q of the lumped circuit in complete analogy to the preceding discussion of distributed circuits (see Eqs. 24.24 and 25.34).

If the elements of the lumped resonance circuit are placed parallel to the driving voltage (Fig. 26.10),

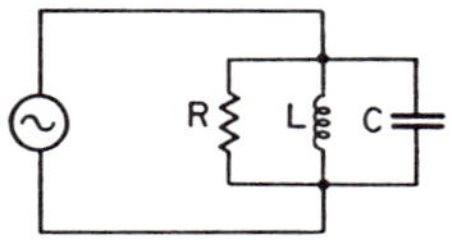

Fig. 26.10. Parallel resonance circuit.

the impedance instead of the admittance becomes a maximum at resonance and the current delivered by a constant voltage source traverses a minimum at ω_0. This situation corresponds to the behavior of a dielectric or a transmission line tuned to antiresonance (see Fig. 24.2).

With the introduction of such equivalent pictures, simulating the response of dielectrics, the task of the macroscopic theory is completed. The deeper probing question as to why dielectrics behave as they do leads from the phenomenological to the molecular theory.

II · MOLECULAR APPROACH

0 · Survey

The interaction between electromagnetic waves and dielectric materials has been treated in Part I from the standpoint of an external observer who describes phenomenologically the effects encountered and believes in the "one world" approach for the electromagnetic spectrum. The two complex parameters, permittivity and permeability, were chosen for a quantitative description of polarization, magnetization, and conduction throughout the frequency region from direct current to X-rays. Conversion formulas and nomographic charts were provided to allow a convenient change-over to alternate sets of parameters such as propagation factor, complex index of refraction, and impedance. The development of the macroscopic theory led automatically to methods of measuring ϵ^* and μ^*. These parameters appear in Maxwell's field theory, not individually but coupled in characteristic combinations. Hence a comprehensive discussion of dielectrics has to deal with their electric as well as their magnetic properties.

This emphasis on a unified approach to electric and magnetic phenomena is again the tenor of the molecular discussion. Here we try to discern the interplay between electrons, atoms, and molecules that causes the macroscopic effects observed. A straightforward admission into the molecular world is provided by a reinterpretation of the field vectors *polarization* and *magnetization*. Macroscopically, these vectors represent the electric and the magnetic dipole moment per unit volume of a dielectric. Microscopically, the moments arise from the additive action of a multitude of elementary moments, and we have to inquire into the origin and coupling of such moments.

This task is taken up in Sec. 1 for polarization. Immediately we find ourselves confronted with three molecular parameters: the number of particles, their polarizability, and the locally acting field. Furthermore, the concept *polarizability* proves to be a generic term, hiding four mechanisms of polarization. Two of them, the electronic and atomic polarizability, concern dipole moments *induced* by the displacement of electrons and nuclei, respectively. A third one, the orientation polarizability, is caused by the rotation of *permanent* dipole moments. In addition, a space charge or interfacial polarizability may result by the piling up of migrating charge carriers.

To simplify the initial treatment we eliminate the unknown local field by substituting, with misgivings, a known local field that increases proportionally to the polarization, the *Mosotti* field. By referring, in addition, to Avogadro's number, we arrive at the Clausius-Mosotti-Lorentz-Lorenz equation (Sec. 2). Here the polarizability is the only remaining molecular parameter and, as long as we concentrate on atoms, it represents unambiguously the electronic polarizability.

For static fields, the classical model of spherical atoms produces a polarizability equal to the volume filled by these spheres, a value somewhat smaller than observed (Sec. 3). For alternating fields, the model of quasi-elastically bound electrons leads to the dispersion formula of classical physics, which properly describes the shape of a spectral line. However, neither the intensity nor the resonance frequency can be derived from this classical picture (Sec. 4). An electron trapped in the Coulomb field of a nucleus has a kinetic energy prescribed by the Virial theorem. It is, classically speaking, a light source which transforms electrostatic energy through kinetic energy into radiation, while the electron spirals into the nucleus. This collapse of the planetary Rutherford atom is not observed; instead, series of spectral lines appear, ruled by the Rydberg-Ritz combination principle. Furthermore, studies of the black-body radiation and of the photoeffect lead to a decisive break with classical physics. Radiation proves to be parceled out in quanta, and Maxwell's theory is therefore one of photon statistics (Sec. 5). On this basis gained by experimental investigations and their interpretation by Planck and Einstein, Bohr's quantum theory (Sec. 6) and wave mechanics (Sec. 7) develop.

Thus, research on the electronic polarization of atoms

leads to a theory of the structure and of the spectra of atoms and to an understanding of the periodic system of the elements (Sec. 8). It also provides an insight into the behavior of atoms in electric fields, the Stark effect (Sec. 9). In accounting for the structure and behavior of atoms, it is of fundamental importance that the angular momenta of the electron clouds and the mechanical spin of the electrons are quantized and produce quantized permanent magnetic orbital and spin moments. That these permanent magnetic moments, coupled to mechanical gyroscopes, take up quantized orientations in relation to the axis of a magnetic field is proved by the Zeeman effect. An external magnetic field produces, in addition, diamagnetism by inducing magnetic moments (Sec. 10). In a complete discussion of the energy level diagram of atoms, the permanent magnetic moments play a much larger role than is foreshadowed by their magnetic interaction energy. The magnetic quantum numbers codetermine, according to the Pauli exclusion principle, the charge distribution of the stationary states and thus their electrostatic energy (Sec. 11). A scrutiny of the interaction between electromagnetic fields and the atoms of quantum mechanics allows the conversion of the dispersion formula of classical physics into its quantum-mechanical counterpart (Sec. 12).

Progressing to the interaction of atoms, we find that, as the partners approach each other, several forces begin to operate in succession: the nonclassical van der Waals attraction caused by fluctuating dipole moments; the classical electrostatic attraction coming into play as the electron clouds begin to overlap; quantum-mechanical pair-bond formation or quantum-mechanical repulsion; and, as a final possibility, electron exchange and ionic bonding. In consequence, the interatomic separation distance of atoms may be characterized by van der Waals, covalent, or ionic radii (Sec. 13). If molecules are formed, quantum mechanics describes the electron states by wave functions which change with the interatomic distance and may either be derived from overlapping atomic orbitals (Heitler-London method) or be composed directly of molecular wave functions (Hund-Mulliken approach). Any possible electronic configuration contributes to the strength of molecular bonds according to the concept of quantum-mechanical resonance (Sec. 14).

When atoms of a single type form symmetrical diatomic molecules, the build-up process can be discussed in close analogy to the formation of atoms. For unlike atoms the electronegativity of the partners differs, and permanent electric dipole moments result (Sec. 15). These permanent moments contribute to the static dielectric constant of gases because the torque of an external field tends to align the dipole axes in the field direction. The contribution is temperature dependent and is classically given by the Langevin function, which describes statistically how thermal agitation counteracts the electric field and tends to maintain a random orientation (Sec. 16).

The structure and dipole moments of polyatomic molecules may be understood by considering the various bonds which operate, the ionic interaction of the partners and the vector addition of dipole moments (Sec. 17). Molecules are characterized by vibration and rotation spectra. For diatomic molecules, these spectra, in a first approximation, can be derived by using the models of the harmonic oscillator and rigid rotator. The treatment of polyatomic molecules is facilitated by referring to the principal moments of inertia and the normal vibrations (Sec. 18). This background suffices for a general discussion of the electronic, atomic, and orientation polarization of gas molecules (Sec. 19).

In spectroscopy, the bandwidth of a spectral line, its Q, decides the amount of fine structure that may be discerned. This bandwidth is determined by radiation damping, Doppler effect, and pressure broadening (Sec. 20). How well the shape of a spectral line can be measured actually depends on the resolving power (or Q) of the analyzing optical instrument. Here the present infrared techniques prove to be definitely inferior. Microwave spectroscopy, on the other hand, using the monochromatic sources and extremely sensitive receivers of the electrical engineer, approaches the best optical interferometers. Microwave spectra are therefore a source of amazingly detailed information concerning the properties of molecules (Sec. 21).

Progressing from gases to liquids and solids we enter regions mapped only in rough outlines. In principle it seems a simple approach to proceed from the rotation spectra of gases by pressure broadening to the relaxation spectra of the condensed phases, but a real transition has not yet been made by theory or by experiment. Debye's relaxation equations for the orientation polarization are based on the ad hoc model of a relaxation circuit (Sec. 22).

When polar molecules in liquids and solids follow the response of an orienting field, the susceptibility, according to Langevin's equation, should obey a Curie law in case the coupling between the dipoles can be neglected, and a Curie-Weiss law if the Mosotti field describes their mutual interaction. Any condensed polar material should become a ferroelectric. This "Mosotti catastrophe" actually occurs only in rare cases. Mathematically it can be avoided by postulating other types of local field expressions (Sec. 23). Why physically the catastrophe does not occur by the spontaneous orientation of permanent moments becomes apparent when we consider the formation and structure of liquids and solids (Sec. 24) and the dipoles in their various surroundings (Sec. 25).

Although dipoles in liquids are still sufficiently mobile to jump statistically over potential walls whenever the activation energy becomes available, in crystals they are structural elements codetermining the lattice and, in general, firmly anchored in place. Properly arranged, they may strikingly manifest their existence in piezoelectricity (Sec. 26). A piezoelectric effect exists only in anisotropic crystals. The macroscopic description of the effect, therefore, requires an involved system of electrical, mechanical, and electromechanical tensor relations. A clear concept of the molecular situation may be obtained by visualizing the crystal as a network of permanent moments subject to the crystal class symmetry (Sec. 27).

Ferroelectricity, the spontaneous alignment of electric dipole moments by mutual interaction, does not take place by the rotation of permanent moments as the Mosotti catastrophe foresees, but by the creation of moments through the ordered displacement of nuclei. The phenomenon, thus far observed in only a few substances, is described in detail for $BaTiO_3$ (Sec. 28).

Switching over from the electrical effects: orientation polarization, ferro- and antiferroelectricity, to their magnetic counterparts: para- and ferromagnetism, we find ourselves confronted with many analogies but also with deep-lying differences. The electrical phenomena arise by ionic bonds *between* atoms; the magnetic phenomena by orbital and spin moments residing *in* atoms and coupled to gyroscopic quantum effects. The various attempts to formulate a satisfactory theory of ferromagnetism illustrate strikingly the dilemma of quantum mechanics, that it can solve its problems only by approximations (Sec. 29). In discussing essential aspects of ferromagnetism for metals and semiconductors, we arrive at one of the most active frontiers of solid state research (Sec. 30).

Up to this point, conduction phenomena played only a minor role in this presentation, whereas actually the effects of migrating charge carriers are observed in many dielectric spectra. An extensive discussion of electric conduction requires a book in its own right. However, the principal aspects of interfacial and space-charge polarization, the fourth mechanism of polarization introduced in Sec. 1, are briefly presented (Sec. 31). And, in a final integrating sweep, the formation and the behavior of mobile charge carriers are surveyed from gases to liquids and solids, and from low fields up to the destruction of dielectrics by electric breakdown (Sec. 32).

1 · Molecular Mechanisms of Polarization

A dielectric material can react to an electric field because it contains charge carriers that can be displaced. In Part I this polarization phenomenon was pictured schematically by the formation of dipole chains which line up parallel to the field and bind countercharges at the electrodes (see I, Fig. 2.1). The density of the neutralized surface charge is represented by the polarization vector

$$\mathbf{P} = (\epsilon' - \epsilon_0)\mathbf{E} = (\kappa' - 1)\epsilon_0\mathbf{E} \quad \left[\frac{\text{coul}}{\text{m}^2}\right] \tag{1.1}$$

(I, Eq. 2.7). Alternatively, the polarization **P** proves to be equivalent to the dipole moment per unit volume of the material (I, Fig. 2.4).

This latter interpretation of **P** provides an entrance from the macroscopic into the molecular world. The dipole moment per unit volume may be thought of as resulting from the additive action of N elementary dipole moments $\bar{\boldsymbol{\mu}}$,

$$\mathbf{P} = N\bar{\boldsymbol{\mu}}. \tag{1.2}$$

The average dipole moment $\bar{\boldsymbol{\mu}}$ of the elementary particle, furthermore, may be assumed to be proportional to the *local electric field strength* $\mathbf{E}'$ that acts on the particle,

$$\bar{\boldsymbol{\mu}} = \alpha\mathbf{E}'. \tag{1.3}$$

The proportionality factor α, called *polarizability*, measures the electrical pliability of the particle, that is, the average dipole moment per unit field strength. Its dimensions are

$$[\alpha] = \left[\frac{\text{sec}^2\,\text{coul}^2}{\text{kg}}\right] = [\epsilon\ \text{m}^3]$$

in the mks, or

$$[\alpha] = [\text{cm}^3] \tag{1.4}$$

in the esu system, respectively.

Equations 1.1 to 1.3 give us the two alternative expressions for the polarization

$$\begin{aligned}\mathbf{P} &= (\kappa' - 1)\epsilon_0\mathbf{E} \\ &= N\alpha\mathbf{E}',\end{aligned} \tag{1.5}$$

linking the macroscopically measured permittivity to three molecular parameters: the number N of contributing elementary particles per unit volume; their polarizability α; and the locally acting electric field $\mathbf{E}'$. This field will normally differ from the applied field $\mathbf{E}$, owing to the polarization of the surrounding dielectric medium. It is the goal of the molecular theories to evaluate these parameters and thus to arrive at an understanding of the phenomenon *polarization* and its dependence on frequency, temperature, and applied field strength.

Such theories are not the result of metaphysical speculation, but the product of experiments and their interpretation. It is beyond the scope of this book to trace in detail the interplay between experiment and theory that led to our present-day ideas about the structure of matter. We have to reverse the approach and to introduce at this point some general concepts which are the outcome of this development and allow us to foresee various mechanisms of polarization.

Matter, electrically speaking, consists of positive atomic nuclei surrounded by negative electron clouds. Upon the application of an external electric field the electrons are displaced slightly with respect to the nuclei; *induced* dipole moments result and cause the so-called *electronic polarization* of materials. When atoms of different types form molecules, they will normally not share their electrons symmetrically, as the electron clouds will be displaced eccentrically toward the stronger binding atoms. Thus atoms acquire charges of opposite polarity, and an external field acting on these net charges will tend to change the equilibrium positions of the atoms themselves. By this displacement of charged atoms or groups of atoms with respect to each other, a second type of *induced* dipole moment is created; it represents the *atomic polarization* of the dielectric. The asymmetric charge distribution between the unlike partners of a molecule gives rise, in addition, to *permanent* dipole moments which exist also in the absence of an external field. Such moments experience a torque in an applied field that tends to orient them in the field direction (see I, Eq. 2.12). Consequently, an *orientation* (or *dipole*) *polarization* can arise.

These three mechanisms of polarization, characterized by an *electronic polarizability* α_e, an *atomic polarizability* α_a, and an *orientation* (or *dipole*) *polarizability* α_d, are due to charges that are locally bound in atoms, in molecules, or in the structures of solids and liquids. In addition, carriers that can migrate for some distance through the dielectric usually exist. When such carriers are impeded in their motion, either because they become trapped in the material or on interfaces, or because they cannot be freely discharged or replaced at the electrodes, space charges and a macroscopic field distortion result. Such a distortion appears to an outside observer as an increase in the capacitance of the sample and may be indistinguishable from a real rise of the dielectric permittivity. Thus we have to add to our polarization mechanisms a fourth one, a *space charge* (or *interfacial*) *polarization*, characterized by a *space charge* (or *interfacial*) *polarizability* α_s.

Assuming at present that the four polarization mechanisms indicated schematically in Fig. 1.1 act inde-

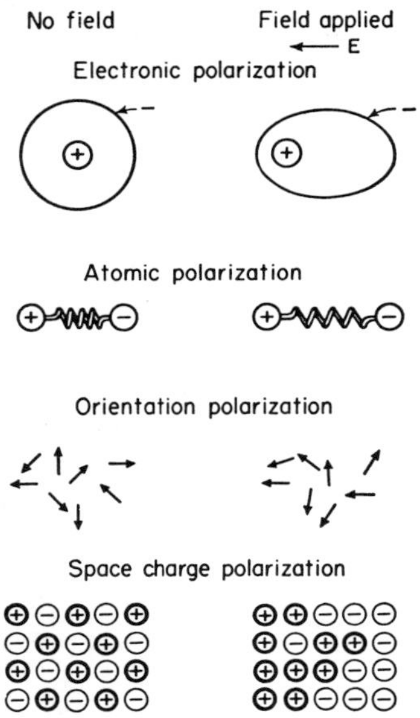

Fig. 1.1. Mechanisms of polarization.

pendently of each other, we may write the total polarizability α of a dielectric material as the sum of the four terms

$$\alpha = \alpha_e + \alpha_a + \alpha_d + \alpha_s, \tag{1.6}$$

where each term again may represent a sum of contributions. How these four effects actually shape the response characteristics of dielectrics will be discussed in subsequent sections.

In Eq. 1.5 we took for granted that the polarizability α is a real quantity. Actually, in alternating fields, a temporal phase shift may occur between the driving field and the resulting polarization, and a loss current component appear (see I, 1), as will be discussed in detail. Thus α becomes complex, and Eq. 1.5 has to be replaced by the more general formulation

$$\mathbf{P} = (\kappa^* - 1)\mathbf{E} = N\alpha\mathbf{E}'. \tag{1.7}$$

2 · The Clausius-Mosotti-Lorentz-Lorenz Equation

Confronted with three molecular parameters (N, α, $\mathbf{E}'$) at once, our tendency will be to concentrate initially on the most instructive one and to eliminate the other two by reasonable approximations. In the present case the parameter α contains the primary information on the electric charge carriers and their polarizing action. Hence we shall try to eliminate N and $\mathbf{E}'$.

We can foresee that the locally acting field $\mathbf{E}'$ will be identical with the externally applied field $\mathbf{E}$ for gases at low pressure where the interaction between the molecules can be neglected. At high pressures, however, and especially in the condensed phases of solids and liquids, the field acting on a reference molecule A may be modified decisively by the polarization of the surroundings. To take this effect into account, we try the following model (Fig. 2.1).

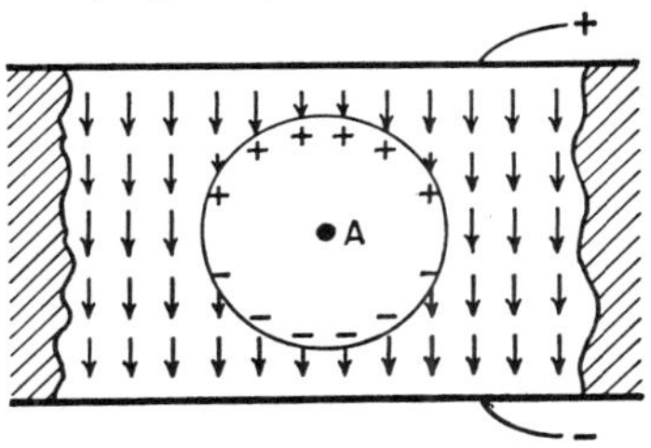

Fig. 2.1. Model for calculation of internal field.

Let our reference molecule A be surrounded by an imaginary sphere of such an extent that beyond it the dielectric can be treated as a continuum. If the molecules inside this sphere were removed while the polarization outside remains frozen, the field acting on A would stem from two sources: from the free charges at the electrodes of the plate capacitor ($\mathbf{E}_1$), and from the free ends of the dipole chains that line the cavity walls ($\mathbf{E}_2$). Actually there are molecules inside the sphere and they are so near to A that their individual positions and shapes have to be considered. This adds an additional contribution $\mathbf{E}_3$ to the local field $\mathbf{E}'$; hence we obtain

$$\mathbf{E}' = \mathbf{E}_1 + \mathbf{E}_2 + \mathbf{E}_3. \tag{2.1}$$

The contribution from the free charges at the electrodes is, by definition, equal to the applied field intensity

$$\mathbf{E}_1 = \mathbf{E}. \tag{2.2}$$

To calculate $\mathbf{E}_2$ we recall that the charge density lining the cavity walls stems from bound charges and is correspondingly determined by the normal component of the polarization vector $\mathbf{P}$ (see I, Eq. 2.5) as

$$\mathbf{P}\cdot\mathbf{n}\,dA = P\cos\theta\,dA \tag{2.3}$$

(Fig. 2.2). Each surface element dA of the sphere contributes at A, according to Coulomb's law, a radial field intensity

$$d\mathbf{E}_2 = \frac{P\cos\theta}{\epsilon_0 4\pi r^2}\,dA. \tag{2.4}$$

For each surface element dA there exists its counterpart, which produces the same vertical but an equal and opposite horizontal field component. Hence only

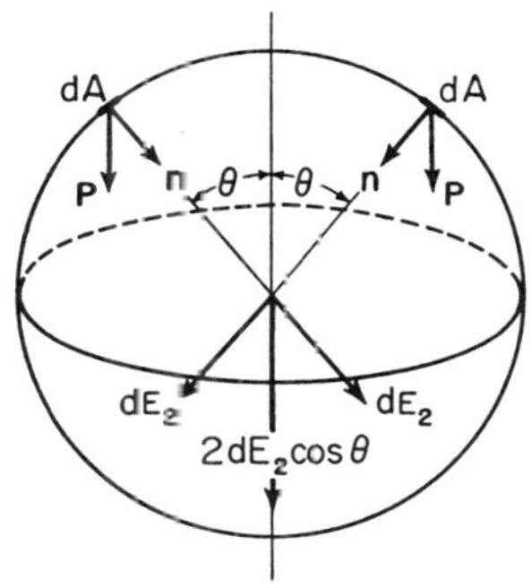

Fig. 2.2. Geometry for calculation of internal field.

the vertical components $d\mathbf{E}_2\cos\theta$ count and create a field intensity

$$\mathbf{E}_2 = \oint_{\text{sphere}} \frac{P\cos^2\theta}{\epsilon_0 4\pi r^2}\,dA, \tag{2.5}$$

orientated parallel to the applied field and strengthening it. By dividing the cavity walls into ring elements

$$dA = 2\pi\sin\theta\,r\,d\theta, \tag{2.6}$$

and integrating over θ, we obtain as the field contribution of the cavity wall charge

$$\begin{aligned}\mathbf{E}_2 &= \int_0^\pi \frac{P\cos^2\theta}{\epsilon_0 4\pi r^2}\,2\pi r^2\sin\theta\,d\theta = \frac{1}{3}\frac{\mathbf{P}}{\epsilon_0}\\ &= \frac{\mathbf{E}}{3}(\kappa' - 1).\end{aligned} \tag{2.7}$$

An evaluation of the field contribution $\mathbf{E}_3$ which arises from the individual action of the molecules inside the sphere requires accurate information on the geometrical arrangement and polarizability of the contributing particles. Even if this information is available, the mathematical treatment may prove prohibitively difficult. In a general way we have recognized the existence of these neighboring molecules already in the calculation of $\mathbf{E}_2$ by assuming that the cavity is scooped out without disturbing the state of polarization of the remaining dielectric. Hence we postulate for the present, as a simple expedient, that the additional individual field effects of the surrounding molecules on the particle at A shall mutually cancel, that is,

$$\mathbf{E}_3 = 0. \tag{2.8}$$

This assumption, first made by Mosotti [1] in 1850, is not altogether a confession of ignorance. It is a reasonable approximation when the elementary particles are neutral and without permanent dipole moment, or when they are arranged either in complete disorder or in cubic or similar highly symmetrical arrays. It allows us to substitute for the unknown molecular parameter E' the expression

$$\mathbf{E}' = \mathbf{E}_1 + \mathbf{E}_2 = \mathbf{E} + \frac{\mathbf{P}}{3\epsilon_0} = \frac{\mathbf{E}}{3}(\kappa' + 2), \tag{2.9}$$

that is, known macroscopic parameters.

By inserting this local *Mosotti field* into Eq. 1.5, we obtain the relation between the *polarizability per unit volume*, $N\alpha$, and the relative permittivity of the dielectric, κ',

$$\frac{N\alpha}{3\epsilon_0} = \frac{\kappa' - 1}{\kappa' + 2}. \tag{2.10}$$

For gases at low pressure, $\kappa' - 1 \ll 1$; hence, $\kappa' + 2$ may be replaced by the digit 3. This is the same as replacing the local field $\mathbf{E}'$ by the applied field $\mathbf{E}$, and Eq. 2.10 simplifies to

$$\frac{N\alpha}{\epsilon_0} = \kappa' - 1 = \chi, \tag{2.11}$$

where χ is the *electric susceptibility of the gas* (cf. I, Eq. 2.8).

Frequently we will refer to an ideal gas under standard conditions (0°C, 760 mm Hg). The number of molecules per unit volume, N, is in this case identical with the *Loschmidt number*

$$N_L = 2.687 \times 10^{25} \quad [\mathrm{m}^{-3}]. \tag{2.12}$$

In other cases, as long as the molecules themselves are the dipole carriers, it will be convenient to eliminate the dependence of the polarization on the density of the material by referring to the *polarization per mole*. The number of molecules per mole is *Avogadro's number*

$$N_0 = \frac{NM}{\rho} = 6.023 \times 10^{23}, \tag{2.13}$$

where M designates the molecular weight in [kg] and ρ the density in [kg m^{-3}], if mks units are used.

Substituting N_0 for N in Eq. 2.10, we obtain the *polarizability per mole* (*molar polarization*),

$$\Pi = \frac{N_0\alpha}{3\epsilon_0} = \frac{\kappa' - 1}{\kappa' + 2}\frac{M}{\rho} \quad [\mathrm{m}^3]. \tag{2.14}$$

This is the famous *Clausius-Mosotti equation*, [2] in which we have reached our goal of eliminating N and E' and retain the polarizability α as the only unknown molecular parameter.

Because of the tendency to treat the electrical and optical frequency range as two separate fields of interest, the same equation was formulated independently for the optical range by Lorentz [3] in Holland and Lorenz [4] in Denmark. In this case, the relative dielectric constant κ' of Eq. 2.14 may be replaced by the square of the index of refraction, n^2, according to the Maxwell relation (I, Eq. 9.11). Thus we arrive at the *Lorentz-Lorenz equation*

$$\Pi = \frac{N_0\alpha}{3\epsilon_0} = \frac{n^2 - 1}{n^2 + 2}\frac{M}{\rho}, \tag{2.15}$$

with Π called the *molar refraction*.

Both equations are identical and not quite general enough, because they assume that α is a real quantity. By replacing the real permittivity or index of refraction with their complex counterparts (see Eq. 1.7), we arrive at the more general formulation of the *Clausius-Mosotti-Lorentz-Lorenz equation*

$$\Pi = \frac{N_0\alpha}{3\epsilon_0} = \frac{\kappa^* - 1}{\kappa^* + 2}\frac{M}{\rho} = \frac{n^{*2} - 1}{n^{*2} + 2}\frac{M}{\rho} \quad [\mathrm{m}^3]. \tag{2.16}$$

In using this equation, we can obtain valuable information on the polarizability α, as we shall see in the following sections. However, we should not be surprised to arrive also at quite erroneous conclusions in case the "near field" $\mathbf{E}_3$ cannot be neglected.

[1] O. F. Mosotti, *Mem. Soc. Ital. 14*, 49 (1850); see also P. Debye, *Polar Molecules*, Dover Publications, New York, 1945.

[2] R. Clausius, *Die mechanische Wärmetheorie*, Vieweg, Braunschweig, 1879, Vol. II, pp. 62 ff.

[3] H. A. Lorentz, *Ann. Physik 9*, 641 (1880).

[4] L. Lorenz, *Ann. Physik 11*, 70 (1880).

3 · Electronic Polarization

The phenomenon of electronic polarization can be observed undisturbed by other effects in *monatomic* gases. Here the elementary particles are nuclei of the positive charge $+Ze$, surrounded by a neutralizing electron atmosphere of the charge $-Ze$; the factor Z, the *order number of the atom,* designates the position of the atom in the periodic system of the elements; the elementary charge e has the value

$$|e| = 1.602 \times 10^{-19} \quad [\text{coul}]. \tag{3.1}$$

An external field $\mathbf{E}$ will exert a force

$$\mathbf{F} = -Ze\mathbf{E} \tag{3.2}$$

on the electron atmosphere and displace its charge center, which previously coincided with that of the nucleus, by a distance d. Here it will be balanced by the Coulomb attraction exercised by the positive nucleus $+Ze$ on the eccentric part of the electron cloud (Fig. 3.1). Let us assume tentatively that the electrons

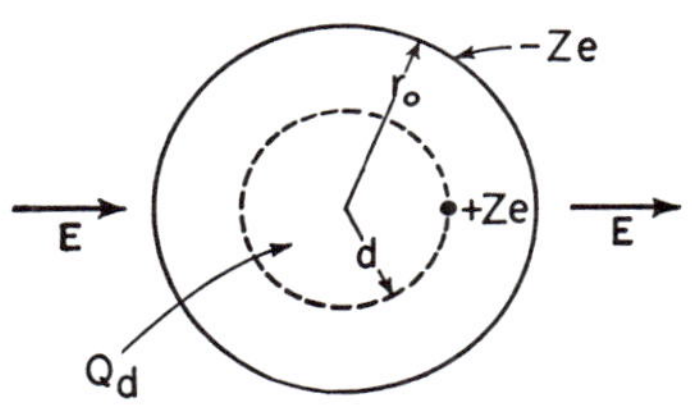

Fig. 3.1. Displacement of electron cloud by external field (overemphasized).

originally form a cloud of constant charge density around the nucleus, confined to a sphere of the radius r_0. After displacement, the eccentric part of the charge, Q_d, is the fraction of the cloud that fills a sphere (radius d) outside the nucleus; that is,

$$Q_d = -Ze\frac{d^3}{r_0^3}. \tag{3.3}$$

It acts as if concentrated at the center of the small sphere, whereas the charge farther out does not exercise a force on the nucleus (see I, Sec. 4). Hence the Coulomb force between the nucleus of charge $+Ze$ and the negative charge Q_d,

$$\mathbf{F}_c = -(Ze)^2\frac{d^3/r_0^3}{\epsilon_0 4\pi d^2}, \tag{3.4}$$

has to balance $\mathbf{F}$.

The external field has induced in the atom a dipole moment

$$\boldsymbol{\mu} = (Ze)\mathbf{d} = \alpha_e\mathbf{E} \tag{3.5}$$

(see Eq. 1.3). From the equilibrium condition

$$\mathbf{F} = \mathbf{F}_c, \tag{3.6}$$

we obtain this dipole moment as

$$\boldsymbol{\mu} = \epsilon_0 4\pi r_0^3\mathbf{E}; \tag{3.7}$$

that is, the electronic polarizability of the atom is

$$\alpha_e = \epsilon_0 4\pi r_0^3. \tag{3.8}$$

The molar polarization of our monatomic gas, defined in Eq. 2.14, becomes

$$\Pi = \frac{N_0\alpha_0}{3\epsilon_0} = N_0\frac{4\pi}{3}r_0^3 \quad [\text{m}^3]. \tag{3.9}$$

This result has the graphic meaning: the molar polarization (polarizability per mole) of a monatomic gas of spherical atoms equals the volume actually filled by these spheres.

Consequently, we have only to know the radius r_0 of our atoms to arrive at numerical results. This radius can be obtained in various ways, for instance, as the van der Waals radius in gaskinetic measurements. It is of the order of one angstrom unit,

$$1\ \text{A} = 1 \times 10^{-10} \quad [\text{m}]. \tag{3.10}$$

For hydrogen, consisting of one proton surrounded by one electron, we find

$$r_0 \simeq 0.53\ [\text{A}] \simeq 0.53 \times 10^{-10} \quad [\text{m}]. \tag{3.11}$$

Since the dielectric constant of vacuum is $\epsilon_0 = 8.854 \times 10^{-12}$ [farad m] (see I, Eq. 8.4), we obtain for the hydrogen atom the electronic polarizability (see Eq. 3.8)

$$\alpha_{eH} = 1.66 \times 10^{-41} \quad [\epsilon\ \text{m}^3]. \tag{3.12}$$

In the unrationalized esu system the factor $\epsilon_0 4\pi$ disappears, and the electronic polarizability is simply equal to the cube of the atomic radius,

$$\alpha_{eH} = r_0^3 = 1.5 \times 10^{-25} \quad [\text{cm}^3]. \tag{3.13}$$

The molar polarization has the value

$$\Pi_H = 37.8 \times 10^{-8}[\text{m}^3] = 37.8 \times 10^{-2} \quad [\text{cm}^3]. \tag{3.14}$$

For atomic hydrogen under standard conditions our simple sphere model thus predicts the susceptibility (see Eqs. 2.11 and 2.12)

$$\chi = \kappa' - 1 \simeq 5.1 \times 10^{-5}. \quad (3.15)$$

This static susceptibility, though small, can be determined with accuracy quasi-statically by a bridge or beat-frequency measurement, in which a gas-filled capacitor of about 10^{-4}-meter plate separation might be used at 20 volts. The maximum field strength of 2×10^5 [volt m^{-1}] thus applied would induce a dipole moment

$$\boldsymbol{\mu} = \alpha_e \mathbf{E} = e\mathbf{d} \simeq 3.3 \times 10^{-36} \quad \text{[coul m]}, \quad (3.16)$$

that is, displace the center of the electron cloud from that of the proton by a distance

$$d \simeq 2 \times 10^{-17} \quad \text{[m]}. \quad (3.17)$$

By measuring the dielectric constant of the gas, the electrical engineer has thus performed the amazing feat of determining this subnuclear length! However, his record for micro-distance measurements loses some of its luster when we consider that the electrical technique employed integrates over the polarization of about 10^{18} atoms. Furthermore, the actual meaning of d is not that of a precise distance.

The purpose of the preceding numerical calculation is to give a feeling for the orders of magnitude involved. A measurement of the electric susceptibility of atomic hydrogen has unfortunately not yet been carried through because the gas consists normally of diatomic (H_2) molecules. However, we can predict[1] that the actual value will be about 4.5 times larger than that just calculated. The concept of a proton surrounded by an electron cloud of radius r_0 and constant density, used above, gives the same result as the Bohr model in which the electron revolves in a circular orbit of the radius r_0 around the proton (see Sec. 6). Quantum mechanics leads to the picture of a more extended electron cloud; and since the distant parts of the electron atmosphere are more weakly bonded to the nucleus, they contribute appreciably to the polarization (Fig. 3.2) in spite of their rapidly decreasing density.

[1] J. H. Van Vleck, *The Theory of Electric and Magnetic Susceptibilities*, Oxford University Press, 1932, pp. 203 ff.

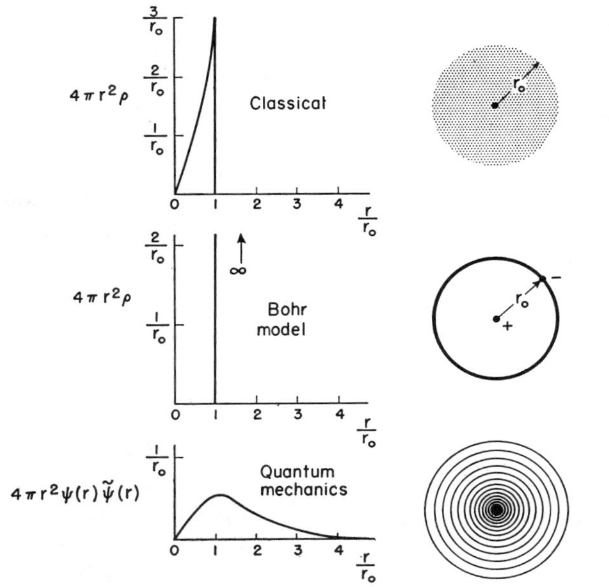

Fig. 3.2. Various models for electron charge distribution in the hydrogen atom: (*a*) electron cloud of uniform density, (*b*) Bohr model, and (*c*) quantum mechanics (normalized to unit electronic charge).

The electric field strength exercised by a proton on an electron at distance r_0 is

$$E_{r_0} = \frac{e}{\epsilon_0 4\pi r_0^2} \simeq 5 \times 10^{11} \quad \text{[volt m}^{-1}\text{]}. \quad (3.18)$$

Compared with such enormous interior fields of atoms any external field at present attainable represents only a minor disturbance. This fact justifies the approximation of Eq. 3.5 that the induced dipole moment is proportional to the inducing field strength.

4 · Anomalous Dispersion and Resonance Absorption

More detailed information about the structure of atoms may be obtained by polarization measurements in alternating fields, that is, by investigating the frequency response characteristics spectroscopically. Not knowing what to expect in detail, we may start with the simple model of electrons quasi-elastically bound to equilibrium positions and reacting to field changes like linear harmonic oscillators.

Transient response

Such an oscillator, displaced in the z-direction from its equilibrium position by an amount z_0 and released, will return to its original position in an harmonic motion prescribed by the force equation

$$m\frac{d^2z}{dt^2} + F_s + fz = 0. \tag{4.1}$$

In the absence of a friction force F_s, the electron would describe an undamped oscillation

$$z = z_0 \cos \omega_0 t \tag{4.2}$$

of the *resonance frequency*

$$\omega_0 = \sqrt{f/m}. \tag{4.3}$$

Actually, friction is produced according to classical theory by the radiation of the linear oscillator as calculated in I, 13. The friction force exerted on an electronic dipole of the charge $Q = e$ by the emission of electromagnetic radiation of the phase velocity $v = c$, is, according to I, Eq. 13.25,

$$F_s = -\frac{\mu_0 e^2}{6\pi c}\frac{d^3z}{dt^3}. \tag{4.4}$$

The energy loss per cycle caused by this radiation is small; hence Eq. 4.4 can be replaced by the approximation

$$F_s \simeq \frac{\mu_0 e^2 \omega_0{}^2}{6\pi c}\frac{dz}{dt}. \tag{4.5}$$

(The friction force, out of phase by 90° in relation to the restoring force, must be an odd differential of z with respect to t.) By defining a friction factor †

$$2\alpha \equiv \frac{\mu_0 e^2 \omega_0{}^2}{m 6\pi c}, \tag{4.6}$$

we rewrite the equation of motion (Eq. 4.1)

$$\frac{d^2z}{dt^2} + 2\alpha\frac{dz}{dt} + \omega_0{}^2 z = 0. \tag{4.7}$$

Its solution is

$$z = z_0 e^{-\alpha t} \cos \omega_0' t, \tag{4.8}$$

where

$$\omega_0' = \sqrt{\omega_0{}^2 - \alpha^2} \tag{4.9}$$

designates the reduced resonance frequency of the damped oscillator. Since for the oscillating electron the attenuation due to radiation damping is low ($\alpha \ll \omega_0$), the difference between ω_0 and ω_0' can usually be neglected.

Without attenuation, the classical electronic oscillator would emit a *monochromatic radiation* of the wavelength

$$\lambda_0 = \frac{2\pi c}{\omega_0} \quad [\text{m}]. \tag{4.10}$$

The radiation damping and other causes (see Sec. 20) broaden this *spectral line* over an infinite range of frequencies as may be seen when we express the oscillator amplitude z of Eq. 4.8 by Fourier integrals.[1] We obtain

$$z = \frac{1}{\sqrt{\pi}}\int_0^\infty a(\omega)\cos \omega t\, d\omega - \frac{1}{\sqrt{\pi}}\int_0^\infty b(\omega)\sin \omega t\, d\omega, \tag{4.11}$$

where the amplitudes of the components are

$$a(\omega) = \frac{z_0}{2\sqrt{\pi}}\left[\frac{\alpha}{\alpha^2 + (\omega - \omega_0)^2} + \frac{\alpha}{\alpha^2 + (\omega + \omega_0)^2}\right],$$

$$b(\omega) = \frac{z_0}{2\sqrt{\pi}}\left[\frac{\omega + \omega_0}{\alpha^2 + (\omega + \omega_0)^2} + \frac{\omega - \omega_0}{\alpha^2 + (\omega - \omega_0)^2}\right]. \tag{4.12}$$

The energy content of the oscillator is proportional to the square of its amplitude; hence it decreases exponentially with time as

$$W = W_0 e^{-2\alpha t} = W_0 e^{-t/\tau}, \tag{4.13}$$

where

$$\tau \equiv 1/2\alpha \tag{4.14}$$

designates the relaxation time of the oscillator. Consequently, the intensity of radiation which corresponds to this energy loss will be distributed over the spectrum proportionally to the square of the Fourier amplitudes, or as

$$\begin{aligned} I &= C[a^2(\omega) + b^2(\omega)] \\ &= \frac{C}{4\pi}\left\{\frac{1}{\alpha^2 + (\omega + \omega_0)^2} + \frac{1}{\alpha^2 + (\omega - \omega_0)^2} + 2\frac{\alpha^2 + (\omega + \omega_0)(\omega - \omega_0)}{[\alpha^2 + (\omega + \omega_0)^2][\alpha^2 + (\omega - \omega_0)^2]}\right\}. \end{aligned} \tag{4.15}$$

Near resonance, only the second term of this expression counts, and, if I_0 designates the intensity of the spectral line at the resonance frequency ω_0, the relative intensity for any other frequency becomes

$$\frac{I}{I_0} = \frac{1}{1 + \left(\frac{\omega - \omega_0}{\alpha}\right)^2}. \tag{4.16}$$

† The attenuation factor α should not be mistaken for the polarizability α; it is unfortunately customary to designate both by the same letter.

[1] See, for example, E. A. Guillemin, *The Mathematics of Circuit Analysis*, John Wiley and Sons, New York, 1949.

The relative intensity falls from its maximum at ω_0 symmetrically to a half value at the frequencies $\omega = \omega_0 \pm \frac{\Delta\omega_h}{2}$ (Fig. 4.1), where

$$\frac{\Delta\omega_h}{2} \equiv \pi\,\Delta\nu_h = \alpha. \tag{4.17}$$

The frequency interval $\Delta\omega_h$ between the half-value points is the *half-width* or *line-breadth of a spectral line,*

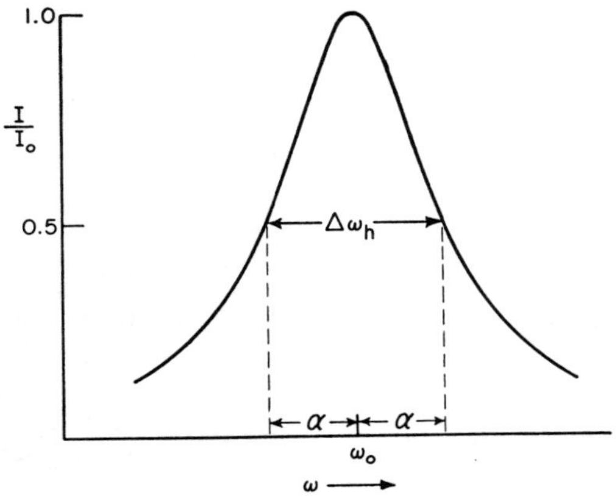

Fig. 4.1. Half width of spectral line.

in analogy to the half-width of a resonator (see I, Fig. 24.7).

According to Eqs. 4.14 and 4.17 the relaxation time of the electronic oscillator is connected to the half-width of the emitted spectral line as

$$\tau = \frac{1}{2\alpha} = \frac{1}{2\pi\,\Delta\nu_h}. \tag{4.18}$$

This expression is identical to the ringing time of a cavity resonator

$$\tau = Q/\omega_0. \tag{I, 24.26}$$

Hence we can compare the qualities of an electronic and a cavity resonator by referring to the *Q of a spectral line,*

$$Q = \frac{\omega_0}{2\alpha} = \frac{\nu_0}{\Delta\nu_h}. \tag{4.19}$$

Steady-state response

If the electronic oscillator is not free but subjected to a driving force $e\mathbf{E}'$ (where $\mathbf{E}'$ represents the locally acting electric field) the law of motion changes from Eq. 4.7 to

$$\frac{d^2z}{dt^2} + 2\alpha\frac{dz}{dt} + \omega_0^2 z = \frac{e}{m}\mathbf{E}'. \tag{4.20}$$

We can rewrite this expression in the form of a differential equation for the polarization $\mathbf{P}$ by assuming that the dielectric is composed of N oscillators per unit volume, each of them contributing an induced electric moment

$$\boldsymbol{\mu} = e\mathbf{z}; \tag{4.21}$$

then (see Eq. 1.2)

$$\mathbf{P} = Ne\mathbf{z}. \tag{4.22}$$

If, in addition, we assume that the locally acting field $\mathbf{E}'$ can be described by the Mosotti approximation (see Eq. 2.9) as

$$\mathbf{E}' = \mathbf{E} + \frac{\mathbf{P}}{3\epsilon_0}, \tag{4.23}$$

the differential equation becomes

$$\frac{d^2\mathbf{P}}{dt^2} + 2\alpha\frac{d\mathbf{P}}{dt} + \left(\omega_0^2 - \frac{Ne^2}{3m\epsilon_0}\right)\mathbf{P} = \frac{Ne^2}{m}\mathbf{E}. \tag{4.24}$$

The effect of the polarization of the surroundings is to lower the resonance frequency of the individual oscillator from ω_0 to

$$\omega_0'' = \sqrt{\omega_0^2 - \frac{Ne^2}{3m\epsilon_0}}. \tag{4.25}$$

The steady-state solution of Eq. 4.24 for a sinusoidal driving field

$$\mathbf{E} = \mathbf{E}_0 e^{j\omega t} \tag{4.26}$$

is

$$\mathbf{P} = \mathbf{P}_0 e^{j(\omega t + \psi)} = \frac{Ne^2/m}{\omega_0''^2 - \omega^2 + j\omega 2\alpha}\mathbf{E}. \tag{4.27}$$

Because of the friction factor 2α, a phase shift ψ occurs between the driving field and the resultant polarization; $\mathbf{P}$ becomes complex. The ratio $\mathbf{P}/\epsilon_0\mathbf{E}$ (cf. Eq. 1.1) determines the complex relative permittivity of the medium in molecular terms as

$$\kappa^* = 1 + \frac{\mathbf{P}}{\epsilon_0\mathbf{E}} = 1 + \frac{Ne^2/\epsilon_0 m}{\omega_0^2 - \omega^2 + j\omega 2\alpha}. \tag{4.28}$$

So far we have assumed that the dielectric contains only one oscillator type. In the more general case of s oscillator types which contribute to κ^* without mutual coupling, the equation may be generalized as

$$\kappa^* = 1 + \sum_s \frac{N_s e^2/\epsilon_0 m_s}{\omega_s^2 - \omega^2 + j\omega 2\alpha_s}. \tag{4.29}$$

For nonmagnetic media,

$$\kappa^* = n^{*2}, \tag{4.30}$$

where n^* is the complex index of refraction of the medium (see I, Eq. 9.14). Equation 4.29 represents the *dispersion formula of classical physics;* N_s designates the *number of dispersion electrons* per unit volume for the oscillator type s.

Far below its resonance frequency ($\omega \ll \omega_s$), each oscillator type adds a constant contribution

$$\frac{N_s e^2/\epsilon_0 m_s}{\omega_s{}^2} \tag{4.31}$$

to the static dielectric constant of the medium, whereas far above the resonance frequency its contribution vanishes. To follow the behavior of the lowest oscillator type r through its resonance region, we lump the effect of vacuum and of the remaining resonator types in a constant contribution

$$A = 1 + \sum_s \frac{N_s e^2/m_s}{\omega_s{}^2} \qquad \text{for } s \neq r. \tag{4.32}$$

Furthermore, we introduce in place of ω the deviation from resonance

$$\Delta\omega \equiv \omega_r - \omega \tag{4.33}$$

as the variable, approximate

$$\omega_r + \omega \simeq 2\omega_r$$
$$\frac{\omega}{\omega_r} \simeq 1, \tag{4.34}$$

and thus rewrite Eq. 4.29 as

$$\kappa^* = n^{*2} = A + \frac{B}{\Delta\omega + j\alpha}. \tag{4.35}$$

The factor B stands for

$$B \equiv \frac{N_r e^2/\epsilon_0 m_r}{2\omega_r}. \tag{4.36}$$

The frequency dependence of the real part of the relative permittivity,

$$\kappa' = n^2(1 - k^2) = A + \frac{B\,\Delta\omega}{(\Delta\omega)^2 + \alpha^2}, \tag{4.37}$$

describes the *dispersion characteristic* of the dielectric medium near resonance (Fig. 4.2).† It rises hyperbolically from the low frequency value $A + \frac{2B}{\omega_r}$ to a maximum

$$\kappa'_{\max} = A + \frac{B}{2\alpha} \tag{4.38}$$

at $\Delta\omega = +\alpha$, falls with a linear slope $-B/\alpha^2$ through the value A at the resonance frequency ($\Delta\omega = 0$), reaches a minimum

$$\kappa'_{\min} = A - \frac{B}{2\alpha} \tag{4.39}$$

at $\Delta\omega = -\alpha$, and then rises again asymptotically to the constant value A for very high frequencies ($\omega \gg \omega_r$).

The absorption characteristic of the dielectric, identified near resonance by the relative loss factor

$$\kappa'' = 2n^2k = \frac{B\alpha}{(\Delta\omega)^2 + \alpha^2}, \tag{4.40}$$

starts from zero at low frequencies, traverses its maximum B/α at resonance and falls again symmetrically to zero at high frequencies. The half-value points $B/2\alpha$ of this bell-shaped absorption characteristic are reached at the deviation from resonance $\Delta\omega = \pm\alpha$, in agreement with the half-width of the emission characteristic (Fig. 4.1).

Since the real dielectric constant and index of refraction rise with increasing frequency over the major part

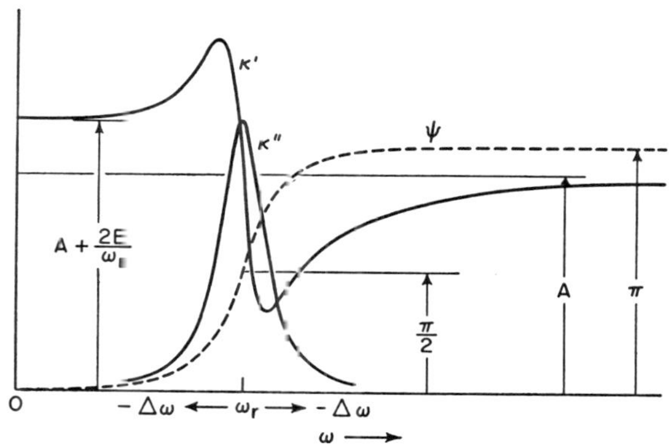

Fig. 4.2. Anomalous dispersion and resonance absorption.

of the dispersion characteristic, this behavior has been called *normal dispersion* in contrast to the *anomalous dispersion* in the half-width region of the spectral line, where the characteristic falls toward shorter wavelengths. This combination of normal and anomalous dispersion is typical for resonance phenomena, as we have pointed out in I, Sec. 26.

For the phase relation between applied field and induced dipole moment near resonance, we obtain the relation

$$\tan\psi = \frac{\kappa''}{\kappa' - A} = \frac{\alpha}{\Delta\omega}. \tag{4.41}$$

At low frequencies ($\omega \ll \omega_r$) the moment follows in phase, at resonance it lags by $\pi/2$, and at very high frequencies by π. In between, it passes the values $\pi/4$ and $3\pi/4$ at half-width points $\Delta\omega = \pm\alpha$. The loss tangent itself,

$$\tan\delta = \frac{\kappa''}{\kappa'} = \frac{\alpha B}{A(\Delta\omega^2 + \alpha^2) + B\,\Delta\omega}, \tag{4.42}$$

rises from zero through the value $B/\alpha A$ at resonance to maximum at $\Delta\omega = -B/2A$ and falls again to zero.

† Cf. I, Fig. 26.9, and the corresponding discussion for the series resonance circuit on p. 30.

The ratio B/α, which characterizes the height of the dispersion and absorption characteristic, has a simple meaning in the case of radiation damping. By introducing the value for α from Eq. 4.6 and for B from Eq. 4.36 we find

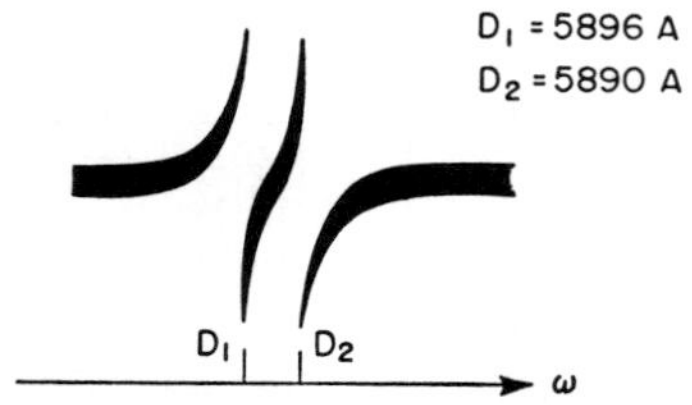

Fig. 4.3. Anomalous dispersion of sodium D lines. (After Wood.[3])

$$\frac{B}{\alpha} = \frac{3}{4\pi^2} N_r \lambda_r^3; \tag{4.43}$$

that is, $13 \dfrac{B}{\alpha}$ equals the number of dispersion electrons contained in a wavelength cube at the resonance frequency.[2]

That the shape of a spectral line corresponds to the prediction of Fig. 4.2, Fig. 4.3 confirms; it shows the dispersion near the D-line doublet of sodium vapor as drawn from an actual photograph.[3]

[2] See M. Born, *Optik*, Springer, Berlin, 1933, p. 478.
[3] R. W. Wood, Physical Optics, Macmillan, New York, 1934.

5 · Various Aspects of Electromagnetic Radiation

Although the dispersion formula of classical physics predicts correctly the general appearance of a spectral line, neither the intensity nor the resonance frequency can be derived from such classical considerations. This becomes apparent when we inquire more seriously into the structure of atoms.

Investigations of Lenard[1] and Rutherford[2] on the scattering of cathode and α rays by matter showed only an extremely small fraction of the *gaskinetic volume* $\frac{4\pi}{3} r_0^3$ of atoms to be impenetrable. Against a background of small, overall scattering there appeared individual deflections of these rays through large angles which indicated the existence of very small mass centers acting with intense fields. Rutherford thus arrived at the basic atom model: a positive nucleus of the charge Ze containing the bulk of the mass and surrounded by a neutralizing atmosphere of Z electrons of the charge $-e$. In contrast to the gaskinetic collision radius of about 10^{-10} [m] the radius of the nucleus according to the scattering experiments[3] proved to be of the magnitude 10^{-14} to 10^{-15} [m]. The factor Z, as van den Broek[4] first postulated, represents the atomic number in the periodic system of the elements.

Rutherford's model corresponds to a planetary system in miniature, in which the electrons move in circular or elliptical orbits around the nuclear sun. According to the classical laws of mechanics and electrostatics such a system is stable when the centripetal force exerted by the Coulomb field of the nucleus balances the centrifugal forces and electrostatic repulsions arising in the rotating electronic system. In the hydrogen atom, a proton of the mass m_+ is circled by one electron of the mass m, where

$$\left.\begin{aligned} m_+ &= 1.6725 \times 10^{-27} \text{ kg} \\ m &= 9.107 \times 10^{-31} \text{ kg} \end{aligned}\right\} \qquad \frac{m_+}{m} = 1836.5\dagger. \tag{5.1}$$

The position of the electron may be indicated by a vector of the length $|\mathbf{r}_0|$, pointing from proton to electron and rotating with an angular velocity

$$\omega = \frac{v}{\mathbf{r}_0} \tag{5.3}$$

as

$$\mathbf{r}_0 = \mathbf{r}_0 e^{j\omega t}. \tag{5.4}$$

This rotation presupposes, according to Newton's law, a central force

$$\mathbf{F} = m \frac{d^2\mathbf{r}_0}{dt^2} = -m\omega^2 \mathbf{r}_0 = -\frac{mv^2}{r_0} \mathbf{r}^0, \tag{5.5}$$

[1] P. Lenard, *Ann. Physik 12*, 714 (1903).
[2] E. Rutherford, *Phil. Mag. 21*, 669 (1911).
[3] See, for instance, H. Geiger and E. Marsden, *Phil. Mag. 25*, 604 (1913); J. Chadwick, *ibid. 40*, 734 (1920).
[4] A. van den Broek, *Physik. Z. 14*, 32 (1913).

† Because this ratio is so large, the proton is considered at rest in the subsequent calculations. An accurate treatment of the motion of electron and proton around their common center of gravity gives precisely the same equations, provided the mass m of the electron is replaced by the *reduced mass* m_r of electron and proton,

$$m_r = \frac{mm_+}{m + m_+}. \tag{5.2}$$

which has to be provided by the Coulomb attraction force (see I, Eq. 3.8)

$$\mathbf{F}_c = -\frac{e^2}{\epsilon_0 4\pi r_0^2}\mathbf{r}^0 \tag{5.6}$$

of the nucleus. The condition of balance

$$\frac{mv^2}{r_0} = \frac{e^2}{\epsilon_0 4\pi r_0^2} \tag{5.7}$$

can be written alternatively by referring to the kinetic energy $\mathcal{E}_{kin}$ and the potential energy U of the electron, as

$$\mathcal{E}_{kin} = -\tfrac{1}{2}U. \tag{5.8}$$

This relation, stating that the kinetic energy of an electron trapped in a Coulomb field equals half its potential energy, is known as the *Virial theorem;* it holds quite generally for electrons moving under the influence of Coulomb fields with velocities very much smaller than light velocity, as Bohr [5] first showed.

In the hydrogen atom of the gaskinetic radius $\mathbf{r}_0 \simeq 0.53$ A the electron rotates, according to Eq. 5.7, with a velocity

$$v = \frac{e}{\sqrt{\epsilon_0 4\pi r_0 m}} \simeq 2.2 \times 10^6 \quad [\text{m sec}^{-1}]; \tag{5.9}$$

hence the condition $v \ll c$ is fulfilled. However, why a particular orbit r_0 should be preferred above others remains unexplained. The classical laws of mechanics and electrostatics allow *any* orbit; if we add the laws of classical electrodynamics, *none* can be stable.

This becomes evident if we visualize the hydrogen atom as a rotating dipole of the moment

$$\boldsymbol{\mu}(t) = -er_0e^{j\omega t}. \tag{5.10}$$

Such a dipole is equivalent to two linear dipoles

$$\boldsymbol{\mu}(t) = -e\left|\,\mathbf{r}_0\,\right|(\mathbf{i}\cos\omega t + \mathbf{k}\sin\omega t) \tag{5.11}$$

that oscillate with a temporal phase shift of 90° in the x- and z-directions, respectively (Fig. 5.1). These linear dipoles, as shown in I, 13, correspond to radiating dipole antennas. An electron trapped in the field of a positive nucleus is, therefore, classically speaking, a *light source.* It transforms electrostatic field energy through kinetic energy into radiation, and, in so doing, it must spiral into the nucleus.

The power radiated per second by two linear dipole antennas into free space is, according to Eq. I, 13.18,

$$2P = \frac{\mu_0\omega^4}{6\pi c}e^2r_0^2 \quad [\text{watt}]. \tag{5.12}$$

[5] N. Bohr, *Phil. Mag. 26*, 1 (1913).

The electron rotates in the r_0 orbit of the hydrogen atom with a frequency

$$\nu = \frac{v}{2\pi r_0} = \frac{\omega}{2\pi} \simeq 6.5 \times 10^{15} \quad [\text{sec}^{-1}]. \tag{5.13}$$

Its kinetic energy in this orbit is

$$\mathcal{E}_{kin} = \tfrac{1}{2}mv^2 \simeq 2.2 \times 10^{-18} \quad [\text{joule}], \tag{5.14}$$

and its energy loss per cycle due to radiation is

$$\frac{2P}{\nu} = \frac{\mu_0\omega^3}{3c}e^2r_0^2 \simeq 3 \times 10^{-23} \quad [\text{joule}]. \tag{5.15}$$

Hence the energy loss per cycle is small, as assumed in Eq. 4.5, but the hydrogen atom would radiate in ca. 10^5 cycles or 10^{-11} sec the total kinetic energy initially available. To replenish this energy the electron has to move nearer to the nucleus in accordance with the mechanical equilibrium condition, Eq. 5.7. This would compel it to circle faster, the wavelength of the emitted light, originally

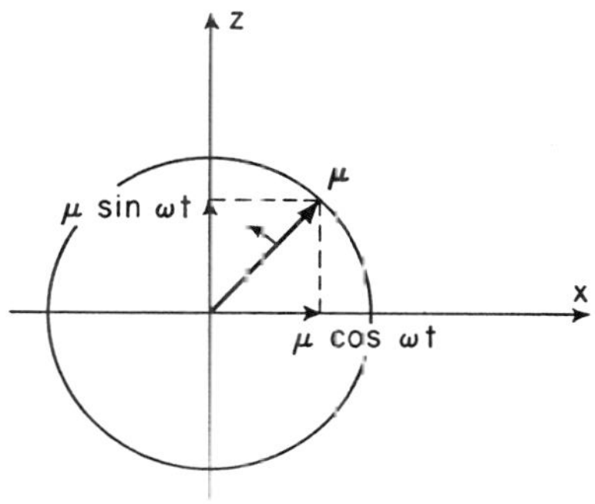

Fig. 5.1. Rotating electric dipole.

$$\lambda_0 = \frac{c}{\nu} \simeq 4.7 \times 10^{-8} \quad [\text{m}], \tag{5.16}$$

would shorten, and the radiation loss per cycle rise with the third power of the frequency. Thus the planetary Rutherford atom must collapse in an extremely short time under emission of a continuous radiation covering a spectral range from λ_0 to extremely short wavelengths.

This obviously does not occur. We are forced to conclude that nonclassical laws come into play in the realm of molecular dimensions, which permit the existence of electrons in selected orbits without radiation. Such entirely new physical concepts outside the range of previous experience cannot be deduced by theoretical speculation; they have to evolve on the basis of an accurate analysis of all available experimental data. In the present case these data concern the various aspects of light radiation.

Extensive spectroscopic studies [6] have established that each atom type is characterized by sets of spectral lines, and that the resonance frequencies in each of these spectral series can be represented as the difference between two spectral terms. This *Rydberg-Ritz combination principle* [7] reduces, in the case of the hydrogen atom, to the *Balmer formula,*

$$\nu = Rc\left(\frac{1}{{n_1}^2} - \frac{1}{{n_2}^2}\right) \quad [\text{sec}^{-1}], \tag{5.17}$$

with n_1 equal 1 or 2 or 3, etc., and n_2 an integer larger than n_1. For $n_1 = 1$, $n_2 = 2, 3, 4 \ldots$, the *Lyman series* of the hydrogen spectrum is obtained, located in the far-ultraviolet (Fig. 5.2); $n_1 = 2$; $n_2 = 3, 4, 5 \ldots$

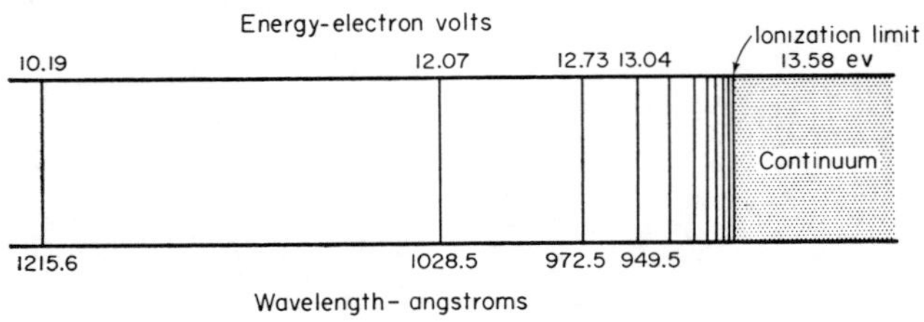

Fig. 5.2. The Lyman series of the hydrogen atom.

corresponds to the *Balmer series* [8] extending from the visible into the near ultraviolet; $n_1 = 3$, $n_2 = 4, 5, 6 \ldots$ to the *Paschen series* located in the infrared, etc. The spectral line with the lowest term values is called the *resonance line* of the atom in question.

Spectroscopists refer, in general, not to the frequency of a spectral line but to its *wave number*

$$\nu' = \frac{\nu}{c} = \frac{1}{\lambda}, \tag{5.18}$$

and therefore write the Balmer formula as

$$\nu' = R\left(\frac{1}{{n_1}^2} - \frac{1}{{n_2}^2}\right). \tag{5.19}$$

The constant R is known as the *Rydberg constant.* The precision of spectroscopic measurements allowed its determination with extreme accuracy as

$$R = 10{,}967{,}758 \quad [\text{m}^{-1}]. \tag{5.20}$$

The resonance line of the hydrogen atom, the first resonance absorption which one would encounter when subjecting an atomic hydrogen gas to alternating fields of higher and higher frequencies, is located in the far ultraviolet region of the electromagnetic spectrum at

$$\nu_r' = R\left(\frac{1}{1^2} - \frac{1}{2^2}\right) = 8.226 \times 10^6 \quad [\text{m}^{-1}]$$

or

$$\lambda_r = \frac{1}{\nu_r'} = 1215.66 \text{ A}. \tag{5.21}$$

This wavelength is about three times as long as that expected from the classical model (see Eq. 5.16).

Additional information on the properties of radiation was obtained from investigations of the continuous spectrum emitted thermally by solids and liquids. The equilibrium radiation that establishes itself in an isothermal cavity enclosed by completely absorbing walls may be observed experimentally through a very small hole in the wall of such a *black body.* The intensity of this *black-body radiation* proved to be a function of temperature and frequency only,[9] and not dependent on the specific properties of the wall material. Obviously, thermodynamics and statistics should be able to calculate the black-body spectrum.

Thermodynamic considerations [10] led to the conclusion that the energy density ρ of the radiation in the cavity should increase with the absolute temperature as

$$\rho = aT^4 \quad [\text{joule m}^{-3}]. \tag{5.22}$$

In addition to this *Stefan-Boltzmann law,*[11] thermodynamics could show that the spectral energy density ρ_λ contained in a wavelength range between λ and $\lambda + d\lambda$ must rise with the fifth power of the absolute temperature, and that the proportionality factor in the spectral density equation can depend only on the product of wavelength and temperature. Thus

$$\rho_\lambda \, d\lambda = T^5 f(\lambda T) \, d\lambda. \tag{5.23}$$

The experimental characteristic of $f(\lambda T)$ plotted as function of the wavelength traverses a maximum at some wavelength $\lambda_{\max}$. At $\lambda = \lambda_{\max}$, the differential of ρ_λ with respect to λ for a fixed temperature is zero; hence $\dfrac{df(\lambda_{\max}T)}{d\lambda} = 0$, or

$$\lambda_{\max}T = \text{constant} \equiv A; \tag{5.24}$$

the wavelength of highest radiation density shifts proportional to $1/T$. Equation 5.24 is known as *Wien's*

[6] See, for example, G. Herzberg, *Atomic Spectra and Atomic Structure,* New York, Prentice-Hall, Inc., 1937.

[7] J. R. Rydberg, *Phil. Mag. 29,* 331 (1890); W. Ritz, *Physik. Z. 9,* 521 (1908); *id., Astrophys. J. 28,* 237 (1908).

[8] The discovery of this series and its term relations by J. J. Balmer [*Ann. Physik. 25,* 80 (1885), *60,* 380 (1897)] marks the beginning of scientific spectroscopy.

[9] G. Kirchhoff, *Berl. Ber.* 1859, p. 216; *Pogg. Ann. 109,* 275 (1860).

[10] See M. Planck, *Theorie der Wärmestrahlung,* Barth, Leipzig, 1906.

[11] J. Stefan, *Wien. Ber. 79,* 391 (1879); L. Boltzmann, *Wied. Ann. 22,* 291 (1884).

displacement law.[12] The values of the Stefan-Boltzmann constant a and of Wien's displacement constant A are

$$a = 7.563 \times 10^{-16} \quad [\text{watt m}^{-3}\ \text{deg}^{-4}], \qquad A = 2.897 \times 10^{-3} \quad [\text{m deg}]. \tag{5.25}$$

There remains the evaluation of the function $f(\lambda T)$ itself; this requires a statistical approach. Classical statistics leads to the *principle* of the *equipartition of energy*, which claims that each degree of freedom of a system in thermal equilibrium contains the same average energy

$$\mathcal{E} = \tfrac{1}{2}kT. \tag{5.26}$$

The factor

$$k = 1.380 \times 10^{-23} \quad [\text{joule deg}^{-1}] \tag{5.27}$$

is the *Boltzmann* or *molecular gas constant*. By applying this principle to the radiation problem, classical physics arrived at theoretical conclusions in complete variance to the experimental facts.

The energy content of the radiation field can be calculated from the number of standing-wave modes that can be accommodated in a black-body cavity since these modes represent the degrees of freedom of the radiation field. In a linear resonator of the length d, n standing waves are possible with nodes at the boundaries; that is,

$$d = n\frac{\lambda}{2}, \qquad \text{with } n = 1, 2, 3 \ldots. \tag{5.28}$$

Consequently, in a wavelength range between λ and $\lambda + d\lambda$ the number of possible wavelengths is

$$dn = \frac{2d}{\lambda} - \frac{2d}{\lambda + d\lambda} \simeq \frac{2d\, d\lambda}{\lambda^2} \qquad \text{for } d\lambda \ll \lambda. \tag{5.29}$$

Since the electromagnetic waves are transversal, each of these dn wavelengths corresponds to two transversal modes, polarized at right angles to each other, and each of these two dn modes may contain $\frac{1}{2}kT$ of electromagnetic energy. The total energy density of the linear radiation field consequently becomes

$$\rho = 2kT\int_0^\infty \frac{d\lambda}{\lambda^2} = \infty \qquad \text{for } T > 0. \tag{5.30}$$

The number of possible modes increases towards infinity with shortening wavelengths; hence no equilibrium is possible between the radiation field and the walls of the container if the classical equipartition principle is valid. Hot bodies would, of necessity, emit a radiation dominated by the highest frequencies until their temperature has fallen to the absolute zero point.

This impasse forced physics to a decisive reorientation. Planck[13] discovered (October, 1900) a function $f(\lambda T)$ which led to a radiation formula in complete agreement with experiment. The *Planck radiation law* for the spectral energy density in the frequency range between ν and $\nu + d\nu$,†

$$\rho_\nu\, d\nu = \frac{8\pi h\nu^3\, d\nu}{c^3\{e^{\frac{h\nu}{kT}} - 1\}} \quad [\text{joule sec}^{-1}\ \text{m}^{-3}], \tag{5.31}$$

represented, however, as was slowly realized, a radical break with the classical ideas. It introduced a universal constant of action, the *Planck constant*

$$h = 6.623 \times 10^{-34} \quad [\text{joule sec}], \tag{5.32}$$

and stated that the energy content of the linear oscillators lining the walls of the black body as well as the energy content of the radiation field itself is parceled in *energy quanta* or *photons* $h\nu$.

Planck drew only the first of these conclusions. Einstein[14] extended the idea of quantization to the radiation field in order to explain the photoelectric experiments of Lenard.[15] Lenard had observed that the *number* of photoelectrons emitted from a metal surface increases proportionally to the light *intensity*, but that the *velocity* of these electrons depended only on the metal used and on the *frequency* of the irradiating light. Classically, we would have expected that the intensity of the electromagnetic field would determine the energy with which the photoelectrons are ejected. By connecting this observation to Planck's law, Einstein arrived at the concept that the energy of the electromagnetic field is contained in *light quanta* $h\nu$. Consequently, if a photoelectron is liberated by the absorption of such a quantum, its energy cannot exceed the amount

$$\tfrac{1}{2}mv^2 = e\mathcal{V} = h\nu - P, \tag{5.33}$$

where the *work function* P represents the binding energy of the metal for electrons.

Einstein's photoelectric equation (Eq. 5.33) suggests a convenient new yardstick for the measurement of energy relations in molecular physics: the *electron volt*

[12] W. Wien, *Sitz. Ber. Akad. Wiss. Berlin*, Feb. 9, 1893, p. 55; *Wied. Ann. 52*, 132 (1894).

[13] M. Planck, *Verh. deut. physik. Ges. 2*, 237 (1900); *Ann. Physik 4*, 553 (1901).

† In switching over from the wavelength to the frequency scale, we have to substitute $d\nu = -\frac{c}{\lambda^2}\,d\lambda$, since $\nu = c/\lambda$.

[14] A. Einstein, *Ann. Physik 17*, 132 (1905).

[15] P. Lenard, *Ann. Physik 8*, 149 (1902).

[ev]. One electron volt represents the energy acquired by an electron in falling unimpeded through a potential difference $\mathcal{V}$ of one volt, or

$$1[\text{ev}] = 1.602 \times 10^{-19} \quad [\text{joule}]. \tag{5.34}$$

Summarizing: Classical physics had come into a conflict with the experimental facts of electromagnetic radiation from which it could not extricate itself. The planetary Rutherford atom was supposed to collapse but proved stable. The radiation of atoms consisted of series of sharply defined spectral lines ruled by the empirical Rydberg-Ritz combination principle. The continuous spectrum of black bodies violated the equipartition principle, and the energy of photoelectrons proved dependent on the frequency and not on the intensity of the light. These last two facts compelled physics to break with its classical continuum ideas and to conclude that radiation is parceled out in light quanta $h\nu$. Maxwell's theory thus loses its absolute validity and becomes a theory of photon statistics. The remaining great problem was to link the quantum theory of radiation to the structure of atoms.

6 · Bohr's Quantum Theory

The first step towards the new theory of atomic structure was made by Bohr.[1] His theory assumes that the electrons of atoms can move in certain stable orbits without radiation (Rutherford atom). It further postulates that electron transitions between these stable energy states of the atom produce the spectral lines by the absorption or emission of light quanta. Since the energy of the system (atom plus photon) has to be conserved, the frequency of the radiation is given by the energy difference between the initial ($\mathcal{E}_1$) and final ($\mathcal{E}_2$) state of the atom as

$$h\nu = \mathcal{E}_1 - \mathcal{E}_2. \tag{6.1}$$

Thus, by representing each spectral line as the difference of two energy terms, *Bohr's frequency condition* (Eq. 6.1) explains immediately the significance of the Rydberg-Ritz combination principle.

In the hydrogen atom, the electron circling the proton at a distance r has, according to the Virial theorem (see Eq. 5.8), the total energy

$$\mathcal{E} = U + \mathcal{E}_{\text{kin}} = \frac{1}{2} U = -\frac{e^2}{\epsilon_0 8\pi r}. \tag{6.2}$$

The frequency condition for an electron transition from an orbit of radius r_2 to an orbit r_1 may therefore be written as

$$h\nu = \frac{e^2}{\epsilon_0 8\pi}\left(\frac{1}{r_1} - \frac{1}{r_2}\right). \tag{6.3}$$

By comparing this expression with the empirical frequency condition of the Balmer formula (see Eq. 5.17)

$$h\nu = hRc\left(\frac{1}{{n_1}^2} - \frac{1}{{n_2}^2}\right), \tag{6.4}$$

we find that the two equations can be alike only if

$$r = n^2 r_0, \tag{6.5}$$

that is, if the radii of the possible orbits increase proportional to n^2, with $n = 1, 2, 3 \ldots$ (Fig. 6.1).

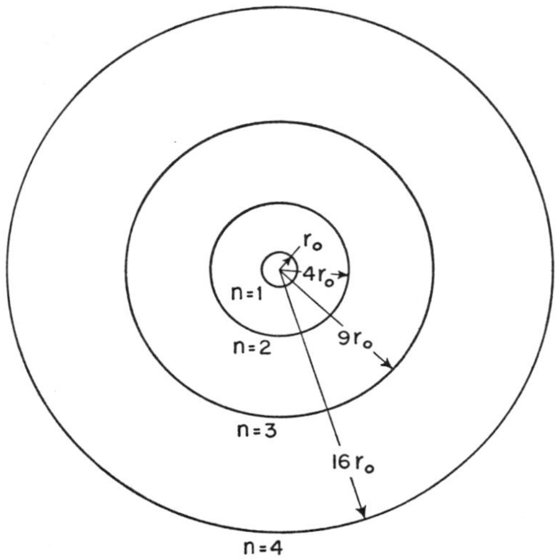

Fig. 6.1. Sequence of circular Bohr orbits of the hydrogen atom.

Equations 6.3 to 6.5 relate the fundamental orbit of the hydrogen atom r_0 to the empirically determined Rydberg constant R (see Eq. 5.20) as

$$r_0 = e^2/\epsilon_0 8\pi hcR. \tag{6.6}$$

A complete theory, however, should be able to calculate r_0 and R directly from fundamental constants. Bohr accomplished this by his *correspondence principle*.

A decisive difference between quantum and classical theory lies in the fact that quantum transitions take place between two energy states of different physical

[1] N. Bohr, *Phil. Mag. 26*, 1, 476, 857 (1913).

properties, whereas the corresponding classical process is concerned with only one set of physical conditions. Thus Bohr's hydrogen atom radiates when the electron jumps from a larger to a smaller orbit, whereas classically the frequency ν is emitted when the electron revolves in a given orbit r with this frequency ν. Bohr now postulated that when the change of physical properties between two quantum states becomes very small, the laws of quantum physics approach asymptotically those of classical physics. This correspondence principle is one of the powerful tools of the quantum theory because it links the atomistic world by extrapolation to the everyday world of macroscopic experience.

Applying the principle to the hydrogen atom, we claim that electron transitions between neighboring, very large orbits ($n \gg 1$, $dn = 1$) produce according to the Balmer formula a quantum frequency

$$\nu_q = Rc\left(\frac{1}{n^2} - \frac{1}{(n+1)^2}\right) \simeq \frac{2Rc}{n^3}, \qquad (6.7)$$

which approaches asymptotically the classical frequency emitted by an electron circling a very large orbit of radius r (see Eqs. 5.9, 5.13, and 6.5),

$$\nu_c = \frac{e}{2\pi n^3}\sqrt{\frac{1}{\epsilon_0 4\pi r_0{}^3 m}}. \qquad (6.8)$$

By postulating for $n \longrightarrow \infty$ the equality

$$\nu_q = \nu_c, \qquad (6.9)$$

the correspondence principle leads thus to a second equation between r_0 and R, which, together with Eq. 6.6, determines the Rydberg constant and the radius r_0 of the hydrogen atom unambiguously and in complete agreement with experience as

$$R = \frac{e^4 m}{\epsilon_0 8h^3 c}$$
$$r_0 = \frac{\epsilon_0 h^2}{\pi m e^2} = 0.528\ \text{A}. \qquad (6.10)$$

A further significant result can now be derived easily. The mechanical stability of an electron in its orbit $r = n^2 r_0$ requires, according to Eq. 5.9, an angular momentum

$$p' = mvr = \sqrt{\frac{e^2 m n^2 r_0}{\epsilon_0 4\pi}}. \qquad (6.11)$$

By introducing the expression for r_0 from Eq. 6.10, we obtain Bohr's famous *quantization condition of the rotator*

$$p' = n\frac{h}{2\pi}, \qquad n = 1, 2, 3 \ldots. \qquad (6.12)$$

The angular momentum of the circling electron must be an integral multiple of $h/2\pi$. The integer n is called the *principal* or *total quantum number* and determines the distance of the electron from the nucleus. The smallest integer, for hydrogen $n = 1$, designates the *fundamental* or *ground state* of the atom; the higher integers correspond to states of higher energy, that is, to *excited* states. The combination $h/2\pi$ is frequently abbreviated as $\hbar$ (h bar).

$$\hbar \equiv h/2\pi. \qquad (6.13)$$

The orbit described by a planet around the sun is in general not a circle but, according to *Kepler's first law*, an ellipse with the sun in one of its two focal points. In the planetary atom model, elliptical orbits have therefore to be admitted (Fig. 6.2). The motion of an

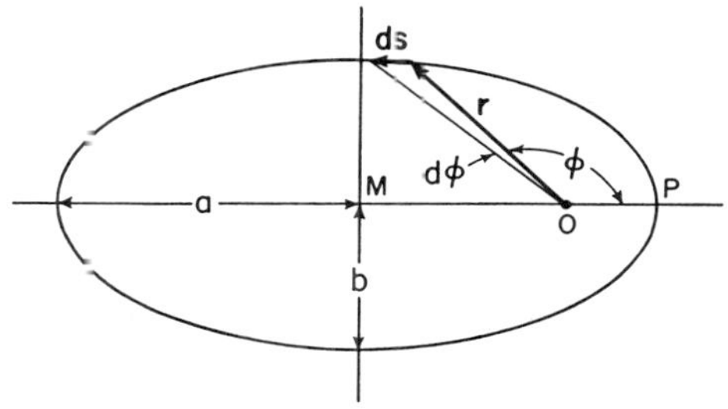

Fig. 6.2. Elliptical orbit.

electron around the positive nucleus in an ellipse is, like the corresponding astronomical situation, a problem of two degrees of freedom since the kinetic energy depends not only on the distance r from the focus but also on the azimuth angle ϕ. If ds designates an element of the orbit, it becomes in these polar co-ordinates (see I, Table 3.2)

$$ds^2 = dr^2 + r^2 d\phi^2; \qquad (6.14)$$

hence the kinetic energy

$$E_{\text{kin}} = \frac{1}{2}m\left(\frac{ds}{dt}\right)^2 = \frac{m}{2}\left\{\left(\frac{dr}{dt}\right)^2 + r^2\left(\frac{d\phi}{dt}\right)^2\right\}. \qquad (6.15)$$

To describe such more general motions, physics uses normally the position co-ordinates q (in this case r and ϕ) and the momentum co-ordinates p, defined as

$$p = \frac{\partial \mathcal{E}_{\text{kin}}}{\partial \dot{q}}, \qquad (6.16)$$

where the dot above the position parameter signifies its derivative with respect to time ($\dot{x} = \frac{dx}{dt}$, etc.). The momentum co-ordinates of the ellipse, according to Eq. 6.15, are

$$p_r = m\dot{r},\ p_\varphi = mr^2\dot{\phi}. \qquad (6.17)$$

The co-ordinate space in which the positions q are plotted as the abscissa and the momenta p as the ordinate is called a *phase space.*

The motion of our electron is a recurrent one; this means that the integral over the motion in phase space, the *phase integral*

$$J = \int p\, dq \tag{6.18}$$

must represent a finite area. According to Bohr's quantum theory, only certain orbits are permissible, selected in the case of circular orbits by the quantum condition Eq. 6.12. In phase space, the selection of certain orbits corresponds to the fact that the phase integral can assume only certain values. Sommerfeld [2] first derived a general rule that the image point of the electron in phase space describes curves of the quantized area

$$\oint p\, dq = nh. \tag{6.19}$$

Equation 6.19 is called *Sommerfeld's quantum condition* (see also Eq. 18.14).

For the elliptical orbits of the hydrogen atom, a system with two degrees of freedom, we obtain thus the two quantum conditions

$$\oint p_r\, dr = n'h, \qquad \int_{\phi=0}^{\phi=2\pi} p_\phi\, d\phi = kh. \tag{6.20}$$

The integers n' and k are designated as the *radial* and the *azimuthal* quantum number. Their sum

$$n = n' + k \tag{6.21}$$

corresponds to the total quantum number of Eq. 6.12 whereas their ratio describes the *numerical eccentricity* of the ellipse (see Fig. 6.2)

$$\epsilon = MO/MP \tag{6.22}$$

as

$$\frac{n'}{k} = \frac{1}{\sqrt{1-\epsilon^2}} - 1$$

or

$$\frac{k^2}{n^2} = 1 - \epsilon^2. \tag{6.23}$$

It can be shown [3] easily that the large and the small half axis of the ellipse are

$$a = r = n^2 r_0, \qquad b = nkr_0. \tag{6.24}$$

[2] See A. Sommerfeld, *Atombau und Spektrallinien*, Vieweg, Braunschweig, 1924, p. 99.

[3] See A. Sommerfeld, *loc. cit.*, pp. 122 ff.

The large half axis of the ellipse proves to be identical with the radius of the circular Bohr orbit of the same principal quantum number (Eq. 6.5). Furthermore, the total energy $\mathcal{E}$ of an electron in its ellipse,

$$\mathcal{E} = -\frac{e^2}{\epsilon_0 8\pi a}, \tag{6.25}$$

depends only on the major half axis a, that is, on the total quantum number n, and is thus identical with that of the corresponding circular orbit (Eq. 6.2). Consequently, to each principal quantum number n belongs one circular Bohr orbit ($n' = 0$; $k = n$) and ($n - 1$) elliptical orbits ($n' = 1, 2, \ldots, n - 1$; $k = n - 1, \ldots, 1$) of the same energy (Fig. 6.3).† This *degeneracy* of the one-electron system, namely, that circular and elliptical orbits have the same energy, makes the hydrogen spectrum especially simple but also atypical.

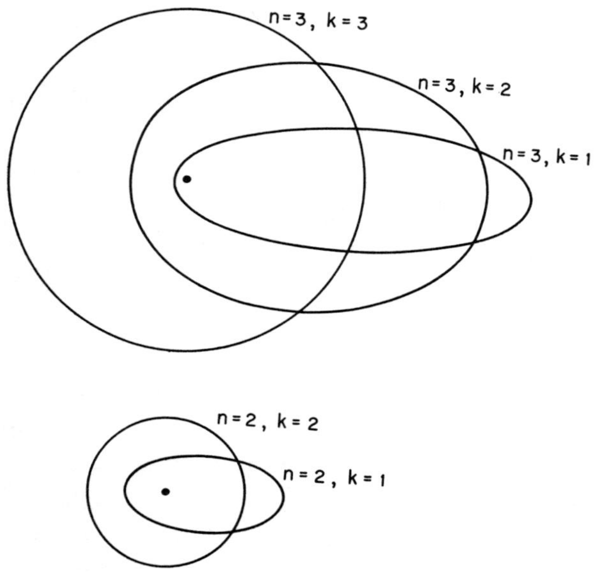

Fig. 6.3. Equivalent circular and elliptical Bohr orbits.

A circular orbit is characterized by the momentum co-ordinates (cf. Eq. 6.17)

$$p_r = 0, \quad p_\phi = mr^2\omega = \text{constant}. \tag{6.26}$$

With these values, the quantum conditions of Eq. 6.20 retransform into Bohr's quantization condition for the rotator (Eq. 6.12).

Bohr's theory proved eminently successful in giving a quantitative explanation of the spectra of atoms and ions with one valence electron and in providing the key for a general interpretation of the structure of atoms

† The case $n' = n$, $k = 0$ is excluded in Bohr's theory, because the electron would have to traverse the nucleus.

and molecules. The quantum nature of the electronic transitions, Bohr's frequency condition, was verified by the experiment of Franck and Hertz [4] (Fig. 6.4). Here it could be directly demonstrated that electrons, when accelerated to a critical kinetic energy (for example, ca. 4.9 ev for mercury vapor), can lose this energy in *one inelastic impact* by exciting an atom to the corresponding energy level; the atom then returns to its original state by the emission of a light quantum, which corresponds in energy to the kinetic energy the electron has lost (for mercury, the spectral line 2536.7 A appears).

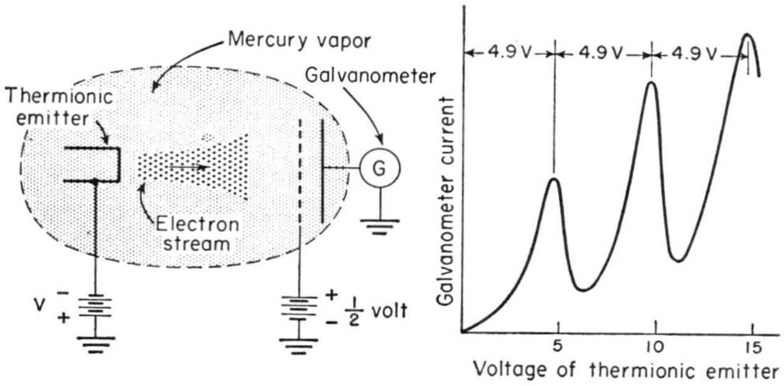

Fig. 6.4. The Franck-Hertz experiment.

[4] J. Franck and G. Hertz, *Ber. deut. physik. Ges. 16*, 457 (1914).

Serious difficulties arose for Bohr's theory, on the other hand, when it was applied to multi-electron systems and to the fine structure of spectral lines. Obviously, the planetary atom with its mysterious stabilization of stationary electron orbits was only an intermediate concept which has to yield to some more deeply founded description.

7 · Wave Mechanics

The preceding development introduced a dualism into the theory of radiation. Although the laws of propagation and interference of electromagnetic energy could be described successfully by a *wave* theory based on Maxwell's equations, the exchange of this energy with matter could be explained only by postulating the existence of light quanta, that is, of *particles* of energy. Since these photons travel with the velocity of light, their mass, energy, and momentum have to be defined on the basis of relativistic mechanics.

Einstein's theory of relativity [1] states that the mass of a particle moving with the velocity v is

$$m = \frac{m_0}{\sqrt{1 - \frac{v^2}{c^2}}}. \tag{7.1}$$

Multiplying this effective mass by c^2 and developing the denominator in a series, we can rewrite the equation as

$$mc^2 = m_0c^2 + \frac{1}{2}m_0v^2 + \frac{3}{8}m_0\frac{v^4}{c^2} + \cdots \tag{7.2}$$

and read it as follows: A particle at rest has the mass m_0 ("rest mass") representing an energy content m_0c^2, which we tap, for example, in the atom bomb. Speeding the particle up to a velocity $v \ll c$ means adding the kinetic energy $\frac{1}{2}m_0v^2$ in accordance with Newtonian mechanics. For a photon which does not possess a rest mass and which travels with the velocity of light the left-hand side of the equation applies. Its energy is

$$\mathcal{E} = h\nu = mc^2; \tag{7.3}$$

hence we may assign to a photon a mass

$$m = \frac{\mathcal{E}}{c^2} = \frac{h\nu}{c^2} = 7.36 \times 10^{-51}\nu \quad \text{[kg]}, \tag{7.4}$$

and a momentum

$$p = mc = \frac{h\nu}{c} = \frac{h}{\lambda}. \tag{7.5}$$

This result can be checked experimentally by the momentum transfer observed in the scattering of X-rays, by the *Compton effect* [2] (Fig. 7.1) (see also Appendix A, II, 1).

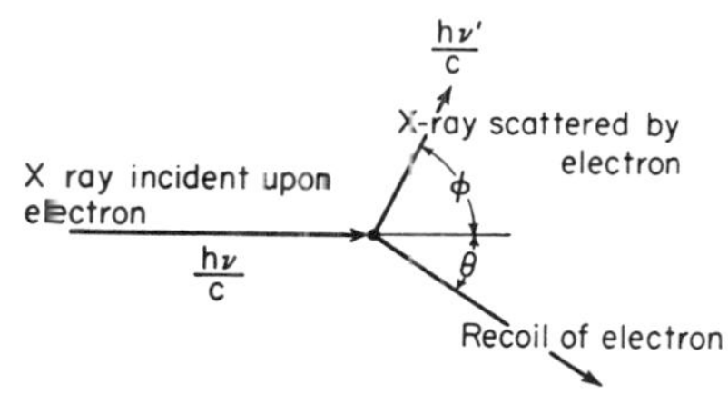

Fig. 7.1. The Compton effect.

According to Eqs. 7.3 to 7.5, the behavior of light may be discussed either from the particle standpoint by the energy and momentum of photons or from the

[1] A. Einstein, *Ann. Physik 17*, 891 (1905).

[2] A. H. Compton, *Phys. Rev. 21*, 715; *22*, 409 (1923).

wave standpoint by the frequency and wavelength of the corresponding electromagnetic wave. De Broglie[3] now postulated that this dualism for photons traveling with the velocity of light also holds for particles of much smaller velocities v and possessing a true rest mass m_0. He assumed that the quantum equations for the total energy and the momentum

$$\mathcal{E} = h\nu \qquad p = mv = \frac{h}{\lambda} \tag{7.6}$$

have universal validity. According to these *de Broglie equations,* the motion of any particle is correlated to a wave phenomenon which statistically prescribes the motion of the particle by its wave patterns.

This revolutionary speculation was soon confirmed by the electron diffraction experiment of Davisson and Germer,[4] who found anomalies in the reflection pattern of electrons from a nickel target which could be traced to the accidental existence of larger single crystals in the target metal. The single crystals produced diffraction spots as in X-ray diffraction, and a corresponding calculation of this interference phenomenon led to a wavelength of the scattered electrons

$$\lambda = \frac{h}{mv} = \frac{h}{p}, \tag{7.7}$$

in agreement with de Broglie's assumption.

Today, electron-diffraction analysis has become a standard research tool, and it seems rather puzzling that this striking phenomenon could elude detection for over twenty-five years. Interference experiments have also been performed with molecular beams of hydrogen and helium[5] and with neutrons.[6] Only the very short wavelengths of heavier particles prevent the demonstration of their wave nature.

In X-ray and electron diffraction the photon and the particle interference patterns appear in complete analogy as a probability distribution of many bullets over some target area. This fact tends to obscure a fundamental difference between electromagnetic waves and de Broglie waves. Electromagnetic waves contain information about the statistical behavior and the energy of *many* photons. The radiation intensity, as expressed by the Poynting vector (see I, Eq. 11.12), represents the average number of photons traversing a unit area per second, the density of the photon stream, times its quantum energy as

$$|\mathbf{S}| = \bar{N}h\nu \quad [\text{watt } m^{-2}]. \tag{7.8}$$

If the number $\bar{N}$ becomes too small, violent statistical fluctuations will take place, and the electromagnetic theory loses its validity. A de Broglie wave, on the other hand, is associated with the location of an *individual* particle. Where it is strong, the particle is likely to be found; where it is weak, the particle will normally not be. The intensity of the de Broglie wave at any point in space represents the probability of finding the particle at that point.

Newtonian mechanics assumed that, once the initial conditions are given, the position and the momentum of a particle are precisely defined at all values of time. It now appears that this claim must be limited to the macroscopic world. Just as geometrical optics is superseded by wave optics, Newtonian mechanics is superseded by the new *wave mechanics* based on the de Broglie equations whenever the wavelength becomes comparable to some significant dimension of the problem.

In the experiences of daily life, where the meter, kilogram, and second are appropriate yardsticks for the dimensions encountered, the wavelength proves completely negligible. From Eqs. 5.32 and 7.7 we obtain for the mass of 1 kg moving with the velocity of 1 m/sec the wavelength

$$\lambda = \frac{h}{1 \times 1} = 6.623 \times 10^{-34} \quad [\text{m}]. \tag{7.9}$$

The small mass of electrons, on the other hand, makes the application of wave mechanics (quantum mechanics) imperative. And it is this wave aspect of electrons which gives the stationary states of the Rutherford atom a reasonable interpretation. If the position of a circling electron is described by a probability wave, it can obviously persist only in orbits in which the de Broglie wave can set up a standing wave pattern instead of destroying itself by interference (Fig. 7.2). By prescribing that the circumference of the circular orbits of the hydrogen atom must correspond to an integral number of wavelengths, we arrive at the stability condition

$$2r\pi = n\lambda, \qquad \text{with } n = 1, 2, 3 \; \ldots . \tag{7.10}$$

If we introduce, in place of the wavelength, the momentum, according to de Broglie's equation, Eq. 7.10

[3] L. de Broglie, *Phil. Mag.* [6], *47*, 446 (1924); *Ann. Physik 3*, 22 (1925).

[4] C. Davisson and L. H. Germer, *Phys. Rev. 30*, 705 (1927); *Proc. Natl. Acad. Sci. 14*, 317 (1928).

[5] F. Knauer and O. Stern, *Z. Physik 53*, 779 (1929); I. Estermann and O. Stern, *ibid. 61*, 95 (1930); I Estermann, R. Frisch, and O. Stern, *ibid. 73*, 348 (1932).

[6] C. G. Shull, E. O. Wollan, W. C. Koehler, *Phys. Rev. 84*, 912 (1951).

reverts immediately to Bohr's quantization condition of the rotator

$$p' = n\hbar. \tag{6.12}$$

Although this outcome is gratifying, it should not be misunderstood. The stringent assignment of an electron of a precise momentum to a precise orbit, as pictured in the Rutherford atom and Bohr's theory, contradicts the spirit of quantum physics. This was realized by Heisenberg,[7] Bohr, and others, before the advent of wave mechanics, through statistical considerations concerning the interaction of particles with the measuring devices of an observer. The existence of Planck's quantum of action introduces an uncertainty in the description of a particle as far as a simultaneous definition of location and momentum are concerned, and consequently a similar uncertainty in its description by de Broglie waves. We shall come back to this *uncertainty principle* (see Sec. 8).

De Broglie's theory had been largely of a qualitative nature. To make use of the wave concept for a quantitative discussion of molecular phenomena requires the formulation of a basic wave equation. This task was first seen and solved by Schrödinger.[8] We have already become familiar with wave equations in the case of the electromagnetic field (see I, Sec. 7). Electric and magnetic plane waves, for example, moving towards the $+x$-direction without attenuation as

$$\mathbf{E} = \mathbf{E}_0 e^{j\left(\omega t - \frac{2\pi}{\lambda}x\right)} \qquad \mathbf{H} = \mathbf{H}_0 e^{j\left(\omega t - \frac{2\pi}{\lambda}x\right)} \tag{7.11}$$

were derived (see I, Eqs. 7.9 to 7.16) as solutions of the differential equations

$$\frac{\partial^2 \mathbf{E}}{\partial x^2} + \left(\frac{2\pi}{\lambda}\right)^2 \mathbf{E} = 0 \qquad \frac{\partial^2 \mathbf{H}}{\partial x^2} + \left(\frac{2\pi}{\lambda}\right)^2 \mathbf{H} = 0. \tag{7.12}$$

In analogy, a de Broglie wave traveling in the $+x$-direction may be described by a wave function

$$\psi = A e^{j\left(\omega t - \frac{2\pi}{\lambda}x\right)} \tag{7.13}$$

which obviously is a solution of the wave equation

$$\frac{\partial^2 \psi}{\partial x^2} + \left(\frac{2\pi}{\lambda}\right)^2 \psi = 0. \tag{7.14}$$

The *frequency* of this wave, according to de Broglie's

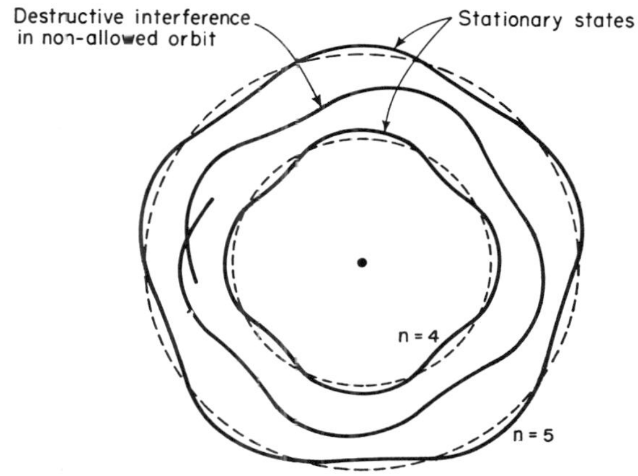

Fig. 7.2. Stationary states and destructive interference.

equations, is given by the *total* energy of the particle

$$\mathcal{E} = h\nu, \tag{7.15}$$

whereas the *wavelength* is prescribed by the particle momentum p or its *kinetic* energy as

$$\lambda = h/p, \qquad \mathcal{E}_{\text{kin}} = \frac{1}{2}mv^2 = \frac{p^2}{2m} = \frac{h^2}{2m\lambda^2}. \tag{7.16}$$

Since the kinetic energy of a particle is the difference of its total and potential energy,

$$\mathcal{E}_{\text{kin}} = \mathcal{E} - U, \tag{7.17}$$

we may substitute this energy difference in Eq. 7.14 for the wavelength and obtain

$$\frac{\partial^2 \psi}{\partial x^2} + \frac{2m}{\hbar^2}(\mathcal{E} - U)\psi = 0. \tag{7.18}$$

Finally, if the particle is free to move in any direction, the space derivative $\frac{\partial^2 \psi}{\partial x^2}$ has to be replaced by the Laplace operator (see I, Eqs. 7.6 and 7.7). Thus the famous *Schrödinger equation* results:

$$\nabla^2 \psi + \frac{2m}{\hbar^2}(\mathcal{E} - U)\psi = 0. \tag{7.19}$$

The solutions of Schrödinger's equation give the pattern of the wave function ψ in space. The square of its absolute value, the product of ψ and its conjugate $\tilde{\psi}$, in a space element dV, represents the probability of finding the particle in that element, as Born[9] first pointed out. The integral of this intensity of the probability wave extended over all possible space must be unity, that is,

$$\oint \psi\tilde{\psi}\, dV = 1, \tag{7.20}$$

[7] W. Heisenberg, *Z. Physik 33*, 879 (1925); *43*, 172 (1927).

[8] E. Schrödinger, *Ann. Physik 79*, 361, 489 (1926).

[9] M. Born, *Z. Physik 37*, 863 (1926).

since the particle, with certainty, must be somewhere in space. This equation is called the *normalization condition* of the de Broglie wave. Obviously it requires that ψ be finite in space; furthermore, the second derivative of ψ with respect to space must be finite because it represents the kinetic energy according to Eq. 7.19. Hence ψ and its first space derivative must be continuous. This is a very important and restricting condition in wave mechanics. It prescribes that, at a boundary, two wave functions have to be joined in such a manner that their amplitudes and slopes are continuous, and is therefore known as the *joining condition.*

If a time dependence of the wave phenomenon has to be considered, we can introduce it into the Schrödinger equation by recalling that

$$\frac{\partial \psi}{\partial t} = j\omega\psi, \tag{7.21}$$

and, according to de Broglie's equations,

$$\omega = \mathcal{E}/\hbar. \tag{7.22}$$

Thus, by substituting

$$\mathcal{E}\psi = -j\hbar \frac{\partial \psi}{\partial t}, \tag{7.23}$$

in the Schrödinger equation and rearranging, we obtain the *Schrödinger wave equation,*†

$$\nabla^2 \psi - \frac{2m}{\hbar^2} U\psi = j \frac{2m}{\hbar} \frac{\partial \psi}{\partial t}. \tag{7.24}$$

† It should be noted that, in contrast to Maxwell's wave equations and to the original derivation of de Broglie's, the Schrödinger equations are nonrelativistic, since they equate $h\nu$ to the classical total energy $\mathcal{E}$ of the particle.

Maxwell's equations have been solved accurately for a very large variety of conditions. Schrödinger applied his equations with complete success to a treatment of the hydrogen atom, the linear oscillator, and various other problems, and they have been tested successfully by many outstanding contributors in progressively more difficult situations. We seem to be justified today in assuming that the wave nature of particles, as formulated, represents a fundamental law of nature, and that the theory of quantum mechanics founded on Schrödinger's equations could give us a quantitatively correct description of the molecular world. But, in contrast to the electromagnetic field equations, Schrödinger's equations can be solved accurately only in very few cases.

The fundamental difficulty lies in the fact that, although the photons of the electromagnetic field do not interact in general, the particles of wave mechanics affect each other with strong forces. Thus an electron moving in the Coulomb field of other charged particles may preserve its total energy and with it the frequency of its de Broglie wave. However, its kinetic energy and thus the wavelength of the electron will vary from point to point in space. In optics we are normally concerned with homogeneous dielectrics; de Broglie waves traverse dispersive media in which the index of refraction varies from point to point. Simultaneously, we face, in a new language, the old difficulty of classical mechanics: the two-body problem can be solved rigorously, the multibody problem can be treated only by approximations.

These are the reasons why quantum mechanics has not progressed faster and why it puts us continuously on guard lest the approximations that are chosen neglect essential aspects of the actual physical situation.

8 · The Structure of Atoms

Quantum mechanics introduces an unaccustomed aspect of *complementarity* into the description of the molecular world. The position of particles is determined by de Broglie probability waves, as interference experiments testify. The stationary states of electrons in atoms correspond therefore to standing wave modes, to *electron clouds* of characteristic sizes and shapes, and the electric charge of the electrons appears smeared out over these probability patterns. In impact experiments, on the other hand, as, for instance, observed in the Wilson cloud chamber or by scintillation, the particle makes its appearance with its concentrated electric charge and a definite momentum. Both the particle and the wave picture describe one and the same physical entity from different standpoints. The two aspects are not contradictory but complementary; the only limitation is that we cannot focus our attention on both of them simultaneously with an arbitrary accuracy.

The existence of an uncertainty in the definition of correlated parameters is a typical feature of any wave phenomenon. A wave function

$$\psi = Ae^{-j\frac{2\pi}{\lambda_0}x}, \tag{8.1}$$

for example, describes a monochromatic wave traveling in the $+x$-direction. If it represents the de Broglie probability wave of a particle, the momentum of this particle is accurately determined by the wavelength λ_0, but the intensity of the wave

$$\psi\tilde{\psi} = A^2 \tag{8.2}$$

is the same from $x = -\infty$ to $x = +\infty$ and thus leaves the position of the particle completely undefined. If we want to localize its position near x_0, we have to quench the probability pattern outside of x_0.

This localization may be achieved by superposing on the constant intensity of the wave a Gaussian error function centered at x_0, that is, by altering the wave function to

$$\psi = A'e^{-\frac{1}{2}\left(\frac{x-x_0}{b}\right)^2} e^{-j\frac{2\pi}{\lambda_0}x} \tag{8.3}$$

with the intensity

$$\psi\tilde{\psi} = A'e^{-\left(\frac{x-x_0}{b}\right)^2}. \tag{8.4}$$

The particle is now confined to a *wave packet* of the half-width †

$$2\,\Delta x = 2b, \tag{8.5}$$

but we have destroyed, in the process of localization, the accuracy with which its wavelength, that is, its momentum, is known. Fourier analysis of such a finite wave train shows, as already discussed in Sec. 4 for the half-width of a spectral line emitted by a damped oscillator, that the intensity of the wave packet plotted on a $2\pi/\lambda$-scale does not peak sharply at $2\pi/\lambda_0$ but spreads around this point in a bell-shaped resonance characteristic. For the wave packet produced by a Gaussian error curve in space, the $2\pi/\lambda$-characteristic is again a Gaussian of the half-width

$$2\Delta\left(\frac{2\pi}{\lambda}\right) = \frac{2}{b} \tag{8.6}$$

(Fig. 8.1). The product of both uncertainties of defining the position and the wavelength of the wave packet has the constant value

$$\Delta x\,\Delta\left(\frac{2\pi}{\lambda}\right) = 1. \tag{8.7}$$

It can be shown that the choice of any other quenching function for the construction of the wave packet leads to even larger uncertainties, since only a Gaussian error function, as a Hermitian function, has the symmetry property of being converted into an identical function by Fourier transformation.

† In this case we define the half-width as the distance between the points at which the intensity has fallen to $1/e$ of its maximum value.

A corresponding discussion can be carried through for a probability wave

$$\psi = Ae^{j\omega t} \tag{8.8}$$

as a function of time and frequency; it leads to the uncertainty relation

$$\Delta t\,\Delta\omega = 1. \tag{8.9}$$

By admitting arbitrary quenching functions we can formulate the *uncertainty principle* that prescribes the

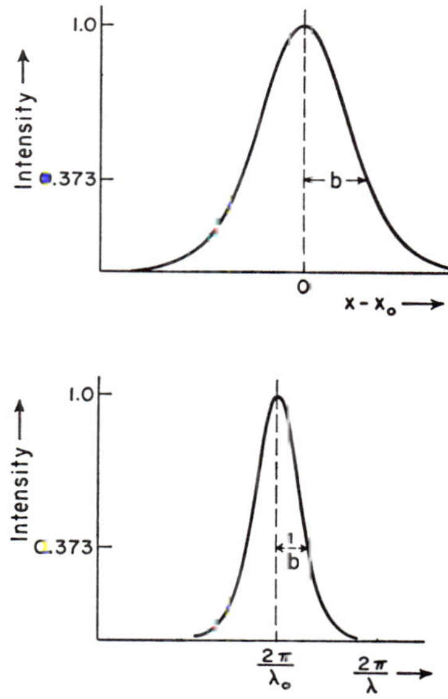

Fig. 8.1. Wave packet constructed with Gaussian distribution in x and $2\pi/\lambda$ scale.

definition of correlated parameters of wave phenomena:

$$\begin{aligned} \Delta x\,\Delta\left(\frac{2\pi}{\lambda}\right) &\geq 1, \\ \Delta t\,\Delta\omega &\geq 1. \end{aligned} \tag{8.10}$$

Quantum mechanics introduces the momentum p for the wavelength and the total energy $\mathcal{E}$ for the frequency according to de Broglie's equations (Eqs. 7.6). Thus the Heisenberg uncertainty principle results:

$$\begin{aligned} \Delta q\,\Delta p &\geq \hbar, \\ \Delta t\,\Delta\mathcal{E} &\geq \hbar. \end{aligned} \tag{8.11}$$

The total energy of a particle at a given time or its momentum at a given location cannot be determined more accurately than Planck's quantum of action allows.

By applying this uncertainty principle to the hydrogen atom, the binding energy of the electron can be obtained.[1] The total energy of the electron at a distance r from the proton is

$$\mathcal{E} = U + \mathcal{E}_{\text{kin}} = -\frac{e^2}{\epsilon_0 4\pi r} + \frac{p^2}{2m}. \tag{8.12}$$

This energy becomes a minimum when r and p assume their smallest possible values. The average distance from the nucleus and the average magnitude of the momentum cannot be smaller than their uncertainties Δ_r and Δ_p, that is, the smallest average values of the parameters are coupled by the uncertainty relation as

$$\bar{r} = \frac{\hbar}{\bar{p}}. \tag{8.13}$$

The energy, expressed as the function of this average distance,

$$\mathcal{E} = -\frac{e^2}{\epsilon_0 4\pi \bar{r}} + \frac{\hbar^2}{2m\bar{r}^2}, \tag{8.14}$$

reaches its minimum for $d\mathcal{E}/d\bar{r} = 0$ at

$$\bar{r}_{\min} = \frac{\epsilon_0 h^2}{\pi m e^2} = 0.528 \text{ A}; \tag{8.15}$$

this minimum average distance is identical with the radius r_0 of the fundamental Bohr orbit (see Eq. 6.10). The total energy in this ground state,

$$\mathcal{E}_H = -\frac{me^4}{\epsilon_0{}^2 8h^2} = -Rhc \simeq -13.5 \quad [\text{ev}], \tag{8.16}$$

represents the *ionization energy* of the hydrogen atom; it serves under the name *Rydberg energy* as a convenient yardstick for the measurement of binding energies.

An actual calculation of the stationary states or the standing-wave modes of the hydrogen atom requires that we introduce the potential energy U of the Coulomb field in the Schrödinger equation (Eq. 7.19). Thus we have to solve the equation

$$\nabla^2\psi + \frac{2m}{\hbar^2}\left(\mathcal{E} + \frac{e^2}{\epsilon_0 4\pi r}\right)\psi = 0 \tag{8.17}$$

under the condition that the wave function ψ be single-valued, continuous, and finite everywhere in space. It proves to have solutions for all positive values of the total energy $\mathcal{E}$, but only for certain negative *eigenvalues* which are identical with those derived from the Bohr theory (see Eqs. 6.4 and 6.10),

$$\mathcal{E}_n = -\frac{Rhc}{n^2}, \qquad \text{with } n = 1, 2, 3 \ldots. \tag{8.18}$$

[1] See F. O. Rice and E. Teller, *The Structure of Matter*, John Wiley and Sons, New York, 1949, p. 7.

This means that, while the electron is trapped in the Coulomb field of the proton, as in a wave guide, only selected standing-wave modes can form. Above the ionization level, however, the electron can move in unbounded space with any energy. Spectroscopically, this is verified by the continuum which sets in at the ionization limit (see Fig. 5.2).

In the hydrogen atom, the electron moves in a central field, Schrödinger's equation has to be solved for spherical co-ordinates, and the wave functions have to be expressed in spherical harmonics.[2] The essential facts of

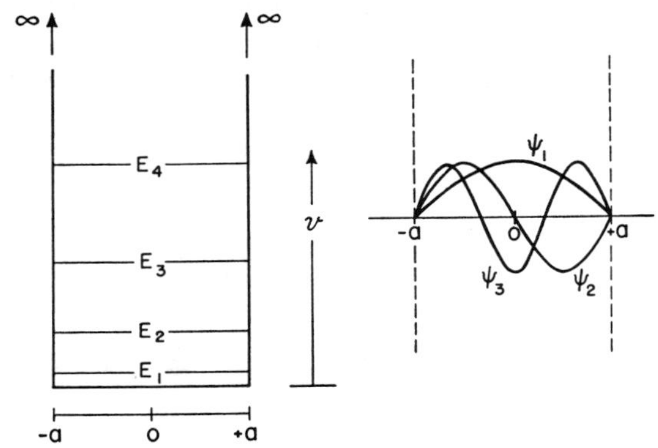

Fig. 8.2. Standing waves in rectangular potential well of infinitely high walls.

the situation, however, can be made clear by adhering to trigonometric functions as previously and referring to the example of a rectangular one-dimensional potential well. In the case of such a well, enclosed by infinitely high walls (Fig. 8.2), an infinite number of energy states exists, characterized by a sequence of simple, standing-wave functions $\cos (m\pi/a)x$ and $\sin (n\pi/a)x$, with $m = \frac{1}{2}, \frac{3}{2}, \frac{5}{2}, \ldots$ and $n = 1, 2, 3 \ldots.$ This situation is completely analogous to the parallel-plane wave guide of I, Sec. 22.

If the walls are cut down to finite height, obviously only a finite number of energy states can be accommodated before the top of the well, the ionization limit, is reached. Above it, the boundary conditions disappear and any energy state is possible (Fig. 8.3). However, on account of the finite wall height, a new aspect enters the situation: the electron has a finite probability of escaping from the well (cf. tunnel effect, Sec. 14). This probability obviously increases the higher the energy state, that is, the nearer the top, and has the consequence that the ψ wave functions are not anymore terminated by a node at the wall. The periodic wave functions in the well have to be joined to exponentially decaying wave functions in the wall according

[2] See, for example, J. C. Slater, *Quantum Theory of Matter*, McGraw-Hill Book Co., New York, 1951, Chapter 5.

to the joining condition (see Sec. 7). Expressed in optical language: the walls cease to be totally reflecting (see I, Sec. 19).

The well of the Coulomb field has some aspects of both situations. Its depth is infinite towards the nucleus but finite towards the outside. Hence, an infinite number

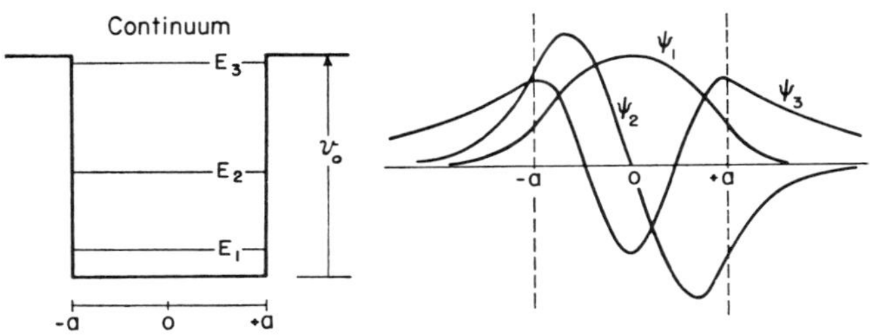

Fig. 8.3. Energy states and wave functions for rectangular well of finite height.

of bound states exists and also an ionization limit followed by a continuum of allowed states. Since the well of the Coulomb field resembles a funnel, infinitely deep below the energy states but of only finite height above them, the states are crowding together towards the ionization limit and not towards the bottom of the well as in Figs. 8.2 and 8.3.

For the ground state of the hydrogen atom, characterized by the total quantum number $n = 1$, the wave function proves to be a simple exponential

$$\psi = Ce^{-r/r_0}. \tag{8.19}$$

The *normalizing factor* C adjusts the integral of the intensity over all space to unity (see Eq. 7.20). The probability of finding the electron, $\psi\tilde{\psi}$, has its maximum in the nucleus ($r = 0$) and falls exponentially towards the outside. The probability, on the other hand, of finding the electron in a spherical shell at a distance between r and $r + dr$ from the proton,

$$P(r)\,dr = C^2e^{-2r/r_0}4\pi r^2\,dr, \tag{8.20}$$

reaches its maximum for $dP(r)/dr = 0$ at the Bohr orbit $r = r_0$ (Fig. 8.4). The wave function in this fundamental state depends only on the distance between the electron and nucleus; it is a spherically symmetrical or *s function.* The electron in such a state is designated as an *s electron.* More specifically, by adding the integer of the fundamental quantum number as a prefix, we denote the electron in the fundamental state of the hydrogen atom as a "1s" electron.

Although Schrödinger's equation has for the energy $\mathcal{E}_1$ of the ground state only this one solution, the first excited state ($n = 2$) with $\mathcal{E}_2 = \frac{1}{4}\mathcal{E}_1$ proves to lead to four different wave functions of the same energy or to be *fourfold degenerate.* One of them is again spherically symmetrical; it is the 2s function

$$\psi_s = \left(\frac{r}{r_0} - 2\right)e^{-r/2r_0}. \tag{8.21}$$

The other three have directional properties, stand normal to each other as

$$\begin{aligned}\psi_x &= xe^{-r/2r_0},\\ \psi_y &= ye^{-r/2r_0},\\ \psi_z &= ze^{-r/2r_0},\end{aligned} \tag{8.22}$$

and are called *p functions.* The various substates belonging to the same principal quantum number n are designated as *orbitals.* For $n = 1$, we had only the 1s orbital; for $n = 2$ we find one 2s orbital and three 2p orbitals. Figure 8.5 schematically illustrates the electron clouds of these s and p orbitals.

The multiplicity of the wave functions and orbitals increases systematically. Nine wave functions belong to the second excited state ($n = 3$); it is *ninefold degenerate* and consists of one 3s, three 3p, and five 3d orbitals. And so it continues, with the general rule

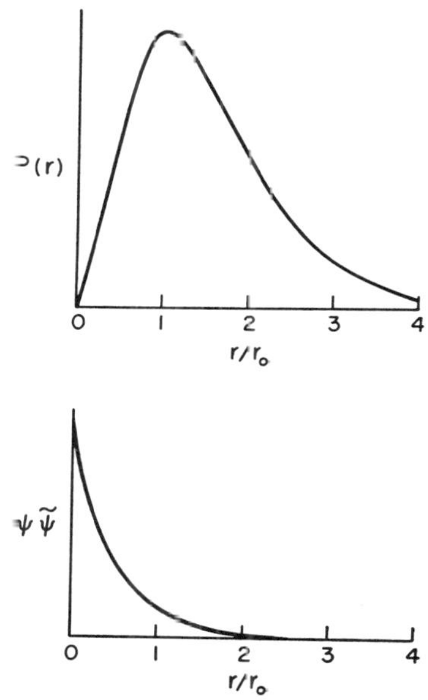

Fig. 8.4. Radial probability function and probability density function of 1s electron in hydrogen atom.

that the state of the fundamental quantum number n is n^2-fold degenerate and that its wave functions and orbitals are composed of n groups, made up by one s, three p, five d, seven f orbitals, etc., adding up to a total of n^2.

Spherical electron clouds, that is, s orbitals, have obviously no angular momenta, but the electron motion in the less symmetrical, $(n - 1)$ configurations leads to definite angular momenta p'. In analogy to the elliptical orbits in Bohr's theory (see Eq. 6.20), these eccentric electron clouds need, in addition to the total quantum number n, a second quantum number, the *azimuthal quantum number* l, for their complete characterization. In slight deviation from Bohr's quantum con-

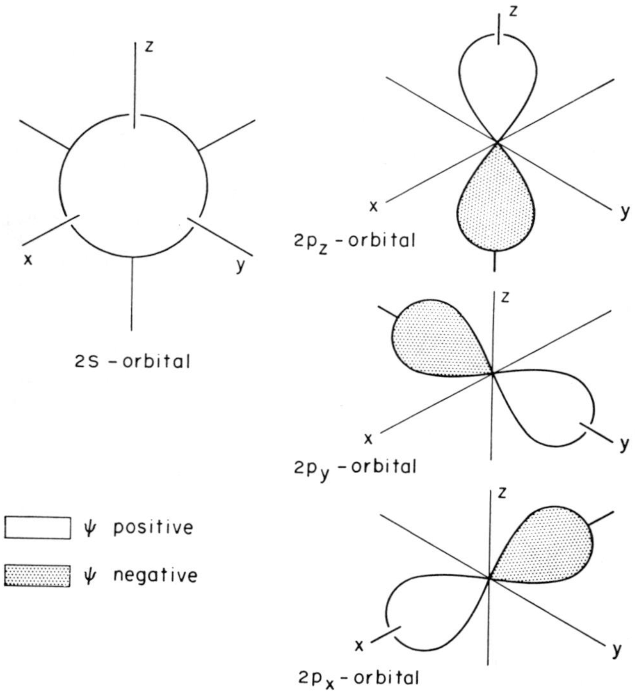

Fig. 8.5. Schematic representation of the angular space dependence of the electron density distribution of s and p orbitals.

dition for the rotator (see Eq. 6.12), wave mechanics leads to the quantization condition

$$p' = \sqrt{l(l+1)}\hbar. \tag{8.23}$$

The azimuthal quantum number assumes the values

$$l = 0, 1, 2, 3 \ldots (n-1), \tag{8.24}$$

where $l = 0$ designates s orbitals, $l = 1$ designates p orbitals, etc.

A net angular momentum of an electron cloud corresponds classically to a d-c current moving in a closed orbit. According to Ampère's circuital law (see I, Eqs. 5.5 and 5.28) such a current creates a magnetic dipole moment. Hence, coupled to the angular mechanical moments p' of electron clouds there must be *permanent magnetic moments*. If an atom containing such electron clouds is brought into a magnetic field, the torque of the field tends to orient the moments in the field direction (see I, Eq. 2.16). However, since the magnetic moments are coupled to mechanical moments, that is, since they point in the axis direction of a gyroscope, the torque of the magnetic field leads to a precession of the momentum vector around the field axis. The *angular momentum of precession* describing this additional periodic motion has to be quantized according to wave mechanics in agreement with Sommerfeld's quantum condition (Eq. 6.19), as

$$p_z' = \textcircled{m}\hbar. \tag{8.25}$$

The new *magnetic quantum number* $\textcircled{m}$ varies from

$$\textcircled{m} = l, l-1, \ldots, -(l-1), -l, \tag{8.26}$$

that is, it prescribes for p-states $(l = 1)$ three orbitals $(1,0,-1)$, for d-states $(l = 2)$ five orbitals $(2,1,0,-1,-2)$, etc. The precession caused by the magnetic field removes the degeneracy of the states and gives each individual orbital its distinguishing energy. Thus, any orbital of the hydrogen atom can be uniquely described by the three quantum numbers n, l, and $\textcircled{m}$. The removal of the degeneracy can be observed experimentally by the splitting of a spectral line into fine-structure components; this is the *Zeeman effect* [3] (Fig. 8.6).

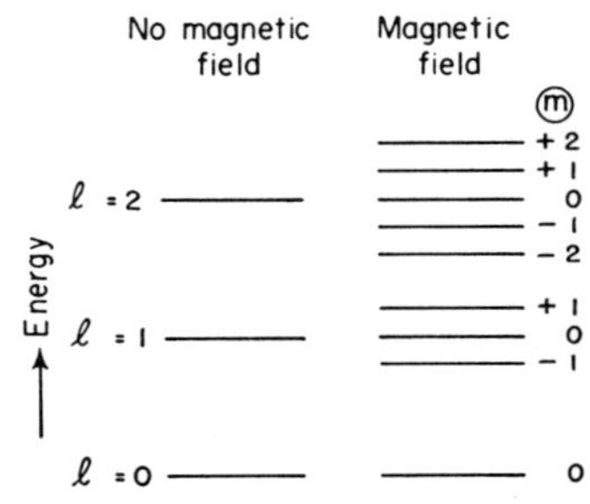

Fig. 8.6. Zeeman effect of a hydrogen-like atom.

Figure 8.7 shows the corresponding electron cloud configurations with their identifying quantum numbers. It may be read as follows: The $\textcircled{m} = 0$ states,

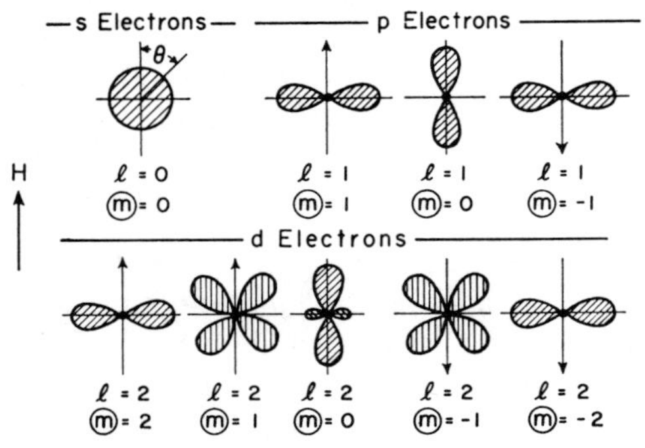

Fig. 8.7. Angular dependence of the electron density $\psi\bar{\psi}$ of some s, p, and d orbitals. (After White.[4])

seen in the field direction, give a circular symmetrical electron cloud distribution resembling an s-state and have no magnetic moment. The $\textcircled{m} \neq 0$ states have

[3] P. Zeeman, *Phil. Mag.* *43*, 226 (1897).
[4] H. E. White, *Phys. Rev.* *37*, 1416 (1931).

magnetic moments, which, in the case of $\textcircled{m} = +$ produce a parallel and for $\textcircled{m} = -$ an antiparallel component. This is indicated by arrows in Fig. 8.7.

The difference between the quantization conditions for the total angular momentum (Eq. 8.23) and the momentum vector in a specific direction (Eq. 8.25) is most important. According to the old quantum theory (Eq. 6.12) the momentum vectors of the three p orbitals, for example, would point parallel, perpendicular, and antiparallel to the external field direction (Fig. 8.8). In quantum mechanics the *components*

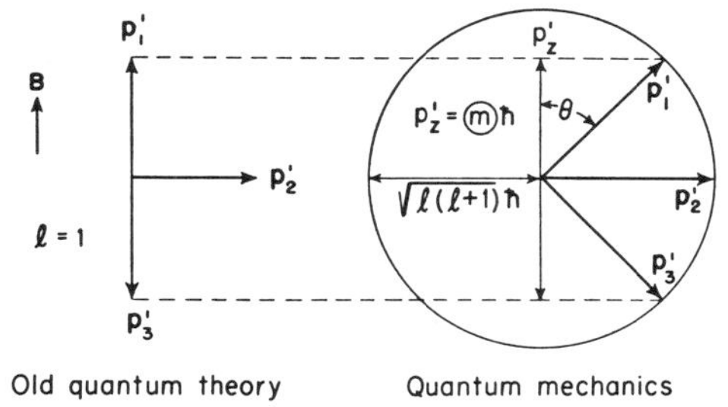

Fig. 8.8. Orientation and magnitude of total angular momentum.

have the same magnitudes as above, but the actual momentum vectors are larger and inclined by an angle θ against the field axis, where

$$\cos\theta = \frac{\textcircled{m}}{\sqrt{l(l+1)}}. \tag{8.27}$$

This situation is a consequence of the uncertainty principle, which does not allow the total angular momentum to point in any one definite direction.

The description of the stationary states of the hydrogen atom can be used to derive a first understanding of the structure of atoms and of the *periodic system of the elements.* According to the Rutherford model, the sequence of atoms in the periodic system arises when we increase the nuclear charge in successive steps of one elementary charge e and simultaneously add for each step one compensating electron to the electronic system. If we would increase the nuclear charge of the hydrogen atom from e to Ze, the potential energy of the Coulomb field would rise to

$$U = -\frac{Ze^2}{\epsilon_0 4\pi r}, \tag{8.28}$$

the Bohr radii would shrink to

$$r_n = \frac{n^2 r_0}{Z}, \tag{8.29}$$

and the eigenvalues of the Schrödinger equation (see Eqs. 8.16 and 8.18) would increase to

$$\mathcal{E}_n = -\frac{Z^2 Rhc}{n^2} = \mathcal{E}_H\left(\frac{Z}{n}\right)^2. \tag{8.30}$$

The type and number of the orbitals, however, would stay the same. To build up the various atoms, we place electrons in these orbitals until the nuclear charge is neutralized.

To justify this procedure we have to assume that the various electrons of an atom move approximately as if a central field acted on each of them. This central field is composed of the Coulomb field of the nucleus, weakened by an average charge of all the other electrons; this average negative charge increases with the distance from the nucleus, since more and more of the electronic charge is encircled according to Gauss's law (see I, Eq. 4.2). Only in a Coulomb field is the energy of circular and elliptical orbits of the same principal quantum number identical (see Eq. 6.25). This degeneracy is removed in a central field; while we operate still with hydrogen-like wave functions, each orbital in multi-electron systems already has a different energy without an external magnetic field.

Before the construction of atoms can begin, a final question has to be settled: How many electrons will an orbital accommodate? The answer to this riddle came from two different sources.

(1) Uhlenbeck and Goudsmit [5] concluded from spectroscopic data that an electron possesses an angular momentum of rotation, an *electron spin,* which has to be quantized according to quantum mechanics (see Eq. 8.23) as

$$p' = \sqrt{s(s+1)}\hbar, \tag{8.31}$$

where $s = 1/2$. The electron spin imparts to an electron a magnetic dipole moment which can orient in a

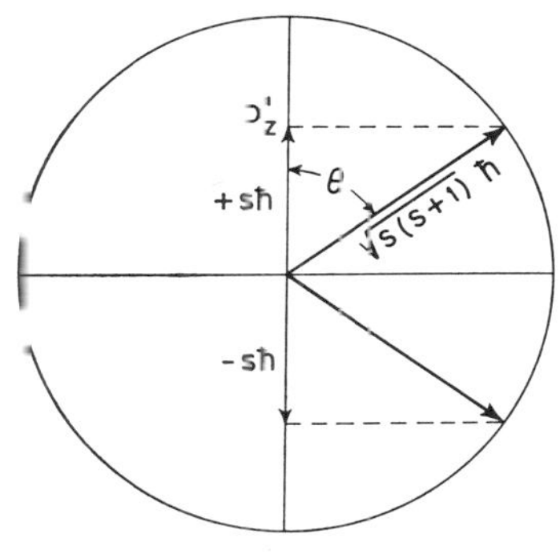

Fig. 8.9. Orientation and magnitude of spin momentum.

magnetic field. The mechanical spin component in the field direction is, in analogy to Eq. 8.25,

$$p_z' = s\hbar \tag{8.32}$$

and may stand parallel or antiparallel (Fig. 8.9).

[5] G. E. Uhlenbeck and S. Goudsmit, *Naturwiss. 13,* 953 (1925); *Nature 117,* 264 (1926).

TABLE 8.1. PERIODIC SYSTEM OF THE

n	Shell	I	II
1	K	1 H Hydrogen 1.0081 1 $^{2}S_{1/2}$	
1 2	K L	3 Li Lithium 6.940 2 1 $^{2}S_{1/2}$	4 Be Beryllium 9.02 2 2 $^{1}S_{0}$
1 2 3	K L M	11 Na Sodium 22.997 2 2 6 1 $^{2}S_{1/2}$	12 Mg Magnesium 24.32 2 2 6 2 $^{1}S_{0}$

n	Shell	Ia	IIa	IIIa	IVa	Va	VIa	VIIa	VIII	VIII
1 2 3 4	K L M N	19 K Potassium 39.096 2 2 6 2 6 1 $^{2}S_{1/2}$	20 Ca Calcium 40.08 2 2 6 2 6 2 $^{1}S_{0}$	21 Sc Scandium 45.10 2 2 6 2 6 1 2 $^{2}D_{3/2}$	22 Ti Titanium 47.90 2 2 6 2 6 2 2 $^{3}F_{2}$	23 V Vanadium 50.95 2 2 6 2 6 3 2 $^{4}F_{3/2}$	24 Cr Chromium 52.01 2 2 6 2 6 4 2 $^{7}S_{3}$	25 Mn Manganese 54.93 2 2 6 2 6 5 2 $^{6}S_{5/2}$	26 Fe Iron 55.84 2 2 6 2 6 6 2 $^{5}D_{4}$	27 Co Cobalt 58.94 2 2 6 2 6 7 2 $^{4}F_{9/2}$
1 2 3 4 5	K L M N O	37 Rb Rubidium 85.48 2 2 6 2 6 10 2 6 1 $^{2}S_{1/2}$	38 Sr Strontium 87.63 2 2 6 2 6 10 2 6 2 $^{1}S_{0}$	39 Y Yttrium 88.92 2 2 6 2 6 10 2 6 1 2 $^{2}D_{3/2}$	40 Zr Zirconium 91.22 2 2 6 2 6 10 2 6 2 2 $^{3}F_{2}$	41 Nb Niobium 92.91 2 2 6 2 6 10 2 6 4 1 $^{6}D_{1/2}$	42 Mo Molybdenum 95.95 2 2 6 2 6 10 2 6 5 1 $^{7}S_{3}$	43 Tc Technetium 99 2 2 6 2 6 10 2 6 6 1	44 Ru Ruthenium 101.7 2 2 6 2 6 10 2 6 7 1 $^{5}F_{5}$	45 Rh Rhodium 102.91 2 2 6 2 6 10 2 6 8 1 $^{4}F_{9/2}$
1 2 3 4 5 6	K L M N O P	55 Cs Cesium 132.91 2 2 6 2 6 10 2 6 10 2 6 1 $^{2}S_{1/2}$	56 Ba Barium 137.36 2 2 6 2 6 10 2 6 10 2 6 2 $^{1}S_{0}$	57 La Lanthanum 138.92 2 2 6 2 6 10 2 6 10 2 6 1 2 $^{2}D_{3/2}$	58 Ce Cerium 140.13 2 2 6 2 6 10 2 6 10 1 2 6 1 2	59 Pr Praseodymium 140.92 2 2 6 2 6 10 2 6 10 2 2 6 1 2	60 Nd Neodymium 144.27 2 2 6 2 6 10 2 6 10 3 2 6 1 2 $^{5}I_{4}$	61 Pm Prometheum 147 2 2 6 2 6 10 2 6 10 4 2 6 1 2	62 Sm Samarium 150.43 2 2 6 2 6 10 2 6 10 5 2 6 1 2 $^{7}F_{0}$	63 Eu Europium 152.0 2 2 6 2 6 10 2 6 10 6 2 6 1 2 $^{8}S_{7/2}$
1 2 3 4 5 6	K L M N O P				72 Hf Hafnium 178.6 2 2 6 2 6 10 2 6 10 14 2 6 2 2 $^{3}F_{2}$	73 Ta Tantalum 180.88 2 2 6 2 6 10 2 6 10 14 2 6 3 2 $^{4}F_{3/2}$	74 W Tungsten 183.92 2 2 6 2 6 10 2 6 10 14 2 6 4 2 $^{5}D_{0}$	75 Re Rhenium 186.31 2 2 6 2 6 10 2 6 10 14 2 6 5 2 $^{6}S_{5/2}$	76 Os Osmium 190.2 2 2 6 2 6 10 2 6 10 14 2 6 6 2 $^{5}D_{4}$	77 Ir Iridium 193.1 2 2 6 2 6 10 2 6 10 14 2 6 9 $^{4}F_{9/2}$
1 2 3 4 5 6 7	K L M N O P Q	87 Fr Francium 223 2 2 6 2 6 10 2 6 10 14 2 6 10 2 6 1 $^{2}S_{1/2}$	88 Ra Radium 226.05 2 2 6 2 6 10 2 6 10 14 2 6 10 2 6 2 $^{1}S_{0}$	89 Ac Actinium 227 2 2 6 2 6 10 2 6 10 14 2 6 10 2 6 1 2 $^{2}D_{3/2}$	90 Th Thorium 232.12 2 2 6 2 6 10 2 6 10 14 2 6 10 2 6 2 2 $^{3}F_{2}$	91 Pa Protoactinium 231 2 2 6 2 6 10 2 6 10 14 2 6 10 2 6 3 2	92 U Uranium 238.07 2 2 6 2 6 10 2 6 10 14 2 6 10 2 6 4 2 $^{5}L_{6}$	93 Np Neptunium 237 2 2 6 2 6 10 2 6 10 14 2 6 10 1 2 6 4 2	94 Pu Plutonium 239 2 2 6 2 6 10 2 6 10 14 2 6 10 2 2 6 4 2	95 Am Americium 241 2 2 6 2 6 10 2 6 10 14 2 6 10 3 2 6 4 2
		s p d f	s p d f	s p d f	s p d f	s p d f	s p d f	s p d f	s p d f	s p d f

ELEMENTS

			III	IV	V	VI	VII	O	Shell	n
								2 He Helium 4.003 2 $^1S_0$	K	1
			5 B Boron 10.82 2 2 1 $^2P_{1/2}$	6 C Carbon 12.010 2 2 2 $^3P_0$	7 N Nitrogen 4.008 2 2 3 $^4S_{3/2}$	8 O Oxygen 16.0000 2 2 4 $^3P_2$	9 F Fluorine 19.00 2 2 5 $^2P_{3/2}$	10 Ne Neon 20.138 2 2 6 $^1S_0$	K L	1 2
			13 Al Aluminum 26.97 2 2 6 2 1 $^2P_{1/2}$	14 Si Silicon 28.06 2 2 6 2 2 $^3P_0$	15 P Phosphorus 30.98 2 2 6 2 3 $^4S_{3/2}$	16 S Sulfur 32.06 2 2 6 2 4 $^3P_2$	17 Cl Chlorine 35.457 2 2 6 2 5 $^2P_{3/2}$	18 A Argon 39.944 2 2 6 2 6 $^1S_0$	K L M	1 2 3
	I_b	II_b	III_b	IV_b	V_b	VI_b	VII_b			
28 Ni Nickel 58.69 2 2 6 2 6 8 2 $^3F_4$	29 Cu Copper 63.57 2 2 6 2 6 10 1 $^2S_{1/2}$	30 Zn Zinc 65.38 2 2 6 2 6 10 2 $^1S_0$	31 Ga Gallium 69.72 2 2 6 2 6 10 2 1 $^2P_{1/2}$	32 Ge Germanium 72.60 2 2 6 2 6 10 2 2 $^3P_0$	33 As Arsenic 74.91 2 2 6 2 6 10 2 3 $^4S_{3/2}$	34 Se Selenium 78.96 2 2 6 2 6 10 2 4 $^3P_2$	35 Br Bromine 79.916 2 2 6 2 6 10 2 5 $^2P_{3/2}$	36 Kr Krypton 83.7 2 2 6 2 6 10 2 6 $^1S_0$	K L M N	1 2 3 4
46 Pd Palladium 106.7 2 2 6 2 6 10 2 6 10 $^1S_0$	47 Ag Silver 107.88 2 2 6 2 6 10 2 6 10 1 $^2S_{1/2}$	48 Cd Cadmium 112.41 2 2 6 2 6 10 2 6 10 2 $^1S_0$	49 In Indium 114.76 2 2 6 2 6 10 2 6 10 2 1 $^2P_{1/2}$	50 Sn Tin 118.70 2 2 6 2 6 10 2 6 10 2 2 $^3P_0$	51 Sb Antimony 121.76 2 2 6 2 6 10 2 6 10 2 3 $^4S_{3/2}$	52 Te Tellurium 127.61 2 2 6 2 6 10 2 6 10 2 4 $^3P_2$	53 I Iodine 126.92 2 2 6 2 6 10 2 6 10 2 5 $^2P_{3/2}$	54 Xe Xenon 131.3 2 2 6 2 6 10 2 6 10 2 6 $^1S_0$	K L M N O	1 2 3 4 5
64 Gd Gadolinium 156.9 2 2 6 2 6 10 2 6 10 7 2 6 1 2 $^9D_2$	65 Tb Terbium 159.2 2 2 6 2 6 10 2 6 10 8 2 6 1 2	66 Dy Dysprosium 162.46 2 2 6 2 6 10 2 6 10 9 2 6 1 2	67 Ho Holmium 163.5 2 2 6 2 6 10 2 6 10 10 2 6 1 2	68 Er Erbium 167.2 2 2 6 2 6 10 2 6 10 11 2 6 1 2	69 Tm Thulium 169.4 2 2 6 2 6 10 2 6 10 12 2 6 1 $^2F_{7/2}$ 2	70 Yb Ytterbium 173.04 2 2 6 2 6 10 2 6 10 13 2 6 1 $^1S_0$ 2	71 Lu Lutecium 175.0 2 2 6 2 6 10 2 6 10 14 2 6 1 $^2D_{3/2}$ 2		K L M N O P	1 2 3 4 5 6
78 Pt Platinum 195.23 2 2 6 2 6 10 2 6 10 14 2 6 10 1 $^3D_3$	79 Au Gold 197.2 2 2 6 2 6 10 2 6 10 14 2 6 10 1 $^2S_{1/2}$	80 Hg Mercury 200.61 2 2 6 2 6 10 2 6 10 14 2 6 10 2 $^1S_0$	81 Tl Thallium 204.39 2 2 6 2 6 10 2 6 10 14 2 6 10 2 1 $^2P_{1/2}$	82 Pb Lead 207.21 2 2 6 2 6 10 2 6 10 14 2 6 10 2 2 $^3P_0$	83 Bi Bismuth 209.00 2 2 6 2 6 10 2 6 10 14 2 6 10 2 3 $^4S_{3/2}$	84 Po Polonium 210 2 2 6 2 6 10 2 6 10 14 2 6 10 2 4 $^3P_2$	85 At Astatine 211 2 2 6 2 6 10 2 6 10 14 2 6 10 2 5 $^2P_{3/2}$	86 Rn Radon 222 2 2 6 2 6 10 2 6 10 14 2 6 10 2 6 $^1S_0$	K L M N O P	1 2 3 4 5 6
96 Cm Curium 242 2 2 6 2 6 10 2 6 10 14 2 6 10 7 2 6 1 2	97 Bk Berkelium 2 2 6 2 6 10 2 6 10 14 2 6 10 8 2 6 1 2	98 Cf Californium 2 2 6 2 6 10 2 6 10 14 2 6 10 9 2 6 1 2							K L M N O P Q	1 2 3 4 5 6 7
s p d f	s p d f	s p d f	s p d f	s p d f	s p d f	s p d f	s p d f	s p d f		

(2) Pauli [6] formulated his *exclusion principle,* which states that each electron in an atom must be described by its individual set of quantum numbers and that no two sets can be alike.† An orbital, as we just discussed, is uniquely described by the three quantum numbers n, l, and ⓜ; an electron in an orbital is characterized, in addition, by its spin component s pointing up or down. Hence, an orbital is filled when it contains two electrons with antiparallel spins.

With this proviso, by filling successively the orbitals of lowest energy, the *Bohr-Stoner arrangement* of the periodic table [7] results (Table 8.1). The periodic character of the system of elements arises because the electrons form consecutive shells while filling the orbitals of lowest energy, according to the Pauli exclusion principle. This sequence of fundamental shells is prescribed by the successive integers of the total quantum number n. Recognized from X-ray spectra [8] before the advent of quantum numbers, they were designated by capital letters, starting with K; therefore, the synonymous designations: $n = 1$ or K-shell; $n = 2$ or L-shell; $n = 3$ or M-shell, etc. In addition to these *main shells,* there are *subshells,* which, when occupied or *closed,* signify electron arrangements of especial stability and inertness. This fact is connected with the shape of the various electron clouds (see Figs. 8.5 and 8.7).

Only the s electrons ($l = 0$) in each main shell, as discussed, correspond to a spherical electron distribution. The $(n - 1)$ configurations characterized by successive integers of the azimuthal quantum number $l \neq 0$ represent less symmetrical wave modes. However, a spherical overall symmetry is restored by summing over the intensities of all modes of the same l. For example, the amplitudes of the three p orbitals ($l = 1$) are proportional to x/r, y/r, and z/r (see Eq. 8.22); the total intensity of the p-states is proportional to the sum of the square of the amplitudes, $(x^2 + y^2 + z^2)/r^2 = 1$, and therefore is independent of angle. Hence, a full occupation of the s and p orbitals does not fill the main shells for $n > 2$, but it creates s,p subshells of spherical symmetry and great stability, as evidenced in the argon, krypton, and xenon cores (see Table 8.1). These s^2,p^6 subshells are the stable *electron octets* of chemistry, first introduced by Kossel [9] and Lewis.[10] By adding the ten d electrons, we arrive again at closed subshells of spherical symmetry but lower stability: the copper, silver, and gold cores. The chemical reactivity of atoms arises from their outer electrons in partly filled subshells, that is, from their *valence electrons.*

For the rare gas helium the K-shell is filled with two electrons in antiparallel spin position. Since this is the $n = 1$ state and an s orbital ($l = 0$), the electron structure of the helium atom can be identified by the symbol $1s^2$, where the exponent signifies the number of electrons occupying the orbital type. The situation is repeated for neon, where filling of the L-shell produces the constellation $1s^22s^22p^6$, and for argon where the occupation of the $3s$ and $3p$ orbitals ($1s^22s^22p^63s^23p^6$) is completed.

The electron clouds of the hydrogen atom illustrate the fact that the s electrons approach the nucleus closest on the average, while the p, d, and f electrons reach farther and farther into space. In fact, the wave function of s states has a maximum in the nucleus (see Fig. 8.4) whereas for all other states the wave function at $r = 0$, that is, $\psi(0)$, vanishes. Hence, we expect that first the s orbital and then the p orbitals will be filled as indicated up to the element argon. For farther outlying levels it becomes a complicated problem to decide which orbital represents the lower energy state, and the sequence may shift as the nuclear charge increases. For the element 19 (K) the $4s$ orbital lies below the $3d$ level and fills first, but at chromium and at copper one $4s$ electron is lost to the $3d$ orbitals. The magnetic behavior of the iron group arises from the unpaired spin moments of the filling $3d$ shell. A similar situation occurs for the rare earth elements, where the $5s$ and $4d$ orbitals compete, and finally in the transuranium elements in a complicated competition between $6d$ and $5f$.

Since the outer electronic configuration of the elements 58 to 71 is so nearly alike, the chemical behavior of these rare earth elements is approximately identical. The shell size, on the other hand, must shrink as the nuclear charge increases, an effect known as *lanthanide contraction.*[11]

During the last fourteen years, nuclear synthesis has added ten new elements to the periodic system: technecium (43), promethium (61), astatine (85), francium (87), neptunium (93), plutonium (94), americium (95), curium (96), berkelium (97), and californium (98). This extension of the sequence of the heaviest elements permits a closer scrutiny of the delayed filling of subshells in the radon core. The electronic distribution for the atoms in the gaseous phase, as shown in Table 8.1, has been proposed by Coryell [12] on the basis of

[6] W. Pauli, Jr., *Z. Physik 31,* 765 (1925).

† For a more general statement of the Pauli principle, see Sec. 11.

[7] E. C. Stoner, *Phil. Mag.* [6], *48,* 719 (1924).

[8] H. G. J. Moseley, *Phil. Mag.* [6], *26,* 1024 (1913); *27,* 703 (1914).

[9] W. Kossel, *Ann. Physik 49,* 229 (1916).

[10] G. N. Lewis, *J. Am. Chem. Soc. 38,* 762 (1916).

[11] V. M. Goldschmidt, *Geochemische Verteilungsgesetze der Elemente,* Oslo. Akad. Wiss., Pt. 7 (1926), Pt. 8 (1927).

[12] C. D. Coryell, *Rec. Chem. Prog. 12,* 55 (1951); *J. Chem. Educ. 29,* 62 (1952).

chemical evidence. It must be kept in mind that in cases of a precarious energy balance between competing levels, the state of aggregation may easily affect the outcome. The information obtained from gas spectroscopy, ionic solutions, salts, and metals can therefore be contradictory.

9 · Atoms in Electric Fields; Stark Effect

The classical model of spherical atoms led us to the conclusion that the electronic polarizability α_e is proportional to the volume of these spheres (Eq. 3.8). By postulating, furthermore, that the atoms react to alternating fields like linear harmonic oscillators, we arrived at the dispersion formula of classical physics (Eq. 4.29). Here the range of the classical approach ended. To gain more precise information, a deeper understanding of the structure of atoms was required. It was provided, as shown in the preceding sections, by spectroscopic studies and their quantum mechanical interpretation.

This more searching inquiry, as always in life, has left us with a much more complex situation. Electronic polarization has become an involved phenomenon integrating over the deformation of various types of electron clouds that are coupled mutually and bound to the nucleus by fields varying with distance in a complicated manner. However, the preceding discussion of the periodic system of the elements suggests that our classical model of spherical atoms will not be far from correct for the closed s^2p^6-cores of the rare gases. These spherical charge distributions of singular stability are found not only in the rare gas atoms but neighboring elements tend also to assume them by taking up additional electrons or by shedding those in excess. Hence *negative ions* are formed by atoms of smaller nuclear charge than the rare gases, and *positive ions* from elements of larger-order number. In this way sequences of similar particles originate such as H^-, He, Li^+, Be^{2+}; O^{2-}, F^-, Ne, Na^+, Mg^{2+}; etc., in which the electronic structure remains identical but the shell radius systematically contracts as the nuclear charge increases in successive steps of one.

Table 9.1. Radii of atoms and ions of rare-gas shell structure (s^2p^6) [1]

He Core		Ne Core		A Core		Kr Core		Xe Core	
		O^{2-}	1.40	S^{2-}	1.84	Se^{2-}	1.98	Te^{2-}	2.21
H^-	2.08	F^-	1.36	Cl^-	1.81	Br^-	1.95	I^-	2.16
He	0.93	Ne	1.12	Ar	1.54	Kr	1.69	Xe	1.90
Li^+	0.60	Na^+	0.95	K^+	1.33	Rb^+	1.48	Cs^+	1.69
Be^{2+}	0.31	Mg^{2+}	0.65	Ca^{2+}	0.99	Sr^{2+}	1.13	Ba^{2+}	1.35
B^{3+}	0.20	Al^{3+}	0.50	Sc^{3+}	0.81	Y^{3+}	0.93	La^{3+}	1.15
C^{4+}	0.15	Si^{4+}	0.41	Ti^{4+}	0.68	Zr^{4+}	0.80	Ce^{4+}	1.01
				V^{5+}	0.59	Cb^{5+}	0.70		

If we calculate classically the electronic polarizability of these particles (see Eq. 3.8) by using the radii of Table 9.1, the curves of polarizability versus order number of Fig. 9.1 result.[2] It illustrates graphically well-known trends. The negative ions with their large size and low nuclear charge are highly polarizable, the positive ones with their smaller size and higher charge much more rigid. In addition, the polarizability of corresponding ions, for example, the negative halogens or positive alkali ions, increases as we progress from the helium to the xenon core.

We may expect that the numerical values calculated from the classical model of finite spheres will turn out to be too small. The electron clouds of quantum mechanics, as discussed in the preceding section, are not rigidly confined but extend with rapidly diminishing density towards infinity (see Eqs. 8.21 and 8.22). Since the field binding the electrons to the nucleus decreases steeply towards the outside, the distant parts of the electron clouds are more pliable and make an appreciable contribution to the polarization in spite of their small charge content. This effect can be calculated in the case of the hydrogen atom where the s electron produces a spherical charge distribution.[3]

[2] See also the semi-empirical data of K. Fajans and G. Joos, *Z. Physik 23*, 1 (1924).

[3] J. H. Van Vleck, *The Theory of Electric and Magnetic Susceptibilities*, Oxford University Press, 1932, pp. 203 ff.

[1] After L. Pauling, *The Nature of the Chemical Bond*, Cornell University Press, Ithaca, N. Y., 1940, p. 346.

The outcome is, as already mentioned in Sec. 3, that the electronic polarizability of H should prove to be about 4.5 times as large as the classical value based on the Bohr radius r_0 (Eq. 3.12).

The static permittivity arises, as the dispersion theory showed (Eq. 4.29), from contributions of all the resonance states of an atom. To understand this situation more clearly, we have to investigate how the resonance states are disturbed by an external electric field.

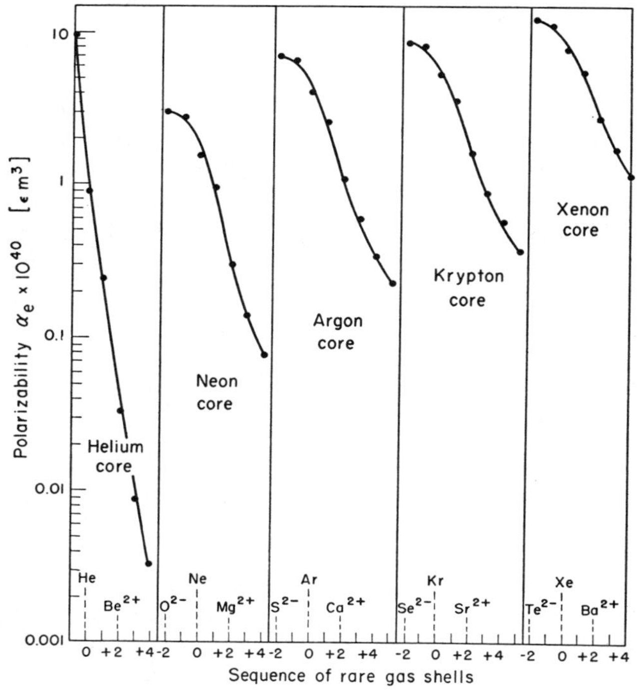

Fig. 9.1. Polarizability of rare gas-type ions (classical calculation).

Such a field is obviously very small in comparison to the central binding field of an atom. Furthermore, its magnitude changes only extremely slowly (*adiabatically*) in comparison to the revolution time of an electron in its Bohr orbit (see Eq. 5.13). Adiabatic disturbances do not produce quantum transitions, that is, do not change the pre-existing quantum numbers of a system (the phase integrals of Sommerfeld's quantum condition, Eq. 6.19) as Ehrenfest [4] first derived (*adiabatic law*). However, the polarizing field creates slight changes in the energy of the electronic states. These perturbations become visible spectroscopically in the splitting of spectral lines under the influence of an electric field, in the *Stark effect*.[5]

Even if the fact is neglected that the electric field distorts the electron cloud structure of an atom, a Stark effect may result. As shown in Sec. 6, a one-electron system like H or He^+ is degenerate: its circular and elliptical orbits of the same principal quantum number are of equal energy. An electric field removes this degeneracy; the various orbits assume, without distortion, different energy values and impart a *fine structure* to the spectral lines. The splitting of the lines increases proportionally to the applied field; this large *linear Stark effect* enabled Stark to make his initial discovery. Normally, a much smaller *quadratic Stark effect* is observed, in which the line splitting increases proportionally to the square of the field strength.

The linear Stark effect of one-electron systems is frequently dismissed as a somewhat abstruse phenomenon. Actually, it is of great interest, since here *permanent electric moments* make their appearance *in atoms*. In consequence, the perturbation energy, given by the product of dipole moment and field strength, varies here proportionally to the field intensity. The quadratic effect in atoms stems from the induced moments of orbital distortions. These moments increase proportionally to the electric field strength, hence an E^2 effect results.

Quantum mechanics, as Schrödinger showed in his third paper,[6] treats the Stark effect of the hydrogen atom by adding to the potential energy $-(e^2/\epsilon_0 4\pi r)$ of the Coulomb field in Eq. 8.17 the potential energy of the electron in the external field. If this field points in the $+z$-direction and $z = 0$ designates the potential plane containing the nucleus, the additional potential energy of the electron in its cloud is $ez\mathbf{E}$, and the Schrödinger equation becomes

$$\nabla^2\psi + \frac{2m}{\hbar^2}\left(\mathcal{E} + \frac{e^2}{\epsilon_0 4\pi r} - ez\mathbf{E}\right)\psi = 0. \qquad (9.1)$$

Schrödinger solved this partial differential equation by the introduction of parabolic co-ordinates in analogy to a treatment of the Stark effect in the old quantum theory by Schwarzschild [7] and Epstein.[8] We shall make use of an early, more graphical approach by Bohr [9] that leads to the same result.

An electron in its elliptical orbit moves much more slowly near its aphelion than near its perihelion. If we imagine the electronic charge smeared out over the segments of the orbit according to the speed of passage, the center of this charge distribution falls at the point S on the long axis (Fig. 9.2). Let the distance from S

[4] P. Ehrenfest, *Ann. Physik 36*, 91 (1911); *51*, 327 (1916).

[5] J. Stark, *Ber. Berlin Akad. Wiss. 47*, 932 (1913); *Ann. Physik 43*, 965, 983 (1914).

[6] E. Schrödinger, *Ann. Physik 79*, 361, 489, 734 (1926).

[7] K. Schwarzschild, *Sitz. kgl. preuss. Akad.*, 1916, p. 548.

[8] P. S. Epstein, *Ann. Physik 50*, 489 (1916).

[9] N. Bohr, *Über die Quantentheorie der Linienspektren*, Vieweg, Braunschweig, 1923, pp. 98 ff.

to the nucleus in the perihelion be designated by the vector $\mathbf{s}$; its magnitude is found to be

$$|\mathbf{s}| = \tfrac{3}{2}\epsilon a, \qquad (9.2)$$

where a represents the major half axis, ϵ the numerical eccentricity of the ellipse (Eq. 6.22), and ϵa the distance from the nucleus to the center of the ellipse. Hence the point S halves the distance from the center to the aphelion.

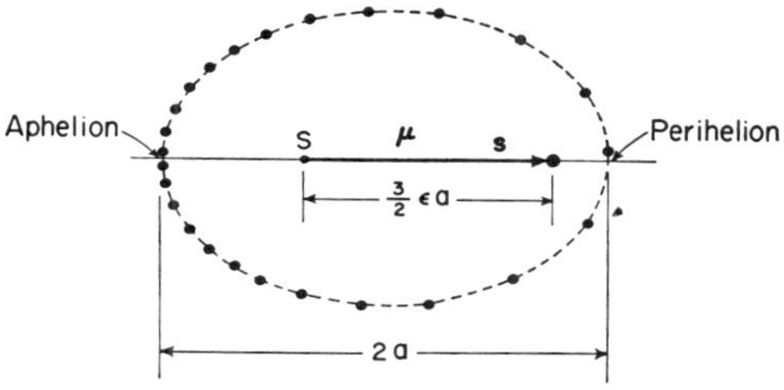

Fig. 9.2. Characterization of elliptical orbit by average dipole moment.

This average charge distribution of the electron allows us to characterize each elliptical orbit by an average electric dipole moment

$$\bar{\boldsymbol{\mu}} = e\mathbf{s}. \qquad (9.3)$$

The existence of this dipole moment does not lead without the superposition of an electric field to a permanent electric moment of the hydrogen atom, because the ellipse can assume any orientation. Thus the overall effect cancels. However, the revolving electron has an angular momentum and corresponds therefore to a gyroscope pointing with its angular momentum axis perpendicular to the electric dipole moment. The application of an external electric field causes a torque on this average dipole moment and thus a precession of the angular momentum vector and of the dipole moment vector around the field axis with a precession frequency [10]

$$\nu_e = \frac{3}{2}\left(\frac{\epsilon_0 a}{m\pi}\right)^{1/2} \mathbf{E}. \qquad (9.4)$$

It is proportional to the electric field strength $\mathbf{E}$ (measured in [volt m^{-1}]) and depends only on the large half axis a of the ellipse.

This half axis is identical with the radius of the circular Bohr orbit of the same principal quantum number n (see Fig. 6.3), that is (see Eqs. 6.5 and 6.10),

$$a = n^2 r_0 = n^2 \frac{\epsilon_0 h^2}{\pi m e^2}. \qquad (9.5)$$

[10] See A. Sommerfeld, *Atombau und Spectrallinien*, Vieweg, Braunschweig, 1924, p. 384.

Hence the precession frequency becomes

$$\nu_e = \frac{3}{2}\frac{\epsilon_0 h}{\pi m e} n\mathbf{E} \simeq 1.93 \times 10^4 n\mathbf{E}. \qquad (9.6)$$

The precession adds to the original term energy $\mathcal{E}_n$ of the hydrogen atom (Eq. 8.18), a perturbation energy $\Delta\mathcal{E}_n$ which has to be quantized as

$$\Delta\mathcal{E}_n = n_e h\nu_e; \qquad (9.7)$$

n_e represents a new *electrical* quantum number.

The actual motion of the electron and the meaning of this quantization are not easily visualized. It is not that the ellipse of Fig. 9.2 remains intact and precesses as such around the field axis; not even the dipole length s, but only the major half axis a remains constant. However, the charge center S remains in a plane perpendicular to the field direction and traverses approximately an ellipse in an harmonic motion of the frequency ν_e around the field axis. The product $s \cos\theta$ remains constant during this motion, where θ is the angle between dipole moment and field axis. Figure 9.3 shows schematically the orbit traversed by the endpoint of the electric moment vector.[11]

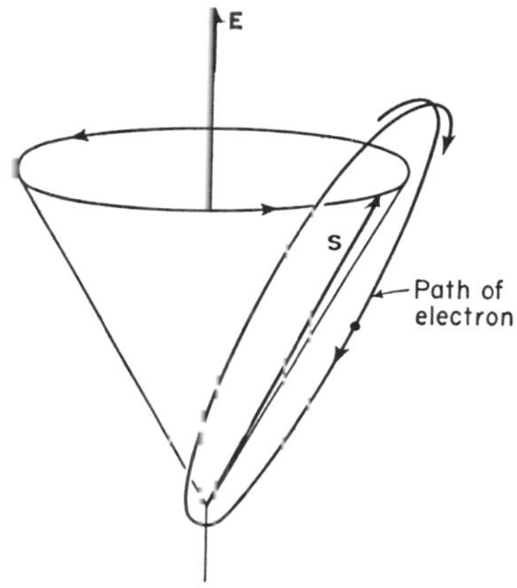

Fig. 9.3. Orbit traversed by endpoint of electric moment vector in linear Stark effect.

The quantum number n_e of the linear Stark effect is related to Bohr's radial quantum number n' of the ellipse (Eq. 6.21). The latter may be expressed as the sum, and n_e as the difference, of two quantum numbers n_a and n_b:

$$\begin{aligned} n' &= n_a + n_b \\ n_e &= n_a - n_b. \end{aligned} \qquad (9.8)$$

Thus the total energy of an electron in the nth quantum state of the hydrogen atom, when an electric field of the

[11] See also R. Ladenburg, *Müller-Pouillets Lehrbuch der Physik*, Vieweg, Braunschweig, 11th ed., Vol. II, 1929, p. 2240.

strength **E** [volt m^{-1}] is superposed, becomes

$$\mathcal{E}_n = -\frac{Rhc}{n^2} + \frac{3}{2}\frac{\epsilon_0 h^2}{\pi m e} n(n_a - n_b)E. \qquad (9.9)$$

For $n_a = n_b$ the average electric dipole moment stands normal to the field direction, and the energy is the same as without field; whereas for $n_a > n_b$ the center S of the electronic charge distribution falls towards the positive and for $n_a < n_b$ towards the negative z-axis.

To illustrate the situation, we consider the Stark splitting of the first line of the Balmer series, H$_\alpha$, which corresponds to a transition between the principal quantum terms $n = 3$ and $n = 2$ (Fig. 9.4). In addition to

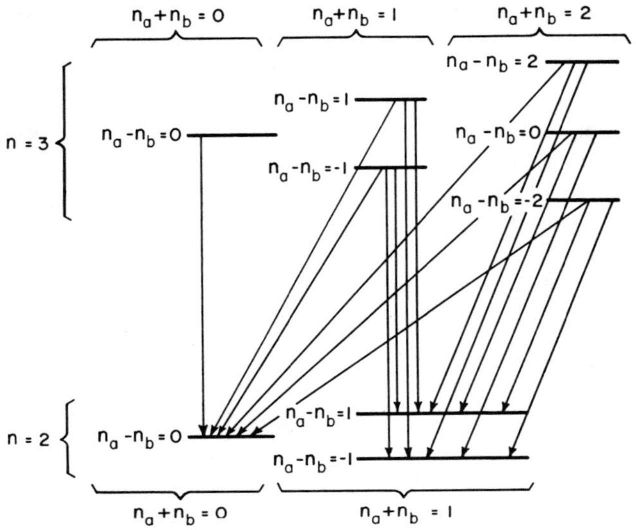

Fig. 9.4. Stark components of H$_\alpha$ line.

the circular orbits ($n_a + n_b = 0$), two elliptical orbits ($n_a + n_b = 1$ and $n_a + n_b = 2$) belong to the upper state, and one ellipse ($n_a + n_b = 1$) to the lower state (see Fig. 6.3). Without field their energy is the same as that of the circular orbits, but with field their energy levels move symmetrically above and below the undisplaced energy level because the electric dipole moment points against or toward the field direction. The splitting of the terms increases proportionally to the field strength and can reach appreciable values, especially for the larger principal quantum numbers n of the higher states. For the term $n = 3$ in a field of 5×10^7 [volt m^{-1}], the energy difference in wave numbers between the outer levels ($n_a = 2, n_b = 0; n_b = 2, n_a = 0$) becomes, according to Eqs. 9.6 and 9.9,

$$\Delta\left(\frac{1}{\lambda_e}\right) = 2 \times \frac{1.93 \times 10^4}{3 \times 10^{10}} n(n_a - n_b)\mathbf{E}$$
$$= 1.28 \times 10^{-6} \times 3 \times 2 \times 5 \times 10^7$$
$$= 384 \quad [\text{cm}^{-1}]. \qquad (9.10)$$

Figure 9.5 shows the relative intensity of the various Stark components.

The *linear* Stark effect of one-electron systems is due to the fact that the undistorted elliptical orbits have average electric dipole moments and that these moments can orient in the external field, and thus produce energy values different from those of the circular orbits (removal or degeneracy). The displacement is symmetrical to the original energy state towards both higher and lower energy values. For strong fields we observe, in addition, a *quadratic* Stark effect which results from the distortion of the orbitals, that is, the induced dipole moments. Since the induced moment is created by the field, its momentum vector will always point towards the field direction; the energy terms are

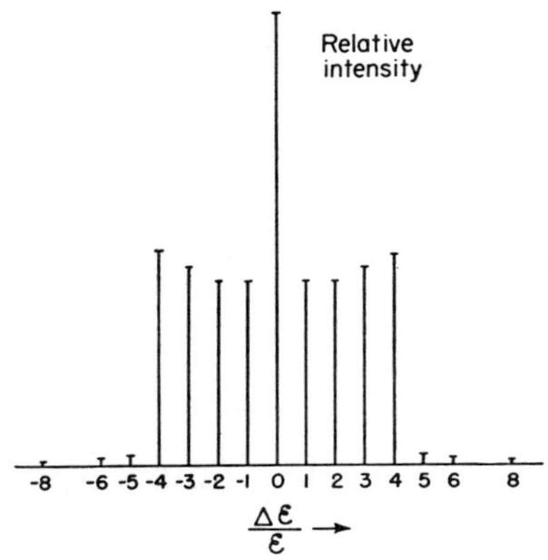

Fig. 9.5. Relative intensity of the sixteen Stark components of the H$_\alpha$ line. (After Sommerfeld.[10])

correspondingly displaced, not symmetrically to the original position, but only downwards. The electrical quantum number ($n_a - n_b$) of Eq. 9.9 appears as the quadratic expression $-(n_a - n_b)^2$, and the total energy of the nth quantum state of the hydrogen atom becomes in this second approximation [12]

$$W_n = -\frac{Rhc}{n^2} + \frac{3}{2}\frac{\epsilon_0 h^2}{\pi m e} n(n_a - n_b)\mathbf{E}$$
$$-\frac{1}{16}\left(\frac{\epsilon_0 h}{\pi m e}\right)^2 n^4\{17n^2 - 3(n_a - n_b)^2$$
$$- 9(k-1)^2 + 19\}\mathbf{E}^2. \qquad (9.11)$$

The quadratic Stark effect depends not only on the principal quantum number n and the electrical quantum number ($n_a - n_b$) but also on the azimuthal quantum number k (Eq. 6.21). In comparison to the linear effect in hydrogen, the quadratic effect is very small: in a field of 5×10^7 volts/m it lowers the term $n = 3$ by about 1.6 [cm^{-1}], whereas the linear splitting, according to Eq. 9.10, amounts to 384 wave numbers.

[12] See R. Ladenburg, *loc. cit.*, p. 2248.

The degeneracy of the one-electron system is due to the fact that the electron moves solely in the Coulomb field of the nucleus. As soon as an external field is superposed, the shape of the various electron clouds has an influence on their energy, and a *fine structure* of the spectral lines results. In multi-electron systems the interaction between the electrons changes the Coulomb field into a central field where the shape of the electron clouds affects their energy (see Sec. 8). This "internal" linear Stark effect removes therefore the degeneracy and creates by itself an energy difference between terms of the same principal quantum number n but different azimuthal quantum numbers l. Hence, an external electric field causes in multi-electron systems, as its first action, a quadratic Stark effect which produces a small additional splitting of the spectral lines. Very strong external fields begin to compete with the internal interaction of the electrons. Thus, in addition to the deformation of the orbitals which generates the induced moments of the quadratic effect, a small linear Stark effect becomes observable even in multi-electron systems, indicating a slight orientation of the average orbital moments against the field axis.

10 · Atoms in Magnetic Fields; Zeeman Effect

Atoms, as the discussion of their structure has shown (see Sec. 8), may possess two types of *permanent magnetic moments:* those causally connected to the mechanical angular moments of the electron clouds (*orbital moments*), and one moment for each electron (*spin moments*). The correlation between angular momentum and magnetic moment is based on the macroscopic observation that a current I circling an area A creates a magnetic field identical to that of a magnetic dipole (see I, Fig. 5.7 and Eq. 5.28)

$$\mathbf{m} = IA\mathbf{n}. \tag{10.1}$$

The electron of Bohr's hydrogen atom, circling the proton ν-times per second in an orbit of the radius r, consequently produces a magnetic moment

$$\mathbf{m} = -e\nu r^2\pi. \tag{10.2}$$

Simultaneously, it has a mechanical angular moment

$$p' = mvr = m\nu 2r\pi r, \tag{10.3}$$

designated by a momentum vector $\mathbf{p}'$ antiparallel to $\mathbf{m}$ (Fig. 10.1). Hence the magnetic and angular moments are interrelated as

$$\mathbf{m} = -\frac{e}{2m}\mathbf{p}'. \tag{10.4}$$

Equation 10.4 is called the *magneto-mechanical parallelism;* it specifies that the magnetic and mechanical moments of circling electrons are interdependent and that their *gyromagnetic ratio* γ is classically

$$\gamma \equiv \frac{|\mathbf{m}|}{|\mathbf{p}'|} = \frac{e}{2m}. \tag{10.5}$$

The angular moment is quantized in Bohr's theory † as (see Eq. 6.12)

$$p' = n\frac{h}{2\pi} = n\hbar, \qquad n = 1, 2, 3 \ldots. \tag{10.6}$$

Hence quantum theory leads to an *elementary magnetic moment,* the *Bohr magneton* ‡

$$|\mathbf{m}_B| = \frac{e}{2m}\hbar = 9.27 \times 10^{-24} \quad [\text{joule/weber m}^{-2}]$$
$$= 9.27 \times 10^{-21} \quad [\text{erg gauss}^{-1}]. \tag{10.7}$$

If the magnetic moment is measured in Bohr magnetons and the angular momentum in units of $\hbar$, the ratio of

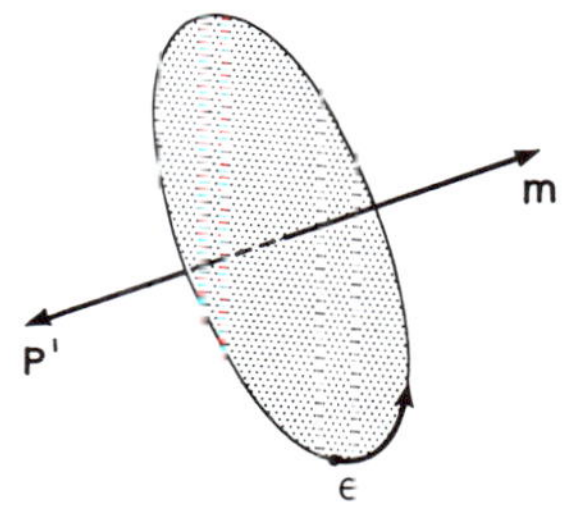

Fig. 10.1. Angular momentum and magnetic moment of orbital electron.

magnetic to mechanical moment (Eq. 10.5), known as the *g factor*, becomes for $n = 1$

$$g = \frac{|\mathbf{m}|}{|\mathbf{p}|}\frac{\hbar}{|\mathbf{m}_B|} = \gamma\frac{\hbar}{|\mathbf{m}_B|} = 1. \tag{10.8}$$

† Note that quantum mechanics prescribes the slightly different condition, Eq. 8.23.

‡ In measurements of the magnetic moments of nuclei the *nuclear magneton* is used as a common unit by substituting the mass of the proton for that of the electron in Eq. 10.7; hence

$$|\mathbf{m}_N| = \tfrac{1}{1836}|\mathbf{m}_B| = 5.05 \times 10^{-24} \text{ [erg gauss}^{-1}].$$

Actual measurements of g on ferromagnetic bars, in which a mechanical moment was produced by a change in magnetization (*Richardson-Einstein-de Haas effect*[1]) or a magnetic moment by rotation (*Barnett effect*[2]), proved that the experimental value was about twice as large. This *magneto-mechanical anomaly* found its explanation in the hypothesis of Uhlenbeck and Goudsmit[3] that an electron itself carries a magnetic moment of 1 Bohr magneton,[4] whereas its mechanical *spin* corresponds to only ½ quantum of action (see Eq. 8.32):

$$p' = \tfrac{1}{2}\hbar. \tag{10.9}$$

Hence for free electrons

$$g = 2. \tag{10.10}$$

Apparently it is this electron spin, without much additional contribution of orbital moments, that is measured in ferromagnetic materials (see Sec. 29).

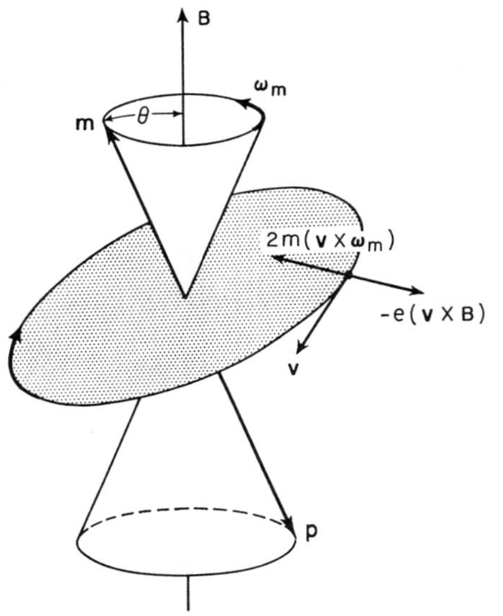

Fig. 10.2. Balance of magnetic and Ceriolis force.

An external magnetic field exercises a torque on the permanent magnetic moments. Since these moments are linked to angular mechanical moments, that is, to a gyroscope, a precession of the momentum vectors around the magnetic field axis results. In a field of the magnetic induction $\mathbf{B} = \mu_0\mathbf{H}$, an electron traversing its orbit with the velocity $\mathbf{v}$ is subjected to a magnetic force (see I, Eq. 10.21)

$$\mathbf{F}_m = -e(\mathbf{v}\times\mathbf{B}) \tag{10.11}$$

(Fig. 10.2). If the electronic orbit precesses around the field axis with an angular frequency $|\boldsymbol{\omega}_m|$, the electron, according to classical mechanics, is simultaneously subjected to a *Coriolis force*

$$\mathbf{F}_c = 2m(\mathbf{v}\times\boldsymbol{\omega}_m). \tag{10.12}$$

The two forces act in opposition and mutually cancel each other, as Larmor[5] first derived, if the orbit precesses around the field axis with an angular velocity

$$\boldsymbol{\omega}_m = -\frac{e}{2m}\mathbf{B} = -\frac{e}{2m}\mu_0\mathbf{H} = -\gamma\mu_0 H. \tag{10.13}$$

The magnetic precession frequency

$$\nu_m = \frac{1}{2\pi}\frac{e}{2m}\mu_0|\mathbf{H}| \simeq 1.76\times 10^4 H \tag{10.14}$$

is called the *Larmor frequency*. The magnetic field $\mathbf{H}$ in our equation is measured in amperes per meter; for a field measured in oerstedts, the numerical value has to be multiplied by a factor $10^3/4\pi$ (see I, Table 8.2).

The external magnetic field, as previously the electric field, represents an adiabatic disturbance. It does not alter the pre-existing quantum numbers of the electronic system but adds to the original energy $\mathcal{E}_n$ a perturbation energy $\Delta\mathcal{E}_n$ that has to be quantized (cf. Eq. 9.7). Thus we obtain

$$\Delta\mathcal{E}_n = -\textcircled{m}h\nu_m = \textcircled{m}\frac{e}{2m}\hbar|\mathbf{B}|, \tag{10.15}$$

where $\textcircled{m}$ represents the *magnetic quantum number* introduced by Eq. 8.25.

The magnetic perturbation energy has a graphic interpretation: the angular momentum of an electron cloud is characterized by the azimuthal quantum number l (Eq. 8.23). This angular momentum produces a magnetic orbital moment

$$\mathbf{m} = \sqrt{l(l+1)}\,\mathbf{m}_B \tag{10.16}$$

where $\mathbf{m}_B$ is the Bohr magneton (Eq. 10.7). Placed in a magnetic field of the induction $\mathbf{B}$, the moment contributes a potential energy

$$U = -\mathbf{m}\cdot\mathbf{B} = -\sqrt{l(l+1)}\,|\mathbf{m}_B|\,|\mathbf{B}|\cos\theta \tag{10.17}$$

that depends on the angle of inclination θ against the field axis. A comparison of Eqs. 10.15 and 10.17, by referring to Eqs. 8.27 and 10.7, shows that the mag-

[1] O. W. Richardson, *Phys. Rev. 26*, 248 (1908); A. Einstein and W. J. de Haas, *Verh. deut. physik. Ges. 17*, 152 (1915).

[2] S. J. Barnett, *Revs. Mod. Phys. 7*, 129 (1935).

[3] G. E. Uhlenbeck and S. Goudsmit, *Naturwiss. 13*, 953 (1925).

[4] Extremely accurate measurements, possible today by combining atomic beam and nuclear resonance techniques, give for the magnetic moment of the electron in the hydrogen atom $\mathbf{m} = (1.001146 \pm 0.000012)\mathbf{m}_B$ [S. H. Koenig, A. G. Prodell, and P. Kusch, *Phys. Rev. 88*, 191 (1952)]; see also N. F. Ramsay, *Nuclear Moments*, John Wiley and Sons, New York, 1953.

[5] J. Larmor, *Phil. Mag. 44*, 503 (1897).

netic perturbation energy is this potential energy, and that the magnetic quantum number corresponds to

$$ⓜ = \sqrt{l(l+1)} \cos\theta. \quad (10.18)$$

Since it can assume only the values

$$ⓜ = l, \quad l-1, \quad \cdots, \quad -(l-1), \quad -l, \quad (8.26)$$

the orbital angular and magnetic moments have to point into fixed angular positions in reference to the field axis (Fig. 10.3).

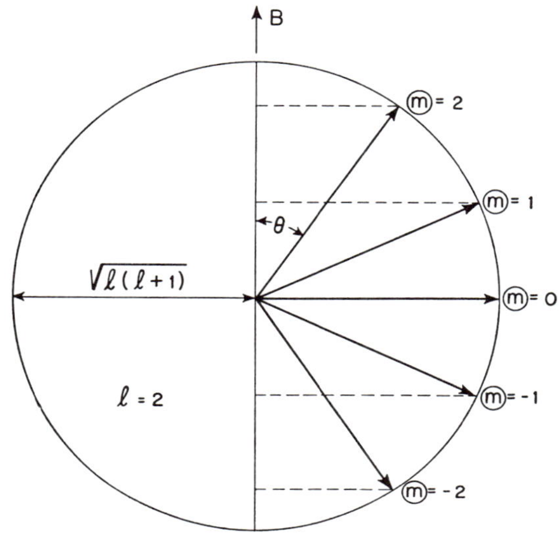

Fig. 10.3. Quantized orientation of magnetic moments.

The magnetic perturbation energy, caused by the interaction of the permanent magnetic moments of atoms with an external magnetic field, produces a magnetic splitting of the energy terms of atoms and thus of the spectral lines; this is the *Zeeman effect.*[6]

There is obviously a close analogy between the Zeeman effect and the linear Stark effect of the preceding section, but also a great difference. By introducing an *elementary electric dipole moment* †

$$|\boldsymbol{\mu}_e| = \frac{3}{2}\frac{\epsilon_0 h^2}{\pi m e} = 12.8 \times 10^{-30} \quad [\text{coul m}], \quad (10.19)$$

we can write the electrical perturbation energy of Eq. 9.9 in the same way as the magnetic perturbation energy of Eq. 10.17 as a dot product of the moment and field

$$U = -n\boldsymbol{\mu}_e \cdot \mathbf{E}. \quad (10.20)$$

The component of the electric moment in field direction is quantized,

$$-n|\boldsymbol{\mu}_e| \cos\theta = n_e, \quad (10.21)$$

and the electric quantum number varies like the magnetic one from l to $-l$,

$$-n_e = \sqrt{l(l+1)} \cos\theta = ⓜ. \quad (10.22)$$

The essential difference is that in the magnetic case the force acts perpendicular to the electron orbit and is of constant magnitude at any point of the circular orbit. In consequence, the electron orbit remains intact during the precession, the magnetic moment is constant, and the cosine of the angle between moment and field is quantized. The angular momentum, and with it the magnetic moment, are conservative. In the electric case the force acts in the plane of the orbit and, since the orbit is an ellipse, the relative effect differs from point to point because the binding force to the nucleus changes. Consequently, the electron orbit does not remain intact but dissolves into a very complicated Lissajou figure; the average electric moment varies during the precession and not the cosine θ but the product of dipole moment and cosine θ is quantized (see Eq. 10.21 and Fig. 9.3).

The Larmor frequencies of the electric and of the magnetic precession, measured in mks units (Eqs. 9.6 and 10.14), are of the same order of magnitude. Since the electric dipole moment depends on the extent of the electron cloud, that is, on the principal quantum number n, there appears in the electric precession frequency Planck's quantum of action. The magnetic frequency, on the other hand, does not give any hint that quantum theory applies because the effect of the Bohr orbit contained in the velocity term $\mathbf{v}$ of Eqs. 10.11 and 10.12 cancels out.

The actual magnetic effect will be more complex than discussed as yet because, in addition to its orbital moment, the electron of the hydrogen atom carries its spin moment of 1 Bohr magneton. Both moments will tend to precess and also act on each other magnetically. For a multi-electron system with numerous orbital and spin moments, the situation may become very complicated.

In treating such cases quantum theory gives a relatively simple answer: not only the individual angular momenta but also their vector sum has to be quantized. Two orbital momenta, l_1 and l_2, may therefore couple to form a resultant vector $\mathbf{L}$ which may take on all integral values from $l_1 + l_2$ to $l_1 - l_2$. (It is understood that the quantum number has to be multiplied with Planck's quantum of action h to give the actual value of the angular momentum.)

As the orbital momentum vectors $\mathbf{l}$ may add to a resultant vector $\mathbf{L}$, the spin momentum vectors $\mathbf{s}$ of the multi-electron systems may add to a quantized vector $\mathbf{S}$. In this case the forces coupling the orbital angu-

[6] P. Zeeman, *Phil. Mag. 43*, 226 (1897).

† The customary unit, 1 debye, is about four times as small (see Eq. 15.1).

lar momentum vectors together are large, the forces coupling the spin vectors together are also large, but the forces coupling the orbital to the spin system are weak. This situation, called *L-S* coupling (or *Russell-Saunders* because it was first recognized by Russell and Saunders),[7] occurs preferentially in lighter atoms. The resultant of L and S is the *inner quantum number* J, which measures the total angular momentum of the electronic system and is again quantized (Fig. 10.4).

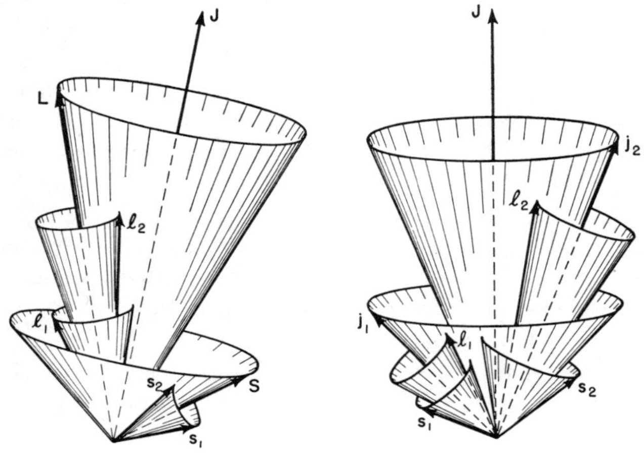

Fig. 10.4. *L-S* coupling and *j-j* coupling of orbital and spin moments.

The opposite extreme is the *j-j coupling*, present in the heaviest atoms. Here the interaction between the orbital magnetic moment and the spin of the same electron is large, hence the **l** and **s** of this electron form a resultant vector **j**, and the **j** vectors of the various electrons couple together to form a resultant **J** that is quantized. There are many intermediate cases between these two extremes, and the patterns of energy levels, even without an external field, may become extremely complicated.

If an external magnetic field is applied, the **J** vector takes up quantized orientations in relation to the field axis. A quantized component

$$\text{Ⓜ} = \sqrt{J(J+1)} \cos\theta \qquad (10.23)$$

is obtained in the field direction in analogy to the magnetic quantum number ⓜ of Eq. 10.18, which described the orientation of a single orbital moment l. The magnetic perturbation energy (see Eq. 10.15) is

$$\Delta\mathcal{E}_J = \text{Ⓜ} gh\nu_L, \qquad (10.24)$$

where the g factor (see Eq. 10.8) designates the ratio of magnetic moment to angular momentum of the atom as a whole, and $g\nu_L$ is the classical precession frequency of the atom. The external magnetic field thus causes a splitting of the multiplet levels of an atom, a Zeeman effect, that can lead to a determination of g and with it to some detailed information about the quantum numbers. Since Landé[8] was the first to analyze in detail the effect of magnetic fields on multiplet structure, the quantity g is often called the *Landé g factor*.[9]

The magnetic splitting may become very complex and vary in type with the magnetic field strength. Sufficiently strong external fields can uncouple the **l** and **s** vectors and orient, for example, the **L** and **S** vectors individually in the external field instead of acting on their resultant **J**. This *Paschen-Back effect*[10] in strong magnetic fields is the analogue of the appearance of a linear Stark effect in strong electric fields for multi-electron systems.

Besides the permanent magnetic moments, *induced magnetic moments* will make their appearance in atoms according to Faraday's induction law (I, Eq. 5.30) as soon as magnetic fields are applied. An electron in its orbit of radius r represents a current loop. A change of the external magnetic flux ϕ through this loop will induce a voltage

$$\oint \mathbf{E}\cdot d\mathbf{l} = -\frac{d\phi}{dt} = -\frac{d}{dt}\int \mathbf{B}\cdot\mathbf{n}\, dA \qquad (10.25)$$

which tends to create a magnetic field opposing the change. For our orbit of the area $r^2\pi$, inclined by an angle θ against the field axis, we obtain

$$E2r\pi = -r^2\pi (\cos\theta)\mu_0 \frac{dH}{dt}. \qquad (10.26)$$

If at $t = 0$ the magnetic field is switched on and reaches at the time t its final value H,

$$\int_0^t E\, dt = -\frac{r\cos\theta}{2}\mu_0 H. \qquad (10.27)$$

The integral on the left side, when multiplied by the electronic charge, represents the action of the induced electric field on the electron, expressed as a change of mechanical moment

$$\int_0^t eE\, dt \equiv -\Delta p = -\frac{er\cos\theta}{2}\mu_0 H. \qquad (10.28)$$

[7] H. N. Russell and F. A. Saunders, *Astrophysical J. 61*, 38 (1925).

[8] A. Landé, *Z. Physik 19*, 112 (1923).

[9] See also J. C. Slater, *Quantum Theory of Matter*, McGraw-Hill Book Co., New York, 1951, pp. 167 ff.

[10] F. Paschen and E. Back, *Ann. Physik 39*, 897 (1912); *40*, 960 (1913).

This change in moment expresses itself as a frequency change of the revolving electron

$$\Delta p = -m2r\pi\, \Delta\nu \tag{10.29}$$

and thus as an induced magnetic dipole moment (cf. Eq. 10.2)

$$\mathbf{m}_i = -er^2\pi\, \Delta\nu = -\frac{e^2r^2}{4m}(\cos\theta)\mu_0 H. \tag{10.30}$$

Since all directions of the orbit in relation to the field axis are equally probable, we obtain, finally, by averaging:

$$\overline{\mathbf{m}}_i = -\frac{e^2r^2}{6m}\mu_0\mathbf{H}$$
$$\simeq -4.7\times 10^{-29}a^2\mathbf{B} \quad [\text{joule/weber m}^{-2}], \tag{10.31}$$

where a represents the radius r, measured in angstroms.

Even a strong magnetic induction **B** of 1 weber/m^2 (10^4 gauss, see I, Table 8.2) acting on an electronic orbit r of 1×10^{-10} [m] ($a = 1$) produces only a weak induced moment $\overline{\mathbf{m}}_i = -4.7\times 10^{-29}$ [amp m^2] $\simeq 5\times 10^{-6}$ Bohr magnetons (cf. Eq. 10.7). This moment gives rise to an additional Larmor precession of the electronic system around the field axis, hence to a quadratic Zeeman effect in analogy to the quadratic Stark effect, but its spectroscopic action can usually be neglected. However, the overall effect of the induced magnetic moments is important; it represents the *diamagnetism* of a material, while the permanent moments cause *paramagnetism* or *ferromagnetism*, as will be discussed later (see Secs. 29 and 30). Diamagnetic measurements, at present relatively neglected, might prove of great value for the study of electronic systems in various states of aggregation.[11]

[11] See, for example, W. R. Myers, *Revs. Mod. Phys.* *24*, 15 (1952).

11 · The Energy Level Diagram of Atoms

The discussion of the preceding three sections has laid the basis for a more comprehensive understanding of the electronic structure of atoms. It has become apparent that the dominating forces shaping the electron clouds are the Coulomb attraction of the nucleus and the electric interaction of the electrons; these forces are of an electrostatic nature. The central field formed by the nucleus and electrons leads to the sequence of fundamental shells, characterized by the principal (fundamental) quantum number n. In the Coulomb field of one-electron systems this quantum number determines the energy of the stationary states nearly completely without reference to the orbital angular momentum of the electron clouds. An external electric field removes this degeneracy and gives terms of the same n, but of different azimuthal quantum numbers l, different positions in the energy scale (linear Stark effect). In multi-electron systems the electrostatic interaction of the electrons changes the Coulomb field into a central field and thus eliminates a priori the degeneracy (internal linear Stark effect). The various (n,l) terms assume different altitudes in the energy level diagram without external action.

Permanent magnetic moments exist, caused by the angular orbital momenta of the electron clouds and the electron spins. These magnetic moments interact, but the direct contribution of their interaction to the overall energy balance is minor. An orbital magnetic moment of 1 Bohr magneton (see Eq. 10.7), coupling to a spin moment of 1 Bohr magneton at a distance r of 1 Ångström unit, produces an interaction energy of approximately

$$U \simeq \frac{\mathbf{m}_B{}^2}{4\pi r^3}\mu_0 \simeq 10^{-23} \quad [\text{joule}]$$
$$\simeq 10^{-4} \quad [\text{ev}], \tag{11.1}$$

whereas the electrostatic binding energy is of the order of 10^1 ev. However, the magnetic coupling of the orbital and spin moments makes a very important indirect contribution to the total energy. The quantum numbers of the angular momenta codetermine, according to the Pauli exclusion principle, the standing-wave modes that may form. Hence, in this roundabout manner, the permanent magnetic moments help prescribe the possible stationary states and with it their electrostatic energy.

The net effect of the magnetic orbital and spin interaction (internal Zeeman effect) is that the (n,l) levels split into sublevels of a characteristic *multiplet structure*. Application of an external electric or magnetic field produces an additional splitting of the various sublevels of a multiplet, that is, a fine structure (quadratic and linear Stark effect, linear and quadratic Zeeman effect) (Fig. 11.1).

In Sec. 8 we introduced the four quantum numbers: n, l, Ⓜ, and s, to specify uniquely each electron of an atom. The magnetic quantum number Ⓜ designates the quantization of an individual orbital momentum

$\sqrt{l(l+1)}$ with respect to an external magnetic field (see Eq. 8.25). We see now that this description was oversimplified; we can dispense with the external field. The internal mutual coupling of the various moments gives the electronic system of an atom a total angular momentum described by the *inner* quantum number J. Its values designate the various energy levels of a multiplet and are derived as the possible resultants (with

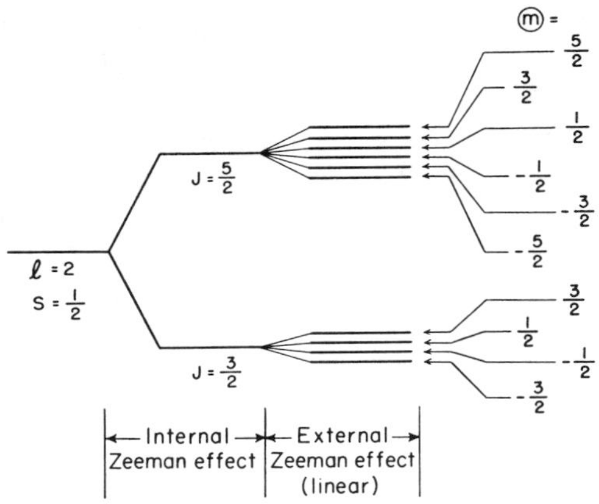

Fig. 11.1. Multiplet structure (internal Zeeman effect) and fine structure (external Zeeman effect) of term $l = 2$, $s = \frac{1}{2}$.

differences of 1) of all the **l** and **s** vectors of the individual electrons (see Fig. 10.4).

Before we lose ourselves in more detailed considerations of the multiplet structure, a reappraisal of the building principles on which the electronic structure of atoms is based may be in order. We know that quantum mechanics applies, that is, that each electron must form its standing-wave pattern. The Pauli exclusion principle prescribes that no two patterns may be identical, that is, no two electrons may occupy the same wave pattern at the same time. This statement is an additional axiom, not derivable from quantum mechanics. A consequence of the Pauli principle is that Fermi statistics applies to electronic problems instead of Maxwell-Boltzmann statistics.

Electrons by themselves cannot be distinguished. If electron 1 is characterized by a wave function ψ_{A_1} and electron 2 by a wave function ψ_{B_2}, the simultaneous existence of electron 1 in state A and electron 2 in state B corresponds to the product of their probability functions, that is, to a new wave function

$$\psi_{A_1B_2} = C\psi_{A_1}\psi_{B_2}. \tag{11.2}$$

There is, however, an equal chance that electron 2 might be found in A and electron 1 in B, that is, that the wave function might be

$$\psi_{A_2B_1} = C\psi_{A_2}\psi_{B_1}. \tag{11.3}$$

Because of this possible exchange of electrons the two states A_1B_2 and A_2B_1 are said to be in *quantum mechanical resonance*,[1] and the actual wave function describing the coexistence of the two states A and B is a linear combination of Eqs. 11.2 and 11.3,

$$\psi = C\psi_{A_1}\psi_{B_2} \pm C\psi_{A_2}\psi_{B_1}. \tag{11.4}$$

The function with the plus sign is called a *symmetric* function since an exchange of the electrons does not alter it. The function with the minus sign is *antisymmetric* because an exchange of the electrons leaves its magnitude unimpaired but changes its sign.

In case the electronic wave functions ψ_A and ψ_B are identical, the antisymmetric wave function is automatically equal to zero. This fact can be used for a simple formulation of the Pauli exclusion principle in the language of quantum mechanics. If we describe each electron of a multi-electron system by an individual wave function ψ which depends on its space co-ordinates x, y, and z, and on its spin, the total wave function of this system must be antisymmetric in the co-ordinates (including spin) of all its electrons. Thus the total wave function becomes automatically zero as soon as two electrons are alike.[2]

The wave functions ψ_A, ψ_B, ... of the individual electrons can be found in a first approximation by *Hartree's method of the self-consistent field.*[3] Here each electron is calculated as moving in a central field produced by the nucleus and all the other electrons. The order number Z of the nucleus appears reduced by a shielding constant S which represents the number of electrons located inward from the electron under consideration. The quantity Z-S takes the place of Z in the potential energy (Eq. 8.28) and in the eigenvalues (Eq. 8.30) of Schrödinger's equation.

The linear combination of these wave functions in forming the actual wave functions of an atom (see Eq. 11.4) expresses the fact that the motions of the electrons are correlated. For instance, if two electrons of parallel spin are close to each other, the antisymmetric wave function, and with it the probability $\psi\bar{\psi}$ of finding the electrons in the corresponding element of space, becomes very small; the electrons stay out of the way of each other. Vice versa, electrons with antiparallel spins characterized by the symmetric wave function of

[1] The concept of quantum-mechanical resonance was first introduced by Heisenberg [W. Heisenberg, *Z. Physik 39*, 499 (1926)] in his treatment of the helium atom. It has become of special importance for molecules, where the exchange energy provides the main strength of the electron-pair bond (see Sec. 14).

[2] See J. C. Slater, *Quantum Theory of Matter*, McGraw-Hill Book Co., New York, 1951, pp. 185 ff.

[3] D. R. Hartree, *Proc. Cambridge Phil. Soc. 24*, 89, 111 (1928); J. C. Slater, *Phys. Rev. 32*, 339 (1928).

their space co-ordinates have a strong tendency to approach each other closely in spite of electrostatic repulsion. This is a completely nonclassical effect which has a great influence on the energy of the electronic states. In the former case the electrostatic repulsion energy between the two electrons is small since they stay away from each other; in the latter case it will be large because the electrons are found near each other. Thus we expect that the wave function which is antisymmetric in the electron co-ordinates (parallel spins) corresponds to a state of lower energy than the symmetric one, because of its smaller repulsion term.

Although the spacing of the sublevels can be obtained only by detailed quantum-mechanical calculations, the multiplet structure as such may be derived relatively simply, at least in the case of L-S coupling (see Sec. 10). The total number of multiplet terms is given by the number of J values that the total angular momentum of the electronic system can assume. If L designates, as above, the resultant angular momentum composed of the individual orbital momentum vectors $\mathbf{l}$ and $\mathbf{S}$ the resultant spin vector composed of the spin momenta s, the total angular moment may vary as

$$J = (L + S),\quad (L + S - 1),\quad \ldots,\quad (L - S). \qquad (11.5)$$

Hence the total number of multiplet terms is $2S + 1$ for $L \geq S$ (Fig. 11.2); it reduces to $2L + 1$ for $L < S$.

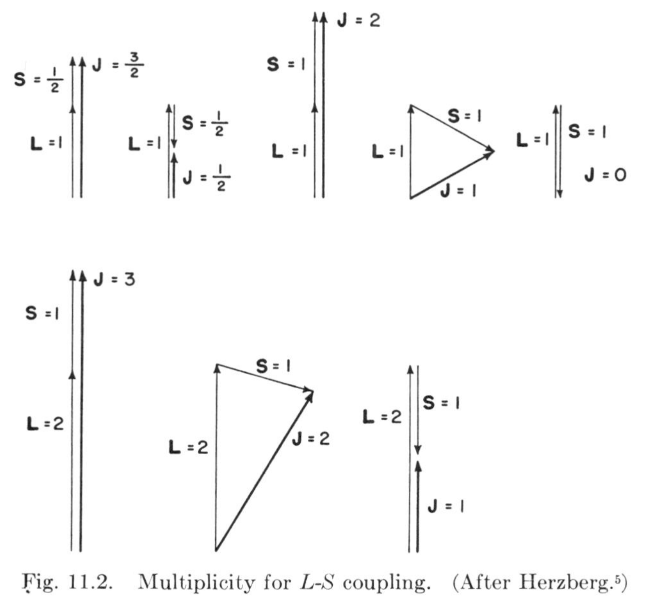

Fig. 11.2. Multiplicity for L-S coupling. (After Herzberg.[5])

The individual sublevels of a multiplet, since they refer to the coupled electron system of the atom as a whole, are designated by capital letters. In analogy to the designation $s, p, d, f, \ldots$ for the orbital moments $l = 0, 1, 2, 3, \ldots$ of the individual electrons (see Sec. 8), the capital letters $S, P, D, F, \ldots$ refer to the orbital moments $L = 0, 1, 2, 3 \ldots$ of the total electronic system. A superscript indicates the possible multiplicity $2S + 1$ and a subscript gives the inner quantum number J of the level.

The vector summation of the orbital and spin momenta is greatly simplified by the fact that the resultant

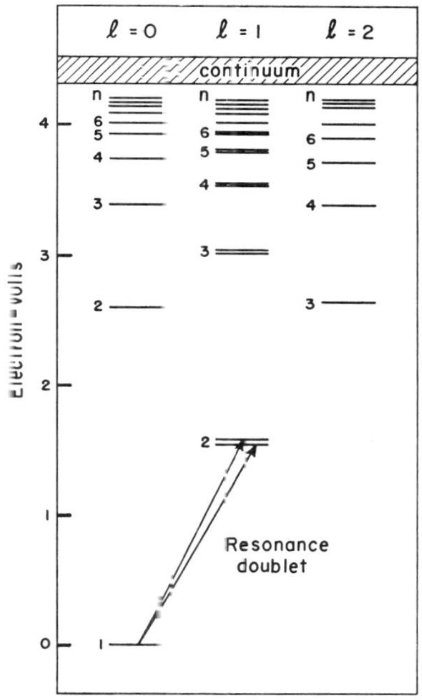

Fig. 11.3. Energy level diagram of potassium atom (doublet splitting shown only for $l = 1$). (After Grotrian.[4])

vectors L and S are zero for closed main shells and subshells (see Sec. 8). Hence only the electrons in uncompleted shells contribute to the angular momentum. The electron spins may line up either parallel (↑↑) or antiparallel (↑↓) to each other. The resultant spin vector $\mathbf{S}$ is therefore a half integral when the number of electrons is odd, and integral when it is even.†

A one-electron system from the standpoint of the multiplet structure is H or an alkali atom ($S = 1/2$); it can have only a doublet term structure ($2S + 1 = 2$). Differently expressed: for a system with one valence electron, L reduces to l, S to s, and the internal quantum number J to

$$j = l + s \quad \text{with} \quad s = \pm\frac{1}{2}. \qquad (11.6)$$

Figure 11.3 shows the characteristic doublet structure

† It is customary, because convenient, to speak in this slightly incorrect way about the moments while referring actually to their components in a given direction (see Fig. 8.9).

for the potassium atom;[4] the D-line doublet of sodium (see Fig. 4.3) in the yellow spectral region corresponds to the resonance transition from $n = 2, l = 1$ to $n = 1, l = 0$.

Proceeding to helium and the alkaline earth atoms, we have two outer electrons that can orientate either parallel ($S = 1$) or antiparallel ($S = 0$); hence triplet and singlet states result. Three electrons outside the atomic core lead to $S = 3/2$ ($\uparrow\uparrow\uparrow$) or $S = 1/2$ ($\uparrow\downarrow\uparrow$), that is, to quartets and doublets, etc. The possible term multiplicities are even for an odd and odd for an even number of valence electrons.

In obtaining the capital letter of the *term symbol*, the possible **L** values of the total orbital angular momentum, we have to consider more carefully whether the contributing electrons belong to the same subgroup (n,l) or to different subgroups. In the latter case of *nonequivalent* electrons all terms are allowed, since the Pauli exclusion principle is already fulfilled. For *equivalent* electrons of identical n and l, on the other hand, fewer terms are possible; the electrons must differ in one quantum number (Table 11.1). Since for an

Table 11.1. Possible multiplet terms for some equivalent and nonequivalent electrons

(See Fig. 11.2)

Electrons	Terms
s^2	$^1S_0$
ss	$^1S_0$, $^3S_1$
sp	$^1P_1$; $^3P_2$, $^3P_1$, $^3P_0$
p^2	$^1S_0$; $^1D_2$; $^3P_2$, $^3P_1$, $^3P_0$
pp	$^1S_0$; $^1P_1$; $^1D_2$; $^3S_1$; $^3P_2$, $^3P_1$, $^3P_0$; $^3D_3$, $^3D_2$, $^3D_1$

empty and for a filled subshell the total angular and the spin moment are zero ($L = S = 0$), the same term configuration is obtained for identical numbers of either electrons or vacancies in a shell (e.g., p^2 and p^4).[5]

The spherical electron clouds of the s electrons have no orbital angular momentum ($l = 0$) in quantum mechanics. The electron of hydrogen or the alkali atoms in its ground state is an s electron, hence the term must remain unsplit because there is no angular moment interacting with the spin vector. In spite of this, the term symbol of the normal state is written as $^2S_{1/2}$ because the superscript refers to the possible multiplicity. The normal state of helium and the alkaline earth atoms is the singlet state $^1S_0$ because a triplet ground state would correspond to two electrons with parallel electron spins in identical orbits, which violates the Pauli principle.† Table 8.1 lists the term symbols of the normal state for the atoms of the periodic system.

The width of the splitting caused by the magnetic interaction between orbital and spin moments varies from fractions of one cm^{-1} in the case of hydrogen to several thousand cm^{-1} for heavy atoms; it increases rapidly with the order number Z. For the doublet of the alkali atoms, for example, the splitting increases with Z^4; and although the lowest ^2P-level doublet is separated by only 0.34 cm^{-1} for lithium ($Z = 3$), the wave number difference has grown to 5540 cm^{-1} for cesium ($Z = 55$).

A typical example of the energy level diagrams of atoms and of the spectral lines resulting from transitions is shown in Fig. 11.4.

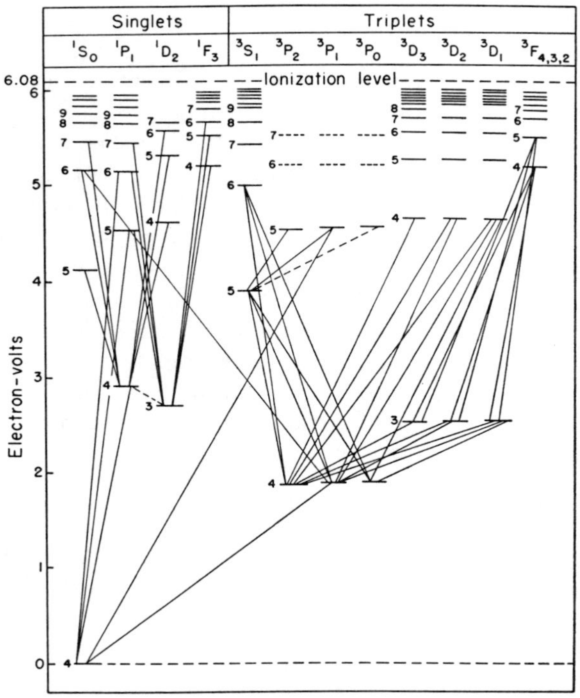

Fig. 11.4. Energy level diagram and transitions for the Ca atom. (After Grotrian.[4])

[4] W. Grotrian, *Graphische Darstellung der Spectren von Atomen und Ionen*, Springer, Berlin, 1928.

[5] See G. Herzberg, *Atomic Spectra and Atomic Structure*, Prentice-Hall, New York, 1937, p. 132.

† There are no intercombinations between the terms of the singlet and triplet system in helium because the electron spins would have to change from parallel to antiparallel. This type of transition is forbidden as inconsistent with dipole radiation. The two independent spectral systems led early investigators to distinguish between two different kinds of helium: *parahelium* (corresponding to the singlet system, superscript 1) and *orthohelium* (corresponding to the triplet system, superscript 3).

12 · Atoms in Electromagnetic Fields; the Dispersion Formula of Quantum Mechanics

In Sec. 4 the response of atoms to electromagnetic waves was treated on a purely classical basis. An atom was pictured as composed of noninteracting electronic oscillators of various resonance frequencies. The steady-state response of these oscillators to an alternating electric field prescribed the complex permittivity of the gas by the dispersion formula of classical physics (Eq. 4.29). The resonance frequencies ω_s of the s-oscillator types had to be assumed, and the intensity of the spectral lines had to be determined empirically. We are now ready to abandon this equivalent representation for a better model.

Let us first settle a question that began to trouble us in the development of the macroscopic theory. We found that the general characterization of a dielectric as the carrier of an electromagnetic field requires the determination of two complex parameters, the permittivity ϵ^* and permeability μ^*. We claimed that fortunately the situation simplifies in practice because, except in ferromagnetics, the magnetization caused by the electromagnetic wave is small as compared to the polarization. Thus μ^* can be replaced by the permeability of free space μ_0 for most practical purposes (see I, Sec. 9). Since we are now familiar with the structure of atoms, we can check this statement.

The polarization $\mathbf{P}$, the electric dipole moment per unit volume, is given, according to Sec. 1, as

$$\mathbf{P} = N\bar{\boldsymbol{\mu}} = N\alpha\mathbf{E}, \tag{12.1}$$

where N represents the number of atoms per unit volume of an atomic gas, $\bar{\boldsymbol{\mu}}$ the average electric dipole moment per atom appearing under the action of the applied electric field strength $\mathbf{E}$, and α the polarizability. In complete analogy we can write the magnetization $\mathbf{M}$, the magnetic dipole moment per unit volume (cf. I, Sec. 2), as

$$\mathbf{M} = N\overline{\mathbf{m}} = N\alpha_m\mathbf{H}, \tag{12.2}$$

where $\overline{\mathbf{m}}$ designates the average magnetic dipole moment caused by the applied magnetic field strength $\mathbf{H}$, and α_m a *magnetizability*. The electric and magnetic susceptibilities follow from I, Eqs. 2.7 and 2.18 as

$$\chi \equiv \kappa' - 1 = \frac{\mathbf{P}}{\epsilon_0\mathbf{E}} = \frac{N\alpha}{\epsilon_0}, \qquad \chi_m \equiv \kappa_m' - 1 = \frac{\mathbf{M}}{\mathbf{H}} = N\alpha_m, \tag{12.3}$$

where κ' and κ_m' are the relative permittivity and permeability of the gas. Hence the ratio of electric to magnetic susceptibility of the gas is given by the ratio of polarizability to magnetizability of the atoms as

$$\frac{\chi}{\chi_m} = \frac{\alpha}{\alpha_m}\frac{1}{\epsilon_0}. \tag{12.4}$$

For the hydrogen atom, if we refer at present to induced moments only and to the Bohr model for an order of magnitude calculation, we obtain from Eq. 3.8 the polarizability

$$\alpha = \epsilon_0 4\pi r_0{}^3, \tag{12.5}$$

and from Eq. 10.31 the magnetizability

$$\alpha_m = \frac{\overline{\mathbf{m}}_i}{\mathbf{H}} = -\frac{e^2 r_0{}^2}{6m}\mu_0. \tag{12.6}$$

Hence

$$\frac{\chi}{\chi_m} = -\frac{24\pi m r_0}{e^2\mu_0} \simeq 8 \times 10^4; \tag{12.7}$$

the electric susceptibility is about 10^5 times that of the diamagnetic susceptibility. Thus, in discussing polarization effects in electromagnetic fields we may, in general, neglect magnetization effects.

We return now to our classical dispersion formula

$$\kappa^* = 1 + \sum_s \frac{N_s e^2/\epsilon_0 m}{\omega_s{}^2 - \omega^2 + j\omega 2\alpha}, \tag{4.29}$$

which represents the relative permittivity of an oscillator gas as a function of the frequency ω of the incident electromagnetic radiation. Each resonance frequency ω_s, according to Bohr's frequency condition (Eq. 6.1), corresponds to a transition between two stationary states of the total energy $\mathcal{E}_i$ and $\mathcal{E}_j$, that is, we have to substitute

$$\omega_s \to \omega_{ij} = \frac{\mathcal{E}_i - \mathcal{E}_j}{\hbar}. \tag{12.8}$$

The stationary states of atoms are prescribed by the energy-level diagram as discussed in the preceding section. However, as Fig. 11.4 already indicates, not all imaginable transitions between these states occur; only selected ones are permitted. Quantum mechanics gives various *selection rules* which specify the possible transitions. They may be expressed as the changes Δ allowed for the various quantum numbers. The principal quantum number n does not underlie any restrictions; any value

$$\Delta n = 0,\ 1,\ 2,\ 3,\ \ldots \tag{12.9}$$

is permissible. Transitions with $\Delta n = 0$ correspond to transitions between multiplet structure levels which may even lie in the spectral region of the electrical engineers. The sequence from $\Delta n = 1$ and up represents the different members of a spectral series (see Eq. 5.17).

For the azimuthal quantum number l the prescription is

$$\Delta l = \pm 1; \tag{12.10}$$

practically, only those states for which the l value differs by one unit can combine. This rule has a graphic physical interpretation: a photon $h\nu$ carries an angular momentum

$$p'_{\text{photon}} = \hbar. \tag{12.11}$$

Hence the selection rule (Eq. 12.10) expresses the law of conservation of angular momentum for the system, atom + photon.

Although the condition Eq. 12.10 must be fulfilled for the electron making the quantum jump, the overall orbital angular momentum L and the total angular momentum J of a multi-electron system can change as

$$\begin{aligned} \Delta L &= 0,\ \pm 1, \\ \Delta J &= 0,\ \pm 1. \end{aligned} \tag{12.12}$$

Furthermore the magnetic quantum numbers obey the selection rules

$$\begin{aligned} \Delta \textcircled{m} &= 0,\ \pm 1, \\ \Delta \textcircled{M} &= 0,\ \pm 1, \end{aligned} \tag{12.13}$$

but $\Delta \textcircled{M} = 0$ is forbidden for $\Delta J = 0$.

These and other special selection rules summarize, in not very lucid prescriptions, a clearly understandable physical situation: (*a*) the system, atom + photon, has to preserve the conservation laws of energy and momentum; (*b*) transitions violating the Pauli exclusion principle are not allowed (see Table 11.2); and (*c*) the atom can radiate and absorb efficiently only when the transition provides an electric dipole moment with which the atom can couple to the electromagnetic radiation field (*dipole radiation*).

In our classical picture the dipole moment was provided by the displacement of the electronic oscillator from its equilibrium position (see Eq. 4.21). In quantum physics, the radiation takes place as a transition between two states $i \longrightarrow j$, and to each such transition we may ascribe a "virtual oscillator" of the resonance frequency ω_{ij}. The dipole moment $\boldsymbol{\mu}_{ij}$ of this oscillator is found by quantum mechanics to be the average dipole moment of the mixed wave functions of the two states

$$\mu_{ij} = \int_{\text{all space}} \psi_i(e\mathbf{r})\tilde{\psi}_j \, d\tau, \tag{12.14}$$

known as the *matrix element of the dipole moment.*

The electromagnetic energy stored and dissipated is proportional to the complex permittivity (see I, Eq. 11.9) and varies with the square of the electric dipole moment (see I, Eq. 13.18). The e^2 factor in the classical Eq. 4.29 contains this dipole moment term, and an accurate quantum-mechanical calculation [1] leads to the substitution expression

$$\frac{e^2}{m} \longrightarrow 2\frac{\omega_{ij}}{\hbar} \,|\, \boldsymbol{\mu}_{ij} \,|^2. \tag{12.15}$$

Finally, one additional change is required to transform the classical dispersion equation into its quantum-mechanical counterpart. In the classical formula the oscillators appear only as absorbers of the incident radiation; in quantum physics, since a transition between two states is involved, the atoms in the lower energy state (N_i) may absorb, whereas atoms in the higher energy state (N_j) may be forced by the radiation to return to the lower energy state. This *forced emission* corresponds classically to special phase relations between oscillator and field and will manifest itself in a reduced absorption. Thus we have to substitute for the number of dispersion electrons

$$N_s \longrightarrow N_i - N_j. \tag{12.16}$$

Carrying out the substitutions indicated in Eqs. 12.8, 12.15, and 12.16 we transform the disperion formula of classical physics (Eq. 4.29) into the *dispersion formula of quantum mechanics:* [2]

$$\kappa^* = 1 + \frac{2}{\epsilon_0\hbar} \sum_{i<j} \frac{\omega_{ij} |\, \boldsymbol{\mu}_{ij} \,|^2}{\omega_{ij}{}^2 - \omega^2 + j\omega 2\alpha} (N_i - N_j). \tag{12.17}$$

[1] See H. Bethe, *Handbuch der Physik*, A. Smekal, Ed., Springer, Berlin, 1933, Vol. 24, Pt. 1.

[2] A first formulation of the dispersion formula of quantum physics was given by H. A. Kramers and W. Heisenberg [*Z. Physik 31*, 681 (1925)]. For a detailed discussion of the "Quantum Theory of Dispersion," see G. Breit, *Revs. Mod. Phys. 4*, 504 (1932); *5*, 91 (1933).

The ratio of quantum mechanical to classical-coupling factor,

$$f_{i\to j} \equiv \frac{2\omega_{ij}|\,\boldsymbol{\mu}_{ij}\,|^2}{\hbar} \bigg/ \frac{e^2}{m}, \qquad (12.18)$$

is called the *oscillator strength* of the transition $i \longrightarrow j$.

The *average life time* τ of the atom in the excited state j before returning to the state i is according to the classical decay law (see Eqs. 4.6 and 4.14), if we replace the permeability μ_0 of free space by $1/c^2\epsilon_0$ (see I, Eq. 8.1),

$$\tau = 4\pi\epsilon_0 \frac{3mc^3}{2e^2\omega_{ij}^2}. \qquad (12.19)$$

The ratio of oscillator strength to life time is designated as the *transition probability of the transition* $i \longrightarrow j$:

$$A_{i\to j} = \frac{f_{i\to j}}{\tau} = \frac{1}{4\pi\epsilon_0} \frac{64\pi^4\nu_{ij}^3|\,\boldsymbol{\mu}_{ij}\,|^2}{3hc^3}. \qquad (12.20)$$

The transition probability, times the quantum $h\nu_{ij}$ emitted, gives the *intensity of the spectral line* per emitting atom, the energy radiated per second

$$P_{j\to i} = \frac{1}{4\pi\epsilon_0} \frac{64\pi^4}{3} \frac{\nu_{ij}^4}{c^3} |\,\boldsymbol{\mu}_{ij}\,|^2 \quad \text{[watt]}. \qquad (12.21)$$

This intensity is identical to that of a classical dipole antenna of the average dipole moment $\boldsymbol{\mu}_{ij}$ emitting a frequency ν_{ij}, as the equivalence of Eq. 12.21 with the dipole radiation equation (I, Eq. 13.18) confirms.

Table 12.1 gives some calculated data on oscillator strength and life time for the hydrogen atom.[1]

Table 12.1. Oscillator strength and life time for some transitions of the hydrogen atom

Wavelength (Å)	Transition	$f_{i\to j}$	τ (10^{-8} sec)
1216	$1s - 2p$ (resonance line)	0.416	0.16
6563	$2p - 3d$	0.694	1.56
18751	$3d - 4f$	1.016	7.3
40532	$4f - 5g$	1.345	23

13 · The Formation of Molecules

Until now we have been concerned with the structure of the electronic atmosphere that surrounds *one* nucleus. We found that the electrons assume energy states which form a complicated pattern of multiplet terms (Sec. 11). These stationary states are prescribed by quantum mechanics as the standing-wave patterns of probability waves. The electrons are trapped by the Coulomb field of the nucleus, and also their mutual interaction is, in the main, of an electrostatic nature. However, the electron clouds may have, in addition, orbital magnetic moments, and the electrons themselves have magnetic spin moments, by which they affect each other through magnetic dipole forces. This magnetic interaction produces a coupling of the orbital with the spin moments, and not only the individual moments but also their vector resultants have to be quantized. Each electron, according to the Pauli principle, is characterized by a unique set of four quantum numbers, but electrons, per se, cannot be distinguished and might therefore be interchanged. The coexistence of several electronic states expresses itself in quantum mechanics as the product of wave functions and the possibility of interchange by linear combinations of these products (see Eq. 11.4). These linear combinations are the stationary states actually realized; the contribution of various possible states to the actual wave function produces an extra coupling effect known as quantum-mechanical resonance.

In extending the discussion to the electron clouds of molecules we may expect that the same considerations remain valid, but that the electronic systems surrounding *several* nuclei will prove appreciably more difficult to treat. For once the systems are of lower symmetry. In diatomic molecules the spherical symmetry of atoms is replaced by a cylindrical symmetry around the molecular axis; in nonlinear, triatomic molecules only one or two planes of symmetry remain; in still more complex structures no symmetry at all may be left. Furthermore, the possible wave modes depend critically on the distance of the nuclei and their mutual orientation in space. To develop concepts for handling such situations, we consider first the interaction of atoms as a function of their nuclear separation.

In approaching each other, atoms will initially undergo an attraction. This was already concluded by van der Waals in his discussion of the gas laws.[1] An *ideal gas*, held under a pressure p in a volume V at a temperature T, obeys the equation of state

$$pV = nRT, \qquad (13.1)$$

[1] J. D. Van der Waals, *Over de Continuiteit van den Gas en Vloeistoftoestand*, Leyden, 1873.

where V/n represents the volume containing a mole of the gas (the mole volume), and R the *gas constant* per mole (see Eqs. 2.13 and 5.27):

$$R = N_0 k = 8.314 \quad [\text{joule mole}^{-1}\ \text{deg}^{-1}] \\ = 1.9864 \quad [\text{calories mole}^{-1}\ \text{deg}^{-1}].\dagger \tag{13.2}$$

This law does not take into account the forces between the gas molecules or the volume V_i that they actually fill. Gases may condense into liquids and solids, hence their molecules must attract each other. Van der Waals took note of this fact by introducing an internal pressure p_i which assists the external pressure p_a. He furthermore referred to the volume actually available as the difference of the volume V_a of the gas container and of the volume V_i. Thus he arrived at the *van der Waals equation for real gases:*

$$(p_a + p_i)(V_a - V_i) = nRT. \tag{13.3}$$

London [2] gave the *van der Waals attraction force* represented by p_i its molecular interpretation. The attraction between atoms is due to their mutual polarization by fluctuating dipole moments. These induced moments arise because the outer electrons of the approaching atoms tend to keep away from each other by electrostatic repulsion. Thus the motion of these electrons in neighboring atoms becomes correlated and a dipole-dipole coupling force results.

Atom 1, developing a moment $\boldsymbol{\mu}_1$, creates in atom 2, at a distance r, a moment

$$\boldsymbol{\mu}_2 = \alpha \mathbf{E}; \tag{13.4}$$

the inducing field strength $\mathbf{E}$ stemming from $\boldsymbol{\mu}_1$ is (see I, Eq. 3.19)

$$E \simeq \frac{|\boldsymbol{\mu}_1|}{\epsilon_0 2\pi r^3}. \tag{13.5}$$

The energy of interaction for two dipoles lined up end to end is found easily by the superposition of their Coulomb potentials (see I, Eq. 10.4) as

$$U_e = -\frac{|\boldsymbol{\mu}_1|\,|\boldsymbol{\mu}_2|}{\epsilon_0 2\pi r^3} = -\frac{|\boldsymbol{\mu}_1|\alpha E}{\epsilon_0 2\pi r^3}. \tag{13.6}$$

For random orientation of the dipoles we obtain an average binding energy [3]

$$U = -\frac{3}{2}\frac{\alpha|\boldsymbol{\mu}_1|^2}{(\epsilon_0 4\pi)^2}\frac{1}{r^6}. \tag{13.7}$$

† 1 calorie (gram-calorie at 4°C) = 4.185 joules.

[2] F. London, *Z. physik. Chem. B11*, 222 (1930); *Z. Physik 63*, 245 (1930); *Trans. Faraday Soc. 33*, 8 (1937). See also H. Margenau, *Phys. Rev. 38*, 747 (1931).

[3] See, for instance, J. C. Slater, *Introduction to Chemical Physics*, McGraw-Hill Book Co., New York, 1939, pp. 363 ff.

This *van der Waals binding energy*, which is proportional to the electronic polarizability α of the atoms and to their mean square dipole moment, is inversely proportional to the sixth power of the interatomic distance. The *van der Waals force* between neutral atoms, the first space derivative of U, is therefore inversely proportional to the seventh power of the distance; it is short-reaching. Electrostatic forces of longer range are obtained between particles carrying permanent dipole moments or even net charges. Such forces are responsible, for example, for the peculiar properties of colloidal systems.

As two atoms draw closer together under the influence of the van der Waals attraction, their electron

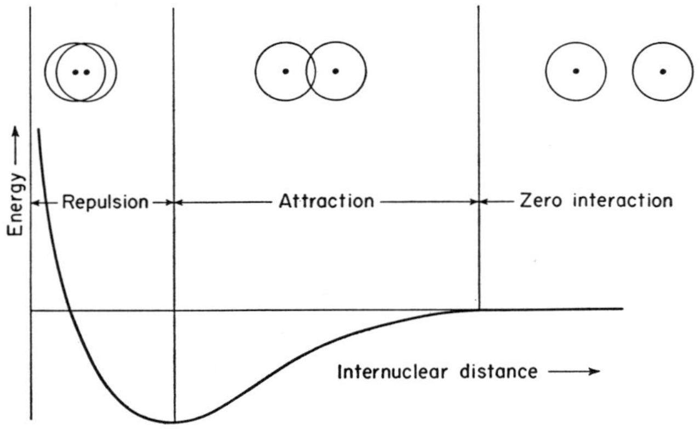

Fig. 13.1. Electrostatic energy between two rigid spherical atoms. (Adapted from Slater.[4])

clouds begin to overlap. According to electrostatics, this causes a true *Coulomb interaction*, and the attraction increases rapidly. Gauss's law shows that a spherical charge distribution acts towards the outside as if the charge were concentrated at the center (see I, Eqs. 3.6 and 4.2). Consequently, a rigid atom, that is, a positive nucleus surrounded by a neutralizing spherical electron cloud of finite radius, exerts no electrostatic force at outside points. However, when we penetrate into the negative charge distribution, only the charge contained in the sphere from the reference point inwards is found to contribute; the effect of the negative charge outside this smaller sphere has to be neglected (see I, Sec. 4). The nearer we move towards the center, the less shielded is the positive charge of the nucleus. Thus we expect a potential energy characteristic as function of the internuclear distance of the partners that starts with zero at large distance, then falls, as first the van der Waals attraction and thereafter the Coulomb attraction between the electron clouds of the atoms and the nuclei of their partners come into play, and finally rises steeply as the repulsion between the nuclei takes over (Fig. 13.1).

Seen from the standpoint of the electrons, the Coulomb potential of each nucleus represents a deep, funnel-shaped trap. When the electron atmospheres begin to overlap, the potential barrier between the nuclei will be lowered (Fig. 13.2).[4] This effect tends to concentrate more electric charge between the nuclei, until finally, for two hydrogen atoms, the quadrupole arrangement ($+e \leftarrow \; -2e \rightarrow e+$) might result. The binding energy of this constellation, as a comparison of its electrostatic potential with that of two separate hydrogen atoms shows, is $-3.5(e^2/\epsilon_0 4\pi r)$ or 1.75 times

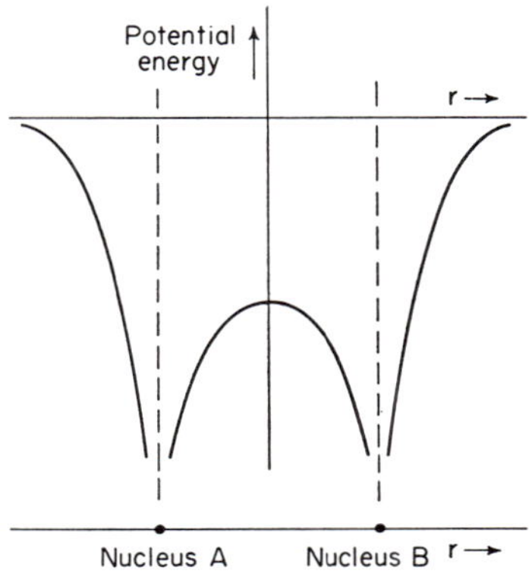

Fig. 13.2. Potential double well formed by two like atoms.

that of the two separate atoms. The Coulomb interaction, by which the electron clouds of atom 1 are attracted to the nucleus of atom 2, and vice versa, depends, in contrast to the van der Waals attraction, on the average and not on a fluctuating field. It produces, in consequence, a large, overall polarization of the electron clouds.

Actually, before this occurs, quantum mechanics interferes through the Pauli exclusion principle, which forbids shared electrons to occupy identical quantum states. Bonding will therefore take place only when the interpenetrating electron clouds can form joint standing-wave patterns. This is simple when the interacting orbitals are partly filled (see Table 8.1); such orbitals may link together by contributing electrons from each partner, thus forming electron pairs with antiparallel spins. Lewis [5] postulated, on chemical grounds, the existence of such *electron-pair bonds* even ten years before the advent of quantum mechanics. The electron-pair bond is the normal *covalent* (*homopolar*) *bond* of chemistry.

If such bonding is not possible, the Pauli principle can be satisfied only if one of the partners goes over into an excited state. Usually the collision energy will not suffice to bring this about, but, as a pre-stage for such an eventuality, the electron clouds tend to avoid each other. Hence, in contrast to the classical prediction, the electron density between the nuclei diminishes and a strong repulsion results. This repulsion balances the van der Waals attraction at a distance of nearest approach, and defines thus a *van der Waals radius* r_v of the atoms. The volume V_i of van der Waals' equation (Eq. 13.3) is the sum of the gas-kinetic volume $(4\pi/3)r_v^3$ of all the molecules contained in V_a.

The van der Waals radii can be determined by various means: by gaskinetic experiment, X-ray or electron diffraction, and, most directly, by the scattering of atomic beams.[6] When electron-pair bonds are formed, the nuclei will obviously approach each other appreciably closer. *Covalent radii* r_c for these bonds have been calculated by Bragg, Goldschmidt, Pauling and Huggins [7], and others from investigations of the structure of homopolar compounds. Furthermore, interacting atoms may give up or take on electrons, thus forming ions of opposite polarity which attract each other by Coulomb forces. Again, a distance of closest approach results when this attraction is balanced by the repulsion called into play by the Pauli principle. By measuring these distances in ionic compounds through X-ray analysis, Goldschmidt [8] first derived tables of *ionic radii* r_i, which were later extended and modified, especially by Pauling.[7] Finally, by measuring interatomic distances in metals, *metallic radii* r_m may be derived; the *metallic bond* is of an intermediate character between covalent and ionic binding. Tables 9.1 and 13.1 summarize typical values for these types of radii.[9]

These values should not be taken too seriously. It is convenient to talk about distinct classes of bonds, but frequently mixed bond types are realized (see Sec. 14). Furthermore, the radius of an atom varies with its environment. For accurate computations it is therefore preferable to refer to tables of interatomic distances. A critical table of this kind has been published by Huggins.[10] Concerning the *hydrogen bond* and its special role see Sec. 24.

[4] J. C. Slater, *Quantum Theory of Matter*, McGraw-Hill Book Co., New York, 1951.

[5] G. N. Lewis, *J. Am. Chem. Soc. 38*, 762 (1916).

[6] I. Amdur, M. C. Kells, and D. E. Davenport, *J. Chem. Phys. 18*, 1676 (1950); *20*, 1620 (1952); H. W. Berry, *Phys. Rev. 75*, 913 (1949).

[7] See L. Pauling and M. E. Huggins, *Z. Krist. 87*, 205 (1934); L. Pauling, *The Nature of the Chemical Bond*, Cornell University Press, Ithaca, N. Y., 1940, pp. 160 ff and p. 346.

[8] V. M. Goldschmidt, *Oslo. Akad. Wiss. 7* (1926), *8* (1927); *Trans. Faraday Soc. 25*, 253 (1929).

[9] Data from L. Pauling, *J. Am. Chem. Soc. 69*, 542 (1947) and Ref. 7; C. A. Coulson, *Valence* Clarendon Press, Oxford, 1952.

[10] M. L. Huggins, *J. Am. Chem. Soc. 75*, 4126 (1953).

Table 13.1. Radii of atoms and ions [9]

Atomic Ion	Covalent r_c	Metallic r_m	Ionic r_i	Van der Waals r_v	Atomic Ion	Covalent r_c	Metallic r_m	Ionic r_i	Van der Waals r_v
H	0.37 A			1.2 A	Co	1.16 A	1.25 A		
Li	1.34	1.55 A			Ni	1.15	1.24		
Li^+			0.47 A		Cu	1.17	1.28		
B	0.90	0.98			Zn	1.25	1.38		
C †	0.77	0.91			Ge	1.20	1.37		
N †	0.73			1.5	As	1.22			
O †	0.74			1.4	Se	1.17	1.40		2.0 A
O^{2-}			1.40		Se^{2-}			1.98 A	
F	0.72			1.4	Br	1.14		1.60	1.95
F^-			1.11		Br^-				
Na	1.54	1.89			Rb	2.11	2.48		
Na^+			0.87		Rb^+			1.48	
Mg	1.36	1.60			Sr	1.91	2.15		
Mg^{2+}			0.65		Sr^{2+}			1.13	
Al	1.25	1.43			Zr	1.45	1.60		
Al^{3+}			0.50		Zr^{4+}			0.80	
Si	1.16	1.32			Ag	1.34	1.44		
Si^{4+}			0.41		Cd	1.41	1.54		
P	1.10				In	1.50	1.66		
S	1.04				Sn	1.40			
S^{2-}			1.84		Sb	1.41			
Cl	0.99			1.8	Te	1.41	1.60		2.2
Cl^-			1.47		Te^{2-}			2.21	
K	1.96	2.35			I	1.33			2.2
K^+			1.18		I^-			1.78	
Ca	1.74	1.97			Cs	2.35	2.67		
Ca^{2+}			0.99		Cs^+			1.69	
Sc	1.44	1.62			Ba	1.98	2.21		
Sc^{3+}			0.81		Ba^{2+}			1.35	
Ti	1.32	1.47			Pt	1.29	1.38		
Ti^{4+}			0.68		Au	1.34	1.44		
V	1.22	1.34			Hg	1.44	1.57		
Cr	1.17				Tl	1.55	1.71		
Mn	1.17				Pb	1.54	1.75		
Fe	1.16	1.26							

† Single bonds; for multiple-bond values see Table 17.1.

Ionic and van der Waals radii represent the atoms as spheres, since the interaction forces have spherical symmetry. Covalent radii, on the other hand, arise, in general, from nonspherical wave functions (see Figs. 8.5 and 8.7); hence covalent bonds are directed into space and the angles between them prescribed by quantum mechanics. By cutting sections from van der Waals spheres in the proper orientation and depth, models are made in which the nuclei can approach each other in the bond directions up to the covalent distance (Fig. 13.3). Such models [11] illustrate the stereostructure of homopolar molecules with great clarity.

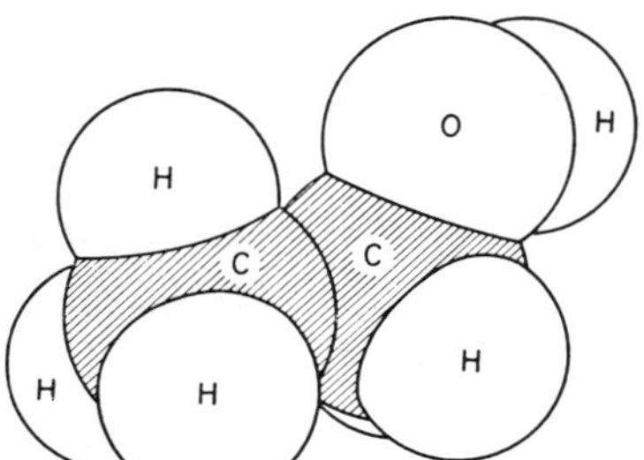

Fig. 13.3. Model of alcohol molecule C_2H_5OH.

When two atoms or molecules collide, the conservation laws of energy and momentum will normally compel the partners to separate again. To keep them together requires the participation of a third body which can take over the excess of momentum (*three-body impact*). However, the particles emerging from a collision need not be identical with the initial ones. The

[11] They were first proposed by M. Magat [*Z. physik. Chem. B16*, 1 (1932)], introduced in Germany by H. A. Stuart (see *Struktur des freien Moleküls*, Springer, Berlin, 1952), and in this country made available as Fisher-Hirschfelder-Taylor atom models (Eimer and Amend, New York).

motion of the nuclei is slow in comparison to the speed with which the electronic systems can rearrange themselves (*Franck-Condon principle*) (see also Appendix AII, Sec. 8).[12] The electronic clouds therefore assume quasi-instantaneously during a collision the wave modes prescribed by the mutual constellation of the nuclei and the total energy of the system. In this transient molecular state nuclei, electrons, and energy may be exchanged and radiation emitted. It is this uncertainty of the outcome and the variety of products that may form in the breakup of such systems that make the chemistry of molecular reactions complicated and exciting.

[12] J. Franck, *Trans. Faraday Soc. 21*, 536 (1925); E. U. Condon, *Phys. Rev. 28*, 1182 (1926); *32*, 858 (1928); *41*, 759 (1932).

14 · Wave Functions of Molecules and the Concept of Quantum-Mechanical Resonance

In deriving the electronic structure of molecules, we have the choice between two antipodean approaches. We may start with the individual atoms as in the preceding section and investigate quantitatively how the wave functions change when the separation distance shrinks to the actual molecular spacings. This is the *Heitler-London method*[1] of overlapping atomic wave functions. Alternatively, we may place, at the outset, the nuclei at their true equilibrium positions and fill molecular states by adding the neutralizing electrons one by one. This *method of molecular orbitals* repeats, on a more complicated scale, the procedure used in building up the electronic structure of atoms (see Sec. 8); it was originated by Hund[2] and Mulliken.[3]

Both approaches are approximations which have their special advantages and shortcomings. The Heitler-London method gives a graphic account of the interaction of atoms, but fails when new molecular orbitals arise. The Hund-Mulliken procedure may describe the electronic structure of the molecule appropriately for the nuclei near their equilibrium positions but fails when, at larger internuclear distances, the atoms begin to act as individuals. Either starting point may lead to the same correct solution, as especially Slater[4] has shown, if the treatment is extended by perturbation methods. However, this refinement requires a prodigious amount of labor in more complicated cases, and only the latest computing machine techniques offer a prospect of producing quantitative results.

[1] W. Heitler and F. London, *Z. Physik 44*, 455 (1927).

[2] F. Hund, *Z. Physik 51*, 759 (1928); *63*, 719 (1930); *Handbuch der Physik, 1933*, Vol. 24, Pt. I.

[3] R. S. Mulliken, *Phys. Rev. 32*, 186 (1928); *33*, 730 (1929); *Revs. Mod. Phys. 4*, 1 (1932).

[4] See J. C. Slater, *Quantum Theory of Matter*, McGraw-Hill Book Co., New York, 1951, Chap. 8.

The show piece of the molecular theory is the hydrogen molecule. It has been treated relatively rigorously from both approaches, and will serve here to develop some essential concepts.

While discussing the wave functions of atoms we applied the Heitler-London method (see Eqs. 11.2 to 11.4) implicitly. We need only to identify A and B with the two hydrogen atoms, and ψ_A and ψ_B with the s-wave functions of their ground states. The fact that each of these functions is simultaneously occupied by one electron is expressed by the product of the wave functions, $\psi_A\psi_B$, and the possibility that the electrons $_1$ and $_2$ can be interchanged by the linear combination of the products: $\psi_{A_1}\psi_{B_2} \pm \psi_{A_2}\psi_{B_1}$. Thus, by compounding the H_2 molecule from the unperturbed wave functions of the two atoms, the Heitler-London approach arrives at the two possible wave functions

$$\psi_{H_2} \begin{cases} = C'(\psi_{A_1}\psi_{B_2} + \psi_{A_2}\psi_{B_1}), \\ = C''(\psi_{A_1}\psi_{B_2} - \psi_{A_2}\psi_{B_1}), \end{cases} \tag{14.1}$$

the one symmetric (+), the other antisymmetric (−) in the electron co-ordinates (exclusive spins).

The total energy of these two molecular states, $\mathcal{E}_\pm$, is found to consist of three contributions:[4] (1) The energy of the two separate atoms, $-2Rhc$ (see Eq. 8.16). (2) A Coulomb energy H_0, that is, the electrostatic interaction energy (4 terms) between the two atoms, where electron cloud 1 surrounds nucleus A and electron cloud 2 nucleus B. These four terms will mutually cancel unless the two electron clouds interpenetrate (see Fig. 13.1). And (3) a *nonclassical exchange energy* H_1 several times larger in magnitude than the Coulomb energy. An accurate calculation leads to the expression

$$\mathcal{E}_\pm = -2Rhc + \frac{H_0 \pm H_1}{1 \pm \alpha^2}. \tag{14.2}$$

The quantity α represents the product of the two atomic wave functions when occupied by the *same* electron and integrated over all space, the so-called *overlap integral*,[5]

$$\alpha = \int \psi_{A_1}\psi_{B_1}\, dV = \int \psi_{A_2}\psi_{B_2}\, dV. \qquad (14.3)$$

The magnitude of this integral should obviously depend on the internuclear spacing, and α proves distance-sensitive as is easily confirmed: the simple exponential wave functions ψ_A and ψ_B (see Eq. 8.19) fall rapidly with the distance from the nucleus A and B, respectively, hence the main contribution to α is made by the parts of the wave functions midway between A and B. For widely separated nuclei the quantity α approaches zero; in the opposite extreme of coalescent nuclei, an He^+ ion is formed, and α represents the overall intensity of its probability wave, which is unity (see Eq. 7.20). Thus α increases from $0 \rightarrow 1$ as the nuclei of the two hydrogen atoms approach each other.

The quantities C' and C'' in Eq. 14.1 are normalizing factors which take care of the condition that the intensity of a probability wave must integrate to unity (Eq. 7.20). Since the wave functions of the hydrogen molecule contain the distance-sensitive factor α, the normalizing factors must also depend on the internuclear distance. The calculation shows that

$$C' = \frac{1}{\sqrt{2(1 + \alpha^2)}}, \qquad C'' = \frac{1}{\sqrt{2(1 - \alpha^2)}}. \qquad (14.4)$$

If the amount of charge of one electron, held jointly by both wave functions, is designated as its exchange charge $-\alpha e$, the exchange energy H_1 is the electrostatic energy of interaction (4 terms) between the two identical, distributed exchange charges and their positive countercharges $+\alpha e$ at the two nuclei.

The exchange energy, plotted as function of the nuclear distance, has the same general trend as the Coulomb energy. Initially, the attraction between the electronic charges and their nuclear countercharges dominates and the energy becomes negative; as the nuclei close in, their mutual repulsion dominates and drives the energy steeply positive.

Since the exchange energy enters the symmetric wave function with a plus sign, it leads to a potential minimum and binding. The symmetric wave function, to satisfy the Pauli principle, has to contain the two electrons with antiparallel spins. Thus the H_2 molecule is formed by the electron pair bond described in Sec. 13.

[5] See, for the relation between overlap integrals and chemical binding, for example, R. S. Mulliken, *J. Am. Chem. Soc. 72*, 4493 (1950).

It should now have become quite clear that the energy of the electron pair bond does not stem from the magnetic interaction energy of the antiparallel magnetic spins but from the electrostatic energy of the quantum-mechanical exchange forces.

The antisymmetric wave function contains the electrons with parallel spins, the exchange energy enters with a minus sign in Eq. 14.2, the hydrogen atoms repel each other, and a nonbonding state without potential minimum results.

The H_2 molecule as a two-electron system can have singlet and triplet states, the former corresponding to antiparallel, the latter to parallel electron spins (see Sec. 11). It is customary to designate the states of molecules by *capital Greek letters* in contradistinction to the capital Roman letters used for the states of atoms. Σ states have no orbital angular momentum around the axis of the molecule. They are realized here because the composing *s*-functions of the atoms have no angular momentum. The symmetric wave function of Eq. 14.1 refers accordingly to a *bonding* singlet Σ state, whereas the antisymmetric function corresponds to an *antibonding* triplet Σ state (Fig. 14.1).

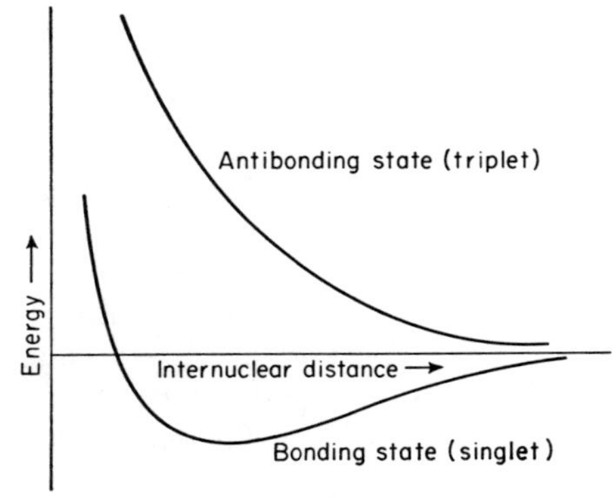

Fig. 14.1. Bonding and antibonding states of the H_2 molecule.

From the standpoint of the molecular orbital method, the discussion begins with the two protons in place and initially only one electron added, that is, with the hydrogen molecule-ion H_2^+. This molecule is the only one for which the wave equation can be solved exactly. The problem can be formulated in ellipsoidal co-ordinates with the nuclei as the foci of the ellipsoids; for this situation the variables can be separated in Schrödinger's equation.[6]

The two nuclei represent electrostatic potential wells with a saddle between them (see Fig. 13.2). An electron, added to one of the protons, can *classically* trans-

[6] P. M. Morse and E. C. G. Stueckelberg, *Phys. Rev. 33*, 932 (1929); P. M. Morse, *ibid. 34*, 57 (1929); E. Teller, *Z. Physik 61*, 458 (1930).

fer to the other proton only by jumping over the potential wall, a process requiring a high activation energy. Quantum mechanics allows the electron to leak through the potential barrier by the so-called *tunnel effect.*

This penetrating of barriers is actually not as absurd as it first seems; it has its analogue in classical physics. In I, Sec. 16, we discussed the case of total reflection and mentioned that the electromagnetic wave penetrates for some distance into the totally reflecting medium 2 before being returned to medium 1. If we slice medium 2 thin enough, some light will leak through; this trick has been used, for instance, in providing illumination for an ultramicroscope. Similarly the probability waves of quantum mechanics leak through potential barriers at a rate that rapidly increases as the barrier shrinks in height or width. As time passes, the electron will be found with increasing probability on the other side.

For the double well of the H_2^+ ion this leakiness implies that, in the stationary state, the electron will be found with equal probability at nucleus A or B. The actual wave function is therefore a linear combination of the two individual s-functions ψ_A and ψ_B,

$$\psi_{H_2^+} \begin{cases} = C_1(\psi_A + \psi_B) \\ = C_2(\psi_A - \psi_B). \end{cases} \qquad (14.5)$$

The (+) function is symmetric, the (−) function antisymmetric in respect to the midpoint of the barrier (Fig. 14.2).

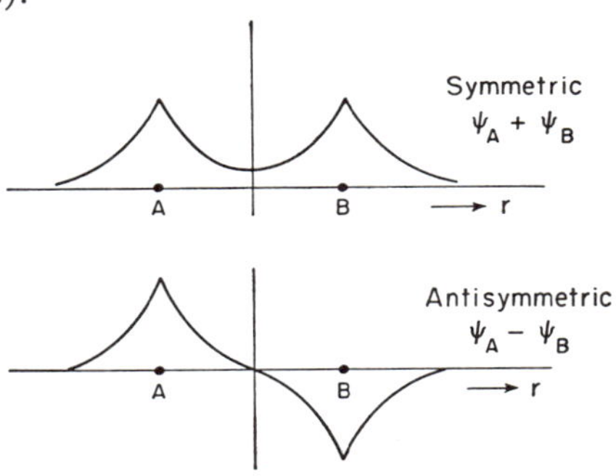

Fig. 14.2. Symmetric and antisymmetric wave functions of H_2^+ ion.

Quite generally every one of the stationary states an electron can assume in a single well splits into two states when the electron is trapped in a double well; one of the doublet states is symmetric, the other, antisymmetric. The closer the coupling, that is, the lower and narrower the barrier, the wider the splitting. †

† The situation of a double well is illustrated in greater detail in Sec. 21 by the example of the inversion vibration of the ammonia molecule (Fig. 21.2).

The antisymmetric state is of higher energy for two reasons: (*a*) it corresponds to the shorter wavelength, hence to a higher kinetic energy (see Eq. 7.6); (*b*) the symmetric wave function is higher at the midpoint between the protons than the individual s-function would be, while the antisymmetric function is here zero. This means that the electronic charge shared between the nuclei in the symmetric case leads to a lowering of the potential energy; the total energy goes through a minimum, and a stable H_2^+ molecule-ion results. Vice versa, in the antisymmetric case the protons are not shielded against each other by a shared charge; they repel each other, and the total energy characteristic, drawn as a function of the internuclear distance, rises without a minimum similar to Fig. 14.1.

If we add a second electron to the H_2^+ system, it also will spread to both nuclei. The lowest energy state results when both electrons are placed in the symmetric wave function with antiparallel spins. This singlet Σ state is thus composed of the product of the symmetric wave functions of the two electrons,

$$\begin{aligned} \psi_{H_2} &= C(\psi_{A_1} + \psi_{B_1})(\psi_{A_2} + \psi_{B_2}) \\ &= C(\psi_{A_1}\psi_{B_2} + \psi_{A_2}\psi_{B_1}) \\ &\quad + C(\psi_{A_1}\psi_{A_2} + \psi_{B_1}\psi_{B_2}). \end{aligned} \qquad (14.6)$$

A comparison with Eq. 14.1 shows that the first part of this equation is identical with the symmetric wave function of H_2 obtained by the Heitler-London method. The second part shows both electrons at the same proton, hence represents the *ionic molecule* H^+H^-. The molecular orbital method, in its first approximation, allows the electrons to move freely throughout the molecule; hence ionic and atomic wave functions carry equal weight in building up the molecular state.

Obviously, this conclusion has to be modified because the mutual electrostatic repulsion of the electrons makes the constellation H^+H^- rather unlikely. A more accurate calculation[7] allows each of the ionic structures to contribute only about 2 percent to the normal state of the molecule. But the principle of decisive importance is that not only one electronic state or even two different types, but all possible electronic states which refer to the same type of spin orientation (same number of unpaired electrons) can contribute to the strength of a molecular bond. This is the concept of *quantum-mechanical resonance* (see also Sec. 11).

We see now that behind this concept, which includes the exchange energy of the Heitler-London model, lies the phenomenon of the leaking over of electrons into

[7] S. Weinbaum, *J. Chem. Phys. 1*, 593 (1933); H. M. James and A. S. Coolidge, *ibid. 1*, 825 (1933).

other wave functions, that is, the tunnel effect. Quantum-mechanical resonance is a coupling phenomenon in some way akin to the energy fluctuations through coupled resonance circuits in which the charge spreads over the various modes of the molecule. Each additional possible constellation with antiparallel spins, that is, each symmetric wave function, adds its exchange energy to the total bond strength with a weight factor depending on the likelihood of its occurrence.

The designation "resonance" has misled people to believe that an *actual* observation of the molecule at various times would disclose it with statistical probability in the various configurations which are in quantum-mechanical interaction. This is not the case. We have to visualize the electronic charge as distributed in an overall probability pattern which can be broken down by mathematical analysis into contributions made by the various wave functions to this one energy state. Each of these functions adds its character to the molecule as a gene does in a genetic relation. Just as a child inherits traits from all its ancestors in a unique mixture and is not expected to be at one moment only its mother and in the next only its great-grandfather, the molecule is a distinct new individual.

15 · Bond Energies and Dipole Moments of Diatomic Molecules

The hydrogen molecule is the prototype of the *symmetrical diatomic molecules*, a class representing an especially simple situation. The nuclei are two *equal* attracting centers with a potential barrier between them. An electron placed in this double well will be found with

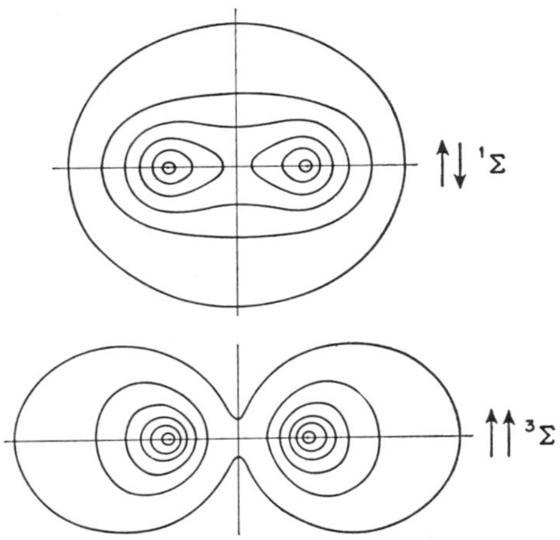

Fig. 15.1. Symmetric and antisymmetric Σ state of hydrogen molecule.

equal probability at one or the other atom, that is, the energy states are of the same height at both sides of the barrier: they are *degenerate*. Each energy state of the individual atom splits into two states, the lower one represented by a symmetrical, the upper one by an antisymmetrical, wave function in respect to the midpoint between the nuclei. The symmetric wave functions designate bonding orbitals; the antisymmetric wave functions, antibonding orbitals (see Sec. 14).

The fact that such a simple sequence of molecular orbitals characterizes the symmetrical diatomic molecule with one electron makes it tempting to apply once more the building principle that served so successfully for the qualitative description of atoms (Sec. 8). The sequence of atoms was obtained by increasing the nuclear charge in steps of one while inserting successively compensating electrons into orbitals derived for an atom containing one electron only. This use of hydrogen-like wave functions proved to be a good first approximation in representing the properties of the periodic system of the elements. Similarly, Hund and Mulliken[1] constructed the sequence of symmetrical diatomic molecules: $H_2 \rightarrow He_2 \rightarrow Li_2 \rightarrow Be_2 \rightarrow$ etc. by increasing the nuclear charge of both atoms in steps of one while simultaneously filling the molecular orbitals derived for H_2^+ with two electrons each of antiparallel spins.

The electrons of atoms were grouped in the K, L, M, ... shells and designated, in addition, as s, p, d, ... electrons according to the angular orbital momentum ($l = 0, 1, 2, \ldots$) of their cloud configuration. Placed in molecular orbitals these electrons are classified as σ, π, δ, ... electrons according to their angular momentum (0, 1, 2, ...) around the molecular axis. The charge density of the σ electrons has cylindrical symmetry around the axis without orbital momentum (Fig. 15.1), whereas the π, δ, etc., orbitals correspond to more complex wave patterns. Since the number of molecular orbitals is twice that of the individual atoms because of the splitting into bonding and antibonding states, all the electrons of the K, L, M, ... shells of the atoms

[1] F. Hund, *Z. Physik 51*, 759 (1928); *ibid. 63*, 719 (1930); *Handbuch der Physik*, Vol. *24*, Pt. I; R. S. Mulliken, *Phys. Rev. 32*, 186 (1928); *33*, 730 (1929); *Revs. Mod. Phys. 4*, 1 (1932).

can be accommodated in corresponding K, L, M, . . . shells of the molecules.

For the H_2 molecule we already have found that the two $1s^1$ electrons of the partners combine in a symmetric wave function with antiparallel spins; that is, a molecular $^1\Sigma_0$ state results resembling in saturation the $^1S_0$ ground state of the He atom (see Sec. 11). The additional electron of a third approaching hydrogen atom would have to be placed in the antibonding orbital of the asymmetric wave function; hence repulsion ensues. Four electrons, provided by two helium atoms, just suffice to fill the lowest bonding and antibonding orbital, that is, the K-shell of an He_2 molecule. The repulsion of the antibonding counteracts the attraction of the bonding state. Since the excited states of the helium atom lie about 20 ev above the ground state, those of the He_2 molecule prove to be inaccessible without artificial excitation. Thus no normal He_2 molecule forms.

The Li_2 molecule is constructed similarly to the H_2 molecule. Its two additional electrons fill the bonding σ orbital of the L-shell ($2\sigma^2$) above the filled bonding orbital ($1\sigma^2$) and antibonding orbital ($1\sigma_a{}^2$) of the K-shell. Since the σ wave function spreads further out for the L-shell than for the K-shell, a greater equilibrium distance and smaller binding energy than for H_2 result.

In the case of Be_2 again a normal molecule will not form if the additional two electrons occupy the antibonding, $2\sigma_a$ state. However, the $2p$ orbitals of the beryllium atom are not so high above its ground state that it might prove impossible to arrive at a stable Be_2 molecule by using p instead of s electrons.

The p orbitals of atoms have directional properties (see Eq. 8.22) and an orbital momentum ($l = 1$). In the diatomic molecule the molecular axis is an axis of cylindrical symmetry for one of the p-electron pairs, while the other two pairs protrude at right angles (Fig. 15.2).[2] Consequently, the symmetrical p pair forms a molecular σ orbital without angular momentum ($p\sigma$), while the other two pairs produce two molecular π states ($p\pi$). To distinguish between the configurations of these two π orbitals, we may visualize one rotating clockwise, the other counterclockwise around the molecular axis with an orbital momentum $\hbar$.

In B_2 we use the first p-electron pair in a bonding ($p\sigma$) orbital, whereas for C_2 and N_2 the bonding ($p\pi$) states will be filled. Hence, N_2 with the electronic constellation $1\sigma^2 1\sigma_a{}^2 2\sigma^2 2\sigma_a{}^2 2(p\sigma^2)2(p\pi^4)$ should be the most stable of these molecules because it has three pairs of bonding electrons. From O_2 to Ne_2 the three remaining antibonding orbitals of the molecular L shell are occupied; hence the binding energy decreases again and becomes zero for neon. A normal Ne_2 molecule does not materialize.

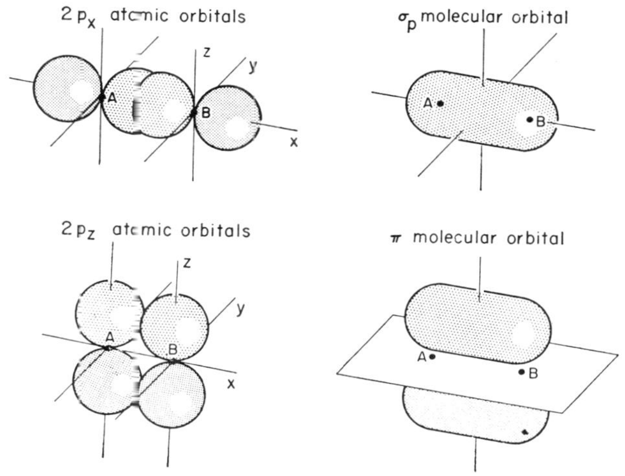

Fig. 15.2. Combination of p orbitals of diatomic molecule. (Adapted in part from Coulson.[2])

Table 15.1 illustrates, in the sequence of binding energies [3] and internuclear distances [4] of the diatomic symmetrical molecules from hydrogen to neon, the build-up scheme just described; it can be extended to the remainder of the periodic system.

Table 15.1. Binding energy and internuclear distance of symmetrical diatomic molecules from H_2 to F_2

Molecule	Binding Energy [3]		Internuclear Distance [4]
H_2	4.47 ev	103.2 kcal	0.741 A
Li_2	1.12	26	2.67
Be_2	?	?	?
B_2	3.0	69	1.59
C_2	3.6	83	1.31
N_2	9.76	225	1.09
O_2	5.08	117.2	1.20
F_2	2.2	72	1.3

This qualitative discussion of symmetrical diatomic molecules is based on simple electron-pair bonds (covalent bonds) without reference to the various combinations of wave functions that are in quantum-mechanical resonance. We know from the preceding section that even ionic states make some contribution to the bond

[2] G. Herzberg, *Molecular Spectra and Molecular Structure. I. Spectra of Diatomic Molecules*, D. Van Nostrand Co., New York, 1950; C. A. Coulson, *Valence*, Clarendon Press, Oxford, 1952.

[3] A. G. Gaydon, *Dissociation Energies and Spectra of Diatomic Molecules*, John Wiley and Sons, New York, 1947; Landolt-Börnstein, *Tabellen*, Springer, Berlin, 1951, Vol. I, Pt. 3, Molekeln II.

[4] H. A. Stuart, *Die Struktur des freien Moleküls*, Springer, Berlin, 1952.

strength, but, because of symmetry, no net electric dipole moment can result. This situation changes decisively for *nonsymmetrical diatomic molecules.* The attraction of the two partners for electrons, their *electronegativity*, is different; hence *permanent electric dipole moments* must be expected and are an important parameter of these molecules. Debye [5] was the first to realize the significance of permanent electric dipole moments in problems of molecular physics. Since the separation of the nuclei is of the order of 10^{-8} cm and the magnitude of the displaced charge comparable to that of one electron ($e = 4.8 \times 10^{-10}$ esu), it proved convenient to express electric dipole moments in the electrostatic system in units of 1×10^{-18} esu, named in honor of P. Debye:

$$1 \text{ debye} \equiv 1 \times 10^{-18} \text{ esu} = 3.33 \times 10^{-30} \quad [\text{coul m}]. \tag{15.1}$$

When this arbitrary unit was chosen, obviously it had not been realized that the linear Stark effect of the hydrogen atom leads to an elementary dipole moment (see Eq. 10.19):

$$|\mu_e| = \frac{3}{2}\frac{\epsilon_0 h^2}{\pi m e} = 12.8 \times 10^{-30} \quad [\text{coul m}]. \tag{15.2}$$

This elementary unit corresponds thus to

$$|\mu_e| \simeq 3.85 \quad [\text{debye}] \tag{15.3}$$

The dipole moments of molecules, like the bond strength, the intermolecular distance, and other parameters, cannot be calculated with accuracy at present, but have to be measured.[6] However, we can make reasonable guesses as to the magnitude of a dipole moment. Let us illustrate two possible approaches by the example of the HCl molecule.

In the symbolism introduced by Lewis [7] and Langmuir [8] the chemist indicates each valence electron by a dot, hence an electron-pair bond by two dots placed between the atoms. The hydrogen atom contributes an $1s^1$ electron, the chlorine atom seven outer electrons ($3s^2 3p^5$, see Table 8.1). Taking an extreme standpoint, we can represent the HCl molecule as either completely ionic (heteropolar) or completely covalent (homopolar):

$$\begin{array}{ccc} & \text{HCl} & \\ \swarrow & & \searrow \\ \text{ionic} & & \text{covalent} \\ [\text{H}]^+\left[:\ddot{\underset{..}{\text{Cl}}}:\right]^- & & \text{H}:\ddot{\underset{..}{\text{Cl}}}:. \end{array} \tag{15.4}$$

In the first case a closed octet shell is formed around the chlorine ion and the partners are held together by Coulomb forces; in the homopolar molecule the octet is completed by an electron-pair bond which binds the two atoms without polarity. Obviously the truth lies somewhere between these extremes and can be found by approximations from either side.

Ionic approach

The nuclear separation distance H—Cl, determined spectroscopically, is $d = 1.28$ A; hence a completely ionic H^+Cl^- molecule would have an electric dipole moment

$$|\mu|_{H^+Cl^-} = ed = 1.6 \times 10^{-19} \times 1.28 \times 10^{-10} \simeq 6.1 \quad [\text{debye}], \tag{15.5}$$

in contrast to its measured moment of ca. 1.12 debye. We have to expect that the proton polarizes the electron cloud of the Cl^- ion and creates an induced moment

$$|\mu_i| = \alpha_e |\mathbf{E}| = \alpha_e \frac{e}{\epsilon_0 4\pi d^2} \tag{15.6}$$

that has to be subtracted from $|\mu|_{H^+Cl^-}$. Hence

$$\mu| = ed\left(1 - \frac{\alpha_e}{\epsilon_0 4\pi d^3}\right). \tag{15.7}$$

The polarizability of Cl^-, according to Fig. 9.1, is $\alpha_e \simeq 6.6 \times 10^{-40}$ [em^3]. With this value the induced moment overcompensates the original one, and the second term in Eq. 15.7 becomes ca. 1.7. The reason for this overcorrection lies obviously in the assumption implied in Eq. 15.6 that the induced moment can be derived from the polarization of the chlorine ion in an homogeneous external field **E**. Actually, the distorting Coulomb field of the proton is strongly inhomogeneous across the Cl^- ion and only a fraction of it induces a moment by acting parallel to the molecular axis. A closer calculation shows that this effect reduces the correction by about 50 percent and thus leads to the magnitude of the dipole moment actually observed.

Covalent approach

The covalent bond between two like atoms was described in Eq. 14.6 by a wave function composed of

[5] P. Debye, *Physik. Z. 13*, 97 (1912).

[6] The most powerful methods for determining the dipole moments of gas molecules are microwave spectroscopy (see Sec. 21) and the molecular beam experiments of Rabi and his co-workers [C. A. Lee, B. P. Fabricand, R. O. Carlson, and I. I. Rabi, *Phys. Rev. 91*, 1395, 1403 (1953)].

[7] G. N. Lewis, *J. Am. Chem. Soc. 38*, 762 (1916).

[8] I. Langmuir, *J. Am. Chem. Soc. 41*, 868 (1919).

atomic and ionic terms. The contribution of the latter to the bond strength proves to be, in general, of minor importance. This situation changes when A and B represent two unlike atoms. The normal covalent bond $A—B$ will be some average of those for the symmetric molecules $A—A$ and $B—B$, and, in addition, the bond will be strengthened by a more or less pronounced contribution of ionic binding energy between the unlike partners. Pauling[9] proposed the relation that the arithmetic mean or, better still, the geometric mean of the two bond-energy values $D(A—A)$ and $D(B—B)$ can be used as representing the energy of the normal covalent bond between the atoms A and B; that is, the difference

$$\Delta = D(A—B) - \tfrac{1}{2}\{D(A—A) + D(B—B)\}, \quad (15.8)$$

or, better,

$$\Delta' = D(A—B) - \{D(A—A)\, D(B—B)\}^{1/2} \quad (15.9)$$

characterizes the ionic energy of the unsymmetrical bond. In this way the ionic bond strength can be measured thermochemically, since the quantity Δ is the heat liberated in the reaction

$$\tfrac{1}{2}A_2 + \tfrac{1}{2}B_2 \rightarrow AB \quad (15.10)$$

in the gaseous state.

The bond energies of diatomic molecules can be obtained experimentally by thermochemical or spectroscopic methods. Thermochemical data refer frequently to the decomposition of polyatomic molecules (H_2O, S_8, etc.) into atoms and give, therefore, not the energy of breaking one individual bond but an average derived from the breaking of all bonds. Table 15.2 lists the bond energies of some important diatomic molecules.

Table 15.2. Bond energies and dipole moments of nonsymmetrical diatomic molecules

Molecule	Bond Energy [3]		Dipole Moment [4]
HF	6.1 ev	145 kcal	1.91 debye
HCl	4.43	102	1.1
HBr	3.60	83	0.8
HI	2.75	63.4	0.38
NO	6.49	150	0.1
CO	11.1	256	0.1
NaI	3.1	71.7	4.9
KCl	4.4	101	6.3

From such bond energies Pauling calculated the extra ionic energy Δ or Δ' of the unsymmetrical bond and succeeded in splitting these values into an expression comprised of the difference of two terms:

$$\Delta(A—B) = 23.06(x_A - x_B)^2 \quad [\text{kcal/mole}]. \quad (15.11)$$

The values x characterize the *electronegativity* of the individual atoms. Figure 15.3 shows Pauling's *electronegativity scale of the elements* based on the parameter x.

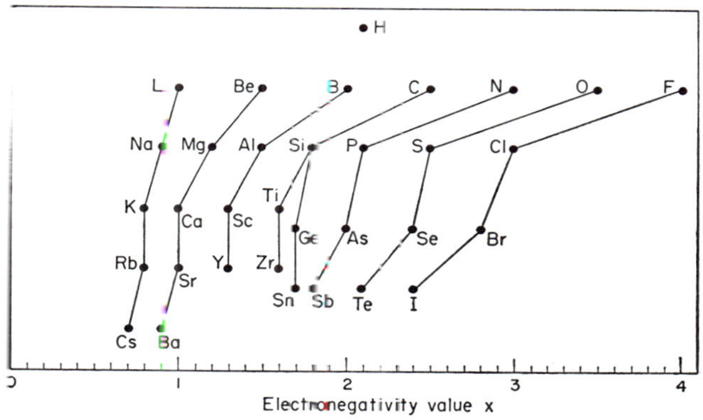

Fig. 15.3. Electronegativity scale of elements. (After Pauling.[9])

That a relation such as Eq. 15.11 may represent reasonably well the ionic part of the bond energy might be foreseen from the fact that the electrostatic interaction energy of two dipoles is proportional to the product of their moments (see Eq. 13.6). Hence, if we assign to a single bond a dipole moment μ_{A-B}, this bond moment should be roughly proportional to the electronegativity difference $x_A - x_B$, as pointed out by Malone.[10] Table 15.2 lists the dipole moments of various nonsymmetrical diatomic molecules.

The electronegativity $x = 2$ in Fig. 15.3 represents approximately the point of division between metals and nonmetals; the metals appear as elements of smaller ($x < 2$) and the nonmetals as those of larger ($x > 2$) electronegativity values.

Returning to the HCl molecule, we find that the covalent approach of Pauling, by which the extra ionic energy of the bond is determined thermochemically, leads to an electronegativity difference in line with the measured dipole moment.

Pauling's x values for the electronegativity of individual atoms neglect, by necessity, the effect of the mutual deformation polarization between two binding atoms. This deformation polarization, as Fajans[11] and his co-workers have emphasized for many years, is of great importance in many chemical problems, for instance, in surface reactions. It may alter the binding energy and dipole moment of molecules appreciably.[12]

[9] See L. Pauling, *The Nature of the Chemical Bond*, Cornell University Press, Ithaca, N. Y., 1940, pp. 58 ff.

[10] J. G. Malone, *J. Chem. Phys. 1*, 197 (1933).

[11] See, for example, K. Fajans, *Ceramic Age 54*, 288 (1949).

[12] See, for instance, the electrostatic calculation of E. S. Rittner for the alkali halide molecules [*J. Chem. Phys. 19*, 1030 (1951)].

16 · Static Dielectric Constants and Dipole Moments of Polar Gases

The dipole moments of Table 15.2 have been determined, in general, by a quasi-static measurement of the polarization of the dielectric as a function of temperature. This method is based on a statistical theory of orientation, first developed for the permanent magnetic moments of paramagnetic substances by Langevin[1] and applied to permanent electric moments by Debye.[2]

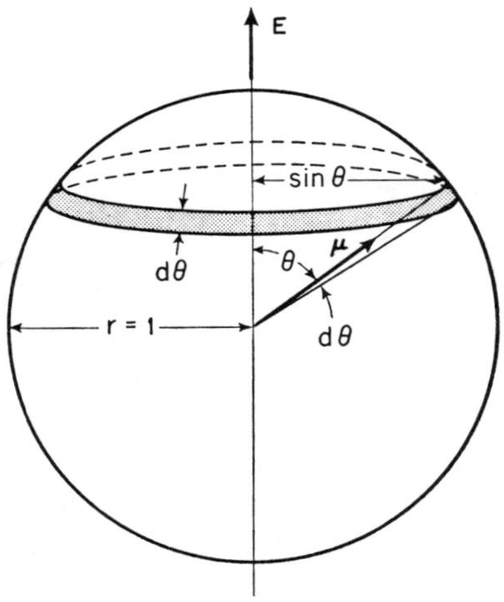

Fig. 16.1. Geometry for calculating average dipole moment.

Molecules carrying a permanent dipole moment suffer a torque in an electric field that tends to align the dipole axis in the field direction (see I, Eq. 2.12). Thermal agitation, on the other hand, tends to maintain a random distribution. The outcome of these counteracting influences is a statistical equilibrium that can be calculated without reference to the actual rotation of the molecules and their interaction as long as the electric field changes so slowly that the equilibrium is reached with certainty.

The assumptions for a simplified calculation are that the permanent dipole moment $\boldsymbol{\mu}$ of the molecules is unaffected by temperature and applied field; that the density of the gas is so low that the dipolar interaction energy is small in comparison to the thermal equilibrium energy kT; and, finally, that the dipoles can assume any direction with respect to the field axis.

[1] M. P. Langevin, *J. Physique 4*, 678 (1905); *Ann. chim. phys. 5*, 70 (1905).
[2] P. Debye, *Physik. Z. 13*, 97 (1912); *Polar Molecules*, Chemical Catalog Co., Inc., New York, 1929.

The potential energy U of a dipole pointing at an angle θ to the field is $-|\boldsymbol{\mu}|\,|\mathbf{E}|\cos\theta$. Consequently, the number pointing at a space angle

$$d\Omega = 2\pi \sin\theta\, d\theta. \tag{16.1}$$

(Fig. 16.1) is, according to Boltzmann's statistics,

$$N = A \exp\left(-\frac{U}{kT}\right) d\Omega$$
$$= A \exp\left(\frac{|\boldsymbol{\mu}|\,|\mathbf{E}|\cos\theta}{kT}\right) d\Omega. \tag{16.2}$$

Each of these dipoles contributes to the orientation polarization the component $|\boldsymbol{\mu}|\cos\theta$ in the field direction; hence it appears to the outside observer as if each molecule carries the average moment

$$\bar{\boldsymbol{\mu}}_d = \frac{\displaystyle\int_0^\pi A \exp\left(\frac{|\boldsymbol{\mu}|\,|\mathbf{E}|\cos\theta}{kT}\right)(\boldsymbol{\mu}\cos\theta) 2\pi \sin\theta\, d\theta}{\displaystyle\int_0^\pi A \exp\left(\frac{|\boldsymbol{\mu}|\,|\mathbf{E}|\cos\theta}{kT}\right) 2\pi\sin\theta\, d\theta}. \tag{16.3}$$

The integration over the space angle θ can be carried out by introducing the parameters

$$x \equiv \frac{|\boldsymbol{\mu}|\,|\mathbf{E}|}{kT}, \qquad \zeta \equiv \cos\theta, \tag{16.4}$$

and rewriting Eq. 16.3 as

$$\frac{\bar{\boldsymbol{\mu}}_d}{\bar{\boldsymbol{\mu}}} = \frac{\displaystyle\int_{-1}^{+1} e^{x\zeta}\zeta\, d\zeta}{\displaystyle\int_{-1}^{+1} e^{x\zeta}\, d\zeta}. \tag{16.5}$$

The denominator of this equation is

$$\int_{-1}^{+1} e^{x\zeta}\, d\zeta = \frac{e^x - e^{-x}}{x}, \tag{16.6}$$

whereas the numerator equals the differential quotient of this denominator after the parameter x, or

$$\int_{-1}^{+1} e^{x\zeta}\zeta\, d\zeta = \frac{x(e^x + e^{-x}) - (e^x - e^{-x})}{x^2}. \tag{16.7}$$

Hence

$$\frac{\bar{\mu}_d}{\bar{\mu}} = \coth x - \frac{1}{x} \equiv L(x). \quad (16.8)$$

The function $L(x)$ which describes the ratio of the average to actual moment of a gas molecule in its dependence on temperature and field strength was first derived by Langevin in the theory of paramagnetism and is therefore known as the *Langevin function.* Figure 16.2 shows the characteristic $L(x)$ as applied by

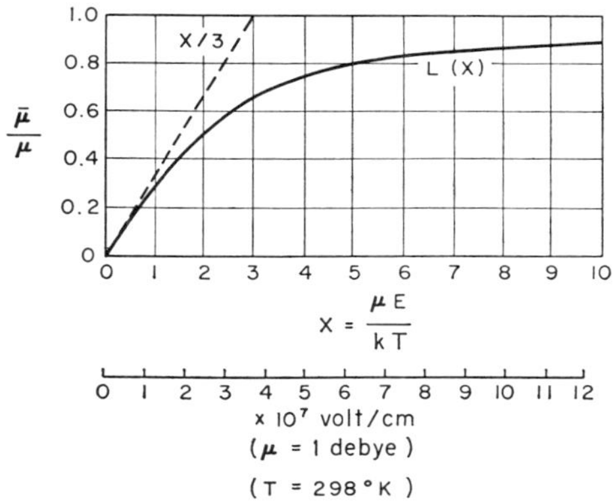

Fig. 16.2. Langevin function of dipole orientation.

Debye for the electric case; the field scale refers to the orientation of polar molecules of unit moment, $\mu = 1$ debye, against the thermal agitation at room temperature.

Obviously very high field strengths are required to produce a deviation from linearity and an approach towards saturation. For the normal case of relatively small fields ($x \ll 1$) the Langevin function can be approximated by its tangent $x/3$; that is, the average moment due to orientation becomes simply

$$\bar{\mu}_d \simeq \frac{|\mu|^2}{3kT}\mathbf{E}. \quad (16.9)$$

The applied field will, in addition, *induce* a moment in the gas molecule by deforming its electron cloud (α_e) and by changing the spacing of the nuclei (α_a). Thus deformation and orientation polarization together produce the total moment per dipole molecule

$$\mu_t = \left(\alpha_e + \alpha_a + \frac{|\mu|^2}{3kT}\right)\mathbf{E}. \quad (16.10)$$

Our derivation assumed a gas of low pressure. The polarizability per unit volume and with it the relative permittivity for static fields, κ_s', follows in this case from Eq. 2.11 as

$$\frac{N\mu_t}{\epsilon_0 \mathbf{E}} = \kappa_s' - 1 = \frac{N}{\epsilon_0}\left(\alpha_e + \alpha_a + \frac{|\mu|^2}{3kT}\right). \quad (16.11)$$

We will designate by κ_∞' the relative dielectric constant due to the induced moments only,

$$\kappa_\infty' \equiv 1 - \frac{N}{\epsilon_0}(\alpha_e + \alpha_a); \quad (16.12)$$

it can be measured at frequencies so high that the orientation polarization has no time to develop. The contribution of the permanent dipoles to the static permittivity then becomes

$$\kappa_s' - \kappa_\infty' = \frac{N|\mu|^2}{\epsilon_0 3kT}. \quad (16.13)$$

The fact that the orientation of the permanent dipole moments is strongly temperature-dependent in contrast to the practically temperature-independent contributions of the induced moments gives a convenient method of determining $\kappa_s' - \kappa_\infty'$ and with it the permanent dipole moment μ of the molecules. To eliminate the effect of varying gas density, we refer to the polarizability per mole Π (see Eq. 3.9) in the formulation for low gas pressure

$$\Pi = \frac{N_0\alpha}{3\epsilon_0} = \frac{\kappa_s' - 1}{3}\frac{M}{\rho}. \quad (16.14)$$

If Π is plotted as a function of $1/T$, we can distinguish easily between polar and nonpolar molecules (Fig. 16.3).

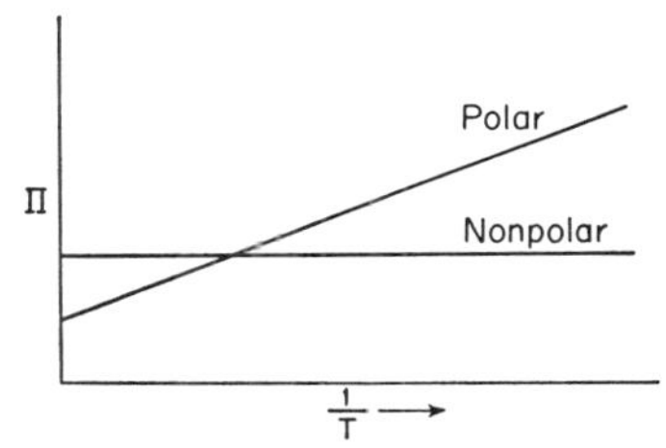

Fig. 16.3. Molar polarization of polar and nonpolar molecules as $f(1/T)$.

For the former a straight line is obtained intersecting the $1/T$ axis at a finite angle, whereas the polarization of a nonpolar gas is temperature-independent, hence produces a horizontal line.

If we write the molar polarization of the gas in the form

$$\Pi = A + \frac{B}{T} \quad [\text{m}^3], \quad (16.15)$$

where

$$A = \frac{N_0}{3\epsilon_0}(\alpha_e + \alpha_a) = 2.27 \times 10^{34}(\alpha_e + \alpha_a) \quad (16.16)$$

contains the effect of the deformation polarization, and

$$B = \frac{N_0}{3\epsilon_0}\frac{|\mu|^2}{3k} = 5.48 \times 10^{56}|\mu|^2 \quad (16.17)$$

that of the orientation polarization, we find the dipole moment as

$$\mu = 4.27 \times 10^{-29}\sqrt{B} \quad \text{[coul m]}$$
$$= 12.7\sqrt{B} \quad \text{[debye]}. \qquad (16.18)$$

Figure 16.4 shows the results of measurements on some

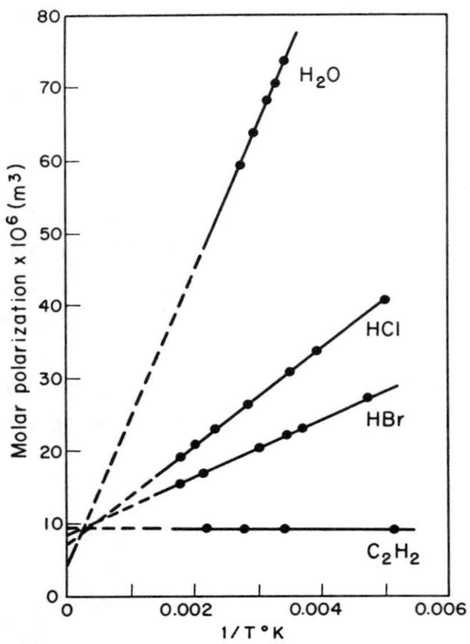

Fig. 16.4. Polarization of polar and nonpolar gases. (Adapted in part from Smyth.[3])

polar and nonpolar gases.[3] Extrapolating the straight lines back to the point $1/T = 0$, we obtain the A values, while their slope determines the B values as

$$B = \frac{\Pi_1 - \Pi_2}{1/T_1 - 1/T_2}. \qquad (16.19)$$

Since the electronic polarization of these compounds can be determined from measurements of the refractive index in the visible region according to the Lorentz-Lorenz equation (Eq. 2.15), the molar polarization Π can be separated into its electronic, atomic, and dipole contributions. As a rule of thumb, the atomic polarization Π_a can be estimated as amounting to about 10 percent of the electronic polarization Π_e for dipole moments greater than 1 debye.[4]

In the preceding discussion we have made the assumption that the permanent dipole moment $\boldsymbol{\mu}$ itself is independent of temperature. This may not be true for more complicated molecules, where the dipole moment is the resultant of group moments and the orientation of such groups restricted by activation energies. In such cases a more involved temperature dependence of the molar polarization arises, as will be shown in Secs. 17 and 25.

[3] C. P. Smyth, *Dielectric Constant and Molecular Structure*, Chemical Catalog Co., New York, 1931.

[4] See R. J. W. Le Fèvre, *Dipole Moments*, Methuen and Co., Ltd., London, 1938.

17 · Polyatomic Molecules

The energy of interaction between *two* atoms, when plotted as a function of their nuclear separation, is represented in the case of covalent bonding by a *potential curve* as shown in Fig. 17.1 (see Sec. 14). Its minimum

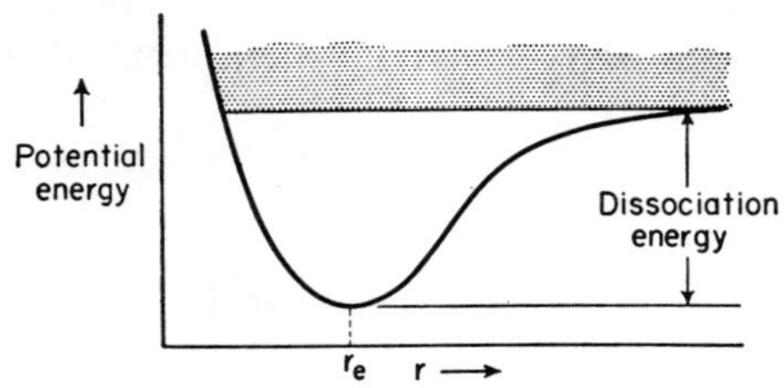

Fig. 17.1. Potential curve of diatomic molecule.

corresponds to the mean distance of separation at the absolute zero of temperature, whereas the depth of this minimum measures the *energy of dissociation, D*. For ionic bonding a similar potential curve is traversed but it leads with increasing internuclear distance to a separation into a positive and negative ion instead of into neutral atoms.

The actual bond between the atoms of a nonsymmetrical diatomic molecule is frequently, as the permanent dipole moment indicates, a complicated mixture of ionic and covalent binding. In the case of the NaCl molecule, for instance, the covalent bond prevails at great distances since the ionization potential of the sodium ($Na \rightarrow Na^+ + e^- - 118$ kcal/mole) is larger than the *electron affinity* of the chlorine atom ($Cl + e^- \rightarrow Cl^- + 86$ kcal/mole). With diminishing nuclear separation, however, the energy of the ionic state decreases more rapidly than that of the covalent one due to the electrostatic attraction of the oppositely charged partners, the potential curves cross, and, near the equilibrium position the ionic state dominates (Fig. 17.2).[1]

[1] J. C. Slater, *Quantum Theory of Matter*, McGraw-Hill Book Co., New York, 1951.

The potential energy of a *polyatomic* molecule consisting of N atoms depends on $3N - 6$ (or $3N - 5$) coordinates (see Sec. 18) and is therefore much more difficult to visualize. However, we know that either definite molecules will form which show small attraction

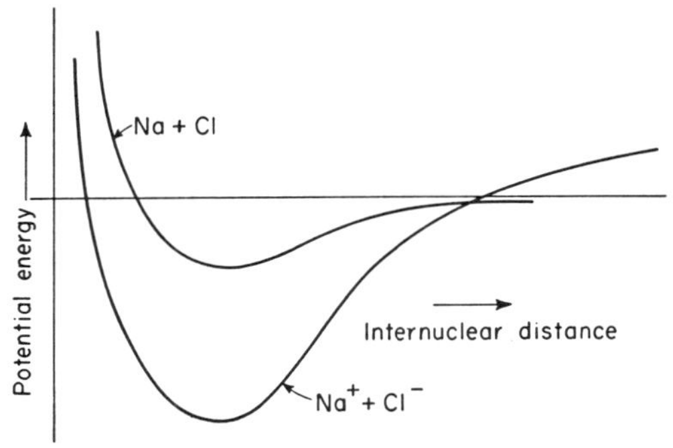

Fig. 17.2. Ionic and covalent state of the NaCl molecule. (Adapted from Slater.[1])

for further atoms or that atoms may flock together without limit and build up the structures of solids. In the former case we can treat the problem as a phenomenon of saturating valence bonds by filling the available orbitals with electron pairs. In solids, on the other hand, no such saturation will take place when more atoms are added. The Coulomb field of the Na^+ and Cl^- ion, for example, can attract more ionic partners and build up the ionic NaCl crystal (Fig. 17.3), or the

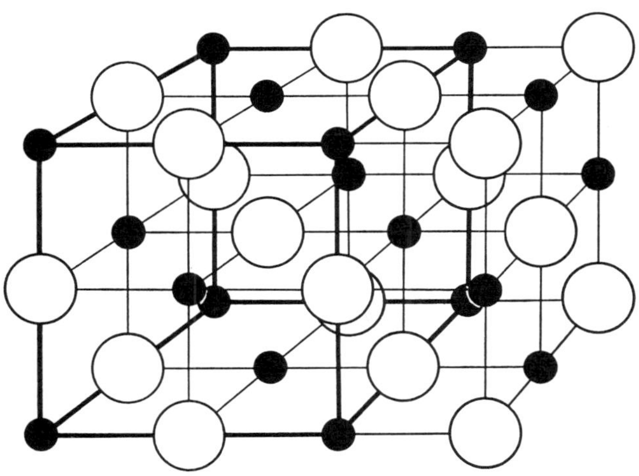

Fig. 17.3. Crystal structure of NaCl. (○ Cl^- ions, ● Na^+ ions, each ion type forming a face-centered cubic lattice.)

metallic bonds between sodium atoms will lead to the formation of metallic sodium (Fig. 17.4) in a packing as dense as the particle size allows (see Sec. 24).

The possible shapes and sizes of polyatomic molecules depend on the types of orbitals that may be used to saturate the chemical valence of the partners by pair-bond formation. To establish these orbitals, we have to go back to the atomic wave functions and to the geometrical patterns of the electron clouds they represent. According to the concept of the exchange energy (see Sec. 14), the strength of a bond between two atoms will increase the more the participating orbitals of the two partners overlap. The integral of the product of the orbitals, taken over all space, the *overlap integral,* is a measure of the bond strength (see Eq. 14.3). This criterion, first introduced by Slater [2] and Pauling,[3] leads to the conclusion that an *s-p* bond is about $\sqrt{3}$ times and a *p-p* bond about 3 times as strong as an *s-s* bond. If both *s-* and *p*-wave functions are available, we might therefore expect that *p* bonds will form, orientated at right angles to each other; but this is not the only possibility. A new type of orbital can be realized by

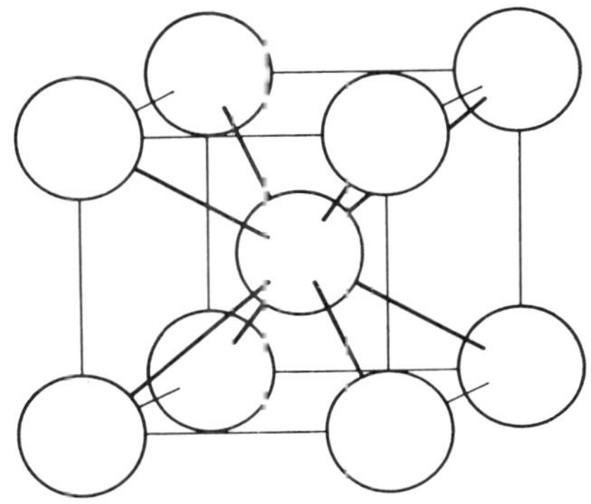

Fig. 17.4. Structure of sodium metal (body-centered, cubic).

a linear combination of the *s-* and *p*-wave functions (see Eq. 11.4), and such *s-p hybrid orbitals* of tetrahedral orientation may be stronger than the original ones (Fig. 17.5).

Let us make the situation clear by the sequence of molecules: $H_2O \rightarrow NH_3 \rightarrow CH_4$. The oxygen atom, with six electrons ($2s^22p^4$) in the L shell, has two p orbitals only half occupied and therefore available for covalent bonding. They can be saturated by forming electron-pair bonds with two hydrogen atoms, and a covalent molecule H_2O with a valence angle of 90° would result (Fig. 17.6). However, according to the electronegativity scale of the elements (Fig. 15.3), the O-H bond is of about 39 percent ionic character, and the measured dipole moment of the water molecule is about 1.85 debye.[4] Thus the hydrogen atoms are partly positively charged and tend to repel one another.

[2] J. C. Slater, *Phys. Rev.* *37*, 481 (1931).
[3] L. Pauling, *Proc. Nat. Acad. Sci.* *14*, 359 (1928); *J. Am. Chem. Soc.* *53*, 1367 (1931).
[4] See G. Birnbaum and S. K. Chatterjee, *J. Appl. Phys.* *23*, 220 (1952).

Hence, the actually measured bond angle is not 90° but ca. 104°.

Similarly, the nitrogen atom with the constellation ($2s^2 2p^3$) has three half-filled *p* orbitals available for

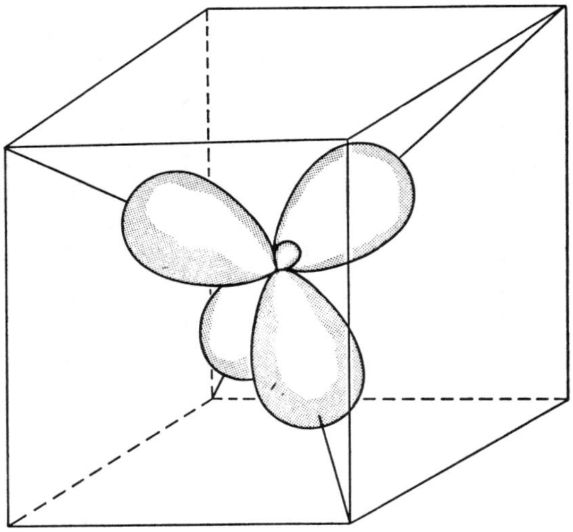

Fig. 17.5. Tetrahedral *s-p* hybrid orbitals.

pair-bond formation; hence it is saturated by forming the ammonia molecule, NH_3. Since the *p* bonds tend to stand at right angles to each other, a pyramidal structure results (Fig. 17.7); but the bond angles are

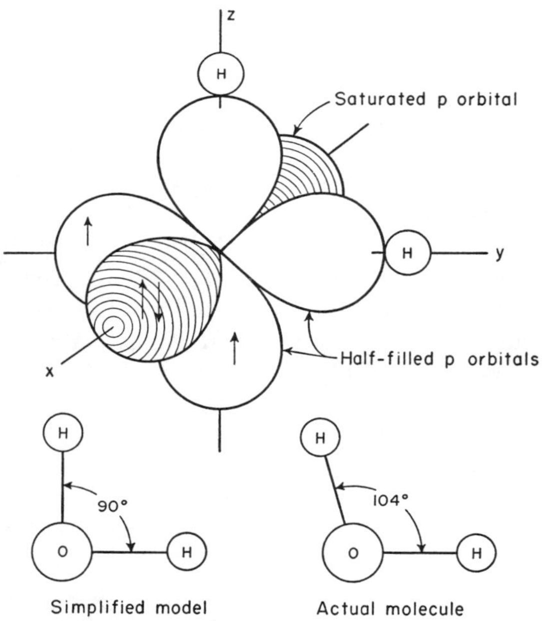

Fig. 17.6. Formation of H_2O molecule by *p* bonds.

again enlarged through electrostatic repulsion of the hydrogen atoms since the permanent dipole moment of NH_3 is 1.46 debyes.

However, there is an alternative explanation for the enlarged bond angles. Instead of using the *p* orbitals of the nitrogen atom, we might *hybridize* its *s*- and *p*-wave functions and bind the hydrogen atoms by *s-p* hybrid bonds to three of the four corners of a regular tetrahedron. The tetrahedral angle H-N-H would be 109° 28′, in relatively good agreement with the actually observed angle of 108°.

Whether *p* bonds and electrostatic repulsion or *s-p* hybrid bonds are mainly responsible for the shape of the ammonia molecule, only an accurate calculation can show. Although the hybrid *s-p* bonds are stronger than *p* bonds, their formation requires an excitation of the *s* electrons which otherwise form a rather stable

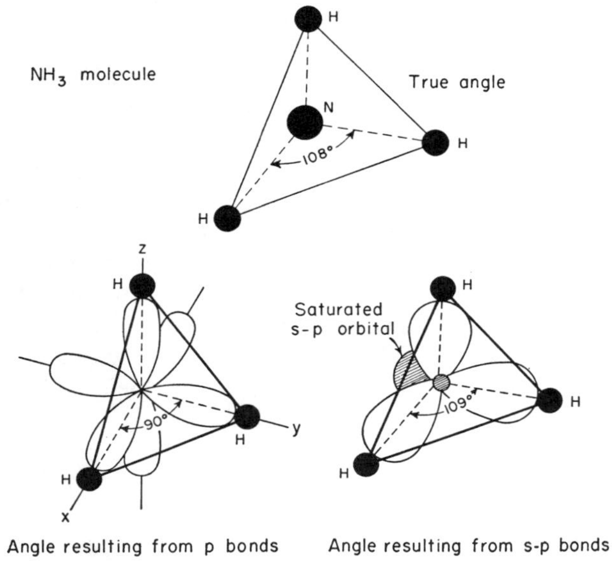

Fig. 17.7. Alternate ways of forming the NH_3 molecule.

$2s^2$ electron pair. The improved bond strength must outweigh the energy loss arising from this atomic excitation in order to make the hybridized form the state of lowest energy.

The situation is not in doubt for the carbon atom. Here, as organic chemists inferred long ago, four equivalent bonds are formed directed towards the four corners of a regular tetrahedron (see Fig. 17.5). This type of *s-p* hybrid orbital is therefore called conveniently a *tetrahedral orbital.* By placing four hydrogen atoms at the four corners, we arrive at methane (CH_4), the mother substance of the aliphatic hydrocarbons.

The existence of four equal tetrahedral orbitals suggests representing the aliphatic carbon atom itself by a tetrahedron. Two such carbon tetrahedra may combine by sharing one corner, that is, forming a single bond while the six remaining free corners can be occupied by hydrogen atoms; in this way the molecule of ethane 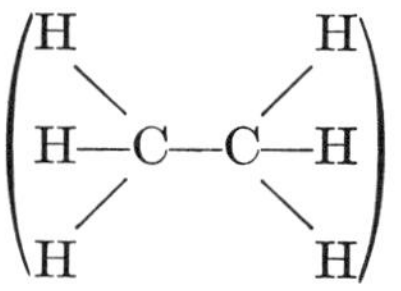 results. By sharing two corners, that is, one edge of their tetrahedron, the two carbon atoms can be joined by a double bond, and only four corners remain for occupation by hydrogen atoms. Thus ethylene 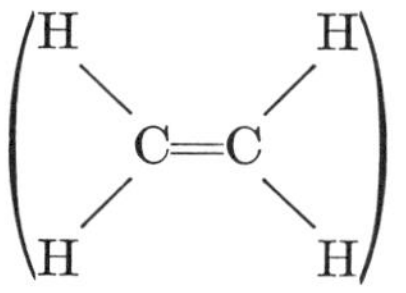 would be obtained with its hydrogen atoms confined to one plane. Finally, by sharing three corners, that is, one tetrahedron face, the carbon triple bond of acetylene (H—C≡C—H) might arise, which places the carbon and hydrogen atoms in a straight line (Fig. 17.8).

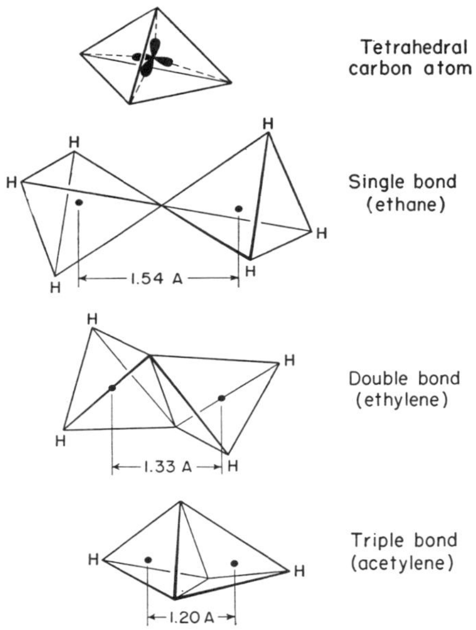

Fig. 17.8. Various ways of joining two tetrahedral carbon atoms.

Actually the situation is not so simple as that. In addition to the tetrahedral *s-p* hybrid bond there exist a *trigonal* and a *digonal* one, and these two bonds are invoked for the formation of ethylene [5] and acetylene.[6] The trigonal or sp^2 type leaves one of the original p orbitals unchanged (for example, p_z), whereas the mixing of s, p_x, and p_y creates three equally strong bonds, directed in the xy-plane at angles of 120° with each other (Fig. 17.9). The digonal bond leaves p_y and p_z unchanged and mixes s and p_x into two equivalent bonds directed at 180°.

The bond strength of the carbon double bond is less than twice the strength of the single bond and that of the triple bond appreciably less than thrice the single

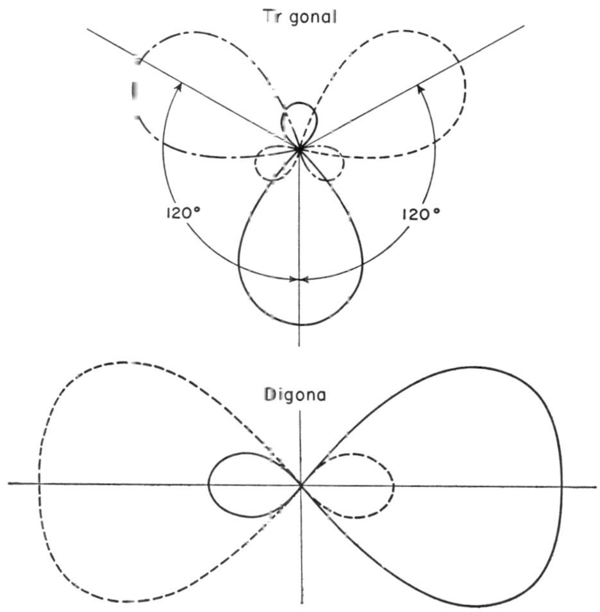

Fig. 17.9. The trigonal and the digonal *s-p* hybrid orbitals. (After Coulson.[6])

bond value (Table 17.1).[7] This instability of the carbon double and triple bonds makes them the favored starting point for many chemical reactions.

Table 17.1. Bond energy and covalent radii of multiple-bonded atoms

Bond	Energy (kcal) C. & S.[6]	Energy (kcal) G.[7]	Covalent radius r_c
C—C	83	84	0.77 A
C=C	146	125	0.67
C≡C	201	181	0.60
N—N	39	55	0.73
N=N	100	104	0.61
N≡N	226	184	0.55
C—O	34	47	0.74
C=O	117	117	0.57

Around a single bond axis, as the model of ethane shows, a more or less *free rotation* of the partners is possible. A double bond, as in ethylene, exerts much greater restraining forces by defining a preferential plane, and the triple bond is still more confining by specifying a preferred direction. The fact that molecu-

[5] E. Hückel, *Z. Physik 60*, 423 (1930).

[6] C. A. Coulson, *Valence*, Clarendon Press, Oxford, 1952, pp. 191 ff; G. E. Coates and L. E. Sutton, *J. Chem. Soc.* (London), 1948, p. 1187.

[7] G. Glockler, *J. Chem. Phys. 19*, 124 (1951).

lar groups may rotate around single bonds is of great importance for stereochemistry; it allows long chain molecules, for example, to change shape with the flexibility of a rope. How far the possibility of free rotation can actually be realized depends on the mutual interference of the rotating groups, on their *steric hindrance.*

The trigonal carbon bond is again invoked to produce the hexagonal coplanar configuration of the benzene molecule, the basic constellation of the aromatic compounds. That the benzene ring molecule (C_6H_6) is a plane molecule, coupled by a mixture of single bonds and double bonds, was proposed by the organic chemists long ago [8] on the basis of various types of valence pictures (Fig. 17.10). According to quantum mechan-

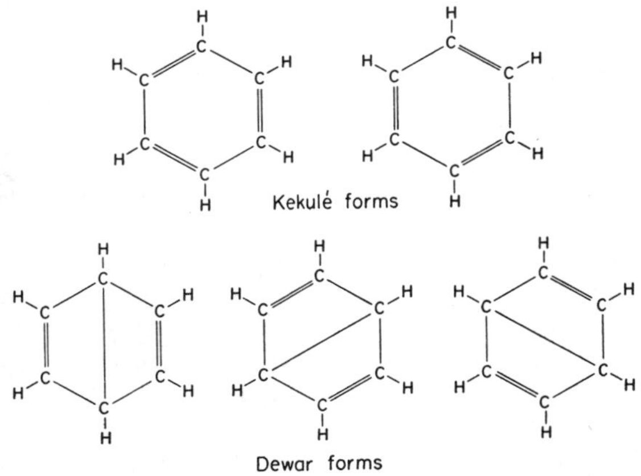

Fig. 17.10. Classical valence pictures of the benzene molecule.

ics there is some truth in all these constellations; they are in quantum-mechanical resonance (see Sec. 14). The total resonance energy contributed to the strength of the carbon-carbon bond beyond that of the single bond value is about 37 kcal/mole. The two Kekulé structures account for nearly 82 percent of the energy, and the three Dewar structures approximately for the remainder.[9]

According to the hybrid bond picture, the s, p_x, and p_y orbitals have combined into three equal coplanar bonds, oriented at 120° to each other and determining the plane of the ring, whereas the p_z bonds protrude in lobes normal to the plane. The exchange between these p_z electrons, which correspond classically to a ring current, gives the extra stability implied by the valence pictures.

For the benzene ring molecule only nonpolar constellations need be considered; dipole molecules such as H_2O have to be described as resonating between covalent and ionic constellations. The four electronic structures of Fig. 17.11 contribute to the bond strength of the water molecule. The completely covalent form

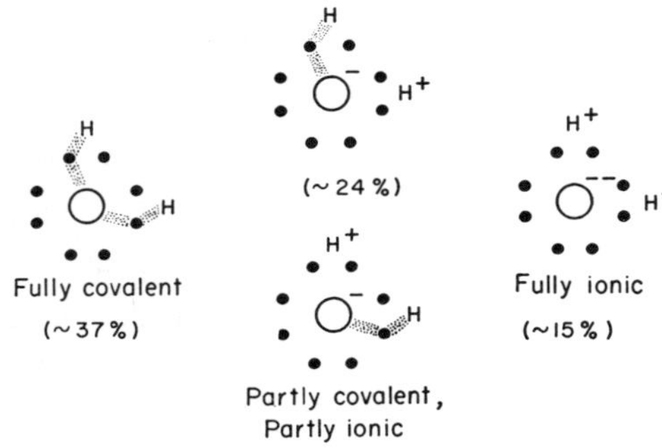

Fig. 17.11. Four bond types of the H_2O molecule contributing in quantum-mechanical resonance. (After Pauling.[9])

has the highest probability, but the two equivalent structures with one ionic bond, taken together, outweigh it, and even the completely ionic form makes a strong contribution.

Polyatomic molecules can be visualized as composed of diatomic groups. Since for diatomic molecules we arrived at quantitative data for bond strengths and dipole moments (see Tables 15.1, 15.2, and 17.1), it is tempting to use this information to evaluate the energy content and dipole moment of polyatomic molecules. Actually we can arrive at a first approximation of the total energy by simply adding arithmetically the individual bond energies. Similarly, we can obtain, in a

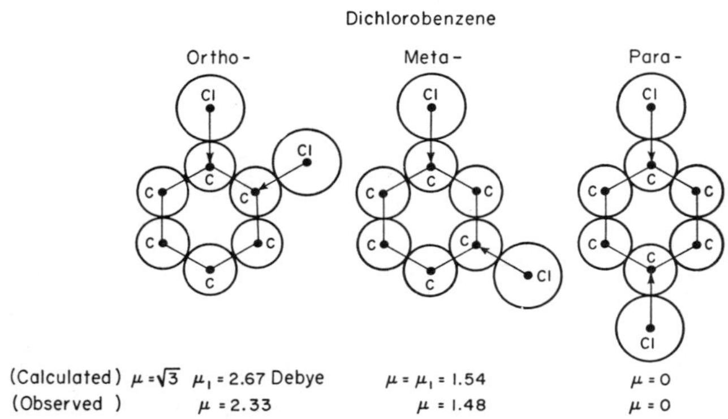

Fig. 17.12. Vector addition of dipole moments.

first approximation, the overall dipole moment of a molecule by a *vector addition* of the individual group moments, as the classical example,[10] dichlorobenzene, illustrates in its ortho, meta and para forms (Fig. 17.12).

[8] A. Kekulé, *Bull. soc. chim. 3,* 98 (1865); *Ann. Chem. 137,* 129 (1866).

[9] L. Pauling and E. B. Wilson, *Introduction to Quantum Mechanics,* McGraw-Hill Book Co., New York, 1935, p. 378; L. Pauling, *The Nature of the Chemical Bond,* Cornell University Press, Ithaca, N. Y., 1942, pp. 128 ff.

[10] See P. Debye, *Polar Molecules,* Chemical Catalog Co., Inc., New York, 1929.

Quantitative agreement, however, should, in general, not be expected from this composite procedure. It misses not only the phenomenon of quantum-mechanical resonance but also the effect of prepolarization which the various groups exercise on the overall electronic structure. It is due to this prepolarization that a substituent can be introduced more readily into certain positions of an aromatic molecule than into other apparently equivalent ones or that polymerization reactions proceed in certain preferential ways. Phenomena of this kind are being studied extensively today by chemists and physicists to learn more about the actual electron distribution and its influence on the kinetics of chemical reactions.

18 · Vibration and Rotation

In problems of *electronic polarization*, the atomic nuclei play, in essence, the role of stationary binding posts for the electron clouds since the nuclear mass outweighs that of the electrons by at least three orders of magnitude. However, when nuclei find themselves bound to partners of comparable mass as in molecules or the condensed phases of liquids and solids, additional types of polarization result from the vibration of nuclei relative to each other (*atomic polarization*) and the rotation of molecules as a whole or of molecular groups around internuclear axes (*orientation* or *dipole polarization*). The classical model for a discussion of these additional modes of motion is the diatomic dumbbell molecule, assumed to oscillate as a *harmonic oscillator* or to rotate as a *rigid rotator* (Fig. 18.1).

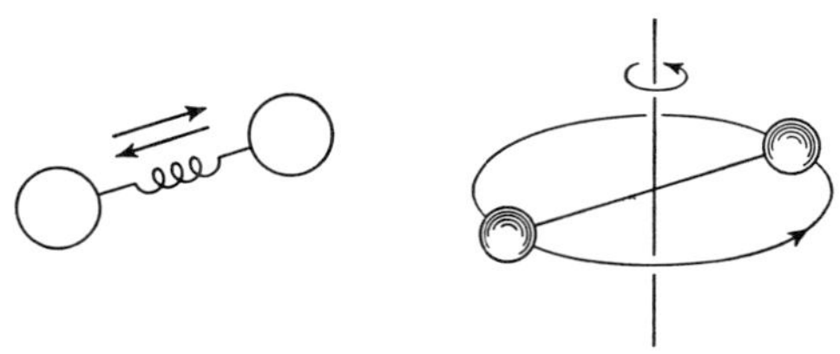

Fig. 18.1. Harmonic oscillator and rigid rotator.

Let r_1 and r_2 be the distances of the mass points m_1 and m_2 from their center C of gravity, and r the mutual distance of the mass centers; that is,

$$m_1 r_1 = m_2 r_2$$

and (18.1)

$$r_1 + r_2 = r.$$

By separating r_1 and r_2 we obtain

$$r_1 = \frac{m_2}{m_1 + m_2} r$$

and (18.2)

$$r_2 = \frac{m_1}{m_1 + m_2} r.$$

The restoring force between the atoms results from a complicated interplay of electrostatic interactions. If it is proportional to the displacement from the equilibrium distance r_e, the two atoms execute harmonic vibrations around their equilibrium positions according to the force equations

$$m_1 \frac{d^2 r_1}{dt^2} = -f(r - r_e)$$

and (18.3)

$$m_2 \frac{d^2 r_2}{dt^2} = -f(r - r_e).$$

Substituting for the distances from the center of gravity the mutual distance, according to Eq. 18.2, we derive from both equations the identical expression

$$\frac{m_1 m_2}{m_1 + m_2} \frac{d^2 r}{dt^2} = -f(r - r_e). \qquad (18.4)$$

By referring to the *reduced mass* of the diatomic molecule,

$$m_r = \frac{m_1 m_2}{m_1 + m_2}, \qquad (18.5)$$

and replacing the internuclear distance r under the differential sign with the displacement distance $r - r_e$ (which is permissible, since r_e is constant), we can rewrite Eq. 18.4 as

$$m_r \frac{d^2(r - r_e)}{dt^2} + f(r - r_e) = 0. \qquad (18.6)$$

The vibrations of the two atoms of the dumbbell molecule are identical to the harmonic vibration of a single mass m_r oscillating with an amplitude equal to the displacement distance. The resonance frequency of this undamped oscillator is (see Eq. 4.3)

$$\omega_{osc} = \sqrt{f/m_r}. \qquad (18.7)$$

Its potential energy

$$U = \tfrac{1}{2}f(r - r_e)^2, \tag{18.8}$$

plotted as function of the internuclear distance, describes a parabola (Fig. 18.2).

In addition to this vibration, the dumbbell molecule, freely suspended in space, may rotate around an axis passing perpendicular to the molecular axis through the

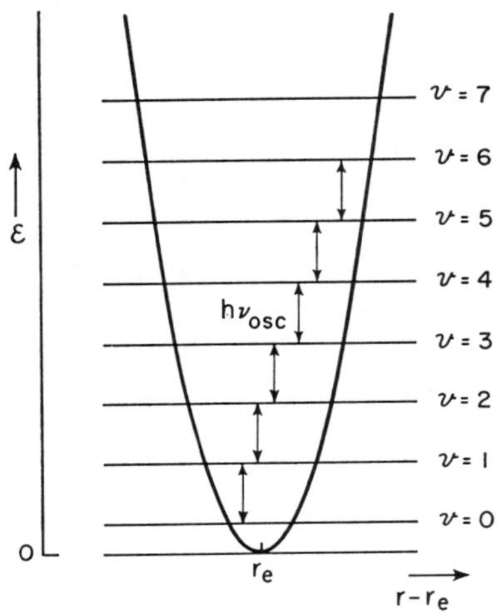

Fig. 18.2. Potential curve and vibration states of harmonic oscillator. (After Herzberg.[1])

center of gravity C. The moment of inertia, I, of the molecule in respect to this axis is

$$I = m_1 r_1^2 + m_2 r_2^2 = m_r r^2, \tag{18.9}$$

and its energy of rotation according to classical mechanics

$$\mathcal{E}_{\text{rot}} = \tfrac{1}{2}I\omega_r^2, \tag{18.10}$$

where $\omega_r = v/r$ represents the angular velocity of rotation. Since the angular momentum of this system is

$$p' = m_r v r = I\omega_r, \tag{18.11}$$

the kinetic energy of this rotator can also be expressed as

$$\mathcal{E}_{\text{rot}} = \frac{p'^2}{2I}. \tag{18.12}$$

A vibrating molecule will be *optically active*, that is, able to emit and absorb electromagnetic radiation according to classical theory (see I, Sec. 13) if its dipole moment *changes* during the oscillation ($\partial\mu/\partial r \neq 0$). The variation of the dipole moment corresponds to the equivalent picture of an alternating current traversing a dipole antenna. The harmonic oscillator has classically only one resonance frequency determined by the force constant f and the two atomic masses (Eq. 18.7). The amplitude of this vibration, and with it the energy, can assume any desired value. The energy radiated per second is proportional to the square of the momentum amplitude (μ_0^2) and to the fourth power of the frequency (see I, Eq. 13.18). It causes radiation damping which will broaden the spectral line (see Sec. 4).

The dumbbell molecule, as a rigid rotator, will be optically active if it carries a permanent moment. Upon such dipole moment (electric or magnetic) the electric or magnetic field can exercise a torque (see I, Sec. 2) and make the molecule rotate classically with any angular frequency. Hence, whereas the classical vibrator emits a single spectral line, the classical rotator emits a continuum extending with rapidly increasing intensity from zero frequency to the explosion frequency of the rotator.

Vibration and rotation may be described in terms of general co-ordinates, a space co-ordinate q and a mechanical momentum co-ordinate p. The co-ordinate q of the vibrator is the displacement distance $r - r_e$ from its equilibrium position; for the rotator it corresponds to the position on the circle of rotation. The momentum p prescribes, according to de Broglie's equations (Eq. 7.6), a wavelength

$$\lambda = h/p. \tag{18.13}$$

Vibration and rotation are recurrent motions. Classically, any free recurrent motion may represent a stationary state. Quantum mechanics admits only states which correspond to standing wave patterns. Forgetting about the uncertainty principle at present, we prescribe that the total path traversed must be an integer n of the wavelength,

$$\oint \frac{dq}{\lambda} = n$$

or

$$\oint p\,dq = nh. \tag{18.14}$$

This is Sommerfeld's quantum condition (Eq. 6.19).

The momentum p of the vibrator depends on the displacement q; that is, the wavelength changes with the position of the mass points. It is shortest at the equilibrium position, where the total energy E has been converted into kinetic energy

$$\mathcal{E}_{\text{kin}} = \frac{1}{2}mv^2 = \frac{1}{2}\frac{p^2}{m}, \tag{18.15}$$

and infinite at the turning points, where only the potential energy (see Eq. 18.8)

$$U = \tfrac{1}{2}fq^2 \tag{18.16}$$

remains. Since the total energy of the harmonic oscillator,

$$\mathcal{E} = \frac{p^2}{2m} + \frac{fq^2}{2}, \tag{18.17}$$

must be constant, the relation between p and q is

$$\frac{p^2}{2m\mathcal{E}} + \frac{fq^2}{2\mathcal{E}} = 1. \tag{18.18}$$

The momentum p, plotted against the space co-ordinate q in the phase space (p,q), represents an ellipse with half axes $\sqrt{2m\mathcal{E}}$ and $\sqrt{2\mathcal{E}/f}$ (Fig. 18.3).

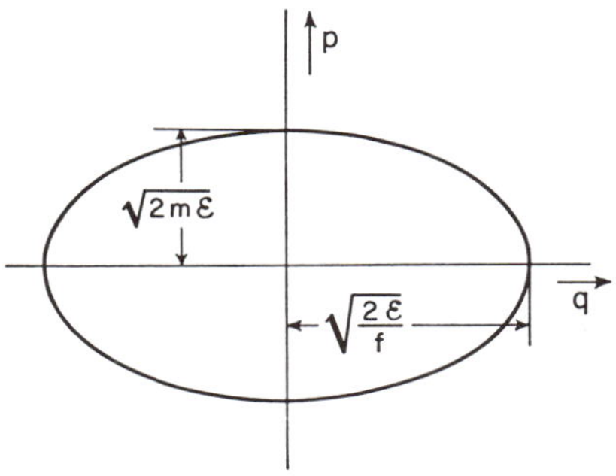

Fig. 18.3. Energy ellipse of harmonic oscillator in phase space.

The phase integral $\oint p\,dq$ of the vibrator is the area of this ellipse and is equal to π times the product of the two half axes. The quantum condition of the linear harmonic oscillator thus becomes

$$\oint p\,dq = 2\pi\mathcal{E}\sqrt{m/f} = vh, \tag{18.19}$$

or, by introducing the resonance frequency from Eq. 18.7,

$$\mathcal{E}_{\text{osc}} = vh\nu_{\text{osc}}, \tag{18.20}$$

where v is the *vibrational quantum number*. Thus the total energy of the linear harmonic oscillator is quantized in integral multiples of the amount $h\nu_{\text{osc}}$.

For the rigid dumbbell rotator, on the other hand, the momentum p is constant; hence the phase integral simplifies to

$$p\oint dq = (mv)(2r\pi) = Jh$$

or (18.21)

$$p' = J\hbar,$$

with J representing the *quantum number of rotation*. Equation 18.21 is Bohr's quantum condition of the rotator, as previously derived (see Eq. 6.12). The energy of the rigid rotator follows from Eq. 18.12 as

$$\mathcal{E}_{\text{rot}} = J^2\hbar^2/2I. \tag{18.22}$$

A comparison of the energy states of the vibrator and rotator (Eqs. 18.20 and 18.22) with those found for the electronic excitation of atoms shows fundamental differences. The stationary energy states of a hydrogen-like atom may be written by generalizing the terms of the Balmer formula (see Eqs. 6.4 and 8.16)

$$\mathcal{E}_e = -R'hc\frac{Z^2}{(n+a)^2}, \tag{18.23}$$

where R' is a modified Rydberg constant, n the principal quantum number, a the so-called Rydberg correction, and Z the charge of the core of the atom (atomic number minus the number of electrons in the closed shells of the core). These quantum states of the atom form a *Rydberg series of terms* (with $n = 1, 2, 3, \ldots$) (Fig. 11.4). As the distance between electron and core increases, the binding energy ($\mathcal{E}_e$ negative) diminishes rapidly according to Coulomb's law. The terms crowd closer and closer together until, at the ionization limit ($\mathcal{E}_e = 0$), the discrete energy states join a continuum of positive energy values. The ionization potential corresponds to the light energy $h\nu$ required to move the electron from its ground state nearest the core to infinity, with the separated electron and ion at rest. The positive energy values $\mathcal{E}_e$ refer to the electron as free and carrying the energy $\mathcal{E}_e$ as unquantized kinetic energy. The discrete negative energy values characterize the electron as *trapped* by the core, and the total energy $\mathcal{E}_e$ is divided equally, according to the Virial theorem (see Eq. 5.8), between potential and kinetic energy.

These Rydberg series of terms, with their rapidly diminishing spacing towards higher excitation energies, are typical for all line spectra of atoms. The linear harmonic oscillator, on the other hand, according to Eq. 18.20 has equally spaced energy levels (see Fig. 18.2), and, for the rigid rotator, the energy of the states even increases quadratically with the quantum number (Fig. 18.4).

According to Bohr's frequency condition (Eq. 6.1), the emission of radiation takes place as the result of a quantum jump from a higher to a lower energy state, and the absorption by the converse process. Which transitions are actually possible is determined by the selection rules of quantum mechanics (see Sec. 12). With atoms, line series like those of Lyman, Balmer, and Paschen appear (Eq. 5.17) because the change of the principal quantum number underlies no restrictions (Eq. 12.9). For the harmonic oscillator the selection rule specifies that the vibrational quantum number v of the two interacting energy states must differ by unity

$$\Delta v = v' - v'' = \pm 1, \tag{18.24}$$

and the same condition applies to the rotational quantum number J:

$$\Delta J = J' - J'' = \pm 1. \tag{18.25}$$

A linear harmonic oscillator should, therefore, as Fig. 18.2 indicates, be characterized by only one spectral line, identical in frequency to that of the classical oscil-

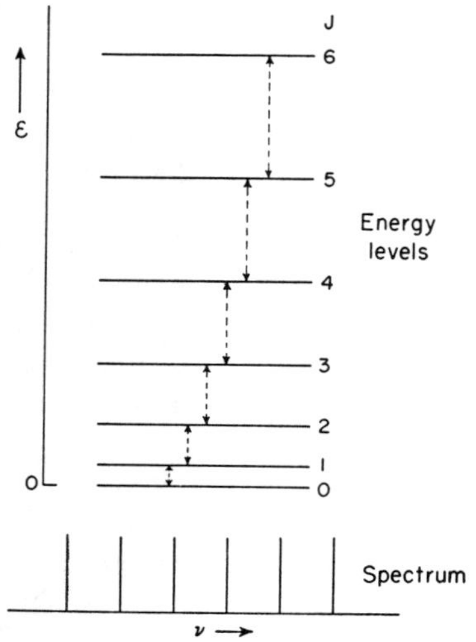

Fig. 18.4. Energy states and spectrum of rigid rotator. (After Herzberg.[1])

lator. The rigid rotator of Fig. 18.4 should show a sequence of lines of the frequencies (see Eqs. 18.11 and 18.21)

$$\nu = \frac{h}{4\pi^2 I} J. \tag{18.26}$$

A classical rotator does not resonate; the quantized rotator has rotational resonance frequencies which increase linearly with the quantum number J.

The simplified treatment, based on Sommerfeld's quantum condition, assumes that the position and the location of the particle can be specified simultaneously with accuracy. This contradicts the spirit of quantum mechanics as expressed in the uncertainty principle (Eq. 8.11). The actual stationary states of the vibrator and rotator are obtained as solutions of the Schrödinger equation.[1] In place of Eqs. 18.20 and 18.22 we thus find the slightly different energy expressions

$$\mathcal{E}_{osc} = h\nu_0(\mathrm{v} + \tfrac{1}{2}) \tag{18.27}$$

and

$$\mathcal{E}_{rot} = \frac{\hbar^2}{2I} J(J+1). \tag{18.28}$$

The result for the rotator agrees with the quantization condition of the angular momentum derived previously (see Eq. 8.23). For the vibrator, Eq. 18.27 postulates that even in the lowest vibrational state, $\mathrm{v} = 0$, that is, at zero absolute temperature, a *zero-point energy* of one half vibrational quantum remains (see Fig. 18.2).

The existence of a zero-point energy for the vibrator, but not the rotator, is required by the uncertainty relation. The position and the momentum of the oscillator at 0°K would be completely defined if a vibrationless state could be realized. For the rotator this is not so, because for a molecule without rotational energy the actual angular position remains still completely undetermined.

Let us check the predictions of the theoretical discussion with the simple example of the HCl molecule. For a thin layer of the gas and small dispersion, a fundamental absorption band is found near 3.5 microns † (Fig. 18.5). Assuming that this is the expected resonance of the vibrator, $\nu_{osc} = 8.65 \times 10^{13}$ sec^{-1}, we obtain from Eq. 18.7, with the reduced mass of the HCl molecule $m_r = 1.63 \times 10^{-27}$ kg, the force constant as $f = 4.8 \times 10^2$ newton/m. To change the internuclear distance by 1 A $= 1 \times 10^{-10}$ [m] against this force constant, would require, according to Eq. 18.8, an energy $U = 2.4 \times 10^{-18}$ joule $= 15$ ev (see Eq. 5.34); this is a reasonable order of magnitude for molecular energies.

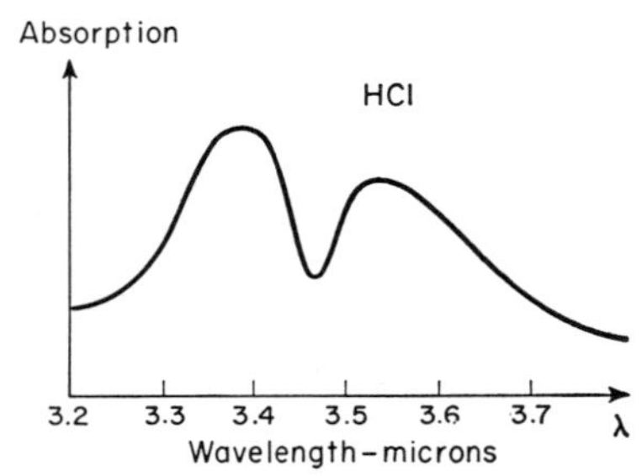

Fig. 18.5. Fundamental vibration band of HCl in near infrared (Bjerrum double band).

In addition to this vibration spectrum in the *near infrared* (wavelengths $<20\ \mu$), there is found an absorption spectrum in the *far infrared* ($>20\ \mu$) fitting the description of the rotation spectrum of HCl (Fig. 18.6).[2] After identifying the rotational quantum numbers J of the transmission minima according to Eq. (18.26), we determine from the rotation frequency of the resonance absorption the moment of inertia of the HCl molecule as $I = 2.71 \times 10^{-47}$ [kg m^2] and then,

[1] See G. Herzberg, *Molecular Spectra and Molecular Structure. I. Spectra of Diatomic Molecules*, D. Van Nostrand Co., New York, 1950.

† Actually, this band shows two maxima (Bjerrum's double band), because the rotation structure consists of two branches (see Fig. 18.7).

[2] Czerny, *Z. Physik 34*, 227 (1925); see also Ref. 1, pp. 57 and 81.

with the reduced mass given above, the internuclear distance from Eq. 18.9 as $r = 1.29$ A.

It is convenient to distinguish between near and far infrared from the standpoint that in the near infrared

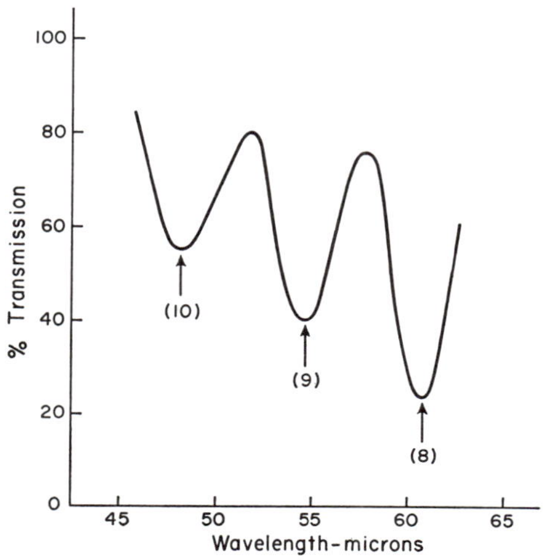

Fig. 18.6. Rotation spectrum of HCl molecule in far infrared. (After Czerny.[2])

the vibration and in the far infrared the rotation spectra of gas molecules are found. The far infrared, still not easily accessible for the experimenter, continues without break into the microwave region.

If the principal absorption band of the HCl gas in the near infrared is observed with high resolution, a fine structure becomes visible (Fig. 18.7). This effect

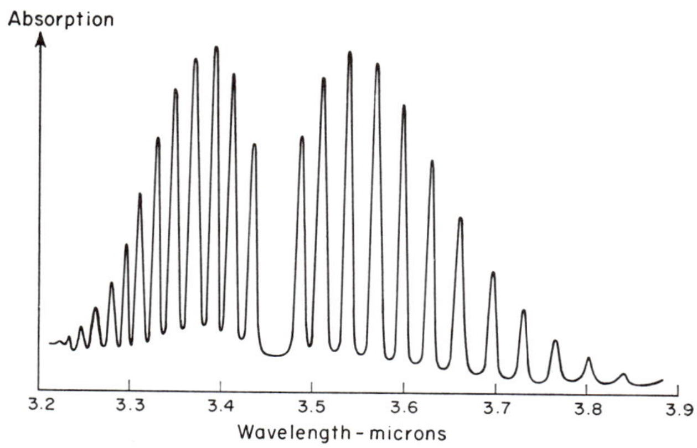

Fig. 18.7. Fundamental vibration band of HCl with resolved rotation spectrum.

is caused by the modulation of the vibration due to the excitation of the various rotation states of the HCl molecule. Since the rotation energy is appreciably smaller than the vibration energy, we may observe a pure rotator but normally not a pure vibrator; vibrating molecules in the gaseous state will be rotating vibrators with a principal energy level diagram as shown in Fig. 18.8.

A more accurate discussion of diatomic molecules has to take into account that actual molecules are neither harmonic oscillators nor rigid rotators. The potential energy of a harmonic oscillator increases indefinitely with increasing displacement from the equilibrium position, whereas a real molecule will dissociate beyond a critical internuclear distance (Fig. 18.9). The true molecule is an anharmonic oscillator, and its potential curve approximates only near the minimum the parabola of Fig. 18.2.[3] To make allowance for this deviation, we can represent the energy of the oscillator by a power series which for all practical purposes may be terminated after the second member. Similarly, the

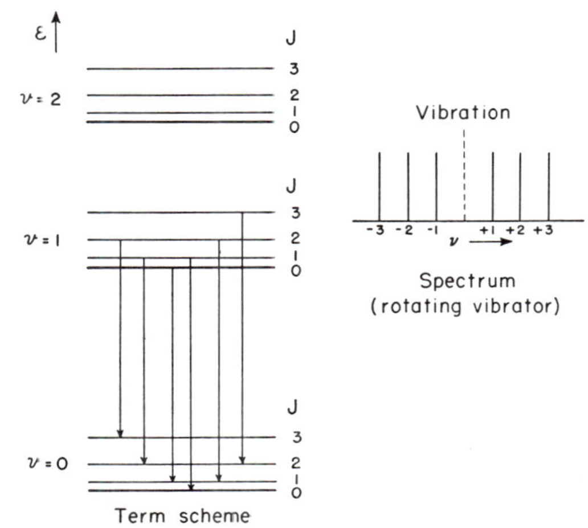

Fig. 18.8. Energy level diagram and spectrum of rotating vibrator.

actual rotator does not have a constant internuclear distance, as assumed, but the separation of the mass points increases slowly under the pull of the centrifugal force. Also, in this case an adequate correction can be made by adding a quadratic term.

An extension of the considerations to polyatomic molecules leads to situations of rapidly increasing complexity, but classical mechanics provides some guiding principles of approach. For a rigid body, three mutually perpendicular *principal axes* exist, which pass through the center of mass and around which the moment of inertia is a maximum or minimum (Fig. 18.10).

[3] A mathematical expression that represents well the potential curve of a diatomic molecule has been given by P. M. Morse, *Phys. Rev.* *34*, 57 (1929) in the form

$$U(r - r_e) = D_e(1 - e^{-\beta(r-r_e)})^2, \tag{18.29}$$

where D_e is the dissociation energy taken from the minimum, and β is a constant. The dotted characteristic in Fig. 18.9 represents such a *Morse curve*.

If these *three principal moments of inertia* are different, the molecule is called an *asymmetric top;* if two are alike, a *symmetric top;* if all three are alike, a *spherical top.* A linear molecule, like the diatomic molecules just discussed, represents a special case of the symmetric top,

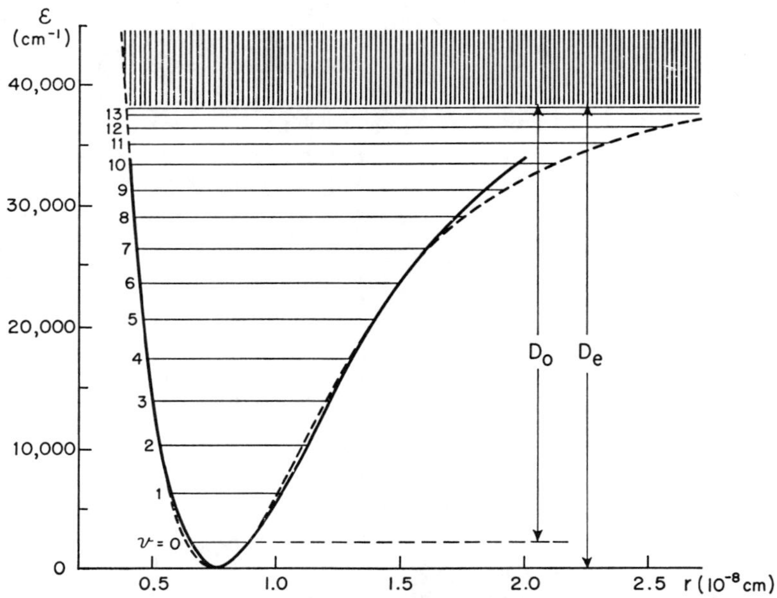

Fig. 18.9. Potential curve and Morse curve (dotted) of H_2 molecule. (After Herzberg.[1])

for which the third principal moment of inertia around the molecular axis is practically zero. A typical spherical top is the methane molecule CH_4; by substituting chlorine for one of the hydrogen atoms, we lower the symmetry to that of a symmetric top; the ethylene

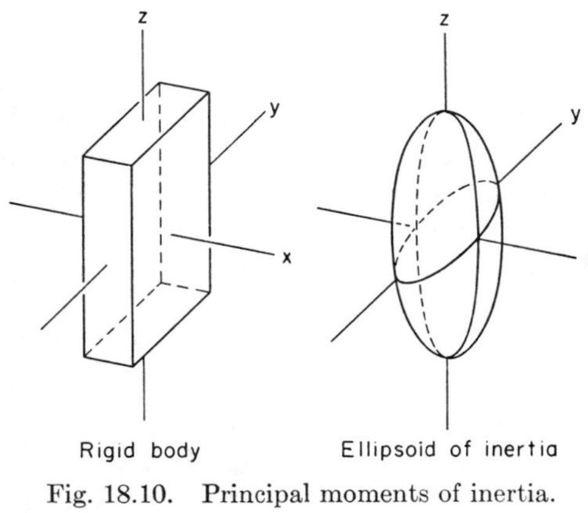

Fig. 18.10. Principal moments of inertia.

molecule, C_2H_4, finally, is a typical example of the asymmetric top class to which the great majority of the polyatomic molecules belong. Each of these classes has typical spectroscopic characteristics.[4]

[4] See G. Herzberg, *Molecular Spectra and Molecular Structure. II. Infrared and Raman Spectra of Polyatomic Molecules*, D. Van Nostrand Co., New York, 1945.

Similarly, a classical analysis of the vibrations of a molecular system leads to a useful systematization. Any complicated oscillation of a system due to harmonic forces can be represented as the superposition of independent simple harmonic motions, of the so-called *normal vibrations.* There are always as many normal vibrations as there are vibrational degrees of freedom.

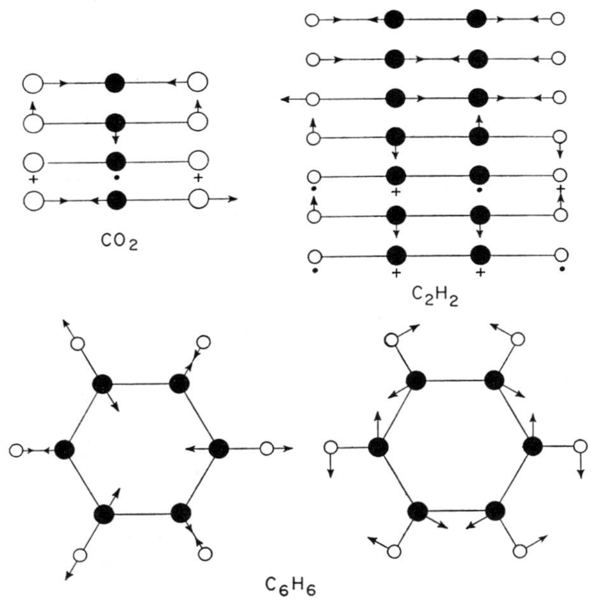

Fig. 18.11. Normal vibrations of some molecules.

The total number of degrees of freedom of a molecule consisting of N mass points is $3N$. Three of these $3N$

degrees of freedom belong to the translation of the molecule as a whole in the x, y, and z direction. Three or, for the linear molecule, two represent the rotation of the molecule around its principal axes as described above. Hence $3N - 6$ or, for the linear molecule, $3N - 5$ vibrational degrees of freedom remain. Figure 18.11 shows normal modes for some typical molecules. The vibrations are obviously of two types: in the one, the valence bonds of a molecule are stretched; in the other, they are bent. We speak therefore of *valence-* or *bond-stretching vibrations* as contrasted with *deformation-* or *bond-bending vibrations*. The former are of appreciably higher frequency, since the head-on motion of the atoms leads to larger restoring forces.†

† For a detailed discussion of molecular spectra see the fundamental monographs by G. Herzberg (Refs. 1 and 4).

19 · Electronic, Atomic, and Orientation Polarization of Gas Molecules

By considering atoms as linear harmonic oscillators, we arrived in Sec. 4 at the dispersion formula of classical physics

$$\kappa^* = 1 + \sum_s \frac{N_s e^2/\epsilon_0 m}{\omega_s^2 - \omega^2 + j\omega 2\alpha}. \tag{4.29}$$

Each resonance frequency ω_0 of an atom is represented in this classical analogue by one special resonator type s. The electric field $\mathbf{E}$ induces in these oscillators an electric moment

$$\boldsymbol{\mu}_i = |\boldsymbol{\mu}| e^{j\psi} = \frac{\mathbf{P}}{N} = \frac{e^2/m}{\omega_0^2 - \omega^2 + j\omega 2\alpha}\mathbf{E}. \tag{19.1}$$

Without attenuation ($\alpha = 0$), the moment stands parallel to the inducing field when its frequency lies below resonance ($\psi = 0$) and antiparallel above resonance ($\psi = -180°$). With attenuation ($\alpha \neq 0$), the discontinuous phase jump of ψ at the resonance frequency is replaced by a continuous phase shift amounting to $-90°$ at ω_0 (Fig. 19.1). The in-phase component of the induced moment below resonance leads to the normal capacitive energy storage ($\kappa' - 1 > 0$); when the moment has an antiparallel component (above resonance), the polarization is negative ($\kappa' - 1 < 0$). The "90° out-of-phase" component of the moment represents energy dissipation since it produces a current in phase with the external field $\left(I \simeq \frac{\partial \boldsymbol{\mu}_i}{\partial t}\right)$. Thus, while the induced moment is a maximum at resonance, its 90° position makes it contribute only to the absorption but not to the dispersion (see also Fig. 4.2).

In static fields the induced moment reduces to (cf. Eq. 4.3)

$$\boldsymbol{\mu}_{i_s} = \frac{e^2/m}{\omega_0^2}\mathbf{E} = \frac{ee\mathbf{E}}{f} = ez; \tag{19.2}$$

it is equal to the dipolar charge times the displacement distance z, where z is given by the ratio of driving force $e\mathbf{E}$ to force constant f. Far above resonance ($\omega \gg \omega_0$) the induced moment falls to zero because the time of a half cycle is too short to permit any charge displacement.

In substituting for the classical formula (Eq. 4.29), the dispersion formula of quantum mechanics (Sec. 12),

$$\kappa^* = 1 + \frac{2}{\epsilon_0 \hbar} \sum_{i<j} \frac{\omega_{ij} |\boldsymbol{\mu}_{ij}|^2}{\omega_{ij}^2 - \omega^2 + j\omega 2\alpha}(N_i - N_j), \tag{12.17}$$

we took cognizance of three facts: (1) the classical resonance absorption corresponds to a quantum transition of the electronic system from a lower (i) to a higher (j) energy state; (2) the number of atoms available for absorption is given by the difference in the population of the two states ($N_i - N_j$), since the atoms in the higher energy state can be forced down into the lower one by light emission; (3) the coupling between the atom and the electromagnetic field is given by the average dipole moment of the product of the wave functions of the two states, the matrix element of the dipole moment, $\boldsymbol{\mu}_{ij}$, taken in the direction of the external field $\mathbf{E}$.

For atoms the higher energy states lie so far above the ground state that the chance of thermal excitation at room temperature (kT ca. 1/40 ev) is negligible. Consequently, the lower energy state i refers to the ground state only and all the atoms are found originally in this state ($N_i = N$, $N_j = 0$). The dispersion formula (Eq. 12.17) applied, for example, to the atomic hydrogen gas therefore claims that the permittivity of this gas in its frequency dependence is prescribed by the frequencies and intensities of the spectral lines of the Lyman series which arises from the ground state (see Sec. 5). On the other hand, we found that the static permittivity of this hydrogen gas is given by the polarizability of the unexcited hydrogen atom (see Sec. 3).

At first glance, it seems difficult to reconcile these two methods of obtaining the static dielectric constant:

by extrapolation from the optical spectrum or by a distortion of the electronic structure of the unexcited atom. The outcome, however, is the same. The optical method considers (in the matrix elements) the dipole moment components in the field direction of the *undisturbed* transitions, whereas the static approach refers to the induced dipole moment of the *perturbed* ground state. In calculating this perturbation, the theory makes use of the same probabilities of transition to the higher energy states which the line spectrum contains.

Progressing from atoms to molecules, we are faced with a much more involved situation. Whereas in atomic gases only the ground state was occupied, for molecules the excited states of vibration and rotation lie low enough for thermal excitation. Atoms, furthermore, are isotropic whereas molecules offer various degrees of anisotropy. Finally, although atoms in their ground state do not act with permanent electric moments, molecules are likely to have such moments.

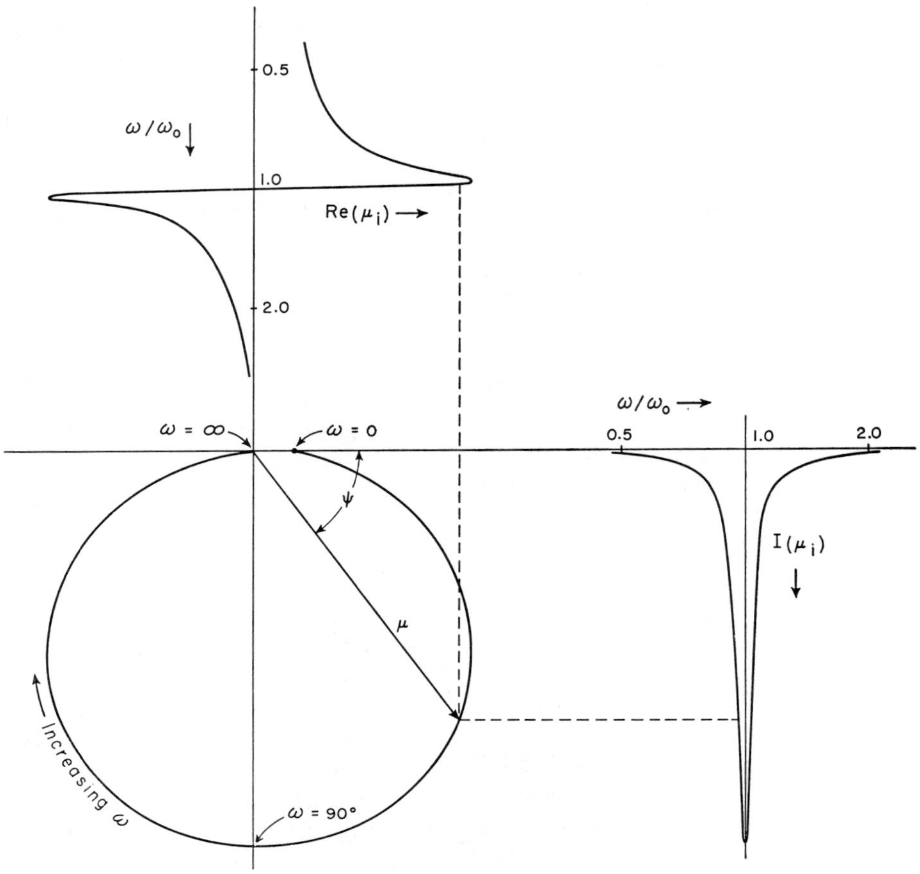

Fig. 19.1. Dispersion and absorption of harmonic oscillator, caused by induced moment μ_i.

The population of various energy states

The inclusion in our considerations of the population of the various energy states is a straightforward procedure. If various energy states $\mathcal{E}_i$ compete for molecules in thermal equilibrium, the fraction of molecules in each of them is given according to Maxwell-Boltzmann statistics as

$$\frac{e^{-\mathcal{E}_i/kT}}{\sum_i e^{-\mathcal{E}_i/kT}}. \tag{19.3}$$

This expression assumes that all energy terms have the same *statistical weight.* Actually, this may be not so; an energy term, as we discussed in Sec. 11, may be degenerate, that is, represent several electronic configurations. In a magnetic field these configurations correspond to different energies: the single term splits into a multiplet of states. Since each of these multiplet terms counts statistically as one state, each degenerate state E_i appears in the calculation with a statistical weight factor g_i equal to its multiplicity. Thus, if N is the total number of molecules, the population of the states N_i and N_j of Eq. 12.17 is, for thermal equilibrium,

$$N_i = Ng_i \frac{e^{-\mathcal{E}_i/kT}}{\sum_i e^{-\mathcal{E}_i/kT}}$$

and (19.4)

$$N_j = Ng_j \frac{e^{-\mathcal{E}_j/kT}}{\sum_i e^{-\mathcal{E}_i/kT}}.$$

The difference in population density of the lower and upper state, since $\mathcal{E}_i - \mathcal{E}_j$ is equal to the absorbed quantum energy $h\nu_{ij}$, may therefore be written as

$$N_i - N_j = \frac{N}{\sum_i e^{-\mathcal{E}_i/kT}} \{g_i e^{-\mathcal{E}_i/kT} - g_j e^{-\mathcal{E}_j/kT}\}$$

$$= N_i \left\{1 - \frac{g_j}{g_i} e^{-h\nu_{ij}/kT}\right\}. \quad (19.5)$$

Let us illustrate the influence of the population density of the various energy states on the intensity of spectral lines by returning to the example of the diatomic molecules of Sec. 18.

We found for the rigid rotator the energy expression

$$\mathcal{E}_{\text{rot}} = \frac{\hbar^2}{2I} J(J+1). \quad (18.28)$$

The selection rule $\Delta J = \pm 1$ prescribes that transitions can take place only to adjacent states. Thus, to make a nonrotating molecule rotate requires an energy input ($J = 0 \rightarrow J = 1$) of $\hbar^2/I$. We may therefore specify with Debye[1] a *critical temperature* T_c for the rotator molecule at which the average thermal energy kT just suffices to start rotation by defining

$$\hbar^2/I \equiv kT_c \quad (19.6)$$

or

$$T_c = \frac{\hbar^2}{kI} = \frac{8.0 \times 10^{-46}}{I} \quad [°\text{K}]. \quad (19.7)$$

For the HCl molecule, with the moment of inertia $I = 2.71 \times 10^{-47}$ [kg m^2], this critical temperature T_c is ca. 30°K, that is, at room temperature many molecules are excited to higher levels. To obtain the relative number of molecules in the various rotation states, we have, in addition, to know the statistical weight g (the multiplicity) of the rotation terms.

The angular momentum of the electron clouds of the hydrogen atom was prescribed by a quantum number l and could assume in a magnetic field $2l + 1$ positions (see Eq. 8.26). Similarly, the angular momentum of the molecular rotator, prescribed by the quantum number J, can assume $2J + 1$ positions. Hence the statistical weight is

$$g = 2J + 1. \quad (19.8)$$

Using this weight factor, we obtain the relative occupation of the various rotational levels in thermal equilibrium by plotting the function (see Eq. 19.4)

$$ge^{-\mathcal{E}/kT} = (2J+1)e^{-(T_c/2T)J(J+1)} = f(J) \quad (19.9)$$

as function of J (Fig. 19.2). Owing to the increasing

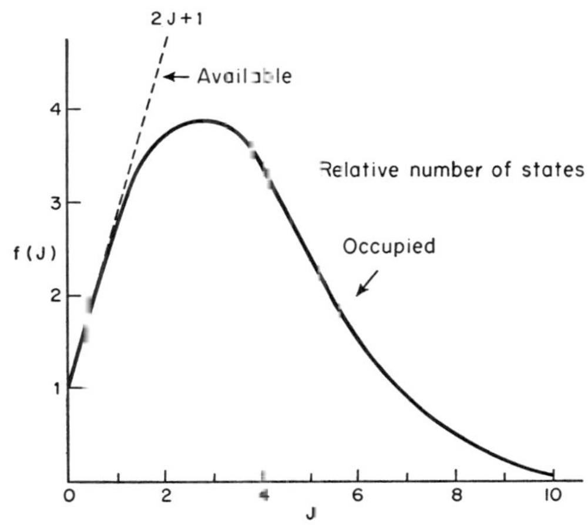

Fig. 19.2. Relative occupation of the various rotation levels of the HCl molecule at room temperature ($T/T_c \simeq 10$).

multiplicity of the higher levels, the number of molecules first increases with J, reaches a maximum at

$$J_{\max} = \sqrt{\frac{T}{T_c}} - \frac{1}{2}, \quad (19.10)$$

and then falls rapidly towards zero.

The tangent on this characteristic, $2J + 1$, indicates equivalent numbers of absorbers in the various energy states. For $J \frac{T_c}{T} \ll 1$, the absorption is very small in spite of the large number of absorbers, because the population of neighboring states is practically equal and the forced emission from the upper state approximately cancels the absorption from the lower. This effect makes the rotation lines very weak in the microwave region. In the high energy range, $J \frac{T_c}{T} \gg 1$, the number of thermally excited molecules becomes very small; hence the absorption falls towards zero due to the scarcity of absorbers. Between these extremes the *band of rotational transitions* traverses a maximum of intensity when the difference in population density of adjacent states becomes most favorable (see Fig. 18.7).

[1] P. Debye, *Polar Molecules*, Dover Publications, New York, 1945, Sec. 33.

The rotational states have a multiplet structure $(2J + 1)$ due to the space quantization of the angular momentum of the rotator in an external electric or magnetic field. This situation has been discussed previously for the Zeeman effect of atoms in Sec. 10. The vibration states of the harmonic oscillator represent only a simple sequence of standing-wave patterns $n\frac{\lambda}{2}$ like the mechanical resonances of a stretched wire or the electrical ones of a transmission line. Thus, in contrast to

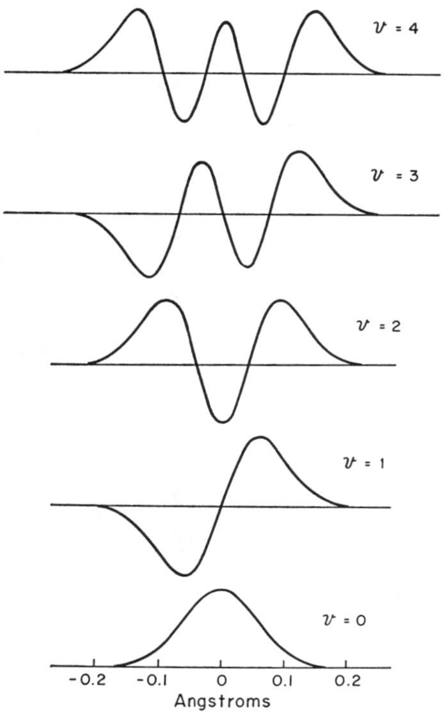

Fig. 19.3. Wave functions of the vibrator (HCl). (After Herzberg.[2])

the wave functions of the rotator, to each quantum state of the vibrator belongs only one wave function (Fig. 19.3); the vibration states have the statistical weight $g = 1$.

In the language of the spectroscopist, the wave numbers of spectral lines

$$\nu' = \frac{1}{\lambda} \quad [\text{cm}^{-1}] \tag{19.11}$$

serve as a measure of the energy of the transitions since

$$\begin{aligned} \mathcal{E} = h\nu = hc\nu' &= 1.985 \times 10^{-23}\nu' \quad [\text{joule}] \\ &= 1.239 \times 10^{-4}\nu' \quad [\text{ev}]. \end{aligned} \tag{19.12}$$

The lowest rotation line of the HCl molecule is found near 0.48 mm or $\nu' = 20.7$, and the fundamental vibration near 3.5 μ or $\nu' = 2886$ cm^{-1} (see Fig. 18.7). At room temperature the lowest rotational state thus lies far below the average thermal energy level $kT \simeq 0.025$ ev $\left(\frac{T_c}{T} \simeq \frac{1}{10}\right)$, whereas the first vibrational state lies appreciably above kT (for HCl at 0.35 ev). The population of even the first vibrational level $(v = 1)$ of HCl amounts therefore to only ca. $1/10^6$ of the nonexcited state in thermal equilibrium at 300°K.

Relation of electronic to atomic polarization

We have seen that the static electronic polarization, caused by a displacement of the electron clouds with respect to the nuclei under the influence of a static electric field, can be found in two ways: either by extrapolation from the undisturbed optical absorption spectrum that arises from the ground state of the atom or molecule, or from the perturbation of the electron clouds of the ground state. Similarly, the atomic polarization, caused by the induced moment arising from a displacement of the atomic masses under the influence of a static field, may be found by extrapolation from the vibration spectrum or from the perturbation of the charge distribution produced by the alteration of internuclear distances.

For the electronic polarization, the optical contribution stemmed from a whole series spectrum of lines since the principal quantum number was unrestricted. For the vibration spectrum, the transitions are limited to neighboring states (Eq. 18.24) and produce, in the case of the harmonic oscillator, even the same spectral line. Furthermore, as just shown for the HCl molecule, the population of the ground state $(v = 0)$ may dominate completely and leave only the transition from the nonvibrating (that is, zero-point vibration) to the first excited state to be considered.

The classical dispersion formula (Eq. 4.29) was based on the model of isotropic harmonic oscillators. A diatomic molecule is an anisotropic oscillator of one instead of three degrees of vibrational freedom. We take account of this anisotropy by inserting a factor 1/3 into Eq. 19.1 for the induced moment; hence

$$\boldsymbol{\mu}_{i_{\text{osc}}} = \frac{e^2/3m}{\omega_0{}^2 - \omega^2 + j\omega 2\alpha}\,\mathbf{E}. \tag{19.13}$$

The static induced moment of the molecule, when lined up in the field direction z, is, as previously (see Eqs. 4.22 and 19.2),

$$\boldsymbol{\mu}_{i_s} = e\mathbf{z}. \tag{19.14}$$

To interpret the dipolar charge e, we have to plot the dipole moment of the molecule as a function of the internuclear distance r. The shape of this curve will depend on the type of bond linking the atoms of the molecule (Fig. 19.4). If the bond is purely ionic, that is, if HCl, for example, dissociated into H^+ and Cl^-, the dipole moment will increase with r to infinity. For the covalent bond between like partners, the dipole moment will remain zero throughout. In the general case of unlike partners, where the molecule dissociates into atoms and not ions, we have a covalent bond with ionic contributions (see Sec. 15). The dipole moment must be zero for coinciding and for widely separated

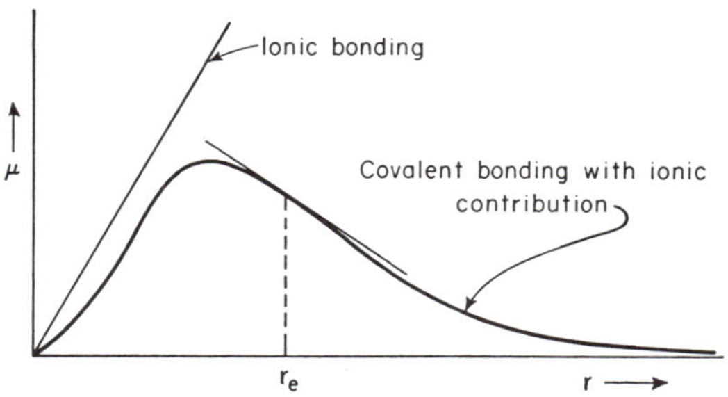

Fig. 19.4. Dipole moment and bond type as function of internuclear distance (r_e for HCl molecule). (After Herzberg.[2])

nuclei and must traverse a maximum at some intermediate distance. At the equilibrium distance r_e, the dipole moment has some value μ_e, and the induced moment will be determined, in a first approximation, by the tangent of the curve at this point as

$$\mu_{i_s} = \frac{d\mu_e}{dr}\,\mathbf{r}. \tag{19.15}$$

Hence the strength of the oscillating charge e of Eq. 19.14 is identical with the space derivative of the dipole moment, $\dfrac{d\mu_e}{dr}$, as already stated in Sec. 18.

The tangent of the dipole curve at r_e consequently can be determined from the intensity of the infrared vibration spectrum. This tangent is usually not steep, since the ionic contribution to the bond energy is minor and the equilibrium distance located near the broad maximum of the dipole curve. The infrared intensity is therefore, in general, relatively weak, and the contribution of the atomic polarization is only a small addition to the electronic polarization. It should be noted that the intensity of the vibration spectrum, since it depends on $\left(\dfrac{d\mu_e}{dr}\right)^2$, gives only the absolute value of the moment change but not its direction. On which side of the maximum the equilibrium point is located must be found from additional information; for HCl, according to Bartholomé,[2] this point lies to the right of the maximum as shown in Fig. 19.4.

In the formulation of quantum mechanics, the induced moment of the linear oscillator is not given simply by the classical expression

$$\mu_{i_{osc}} = \frac{d\mu_e}{dz}\,z, \tag{19.16}$$

but found (see Eq. 12.14) as the average dipole moment of the mixed wave functions of the lower (v') and upper (v'') state

$$\mu_{i_{v'v''}} = \frac{d\mu_e}{dz}\int_{z=-\infty}^{z=+\infty} \psi_{v'} z \tilde{\psi}_{v''}\,dz. \tag{19.17}$$

By introducing this matrix element of the dipole moment and the factor 1/3 into Eq. 12.17, we change from the dispersion formula of electronic to that of atomic polarization (see also Appendix A, II, Sec. 3).

Rotation spectrum and orientation polarization

Electronic and atomic polarization are due to the *displacing force* exercised by the external field upon the electrons and atoms, and measure the *induced* moments thus created. The orientation polarization, in contradistinction, arises from the *torque* action of the field on the pre-existing *permanent* moments of the molecules. We have shown that the contributions of the electronic and atomic polarization to the static dielectric constant can be obtained by extrapolation from the electron excitation and the vibration spectrum according to the dispersion formula of quantum mechanics. We will find for the rotation spectrum that this extrapolation also holds but that only the first rotation line contributes to the static orientation polarization.

Let us consider first the classical situation (Fig. 19.5). A dipole molecule in a homogeneous field will

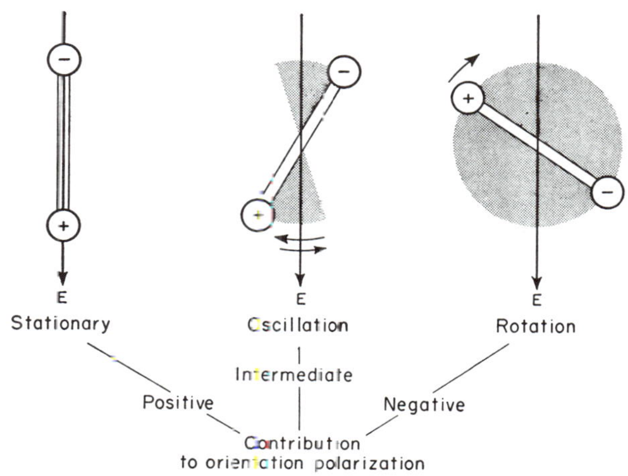

Fig. 19.5. Classical contribution of dipole molecule to orientation polarization.

[2] E. Bartholomé, *Z. physik. Chem.* B*23*, 131 (1933); see also G. Herzberg, *Spectra of Diatomic Molecules*, Van Nostrand, New York, 1950, p. 96.

turn parallel to **E** if no additional forces act; the orientation polarization of polar molecules at rest is always positive. If the molecule can oscillate around an axis normal to the field, its positive contribution rapidly diminishes with increasing amplitude, since it passes through the parallel position with maximum kinetic energy and spends most of its time at the turning points. The contribution will become clearly negative when the oscillation evolves into a slow rotation because the molecule traverses the antiparallel position at a much slower speed than the parallel one. This negative polarization rapidly diminishes with increasing rotational speed and approaches zero when the kinetic energy $\mathcal{E}_{\text{rot}}$ becomes much larger than the potential energy U of the dipole in the field. When the axis of rotation is oriented parallel to the field, the dipole moment of the dumbell molecule stands perpendicular to **E** and does not contribute to the polarization. Intermediate axis positions can be resolved into a parallel and a perpendicular component. Hence classically rotating molecules will not contribute to the orientation polarization as long as

$$\frac{p'^2}{2I} \gg -|\boldsymbol{\mu}|\,|\mathbf{E}|\cos\theta. \tag{19.18}$$

Quantum physics comes to the same result: the angular momentum p' of the diatomic molecule is quantized in the electric field but stands normal to the dipole moment. Hence, whereas the mechanical momentum points in certain field directions, the dipole moment is not oriented and does not contribute to the polarization. In consequence, no linear Stark effect (see Sec. 9) is observed for the rotation lines of these linear dipole molecules. However, a second-order effect exists that causes a line splitting. This is not the normal quadratic Stark effect based on dipole moments induced by the electric field, which can cause only a lowering of the energy terms because such moments point towards the field. The multiplet splitting of the rotation lines is proportional to the square of the permanent moment and of the field strength ($\boldsymbol{\mu}^2\mathbf{E}^2$) and raises some levels while lowering others (Fig. 19.6).[3] The rotating molecule, graphically speaking, makes wobble motions, in which the states shifted towards higher energy correspond to an antiparallel contribution of the moment and the depressed levels to a parallel one. These contributions accurately balance each other for all higher rotation levels. Only the dispersion effect stemming from the transition from the nonrotating to the first rotating state remains and gives, extrapolated to zero frequency, the static orientation polarization.[4]

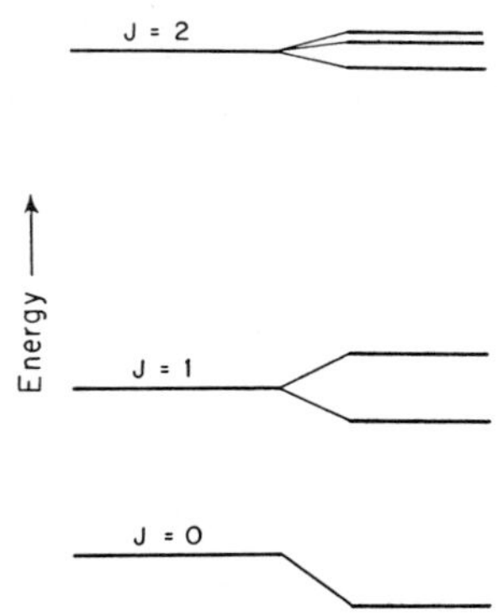

Fig. 19.6. Stark effect splitting of rotation levels of linear dipole molecule. (After Debye.[1])

[3] See P. Debye, *Polar Molecules*, Chemical Catalog Co., New York, 1929, Sec. 29.

[4] For a detailed theory of the Stark effect in rotational spectra, see J. H. Van Vleck, *Theory of Electric and Magnetic Susceptibilities*, Clarendon Press, Oxford, 1932.

20 · The Bandwidth of Spectral Lines

The dispersion formulas of classical and quantum physics (Eqs. 4.29 and 12.17) contain a damping factor 2α which determines the bandwidth (half-width or linebreadth) of the spectral line, that is, the Q or the ringing time τ of the resonator. In Sec. 4 we derived the relations

$$2\alpha = \Delta\omega_h = \frac{1}{\tau} = \frac{\omega_0}{Q}, \tag{20.1}$$

where $\Delta\omega_h$ is the frequency interval between the half-power points of the bell-shaped resonance characteristic (see Fig. 4.1). As the cause of the line broadening we assumed radiation damping and calculated, on the basis of the classical theory of dipole radiation, the damping factor as (see Eq. 4.6)

$$2\alpha = \frac{\mu_0{}^2 e^2 \omega_0{}^2}{m 6\pi c}. \tag{20.2}$$

In the language of atomic physics, the time τ represents the average *natural lifetime* of an atom or molecule in an excited energy state before it returns to a lower state under the emission of a photon $h\nu_0$. If, for example, we calculate on this basis the bandwidth of the green mercury line, located at

$$\lambda_0 = 5461 \text{ A}$$

or (20.3)

$$\nu_0 = 5.494 \times 10^{14} \quad [\text{sec}^{-1}],$$

we obtain

$$\Delta\nu_h = \frac{\Delta\omega_h}{2\pi} = 1.2 \times 10^6 \quad [\text{sec}^{-1}],$$

$$\tau = 1.34 \times 10^{-7} \quad [\text{sec}], \tag{20.4}$$

$$Q = 4.6 \times 10^8.$$

Compared with the optimum Q values of about 10^6 that an electrical engineer can realize in his cavity resonators (see I, Sec. 24), the physicist obviously possesses in his light sources oscillators of very superior monochromacy.

Unfortunately, however, our calculation was misleading, as a comparison between theory and experiment shows. The monochromacy of the radiation can be tested by an interference experiment. Reflecting the light from the front and back surface of a loss-free dielectric layer of the thickness d (for example, air), we may form a standing-wave pattern (see I, Sec. 18). The reflected light intensity, observed as function of the layer thickness, exhibits a sequence of interference maxima and minima. For a truly monochromatic light source, emitting only the wavelength λ_0, the thickness d might be increased ad infinitum without causing a deterioration in the definition of these interference stripes. If the light source emits two wavelengths, λ_1 and λ_2, corresponding to the half-power points of the spectral line, both lines would create their individual interference systems. Since the difference in wavelength of the two lines is very small,

$$\lambda_1 \simeq \lambda_2 \simeq \lambda_0, \qquad \lambda_1 - \lambda_2 = \Delta\lambda_h, \tag{20.5}$$

the stripe systems initially overlap. When the layer thickness is increased to $q\lambda_0/2$ wavelengths, the maxima of the first system fall on the minima of the second and the interference is blurred. As the dielectric layer reaches a thickness $q\lambda_0$, the patterns coincide again perfectly and the interference phenomenon is restored to its original clarity (Fig. 20.1). Thus by measuring the layer thickness from consonance to consonance, we obtain the wavelength difference of the two lines as

$$\Delta\lambda_h = \lambda_0/q. \tag{20.6}$$

Actually, the bandwidth $\Delta\lambda_h$ of our spectral line is filled with a continuum of additional wavelengths; hence a continuous sequence of additional stripe systems is interspersed between the interference maxima produced by λ_1 and λ_2. The position of restored coincidence is therefore not characterized by a perfect blending of two-stripe systems but by a uniformly illuminated field of view without stripes: beyond this distance of first consonance, $q\lambda_0$, stripe systems appear again but they are blurred and hard to see.

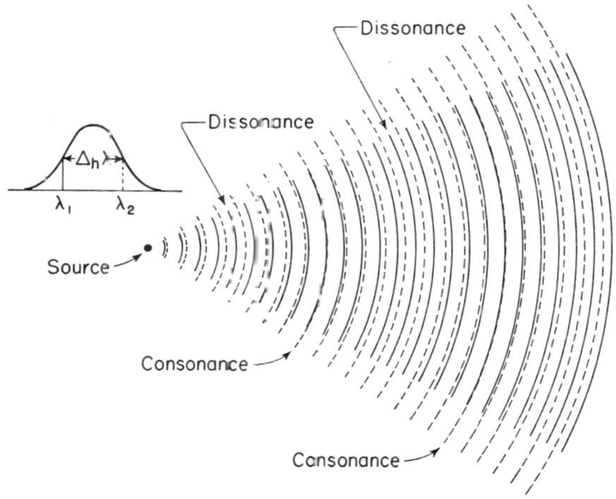

Fig. 20.1. Dissonance and consonance of two systems of interference stripes.

The integer q of Eq 20.6 which measures the sharpness of a spectral line by interference, is, according to Eq. 20.1, identical with the quality factor Q of the electrical engineer,

$$q = Q. \tag{20.7}$$

Consequently, it should be possible to observe interference stripes with the green mercury line † over a consonance distance

$$Q\lambda_0 \simeq 252 \quad [\text{m}]. \tag{20.8}$$

Fabry and Perot [1] performed this interference experiment using a water-cooled, low-pressure, d-c–discharge lamp as their source, and obtained interference over a distance of only 790,000 wavelengths, or 43 cm. Thus the Q of this light source proves to be about 600 times smaller than predicted on the basis of radiation damping. Obviously, we have overlooked some factors which contribute decisively to the line width.

One tacit assumption of our calculation has been that the light source is at rest in relation to the observer. Actually, when atoms or molecules radiate, this condition is usually not fulfilled. The particles of a gas are in thermal agitation, and the probability that they have a velocity component between ζ and

† Actually, the green mercury line consists of several components, and our discussion refers to the width of an individual component.

[1] C. Fabry and A. Perot, *Ann. chim. phys.* [7] *12*, 459 (1897).

$\zeta + d\zeta$ toward or away from the observer, equals, according to Maxwell-Boltzmann statistics,

$$dw = \sqrt{\frac{m}{2\pi kT}}\, e^{-m\zeta^2/2kT}\, d\zeta. \tag{20.9}$$

This relative motion between the radiating particle and the measuring device produces a *Doppler effect.*[2] The observer will note, instead of the frequency ν_0, the frequencies

$$\nu = \nu_0\left(1 \pm \frac{\zeta}{c}\right). \tag{20.10}$$

By replacing the variable ζ in the distribution law with the observed frequency ν, we may rewrite Eq. 20.9, since

$$\zeta = \frac{c}{\nu_0}(\nu - \nu_0), \qquad d\zeta = \frac{c}{\nu_0}\, d\nu, \tag{20.11}$$

as

$$dw = \sqrt{\frac{m}{2\pi kT}}\, e^{-\frac{m}{2kT}\frac{c^2}{\nu_0^2}(\nu-\nu_0)^2}\, \frac{c}{\nu_0}\, d\nu. \tag{20.12}$$

The intensity distribution of the radiation over the various frequencies images this velocity distribution. Identifying with I_0 the intensity for the undisplaced frequency ν_0 at the center of the line, we expect that the intensity will fall towards both sides due to the Doppler effect, as

$$I = I_0 e^{-\frac{m}{2kT}\frac{c^2}{\nu_0^2}(\nu-\nu_0)^2}\, \frac{c}{\nu_0}\, d\nu. \tag{20.13}$$

It drops to $1/e$ or 0.368 of its maximum intensity at a frequency

$$\nu = \nu_0 \pm \frac{\nu_0}{c}\sqrt{\frac{2kT}{m}}; \tag{20.14}$$

the frequency difference between these half-width points may be designated as the bandwidth of the line,

$$\Delta\nu_h = \frac{2\nu_0}{c}\sqrt{\frac{2kT}{m}}, \tag{20.15}$$

produced by *Doppler broadening.*

Let us assume that the water-cooled mercury lamp of Fabry and Perot had a gas temperature T of about 300°K. The weight of the mercury atom is $m_{\text{Hg}} = 3.34 \times 10^{-25}$ [kg], and the molecular gas constant $k = 1.380 \times 10^{-23}$ [joule deg^{-1}]. The half-width of the green mercury line due to the Doppler effect then becomes

$$\Delta\nu_h = 6 \times 10^8 \quad [\text{sec}^{-1}]; \tag{20.16}$$

it is about 500 times larger than that produced by radiation damping (see Eq. 20.4). This is the magnitude indicated by the outcome of the Fabry-Perot experiment.

It is possible to avoid the Doppler effect by using molecular beams as the light source and directing them perpendicularly to the direction of observation.[3] If we exclude this special case, the light sources presented by the finest spectral lines of the physicist are of about the same Q as the best cavity resonators of the electrical engineer.

In the preceding discussion we have made one additional tacit assumption: that excited atoms and molecules can live out their natural life span τ undisturbed. Actually, they will collide with other atoms and molecules or with the walls of the container and be forced to radiate after a normally much shorter collision time τ_c. Fourier analysis shows that the shorter the pulse, the wider the bandwidth; hence a *collision-* or *pressure-broadening* of spectral lines will result.

The order of magnitude of τ_c can be derived by a simple gaskinetic consideration as the ratio of the free path (the average distance $\bar{l}$ traversed between collisions) and of the average velocity $\bar{v}$ of the particle:

$$\tau_c = \bar{l}/\bar{v} \tag{20.17}$$

The free path is inversely proportional to the number N of particles per unit volume and to the target area $4\pi r^2$ that each of them offers. A more accurate calculation gives

$$\bar{l} = \frac{1}{\sqrt{2}\, N 4\pi r^2}, \tag{20.18}$$

where r designates the *collision radius* of the particle in question.

To obtain the average velocity, we recall that each particle has three degrees of freedom and that the kinetic energy associated with each degree of freedom equals, according to the equipartition law of classical physics, $\frac{1}{2}kT$. Hence

$$\tfrac{1}{2}mv^2 = \tfrac{3}{2}kT \tag{20.19}$$

and, as a somewhat more detailed calculation shows,

$$\bar{v} = \sqrt{\frac{8kT}{\pi m}}. \tag{20.20}$$

Thus

$$\tau_c = \frac{1}{N(4r)^2}\sqrt{\frac{m}{\pi kT}}. \tag{20.21}$$

[2] C. Doppler, *Abh. kgl. Böhmisches Ges. Wiss.* [5] *2*, 465 (1842).

[3] See H. Kuhn, *J. Sci. Instr.* *23*, 249 (1946).

Assuming for mercury a collision radius $r \simeq 1.5$ A and for the water-cooled mercury lamp a pressure of $\frac{1}{10}$ mm at 300°K, that is (cf. Eq. 2.12),

$$N = \frac{N_L}{7600} \times \frac{273}{300} \simeq 3.2 \times 10^{21} \quad [\text{m}^{-3}], \qquad (20.22)$$

we obtain a free path $\bar{l} \simeq 7.8 \times 10^{-4}$ [m], an average velocity for the mercury atom of $\bar{v} \simeq 177$ [m sec^{-1}], hence a collision time $\tau_c \simeq 4.4 \times 10^{-6}$ [sec]. Thus, if the effect of the collisions is that the natural lifetime τ of the excited atom must be replaced by the collision time τ_c, we expect, according to Eq. 20.1, a bandwidth due to collision broadening

$$\Delta\nu_h = \frac{1}{2\pi}\frac{1}{\tau_c} \simeq 3.6 \times 10^4 \quad [\text{sec}^{-1}]. \qquad (20.23)$$

In our example of the mercury lamp, the natural lifetime is about thirty-three times shorter than the collision time; hence the effect of pressure broadening can be neglected. At higher pressures this is not the case, but the Doppler effect will normally dominate the line width in the visible and ultraviolet region. For a convenient comparison of the broadening by radiation, Doppler effect, and collision we can write the three contributions in terms of the relative line width [4] as

$$\frac{1}{Q} = \Delta_h = \frac{\Delta\nu_h}{\nu_0} = \underbrace{1.17 \times 10^{-14}\frac{1}{\lambda_0}}_{\text{radiation damping}} + \underbrace{2.7 \times 10^{-8}\sqrt{\frac{T}{M}}}_{\text{Doppler effect}} + \underbrace{1.88 \times 10^{-9}\sqrt{\frac{T}{M}}\frac{\lambda_0}{\bar{l}}}_{\text{collision broadening}}, \qquad (20.24)$$

where λ_0 is the resonance wavelength in [m], T the absolute temperature, M the mole weight in [kg], and $\bar{l}$ the free path in [m]. In the case of an atomic hydrogen gas at 760 mm pressure and 0°C we would find for the resonance wavelength $\lambda_0 = 1.215 \times 10^{-7}$ [m] (see Eq. 5.21), and for $\bar{l} = 8.4 \times 10^{-7}$ [m], that

$$\Delta_h \simeq 1 \times 10^{-7} + 1.4 \times 10^{-5} + 3.7 \times 10^{-7}. \quad (20.25)$$

The Doppler effect still dominates, but the effect of pressure broadening is already 3.7 times the attenuation caused by radiation damping. At long wavelengths, the influence of pressure broadening will clearly become the decisive factor.

[4] See M. Born, *Optik*, Springer, Berlin, 1936, Sec. 86.

21 · Microwave Spectroscopy

How well the shape of a spectral line can be measured depends frequently not on its own bandwidth but on the resolving power of the analyzing optical instrument. This *resolving power*, defined by the physicist as the ratio of the average wavelength λ of the spectral region in question to the observable wavelength difference $\Delta\lambda$, is identical with the Q introduced by the electrical engineer:

$$Q = \frac{\lambda}{\Delta\lambda} = \frac{\nu}{\Delta\nu}. \qquad (21.1)$$

The Q of the instrument must obviously be larger than that of the spectral line, before finer details can be discerned.

In the visible and ultraviolet the resolving power is given by the dispersion which the prisms, gratings, or interferometers can provide. For the region of the green mercury line, a flint-type prism of 10-cm base realizes a Q of about 10^4; a grating of 100,000 lines produces in its third order a Q of about 3×10^5; a Lummer-Gehreke glass plate of 1 cm thickness may reach a Q of about 10^6; and with the best Perot-Fabry air-plate interferometer a Q up to 7×10^6 has been obtained.[1]

In the infrared, the resolving power of the prism or grating is not the decisive factor, but more stringent limitations are imposed by the nature of the heat radiation and the means for its detection. The low intensity of thermal radiation and the relatively insensitive bolometer or thermocouple receivers require wide slits. In addition, disturbing radiation emanates from all parts of the equipment and produces a high noise level. Thus the overall Q of the equipment may decrease in the far infrared to 10^1, and the experimental difficulties encountered in this wavelength region have all but discouraged the investigation of the rota-

[1] For a detailed discussion of the resolving power of interferometers, see, for example, G. Hansen, *Handbuch der physikalischen Optik*, Barth, Leipzig, 1927, pp. 185 ff.

tion spectra themselves. The main sources of our knowledge on rotating molecules were, until recently, the vibration-rotation spectra in the near infrared (cf. Fig. 18.7) and the Raman spectra observed with high dispersion in the visible and ultraviolet region.[2]

The microwave techniques of radar, developed during World War II, have altered this situation radically. Very stable new tubes such as the klystron oscillators have made a microwave spectroscopy possible based on monochromatic sources of accurately adjustable frequency and phase and on extremely sensitive detectors rather than on dispersive instruments and broad-band light sources. Interference techniques thus come into their own again and allow us to measure wavelengths and frequencies electronically to seven significant figures and to analyze line shapes with corresponding accuracy. The microwave spectroscope, operating at present mainly in the region of 20 to 0.5 cm surpasses by nearly a millionfold the resolving power of the grating instruments operating in the far infrared, and reaches that of the best optical interferometers.[3] A typical electronic spectroscope is shown in Fig. 21.1.

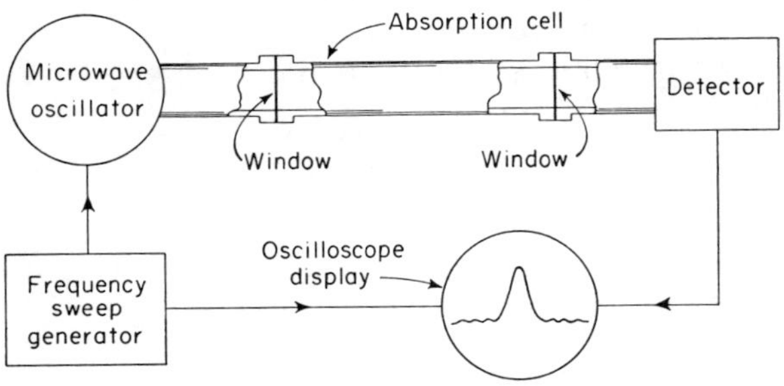

Fig. 21.1. Electronic spectroscope.

The information obtained by such precise measurements of the structure of molecular lines is of a surprising variety. Accurate data result as to moments of inertia, bond distances and bond angles, dipole moments and collision frequencies, nuclear magnetic dipole and electric quadrupole moments, isotopic masses, and the internal electric fields of molecules. The limitations of the method are that microwave absorptions are, in general, weak (see Sec. 19) and that only heavy molecules have rotation lines in this spectral range. However, we can observe a host of additional transitions

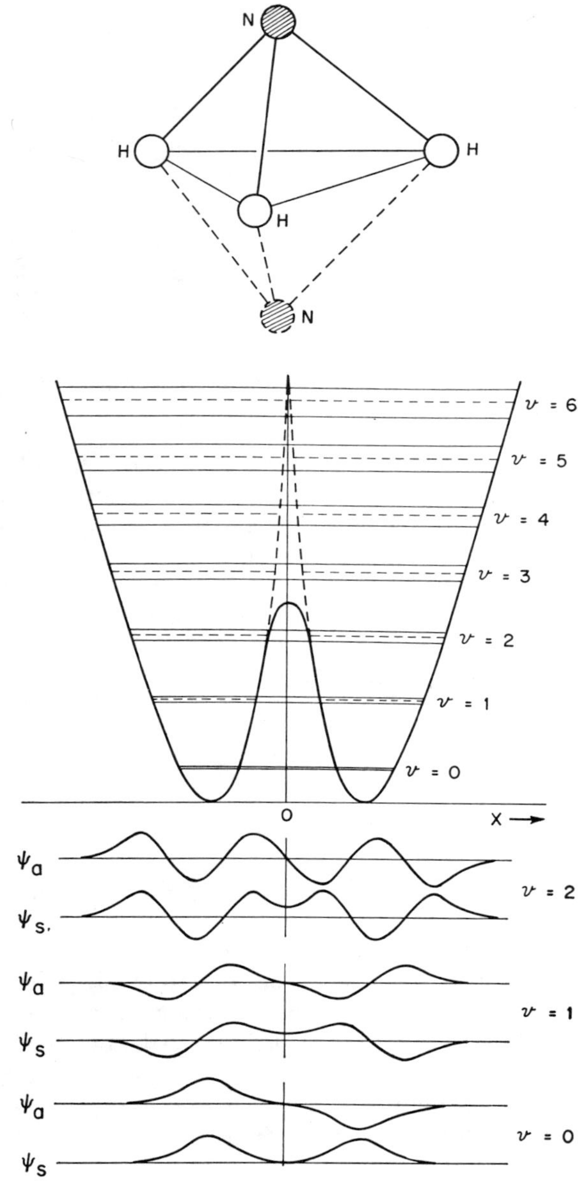

Fig. 21.2. Inversion and potential double well of NH_3 molecule. (After Herzberg.[8])

between fine-structure as well as hyperfine-structure levels by Stark and Zeeman effect measurements on atoms [4] and molecules. Furthermore, nonplanar mole-

[2] In the *Raman effect* [C. V. Raman, *Indian J. Physics 2*, 387 (1928)], the spectrum of the scattered light is observed. It shows, in addition to the original frequency of the incident light (Rayleigh scattering), weak lines displaced by distances corresponding to the vibrations and rotations.

[3] For surveys of microwave spectroscopy, see, for example, D. K. Coles, *Advances in Electronics*, Academic Press, New York, 1950, Vol. II, pp. 299 ff; E. B. Wilson, Jr., *Trans. Faraday Soc. 9*, 108 (1950); Conference on Microwave Spectroscopy, *Ann. New York Acad. Sci. 55*, 743 (1952); W. Gordy, *Physics Today*, December, 1952, p. 5; W. Gordy, W. V. Smith, and R. Trambarulo, *Microwave Spectroscopy*, John Wiley and Sons, New York, 1953.

[4] See, for example, the famous hydrogen experiment by W. E. Lamb and R. C. Retherford, *Phys. Rev. 72*, 241 (1947) [see also H. A. Bethe, *Phys. Rev. 72*, 339 (1947)], which led to a definite disagreement with modern quantum theory because the interaction between the electron and the radiation field had not been considered.

cules like ammonia can exist in two equivalent structures, as Fig. 21.2 illustrates. The "flip-flop" or "inversion" frequency between the two constellations may produce spectra in the microwave region.

The inversion spectrum of ammonia occupies a unique position in microwave spectroscopy. It was the first and only microwave spectrum of gases measured before the war (in the pioneering, semi-optical experiment of Cleeton and Williams [5] at 1.25 cm). It has been reinvestigated by numerous observers with the high resolving power of modern postwar techniques.[3] Also the deutero-ammonias have been studied in detail.[6] The ammonia absorption may therefore serve as a classical example to demonstrate what the microwave spectroscopist sees and how he proceeds in his interpretations.[7]

The absorption near 1.25 cm wavelength (0.8 cm^{-1} in wave numbers) in ammonia gas is caused by the possibility of turning the NH_3 molecule "inside out," that is, carrying the nitrogen atom through the plane of the hydrogen atoms to its image position. The original and the inverted configuration represent, from the standpoint of the nitrogen atom, two identical potential wells separated by an energy barrier (Fig. 21.2).[8] This constellation of a double well has been discussed already in the case of the H_2^+ ion (see Fig. 14.2). We have seen that each energy state of a particle in a single well splits into two states in a double well because the two configurations are in quantum-mechanical resonance. We expect, therefore, that the vibration spectrum of the ammonia molecule shows the same splitting of each term in a symmetrical and an antisymmetrical sublevel and will lead to a doublet structure of its lines. This structure has indeed been found in the far infrared absorption spectrum of NH_3 (Fig. 21.3).[9]

The average time τ a particle requires to transfer from one side of the potential barrier to the other by tunnel effect, is inversely proportional to the energy difference of the two sublevels. The probability of transfer, and with it the doublet splitting, is very small

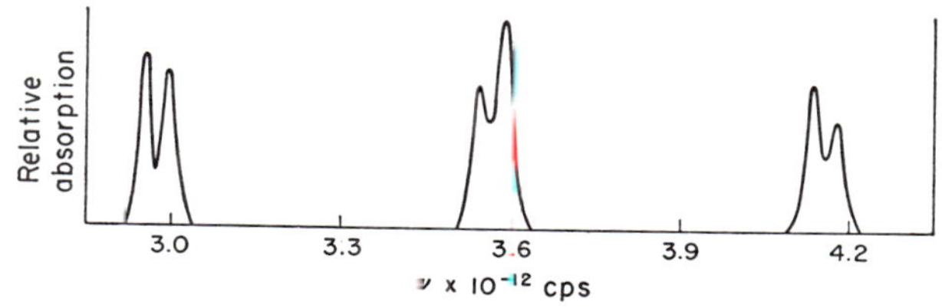

Fig. 21.3. Infrared doublet structure of NH_3 spectrum [path length 1 cm, pressure 8 cm at (4), 10 cm at (5), 7.3 cm at (6)]. (After Wright and Randall.[9])

for levels which lie low in the potential well, but increases rapidly when, with increasing vibration quantum numbers v, the top of the potential barrier is approached. The absorption near 1.25 cm corresponds to the transition between the two sublevels of the doublet for the zero-vibration state (ground vibration), $v = 0$.

When this absorption line is investigated with the high resolution of the microwave spectroscope, an extensive fine structure is observed [10] (Fig. 21.4). Such a

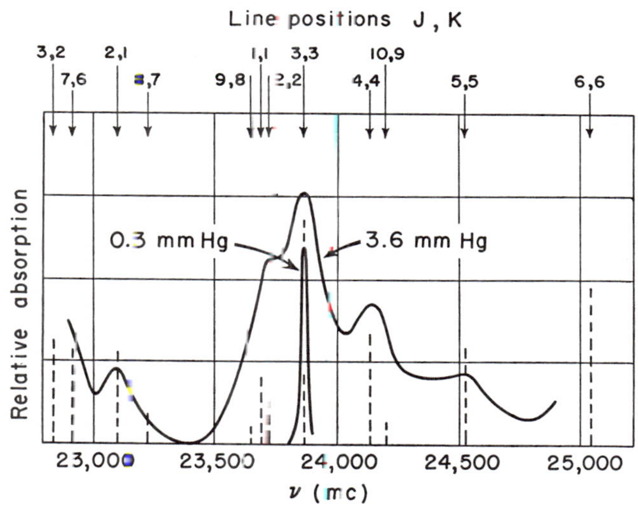

Fig. 21.4. Fine structure of NH_3 spectrum in microwave range. (After Townes.[10])

structure would be expected since the ammonia molecule can rotate around its symmetry axis as well as around an axis perpendicular to the symmetry axis (Fig. 21.5). These rotations lead to slight centrifugal distortions of the molecule in two opposite ways: the

[5] C. E. Cleeton and N. H. Williams, *Phys. Rev. 45*, 234 (1934).

[6] M. T. Weiss and M. W. P. Strandberg, *Phys. Rev. 83*, 567 (1951); K. A. Sawyer and J. B. Kierstead, *M.I.T. Res. Lab. Electronics Tech. Rep. 188* (1951).

[7] Two microwave absorptions troublesome for radar transmission were the first gas spectra encountered with modern equipment during the war: a water vapor intercombination line between two rotation states at 1.35 cm, predicted by Van Vleck [J. H. Van Vleck, *Phys. Rev. 71*, 425 (1947)] and measured with various techniques [see R. H. Dicke, R. Beringer, R. L. Kyhl and A. B. Vane, *Phys. Rev. 70*, 340 (1946); C. H. Townes and F. R. Merritt, *Phys. Rev. 70*, 558 (1946); G. E. Becker and S. H. Autler, *Phys. Rev. 70*, 300 (1946)], and a magnetic dipole absorption of the paramagnetic O_2 molecule between 4 and 6 mm first observed by Beringer [R. Beringer, *Phys. Rev. 70*, 53 (1946)] and calculated by Van Vleck [J. H. Van Vleck, *Phys. Rev. 71*, 413 (1947)].

[8] D. M. Dennison and G. E. Uhlenbeck, *Phys. Rev. 41*, 313 (1932). Figure from G. Herzberg, *Infrared and Raman Spectra of Polyatomic Molecules*, Van Nostrand, New York, 1945.

[9] N. Wright and H. M. Randal, *Phys. Rev. 44*, 391 (1933).

[10] W. E. Good, *Phys. Rev. 70*, 213 (1946); C. H. Townes, *Phys. Rev. 70*, 665 (1946); B. Bleaney and R. P. Penrose, *Nature 157*, 339 (1946); *Proc. Roy. Soc. (London)*, A*189*, 358 (1947).

rotation around the symmetry axis, characterized by an angular momentum quantum number K, tends to flatten the pyramidal molecule, hence to decrease the potential barrier and increase the inversion frequency; vice versa, the rotation around a perpendicular axis

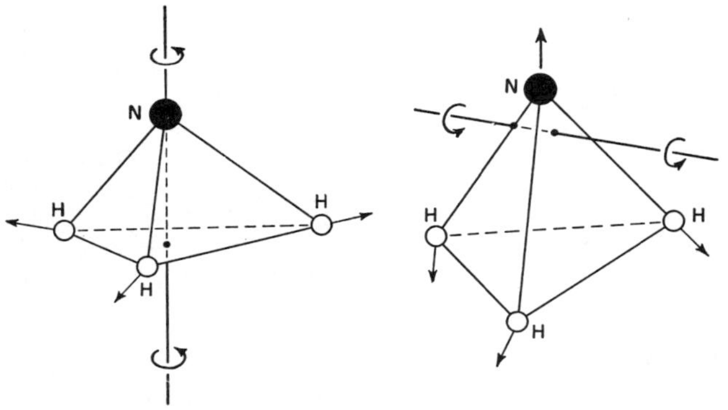

Fig. 21.5. Rotation and centrifugal distortion of NH_3 molecule.

steepens the pyramid, raises the potential hill, and reduces the splitting and with it the inversion frequency. The angular momentum in the latter case is prescribed by the combination of two quantum numbers J and K as $(J^2 + J - K^2)^{1/2}$.

About thirty fine structure lines have been observed which correspond to flip-flop frequencies at various rotation states of the NH_3 molecule. The doublet splitting of the nonrotating molecule in its vibrational ground state equals 0.66 wave number. For the different rotation states of the molecule the inversion lies at the wave numbers

$$\nu' = 0.66 - 0.0011(J^2 + J - K^2) + 0.0005K^2 \quad [\text{cm}^{-1}]. \tag{21.2}$$

The allocation of quantum numbers (J, K) to the various lines (see Fig. 21.4) becomes possible if we observe the intensity of the lines and compare this with the statistical weight of the terms and the transition probabilities theoretically predicted. For example, levels $J = K$ have both a large statistical weight and a high transition probability, and lines with K a multiple of 3 are especially preferred. This leads to the identification of the two strongest lines as $(J,K) = (3,3)$ and $(6,6)$, and so on.

The observed splittings of the lines allow us to determine the separation of the two minima of the potential well and the height of the potential barrier between them.[11] The height of the NH_3 pyramid, which equals one half the separation distance of the two minima, is found as 0.38 A; the pyramid is thus quite flat, in agreement with the large bond angle found for the ammonia molecule (see Fig. 17.7). The height of the potential hill, the activation energy of inversion, proves to be 2076 cm^{-1} or about 0.26 [ev] (cf. Eq. 19.12). The flip-flop frequency of the vibrational ground state, $\nu = 0.66 \times 3 \times 10^{10} = 2 \times 10^{10}$ [sec^{-1}], gives the average time of inversion as

$$\tau = \frac{1}{2\nu} \simeq 2.5 \times 10^{-11} \quad [\text{sec}]. \tag{21.3}$$

When an individual rotation line of the fine structure spectrum (see Fig. 21.4) is carefully examined as a function of the ammonia gas pressure, a hyperfine structure may be discerned (Fig. 21.6). This structure proves to be due to the coupling of the electric quadrupole moment of the nucleus N^{14} with the electronic structure of the ammonia molecule. Such nuclear electric moments can arise because the positive charge of the nucleus need not be spherical in shape. Nuclei may have angular moments and thus magnetic dipole moments. The spin axis must be an axis of rotational charge symmetry, but the charge may be distributed in the form of an ellipsoid, either prolate or oblate (Fig.

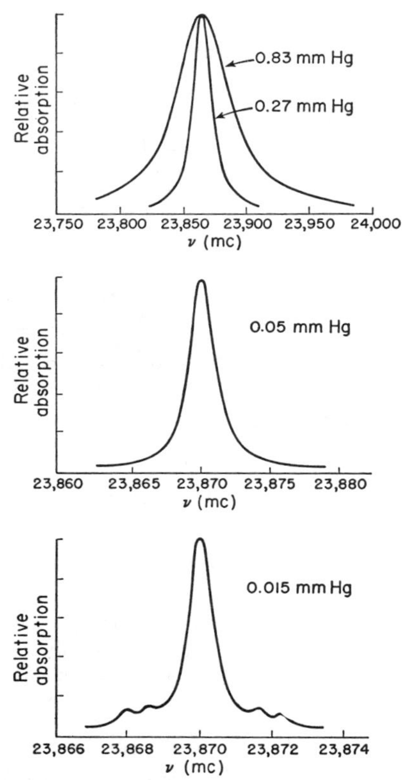

Fig. 21.6. Hyperfine structure of rotation line of NH_3 molecule. (After Good [10] and Townes.[10])

[11] See G. Herzberg, *Molecular Spectra and Molecular Structure. I. Diatomic Molecules*, D. Van Nostrand, New York, 1950, pp. 221 ff.

21.7). Such a charge distribution can be regarded as the superposition of a spherical charge and of an electric quadrupole. By convention, the prolate type of nucleus

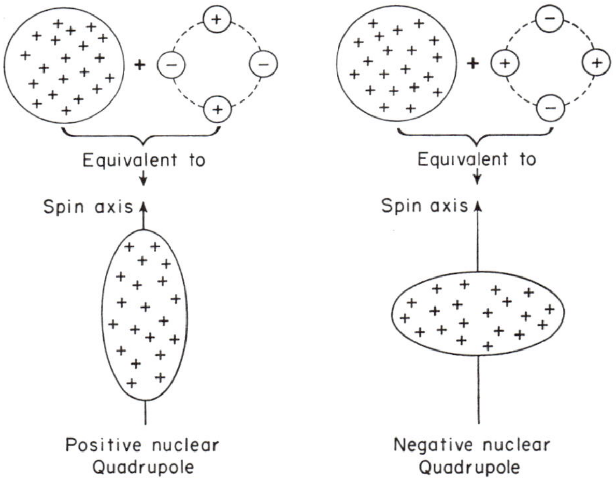

Fig. 21.7. Nuclear quadrupole moment.

is said to have a positive quadrupole moment. The N^{15} nucleus has no quadrupole moment, and the hyperfine structure of Fig. 21.4, in consequence, disappears. In addition to the hyperfine structure due to the nuclear quadrupole, which produces line shifts of the order of megacycles, a much smaller shift of the order of kilocycles results due to the coupling of the magnetic moment of the nucleus with that of the rotating molecule.[6]

The masses of the nuclei enter into the vibration and rotation frequencies; consequently, by replacing N^{14} with N^{15}, or hydrogen with deuterium, large changes in the inversion frequencies can be produced. This *isotope effect* causes the NH_3 absorption at ca. 24,000 megacycles to move to a more than ten times smaller frequency for ND_3 (ca. 2000 megacycles). An observation of such isotope shifts gives the additional equations needed for the evaluation of the force constants of the molecule.

By replacing hydrogen partly with deuterium, the symmetric-top molecules NH_3 are converted into the asymmetric rotors NH_2D and ND_2H. This changes the character of the microwave spectrum appreciably. In the inversion spectrum of the degenerate symmetrical top no rotational transitions occur, whereas in the microwave spectrum of the partly deuterized molecules inversion and rotation transitions take place simultaneously.[6] These more complicated spectra can give information on the sign of the nuclear electric quadrupole moment, which turns out to be negative (oblate) for N^{14}. Furthermore, additional data on the centrifugal distortion of the molecule result.

The Stark splitting of rotational energy levels by an external electric field in the microwave region (Fig. 21.8)[12] offers a new method for the determination of molecular electric dipole moments which has a number of advantages over older methods.[13] The moment is determined for a certain vibrational and rotational

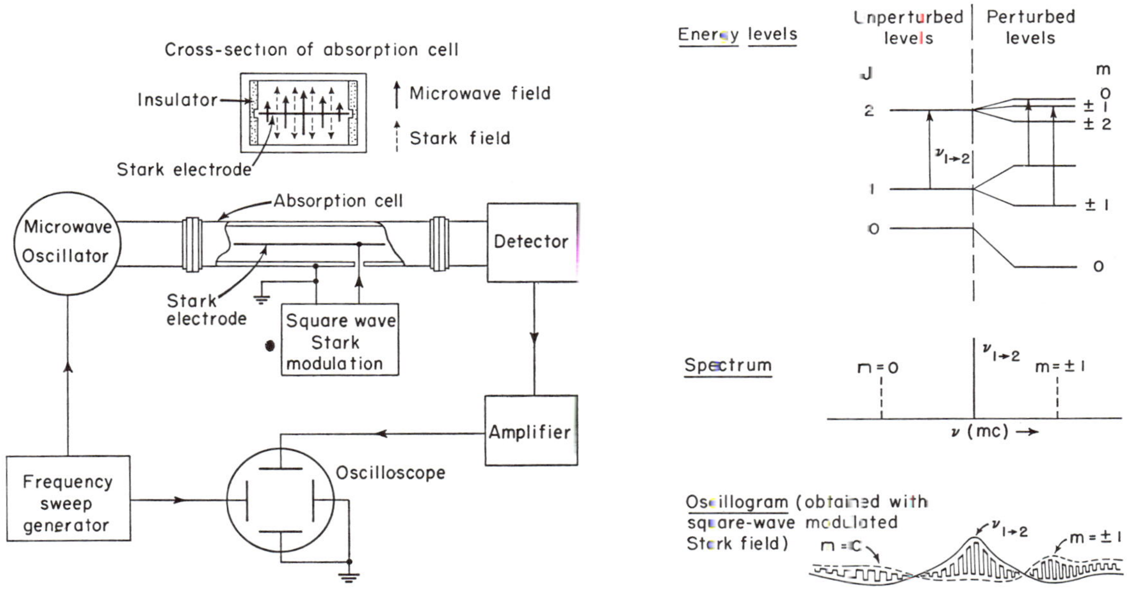

Fig. 21.8. Stark splitting of rotational levels (After Strandberg, Wentink, and Kyhl.[12])

[12] M. W. P. Strandberg, T. Wentink, and R. L. Kyhl, *Phys. Rev. 75*, 270 (1949).

[13] See, for example, K. B. McAfee, Jr., R. H. Hughes, and E. B. Wilson, Jr., *Rev. Sci. Inst. 20*, 821 (1949).

state of the molecule and not as an average of all states. Impurities are no longer of importance since the spectrum of a specific molecule is examined. Microwave measurements on Stark effects are made at gas pressure of only about 0.01 mm Hg, hence many substances can be studied in the vapor state without danger of association. Accuracies to 0.01 debye can be obtained,[14] while a measurement of the dipole moment from the temperature dependence of the dielectric constant (see Sec. 16) is, in general, not more accurate than 0.1 debye. Stark effect measurements on the NH_3 inversion spectrum have led to the permanent moment $\mu = 1.45$ [debye].[15]

This superior accuracy of microwave spectroscopy, however, does not imply that direct electric measurements of dipole moments have become unnecessary. The Stark effect technique is applicable only for simple molecules where the theory can be worked out. Furthermore, it does not extend to the liquid and solid phases, where the resonance spectra of rotation have been destroyed (see Secs. 22 ff.).

[14] See, for example, R. G. Shulmann, B. P. Daily, and C. H. Townes, *Phys. Rev. 78*, 145 (1950).

[15] See P. Kisliuk and C. H. Townes, Molecular Microwave Spectra Tables, *J. Res. U. S. Natl. Bur. Stand. 44*, 611 (1950).

Microwave measurements, finally, allow a detailed study of the pressure broadening of spectral lines. As a first result of observations, as in Fig. 21.6, a comparison is obtained between the gaskinetic collision time τ_c (see Eq. 20.21) and the actual average time interval between collisions. This time proves to be about fifteen times shorter than the kinetic theory predicts; obviously, the large dipole moment of the ammonia molecule has increased its collision radius appreciably. At higher pressures the collision damping becomes so strong that not a symmetrical but an asymmetrical broadening of the line results. The resonance peak shifts to lower frequencies [16] as already foreseen by the simple oscillator theory (see Eq. 4.9).

Microwave spectral lines like the 3,3 line of ammonia at 23,870 megacycles are under consideration as primary frequency standards and for the control of "atomic clocks" which might compare atomic time with sidereal time.[17] Our present frequency standards are based on "quartz clocks" operating by piezoelectric resonance (see Sec. 26).

[16] B. Bleany and J. H. N. Loubser, *Proc. Phys. Soc. (London) A63*, 483 (1950).

[17] H. Lyons, *Phys. Rev. 74*, 1203 (1948); *Ann. N. Y. Acad. Sci. 55*, 831 (1952).

22 · Pressure Broadening and Debye's Relaxation Equation

In discussing the frequency dependence of the permittivity of gases, we started with the anomalous dispersion and resonance absorption of linear harmonic oscillators (see Sec. 4). The dispersion formula of classical physics thus derived predicted correctly the general shape of spectral lines. It was impossible, however, to foresee the intensity and frequency location of these lines from classical considerations because the existence of stationary energy states of atoms and molecules is a nonclassical phenomenon. Only by passing from Newtonian mechanics to quantum mechanics was an interpretation obtained of the stationary states of electronic excitation, vibration, and rotation as standing-wave patterns of the probability waves, and of the spectral lines as transitions between the various energy levels. The intensity of the lines proved to be given by the statistical weight of the terms (their multiplet structure) and by the transition probabilities (see Secs. 12 and 19). The line width results from radiation damping, Doppler effect, and collision broadening (see Sec. 20); for higher pressures, collision damping becomes the decisive factor. Since the density of the condensed phases corresponds to that of normal gases under several thousand atmospheres of pressure, the pressure broadening of spectral lines has to be considered as one important aspect in the discussion of the dielectric absorption of solids and liquids.

That a rapid sequence of interrupting collisions will make the formation of discrete quantum states impossible can be derived immediately from the uncertainty relation (Eq. 8.11):

$$\Delta\mathcal{E}\,\Delta t \geq \hbar. \tag{22.1}$$

If the time interval Δt equals the natural life span τ of the quantum state (Δt = maximum), the energy state $\mathcal{E}$ is defined with optimum sharpness ($\Delta\mathcal{E}$ = minimum). If, on the other hand, a disrupting collision takes place during every vibration or rotation cycle ($\Delta t = 1/\nu$), the uncertainty in the definition of the energy state becomes

$$\Delta\mathcal{E} = \hbar\nu, \tag{22.2}$$

that is, of the order of magnitude of the total quantum energy. The resonance state in this case has vanished, and the spectral line has broadened into a continuum; the resonator, classically speaking, has been overdamped.

In Sec. 20 we treated the case of pressure broadening by substituting in the damping factor of Eq. 4.14

$$2\alpha = 1/\tau, \tag{22.3}$$

the time interval τ_c between collisions in place of the natural lifetime τ of the quantum state. This seems justified if many *weak collisions* constitute a continuous drain on the energy of the oscillator without causing abrupt changes in amplitude and phase. In contrast, the pool game of gas kinetics visualizes frequently the interaction of molecules as a statistical sequence of *strong collisions*, each of them altering the energy and momentum of the partners violently. Under such conditions, the amplitude and phase of the classical oscillator varies discontinuously, and not the steady-state solution (Eq. 4.27) of the force equation applies but the general solution composed of steady-state and transient terms.

The dispersion formula of classical physics pictures, electrically speaking, the resonating atoms or molecules in the gaseous state as *L*, *R*, *C* circuits shunted by a capacitance (*a*). The classical approach to the treatment of dipole molecules in the condensed phases of liquids and solids is to consider the polar molecules as rotating in a medium of dominating friction.[1] We postulate, in effect, that the acceleration term can be neglected, that is, that the *L*, *R*, *C* equivalent circuit reduces to an *R*, *C* circuit (*b*), (Fig. 22.1). In this extreme case of weak-collision broadening we thus substitute by hunch for the resonance spectrum of Fig. 4.2 the relaxation spectrum of I, Fig. 26.5, with its complex permittivity (see I, Eq. 26.14),

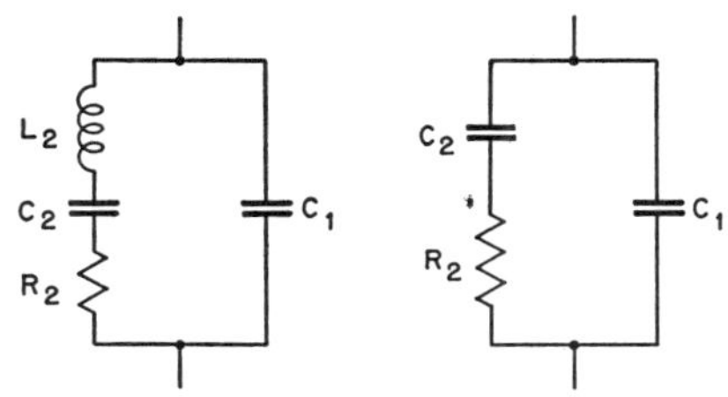

(a) Resonator (b) Rotator in medium where friction term dominates

Fig. 22.1. Equivalent circuits of free resonator molecule and of rotator molecule in medium of dominating friction.

$$\kappa^* = \kappa_\infty' + \frac{\kappa_s' - \kappa_\infty'}{1 + j\omega\tau}. \tag{22.4}$$

[1] P. Debye, *Polar Molecules*, Chemical Catalog Co., New York, 1929, Chap. V.

Now only the task remains to reinterpret the static and optical permittivities κ_s' and κ_∞', and the relaxation time τ by molecular quantities.

The optical dielectric constant κ_∞', represented in the electric circuit analogue by the by-pass capacitor, contains the contribution of the induced moments

$$\boldsymbol{\mu}_i = (\alpha_e + \alpha_a)\mathbf{E}' \tag{22.5}$$

caused by the electronic and atomic polarizabilities of the molecules (the deformation polarization). The static dielectric constant κ_s' contains, in addition, the orientation polarization of the permanent moments $\boldsymbol{\mu}$. According to the statistical theory of Sec. 16, when the electric field energy is small in comparison to the thermal energy of agitation, each dipole molecule contributes an average moment (see Eq. 16.9)

$$\bar{\boldsymbol{\mu}}_d = \frac{\boldsymbol{\mu}^2}{3kT}\mathbf{E}'. \tag{22.6}$$

If we could assume that E′ represents the applied field E, as in gases at low pressure, we would immediately obtain the static and optical dielectric constants as given in Eqs. 16.11 and 16.12 and thus from Eq. 22.4 the complex permittivity

$$\kappa^* = 1 + \frac{N}{\epsilon_0}\left\{(\alpha_e + \alpha_0) + \frac{\boldsymbol{\mu}^2}{3kT}\frac{1}{1 + j\omega\tau}\right\}. \tag{22.7}$$

Actually, the applied field **E** will have to be replaced by some local field **E**′, for example, the Mosotti field of Eq. 2.9:

$$\mathbf{E}' = \mathbf{E} + \frac{\mathbf{P}}{3\epsilon_0} = \frac{\mathbf{E}}{3}(\kappa^* + 2). \tag{22.8}$$

In this case we obtain for the polarizability per unit volume not the $\kappa^* - 1$ of Eq. 22.7, but the more involved equation

$$\frac{N\alpha}{3\epsilon_0} = \frac{\kappa^* - 1}{\kappa^* + 2}$$
$$= \frac{N}{3\epsilon_0}\left\{(\alpha_e + \alpha_a) + \frac{\boldsymbol{\mu}^2}{3kT}\frac{1}{1 + j\omega\tau}\right\}. \tag{22.9}$$

This modified polarizability has the static and the optical value

$$\frac{\kappa_s' - 1}{\kappa_s' + 2} = \frac{N}{3\epsilon_0}\left\{(\alpha_e + \alpha_a) + \frac{\boldsymbol{\mu}^2}{3kT}\right\},$$
$$\frac{\kappa_\infty' - 1}{\kappa_\infty' + 2} = \frac{N}{3\epsilon_0}(\alpha_e + \alpha_a). \tag{22.10}$$

By introducing the expressions on the left into Eq. 22.9,

we return to the formulation for the complex permittivity given in Eq. 22.4,

$$\kappa^* = \kappa_\infty' + \frac{\kappa_s' - \kappa_\infty'}{1 + j\omega\tau_e}, \tag{22.11}$$

with the new time constant

$$\tau_e = \tau \frac{\kappa_s' + 2}{\kappa_\infty' + 2}. \tag{22.12}$$

Hence, by replacing the applied field **E** with the Mosotti field **E**$'$ the interpretation but not the shape of the relaxation spectrum has changed. Only the relaxation time has lengthened from τ to τ_e, and the molecular meaning of the static and optical dielectric constants has been altered by the introduction of the denominators $\kappa_s' + 2$ and $\kappa_\infty' + 2$ in Eq. 22.10. Similarly, any other type of local field expression will not change the shape of the absorption spectrum as long as the motion of the dipole molecule is represented by a first-order differential equation.

The remaining problem is the molecular interpretation of the relaxation time τ. According to the assumption of dominating friction, we have to picture the polar molecules as rotating under the torque **T** of the electric field with an angular velocity $\frac{d\theta}{dt}$ proportional to this torque, or

$$\mathbf{T} = \zeta \frac{d\theta}{dt}. \tag{22.13}$$

The friction factor ζ will depend on the shape of the molecule and on the type of interaction it encounters. If one visualizes the molecule as a sphere of the radius a, rotating in a liquid of the viscosity η according to Stokes's law,[2] classical hydrodynamics leads to the value

$$\zeta = 8\pi\eta a^3. \tag{22.14}$$

In a static field, the spherical dipole carriers will have a slight preferential orientation parallel to this field and thus contribute the average moment of Eq. 22.6. A sudden removal of the external field will cause an exponential decay of this ordered state due to the randomizing agitation of the Brownian movement. The relaxation time τ (or τ_e) measures the time required to reduce the order to $1/e$ of its original value. Debye[1] was able to calculate this time statistically by deriving the space orientation under the counteracting influences of the Brownian motion and of a time-dependent electric field and found

$$\tau = \frac{\zeta}{2kT}. \tag{22.15}$$

[2] G. Stokes, *Trans. Cambridge Phil. Soc. 9*, 8 (1851).

Combining Eqs. 22.14 and 22.15, Debye obtained for the spherical molecule, if it behaves like a ball rotating in oil, the relaxation time

$$\tau = \frac{4\pi a^3 \eta}{kT} = V \frac{3\eta}{kT}. \tag{22.16}$$

The time constant is proportional to the volume V of the sphere and to the macroscopic viscosity of the solution. Water at room temperature has a viscosity $\eta = 0.01$ poise; with a radius of ca. 2 A for the water molecule, a time constant of $\tau \simeq 0.25 \times 10^{-10}$ sec results.

Figure 22.2 shows that indeed the relaxation time of water is located near the wavelength of 1 cm. The agreement, however, is somewhat marred by the realization that experimentally we have determined

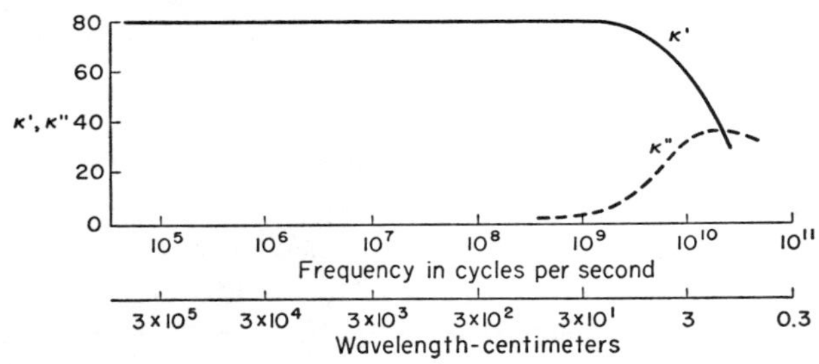

Fig. 22.2. Relaxation spectrum of water at room temperature.

$\tau_e \simeq 20\tau$ instead of τ itself. Obviously, neither the sphere model nor the Mosotti field should be taken too seriously; the essence of Debye's approach is to postulate that the orientation of polar molecules in liquids and solids leads spectroscopically to a simple relaxation spectrum.

The *Debye equation* (Eq. 22.11) may be written in various forms, which have their special merits for the evaluation of experimental characteristics. Separating it into its real and imaginary parts, we obtain the standard version

$$\kappa' - \kappa_\infty' = \frac{\kappa_s' - \kappa_\infty'}{1 + \omega^2\tau_e^2},$$

$$\kappa'' = \frac{(\kappa_s' - \kappa_\infty')\omega\tau_e}{1 + \omega^2\tau_e^2}, \tag{22.17}$$

$$\tan\delta = \frac{\kappa''}{\kappa'} = \frac{(\kappa_s' - \kappa_\infty')\omega\tau_e}{\kappa_s' + \kappa_\infty'\omega^2\tau_e^2}.$$

If we introduce as a new variable[3]

$$z = \ln \omega\tau_e, \tag{22.18}$$

[3] See H. Fröhlich, *Theory of Dielectrics*, Clarendon Press, Oxford, 1949, p. 73.

Eqs. 22.17 may be rewritten in a normalized form

$$\frac{\kappa' - \kappa_\infty'}{\kappa_s' - \kappa_\infty'} = \frac{1}{1 + e^{2z}} = \frac{e^{-z}}{e^z + e^{-z}},$$

$$\frac{\kappa''}{\kappa_s' - \kappa_\infty'} = \frac{1}{e^z + e^{-z}}, \tag{22.19}$$

$$\frac{\tan\delta}{\kappa_s' - \kappa_\infty'} = \frac{1}{\kappa_\infty' e^z + \kappa_s' e^{-z}}.$$

Figure 22.3 shows this logarithmic plot of the dispersion and absorption characteristic, and, added to them,

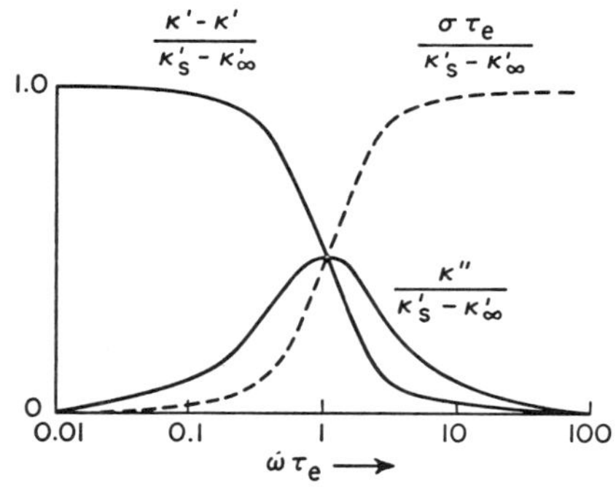

Fig. 22.3. Dielectric constant, loss factor, and conductivity of simple relaxation spectrum in normalized form.

as a third curve, the relative dielectric conductivity (see I, Eq. 1.16)

$$\sigma = \omega\kappa'' \tag{22.20}$$

in the normalized form

$$\frac{\sigma\tau_e}{\kappa_s' - \kappa_\infty'} = \frac{e^z}{e^z + e^{-z}}. \tag{22.21}$$

The conductivity curve is the mirror image of the κ' characteristic; that is, the orientation polarization leads to a constant maximum conductivity contribution beyond the range of the dispersion region.

It may seem surprising that, after the polarizing action of the permanent dipoles has disappeared, their existence is still noted with full force as a conduction effect, but the explanation is simple. As the frequencies range so high that the molecules have no time to turn, we do not notice that the two opposite dipole charges are coupled together; their effect on the conduction is therefore the full contribution of two ions of opposite polarity moving in the electric field according to Ohm's law.

Figure 22.3 shows clearly the frequency spread of the dispersion phenomenon. According to the decade scale, it is practically limited to one decade for κ' and to two decades for κ'' above and below the center frequency. One further graphical representation of the Debye equation proves of value in analyzing and extrapolating experimental data. If we plot κ'' against κ' in the complex plane, points obeying the Debye equation fall on a semicircle with its center at $\frac{\kappa_s' + \kappa_\infty'}{2}$ (Fig. 22.4), as Cole and Cole [4] first pointed out. This becomes evident when we rewrite Eq. 22.11 in the form

$$(\kappa^* - \kappa_\infty') + j(\kappa^* - \kappa_\infty')\omega\tau_e = \kappa_s' - \kappa_\infty'. \tag{22.22}$$

The first member on the left side corresponds to a vector **u**; the second member, as the factor j indicates, adds perpendicular to it and represents a vector **v**; the sum is the diagonal of the circle.

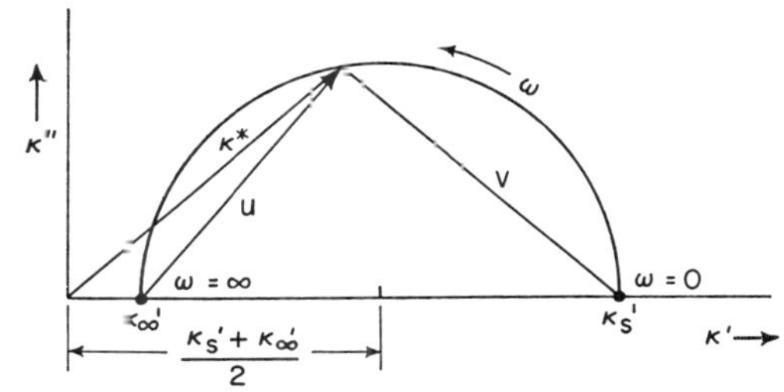

Fig. 22.4. Cole-Cole circle diagram of κ^* in complex plane.

The loss factor κ'' reaches its maximum at the critical frequency

$$\omega_m = 1/\tau_e, \tag{22.23}$$

that is, at the critical wavelength

$$\lambda_\tau = 2\pi c\tau_e \tag{22.24}$$

at which the dipole polarization has fallen to its half value. Furthermore

$$\kappa''_{max} = \frac{\kappa_s' - \kappa_\infty'}{2} \equiv \frac{S}{2}. \tag{22.25}$$

The relaxation time and the contribution S of the orientation polarization to the permittivity can be determined by these relations from the absorption characteristic of a dielectric as long as the Debye equation is valid.

Finally, from Eq. 22.17 we obtain the linear equation in κ' versus $\omega\kappa''$, and κ' versus κ''/ω:

$$\kappa' = \kappa_s' - \omega\tau_e\kappa'',$$
$$= \kappa_\infty' + \frac{\kappa''}{\omega\tau_e}, \tag{22.26}$$

the first of which may be rewritten

$$\epsilon' = \epsilon_s' - \sigma\tau_e. \tag{22.27}$$

Thus the static and optical dielectric constants are given by the intersection, and τ_e by the slope of a

[4] K. S. Cole and R. H. Cole, *J. Chem. Phys.* *9*, 341 (1941).

straight line, as long as the simple relaxation equation applies.

The theory has confronted us with two extreme cases: the classical resonance absorption [5] in a quantum-mechanical reinterpretation and the Debye relaxation absorption. It is tempting to link the two cases graphically by postulating that pressure broadening gradually converts a resonance into a relaxation spectrum (Fig. 22.5). Also mathematically the semblance of a unified theory can be achieved, as the interesting attempt of Van Vleck and Weisskopf [6] shows (extended by Van Vleck and Margenau [7]). However, it seems questionable how far it makes physical sense to force a marriage between two concepts based on entirely different assumptions. The original resonance picture deals with oscillators that are effectively uncoupled and suffer only now and then disturbing impacts. The force of the applied field induces moments, and the frequency of the field changes these moments in amplitude and phase. The relaxation picture deals with rotators in a state of perpetual impact. The field produces a torque on permanent moments, and the orientation of these moments begins to lag as the frequency increases. There is no classical resonance absorption for the rotator, and the replacing of induced moments by permanent moments in the resonance equation does not change the original meaning of the frequency dependence, which is not compatible with that of the relaxation equation. We should expect that the pressure broadening of rotation spectra does not lead to a unique result as Fig. 22.5 implies, but to a diversity of situations in keeping with the great variety of surroundings which the solid and liquid state can offer. The Debye case is one simple prototype of the behavior of polar molecules in condensed phases; we shall encounter others in the subsequent sections.

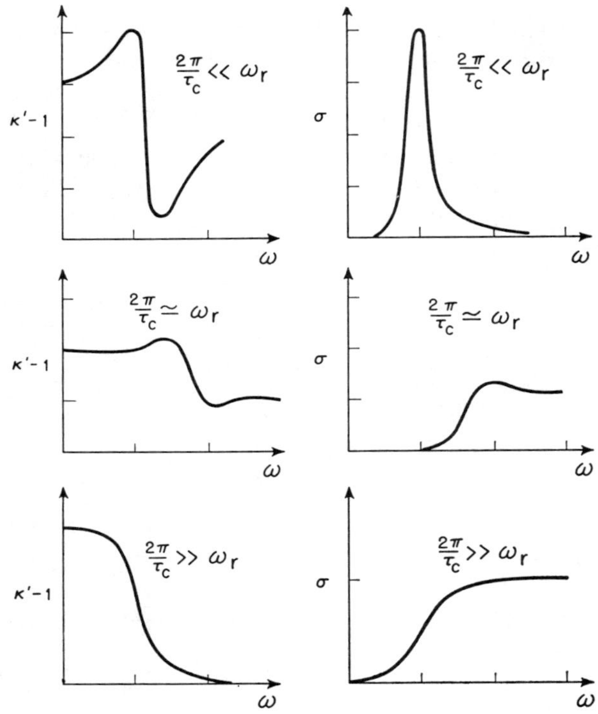

Fig. 22.5. Assumed pressure broadening of resonance into relaxation spectrum (collision frequency very much smaller, equal, or much larger than the rotation frequency).

[5] The classical dispersion equations were first derived by Ketteler and Helmholz [see E. Ketteler, *Wied. Ann. 49,* 382 (1893)] and developed into an electron theory of dispersion by Lorentz (see H. A. Lorentz, *Proceedings Amsterdam Academy,* 1897–1898).

[6] J. H. Van Vleck and V. F. Weisskopf, *Revs. Mod. Phys. 17,* 227 (1945).

[7] J. H. Van Vleck and H. Margenau, *Phys. Rev. 76,* 1211 (1949).

23 · The Mosotti Catastrophe and the Local Field

The assumption of a simple relaxation spectrum fits satisfactorily the frequency response of a number of dielectrics, especially of dilute solutions of polar materials in nonpolar solvents, when the shape of the dipole molecules is approximately spherical. This fact, however, should not be construed as a confirmation of the special Debye equation (Eq. 22.11), which is based on the Mosotti field (Eq. 2.9). By specifying this type of local field we have implicitly resigned ourselves to catastrophic consequences.

This becomes apparent when we return to the defining equation for the polarization (Eq. 1.5)

$$\mathbf{P} = (\kappa' - 1)\epsilon_0\mathbf{E} = N\alpha\mathbf{E}'. \qquad (23.1)$$

By introducing for the local field $\mathbf{E}'$ the Mosotti field

$$\mathbf{E}' = \mathbf{E} + \frac{\mathbf{P}}{3\epsilon_0}, \qquad (23.2)$$

we obtain for the polarization and the electric susceptibility the expressions

$$\mathbf{P} = \frac{N\alpha\mathbf{E}}{1 - \dfrac{N\alpha}{3\epsilon_0}},$$

$$\chi \equiv \kappa' - 1 = \frac{\mathbf{P}}{\epsilon_0\mathbf{E}} = \frac{N\alpha/\epsilon_0}{1 - \dfrac{N\alpha}{3\epsilon_0}}. \tag{23.3}$$

Hence, when the polarizability term of the denominator $N\alpha/3\epsilon_0$ approaches 1, the polarization and susceptibility must approach infinity.

Obviously, this catastrophe must occur for any polar material when the dipole molecules contribute an orientation polarizability

$$\alpha_d = \mu^2/3kT. \tag{23.4}$$

As the temperature falls, a critical temperature T_c is reached where

$$\frac{N\alpha}{3\epsilon_0} = \frac{N}{3\epsilon_0}\left(\alpha_e + \alpha_a + \frac{\mu^2}{3kT_c}\right) = 1. \tag{23.5}$$

This condition determines the critical temperature as

$$T_c = \frac{N\mu^2}{9\epsilon_0 k}\,\frac{1}{1 - \dfrac{N(\alpha_e + \alpha_a)}{3\epsilon_0}}. \tag{23.6}$$

If we can forget about the deformation polarization ($\alpha_e + \alpha_a \ll \alpha_d$), Eqs. 23.3 may be rewritten, by introducing T_c, as

$$\mathbf{P} = \frac{3T_c}{T - T_c}\,\epsilon_0 E,$$

$$\chi = \frac{3T_c}{T - T_c}. \tag{23.7}$$

Analogous expressions appear in the theory of ferromagnetism (see Sec. 29) for the magnetization and magnetic susceptibility caused by the orientation of magnetic dipoles. The linear dependence of $1/\chi$ on $T - T_c$ is known as the *Curie-Weiss law* [1] and T_c as the *Curie temperature.*

Let us explain in words the gist of the situation. The statistical theory with its Langevin characteristic (Fig. 16.2) showed that nearly unattainable external fields are required to enforce more than a slight orientation of the individual dipoles against the randomizing action of the Brownian movement. However, if the dipoles interact by a local field of the Mosotti type, a situation arises of the sort experienced by Munchausen's soldier who lifted himself out of the swamp by his own boot straps. The local field increases the polarization, and the polarization, in turn, increases the local field. Above the Curie temperature the counteraction of thermal agitation is able to maintain a highly disordered state. At T_c the randomizing effect of the temperature is overcome, *spotaneous polarization* sets in, and the dipoles line up in parallel arrays due to their long-range forces of interaction. Any polar substance, according to this simple theory, should, at some critical temperature, become a *ferroelectric material.*

This would be a real catastrophe, as the example of water may illustrate. We calculate its critical temperature according to Eq. 23.6 by determining first the molar deformation polarization of water vapor from Fig. 16.4; extrapolation of the straight line to $1/T = 0$ gives $\Pi_{e+a} \simeq 5 \times 10^{-3}$ [m^3]. We assume that in the liquid phase the molecules contribute about the same deformation polarization. Since the number N of molecules per cubic meter of liquid water is ca. $\frac{1}{3} \times 10^{29}$ and Avogadro's number $N_0 = 6.023 \times 10^{23}$, we find (see Eq. 2.14)

$$\frac{N(\alpha_e + \alpha_a)}{3\epsilon_0} = \Pi_{e+a}\frac{N}{N_0} \simeq 0.28 \quad [\mathrm{m}^3]. \tag{23.8}$$

The dipole moment of the water molecule is $\mu \simeq 1.8$ debye $= 1.8 \times 3.33 \times 10^{-30}$ [coul m], and the gas constant per molecule $k = 1.38 \times 10^{-23}$ [joule deg^{-1}]; thus

$$T_c = \frac{N|\mu|^2}{9\epsilon_0 k}\,\frac{1}{1 - 0.28} \simeq 1520°\mathrm{K}. \tag{23.9}$$

Hence water should solidify by spontaneous polarization at high temperature, making life impossible on this earth!

Ferroelectrics seem to be rare, and the Mosotti catastrophe usually does not take place. We are spared because the local field $\mathbf{E}'$ is, in general, not the simple field assumed in Eq. 2.3. Permanent dipoles tend to lose their individual freedom of orientation in condensed phases through association and steric hindrance. Their interaction with their surroundings has to be taken into account; the near field $\mathbf{E}_3$ of Eq. 2.1 cannot be neglected.

To save face, the theory can try to construct a plausible model that does not lead to a disappearing denominator. Onsager [2] was the first to evade the difficulty by replacing the mathematical cavity of Fig. 2.1 and the Mosotti field in its center with a real cavity of molecular size and a dipole in its center (Fig. 23.1). By assuming that the molecule occupies a sphere of radius R, that its polarizability is isotropic, and that

[1] P. Weiss, *J. physique 6,* 667 (1907).

[2] L. Onsager, *J. Am. Chem. Soc. 58,* 1486 (1936).

its interaction with the surroundings can be described by the polarizing action of its dipole field, he arrived at a simple electrostatic model that avoids the Mosotti catastrophe.

In I, Sec. 10, we have already calculated the electric field arising in the interior of a sphere of the permittivity ϵ_2', when embedded in a medium ϵ_1', and subjected to a homogeneous field $\mathbf{E}_0$ (see I, Eq. 10.13). For an empty cavity ($\epsilon_2'/\epsilon_0 = 1$) this field is

$$\mathbf{E}_c = \frac{3\kappa_1'}{2\kappa_1' + 1}\mathbf{E}_0. \qquad (23.10)$$

In addition to this *cavity field* due to external sources there acts on the dipole molecule inside the Onsager

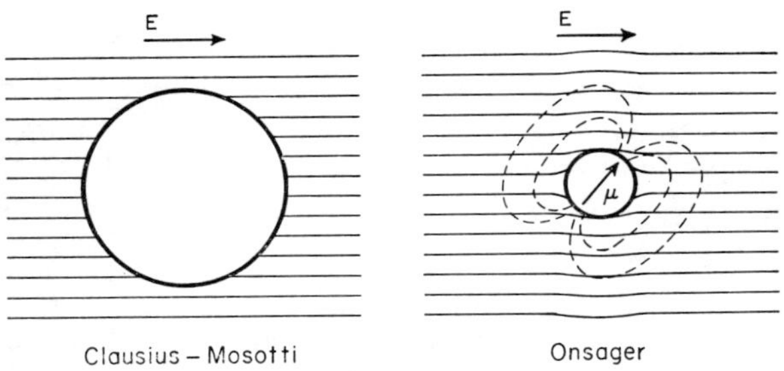

Fig. 23.1. The Clausius-Mosotti and the Onsager cavity. (After van Vleck.[3])

cavity a *reaction field* $\mathbf{E}_r$ due to the polarization caused by the moment $\boldsymbol{\mu}$ of the molecule in the surrounding medium. This field can be calculated from the solution of the Laplace equation (I, Eq. 10.7) with the boundary conditions of Eq. 10.8 when we replace the additional boundary conditions, Eqs. 10.9 and 10.10, with the assumptions: (*a*) $E_0 = 0$ for $r \to \infty$, that is, no external field is applied; and (*b*) $C/r^2 = -(|\mu|/\epsilon_0 4\pi r^2)$ for $r \to 0$, that is, a point dipole is placed in the center of the sphere. The coefficient D in Eq. 10.7 represents now the reaction field and becomes

$$\mathbf{E}_r = \frac{\boldsymbol{\mu}}{4\pi R^3}\, 2\,\frac{\kappa_1' - 1}{2\kappa_1' + 1} = \frac{\boldsymbol{\mu}}{V}\frac{2}{3}\frac{\kappa_1' - 1}{2\kappa_1' + 1}, \qquad (23.11)$$

where $\boldsymbol{\mu}$ is the dipole moment of the molecule and V the volume of the molecular cavity.

The difference between the old Mosotti or Lorentz model and the Onsager model is apparent from Fig. 23.1. The cavity in the Lorentz model is a mathematical fiction, and the field lines in consequence are undeflected. In the Onsager model a real cavity is assumed; therefore the lines of force have cognizance of its existence and are bent by it. The Lorentz sphere contains a large number of molecules while in the Onsager approach it has shrunk to the volume of a single dipole molecule. The center of the Lorentz cavity is marked by a mathematical point and that of the Onsager cavity by a point dipole which polarizes the surrounding medium. This polarization, acting back on the dipole in the cavity, creates the reaction field. When the dipole in the sphere stands parallel to the applied field, the external field lines in the Onsager model, resulting from the superposition of the applied field and the dipole field, must straighten out to those of the Lorentz model. This is true only when the volume of the Onsager cavity is chosen equal to the molecular volume $V = 1/N$, where N is the number of molecules per unit volume.

The reaction field must be always parallel to the elementary dipole, hence can have no effect in orienting it. Only the cavity field $\mathbf{E}_c$ exercises a torque on the dipole. Here, according to Onsager, the Mosotti-Lorentz theory has made its mistake by introducing the Mosotti field of Eq. 23.2 as the locally acting field. This field is, in essence, the sum of the cavity field and of the mean value of the reaction field. It is therefore too large and causes, by including a part of the reaction field, the dipole molecule to orient itself by its own boot straps.[3]

By introducing the cavity field $\mathbf{E}_c$ instead of the Mosotti field in Eq. 23.1, the feedback interaction between $\mathbf{E}'$ and $\mathbf{P}$, and with it the catastrophe, disappears. The Mosotti field rises in media of high permittivity proportional to κ' (see Eq. 2.9), the cavity field of Eq. 23.10 approaches the constant value $\frac{3}{2}\mathbf{E}$. Hence, the Onsager treatment returns us approximately to the situation of Eq. 22.7, where the local field is identified with the applied field and the relaxation time is again τ instead of τ_e. The actual expressions derived by Onsager become more involved due to the deformation polarization which bestows on the cavity an internal index of refraction n (an optical dielectric constant $\kappa_\infty' = n^2$) different from that of vacuum. In fact, this optical dielectric constant is identical with that of the Mosotti-Lorentz model as given in Eq. 22.10, because the induced moments stand parallel to the applied field (see also Appendix A, II, Sec. 7).

The Onsager calculation of the local field has the merit of avoiding the Mosotti catastrophe by considering the long-range dipolar interaction between a polar molecule and its isotropic surroundings. This approach gives relatively good results as long as the surroundings of the dipole act like a smeared-out isotropic medium. Figure 23.2 makes this strikingly clear in a plot of the static dielectric constant of polar liquids versus $N\boldsymbol{\mu}^2/kT$. This representation, due to R. H. Cole,[4] is based on the

[3] See J. H. Van Vleck, *Ann. N. Y. Acad. Sci. 40*, 289 (1940).

[4] The author is indebted to Professor R. H. Cole for this as yet unpublished information.

following consideration: Debye's theory of the static dielectric constant (Eq. 22.10) leads to the expression for the dipole moment of a polar molecule

$$\mu^2 = \frac{9\epsilon_0 kT}{N} \frac{3(\kappa_s' - \kappa_\infty')}{(\kappa_s' + 2)(\kappa_\infty' + 2)}. \qquad (23.12)$$

Onsager's theory, which is a theory for static fields only, gives the modified equation [5]

$$\mu^2 = \frac{9\epsilon_0 kT}{N} \frac{(2\kappa_s' + \kappa_\infty')(\kappa_s' - \kappa_\infty')}{\kappa_s'(\kappa_\infty' + 2)^2}. \qquad (23.13)$$

Rearranged, it leads to the expression for the static dielectric constant

$$\kappa_s' = \frac{N\mu^2}{kT} \frac{(\kappa_\infty' + 2)^2}{18\epsilon_0} + \kappa_\infty'\left(1 + \frac{\kappa_\infty'}{\kappa_s'}\right), \qquad (23.14)$$

that is, in a first approximation to a straight-line relation between κ_s' and $N\mu^2/kT$. This relation is quite well fulfilled for nonassociated polar liquids, as Fig. 23.2 shows. However, for liquids in which the hydrogen atoms form bonds between the molecules and therefore produce a short-range order, the equation fails. The Debye as well as the Onsager model, based on the macroscopic concept of a dielectric continuum, have to be replaced for ordered substances by local field calculations based on some structure analysis of the individual dielectric. This arduous task was started successfully by Kirkwood and his co-workers [6] (see Sec. 25).

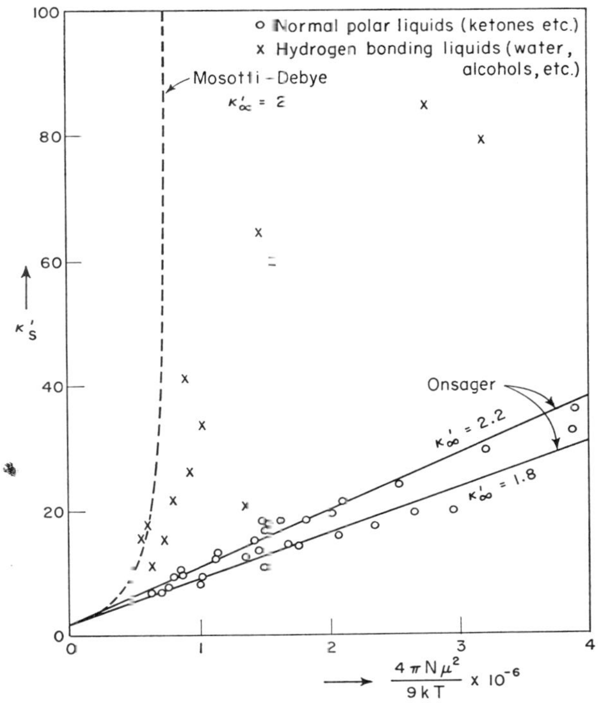

Fig. 23.2. Confirmation of the Onsager theory for nonassociated polar liquids. (After R. H. Cole.[4])

[5] An extensive treatment of the local field equations and their applications for dipole moment determinations is given in C. J. F. Böttcher's *Theory of Electric Polarization*, Elsevier Publishing Co., New York, 1952.

[6] J. G. Kirkwood, *J. Chem. Phys.* 7, 911 (1939).

24 · Formation and Structure of Liquids and Solids

Atoms form because stable positive nuclei and negative electrons exist and the Coulomb field of the nuclei traps electrons, until the nuclear charge is neutralized. The electron clouds of atoms assume distinct structures, because, in addition to electrostatic and magnetic interaction, the wave nature of the electrons restricts their whereabouts. The probable charge distribution of each electron must correspond to a standing-wave pattern in accord with the Schrödinger equation, and these individual wave patterns must differ in their set of quantum numbers, as the Pauli principle postulates.

When atoms combine to molecules, three limiting cases may be distinguished (see Sec. 13):

(1) The atoms keep their identity but their electron clouds perturb each other and establish mutual phase relations; attraction by mutual polarization results and leads to van der Waals' binding.

(2) The atoms lose their identity by merging electrons in a joint electronic system; the new molecules formed by covalent bonding are the basic entities in further molecular processes.

(3) The atoms may alter their identity by the exchange of electrons, in order to form electronic systems of higher overall stability; positive and negative ions result, acting on each other with the ionic bonds of Coulomb attraction.

Actual molecules, as the concept of quantum-mechanical resonance implies, represent, in general, mixtures of these three basic types. Since the bonds formed by mechanisms 2 and 3 are appreciably stronger than those resulting from a van der Waals type of attraction, they are called *primary* in contrast to the weaker *secondary* bonds. This distinction is useful, but frequently we observe intermediate cases.

An intermediate bond type of special importance for the intra- as well as the intermolecular bonding of numerous inorganic and organic compounds is the *hydrogen bond* (cf. Fig. 23.2). Here a hydrogen atom bridges between two, in general identical, electronegative neighbors, for example, F—H····F, O—H····O, N—H····N. In this sequence the bond strength of the bridge decreases from ca. 7 to ca. 2 kcal. The bond type should probably be visualized as resulting from the strongly ionic character of the covalent molecules involved (see Figs. 17.6 and 17.11) and the close approach to the small hydrogen atom. Thus the hydrogen bridge is akin to a dipole bond. It determines, for example, not only the short-range order in water (Fig. 25.4) and the structure of ice, but plays, as Pauling especially pointed out, an important role in the structure of complex biological molecules such as the proteins. Since the bond can be broken easily at elevated temperatures, its influence explains the sensitivity and variability of such molecules. The minimum bond distance [1] changes from 2.5 A for O—H····O, to 2.60 A in N—H····O, 2.94 A in N—H····N, and 3.01 A in N—H····Cl.

The general existence of van der Waals' attraction causes any gas to condense at absolute zero. How far the agglomeration proceeds at higher temperatures is a question of competition between the binding energy of the partners and the disrupting action of thermal agitation. At sufficiently high temperatures and low pressures all interatomic bonds will be broken and the statistical equilibrium shifted completely towards the side of atoms. As the temperature decreases, the atoms will commence to associate. Where this condensation stops depends on the type of bond that forms.

If electrons are shared in covalent pair bonds, the capacity of the partners for the formation of additional primary bonds diminishes and becomes exhausted when their valence is saturated. Thus characteristic molecules emerge with little tendency for adding further atoms. The temperature must be lowered drastically before a condensation to liquids or solids by secondary bonds occurs.

When ionic molecules are formed no such saturation of valency takes place. The far-reaching Coulomb forces can attract partners without limit until the random structure of glasses or the organized structure of ionic crystals results. Similarly, the atoms of metals can conglomerate indefinitely by an overall sharing of electrons, thus forming liquids and solids without distinct molecular groupings. The *metallic* bond is a primary bond of a mixed covalent-ionic character.

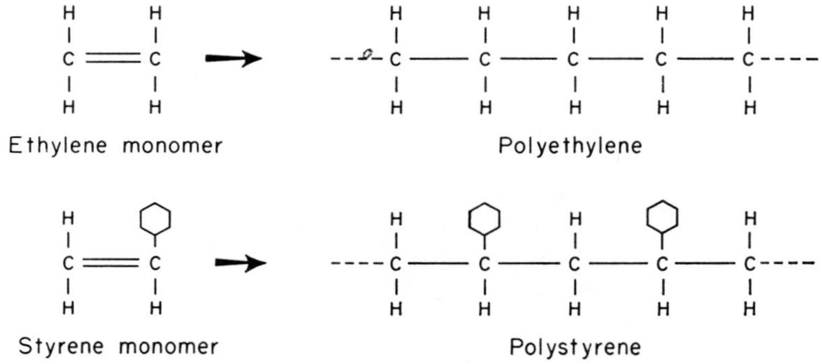

Fig. 24.1. Formation of polyethylene and polystyrene (schematic).

A condensation phenomenon leading to intermediate stages between small molecules and limitless aggregates is the *polymerization* of molecules. Here a chemical reaction is triggered off by heat, light, or some other catalyzing agent and proceeds to build individual macromolecules, frequently with explosive speed, by the formation of primary covalent bonds. The process terminates when a false step in the chain reaction fails to regenerate the required activation energy or adds a foreign building stone which cannot propagate the reaction.

The formation of such macromolecules can proceed by *polyaddition* or *polycondensation*. In the former case the compound is formed by the repeated addition of a *monomer* (M), as indicated symbolically by the formula

$$nM \rightarrow (M)_n. \qquad (24.1)$$

Polyethylene and polystyrene are composed of linear chain molecules of this type (Fig. 24.1), where n may exceed the value 10^4, and the molecular weight (as compared to hydrogen) may range in the order of millions. Polycondensation products are formed by the repeated reaction (condensation) of small molecules; water molecules or other side products are eliminated during the process. The reaction between phe-

[1] J. Donahue, *J. Am. Chem. Soc. 56*, 502 (1952).

nol and formaldehyde (Fig. 24.2),[2] by which Baekeland[3] produced the first synthetic resin of practical importance ("bakelite"), is a famous example in kind. The formation of polysilicones, at present in the forefront for engineering applications at high temperatures, is an example of another class of polycondensation products.

The various polymers may be classified from a different point of view. Polymerization may proceed from one or several reactive spots of a molecule and

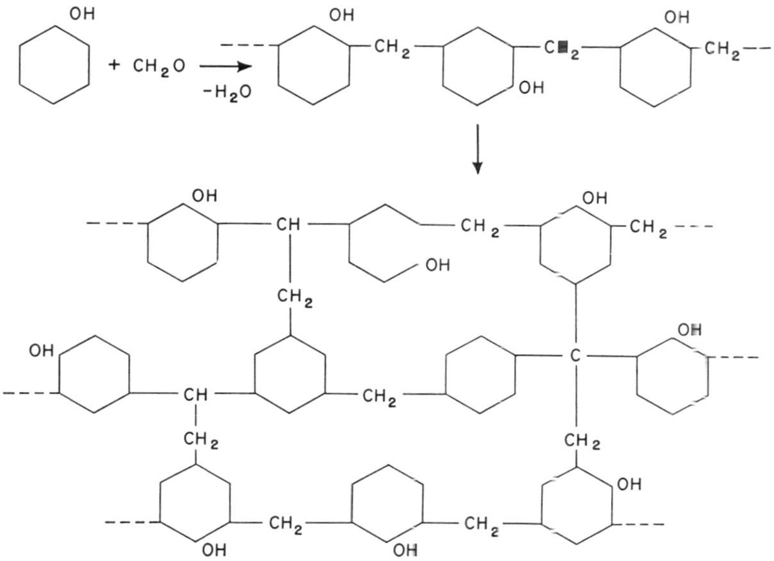

Fig. 24.2. Formation of phenol-formaldehyde polymer. (After Staudinger.[2])

thus lead to linear chains, two-dimensional networks, or three-dimensional lattice structures. Referring to these end products instead of to the chemical reaction forming them, we may distinguish between *thermoplastic* and *thermosetting* compounds. The thermoplastic polymers consist of macromolecules constructed with primary bonds but adhering to each other with weaker secondary bonds. Heating or solvent action will break these secondary bonds preferentially, the material melts or dissolves, while the macromolecules themselves stay intact and can be reassembled; the plastic can be remolded. In the thermosetting materials (like bakelite) the whole structure is knit together by *cross-linking* primary bonds. Heating will therefore not melt the polymer but destroy it by decomposition when a critical temperature range is reached; the plastic cannot be remolded.

Typical prototypes of one-, two-, and three-dimensional polymers are the linear paraffin chain, the two-dimensional network of graphite, and the three-dimensional lattice of diamond (Fig. 24.3). In the saturated linear paraffins (*aliphatic* hydrocarbons), each carbon is linked to two neighboring carbon atoms by single bonds; the methylene radical $-\overset{\displaystyle H}{\underset{\displaystyle H}{\overset{|}{\underset{|}{C}}}}-$ represents the basic chain unit. The graphite sheets can be visualized as developing by the polymerization of benzene rings, the mother substance of the *aromatic* hydrocarbons; each carbon atom is linked to three carbon neighbors by primary bonds. Finally, the diamond structure, where carbon is linked to carbon by four equal bonds, may be visualized as arising from the polymerization of cage-like molecules. Such molecules were first discovered in Galician petroleum and are now becoming of increasing importance in polymer chemistry.[4]

A typical example of the diversity of condensation stages a material may traverse upon cooling is offered by the element sulfur.[5] The sulfur atom, with an electronic constellation $3s^23p^4$ beyond the neon core (see Sec. 8), has two half-filled p orbitals available for covalent bonding. As the atomic gas begins to condense at high temperature, first diatomic S═S molecules form, as can be seen clearly in the sequence of vibration maxima of the optical absorption spectrum. More energy, however, can be gained by the formation of

[2] J. J. P. Staudinger, *British Plastics 25*, 160 (1952).

[3] L. H. Baekeland, *Ind. Eng. Chem. 1*, 149 (1909).

[4] See V. Prelog and R. Seiwerth, *Ber. 74*, 1644, 1769 (1941); W. Nowacki and K. W. Hedberg, *J. Am. Chem. Soc. 70*, 1497 (1948).

[5] See A. von Hippel, *J. Chem. Phys. 16*, 372 (1948).

two single bonds in place of one double bond, hence a tendency exists for the formation of chain molecules —S—S—S—S. . ; and indeed, such molecules appear as the temperature decreases. A sequence of eight sulfur atoms in a chain can close into an S_8-ring molecule of

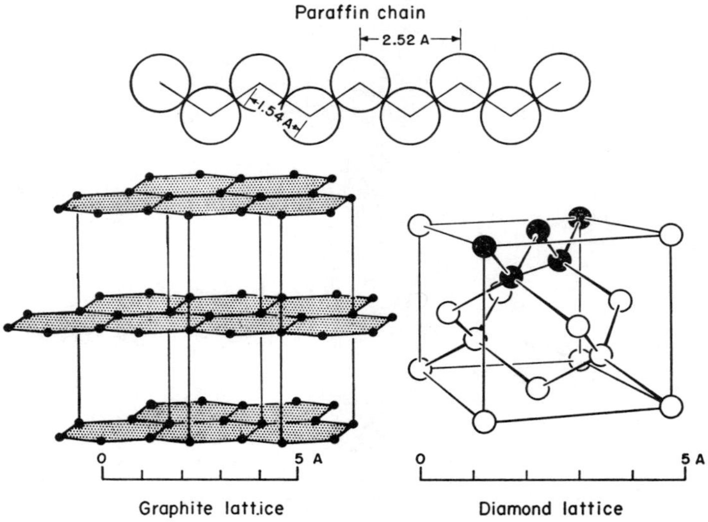

Fig. 24.3. One-, two-, and three-dimensional carbon networks.

greater stability (Fig. 24.4). Thus, in addition to chain segments of statistical length, these ring molecules begin to make their appearance.

The next stage of condensation occurs when the temperature is low enough to allow the secondary bonds of the van der Waals type to act. Now liquid sulfur is

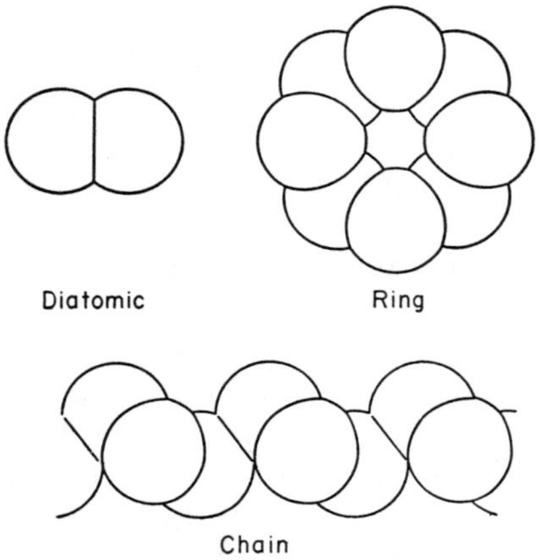

Fig. 24.4. Various types of sulfur molecules.

formed which may consist of ring molecules, chain molecules, or mixtures thereof. A viscosity experiment [6] indicates that sulfur not far above its melting point consists of a liquid of low viscosity, composed of S_8 rings. As the temperature is raised, the viscosity suddenly increases from about 10 centipoises to over 900 poises in a temperature interval of only 20°C (Fig. 24.5). The ring molecules are breaking open and polymerize into long chains which impede the flow of the liquid. Further temperature rise reduces the average chain length; hence the viscosity decreases again.

Rapid quenching of the liquid sulfur from its high viscosity range produces the disorganized chain structure of *plastic sulfur*, a true thermoplastic polymer.

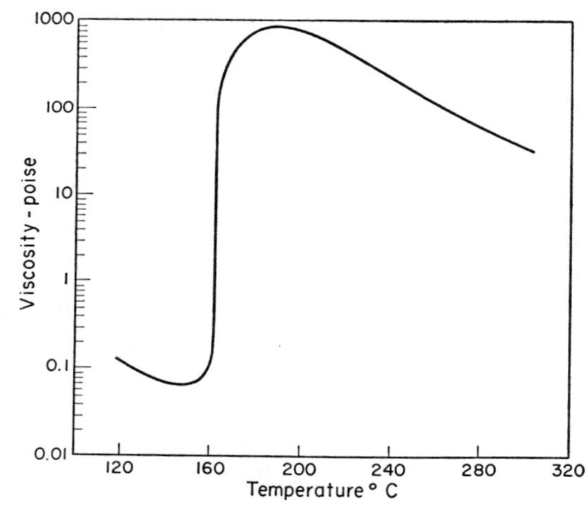

Fig. 24.5. Viscosity characteristic of sulfur, indicating breaking of ring molecules and polymerization. (After Bacon and Fanelli.[6])

Slow cooling leads first to a monoclinic and finally to a rhombic crystal structure. The rhombic [7] and probably also the monoclinic modification is built of S_8-ring

[6] R. F. Bacon and R. Fanelli, *J. Am. Chem. Soc. 65*, 639 (1943).

[7] B. E. Warren and J. T. Burwell, *J. Chem. Phys. 3*, 6 (1935); J. T. Burwell, *Z. Krist. 97*, 123 (1937).

molecules; the plastic variety reconverts to rhombic sulfur upon standing.

The example, sulfur, drives home forcefully three facts of general validity:

(*a*) The state of condensation of a material at any given temperature should be described as a statistical equilibrium.

(*b*) Materials of identical chemical composition may exist at the same temperature in different modifications (*polymorphism*).

(*c*) The prehistory of a sample may be an essential factor in determining its state of condensation and thus its properties; for example, high-temperature equilibria may be frozen-in by quenching.

In the preceding discussion we emphasized the parameter *temperature* as the main physical variable because significant temperature changes can be made easily. Of equal importance, as thermodynamics and statistics show, is the parameter *pressure*. That pressure as well as temperature decides the state of aggregation of molecules, becomes immediately apparent in the pressure-density characteristics of real gases. CO_2, for example, liquefies at room temperature at about 60 atmospheres of pressure, whereas an ideal gas would require more than a thousand atmospheres to reach the same packing density. We have to be aware that we deal here with association phenomena in which the size of the aggregates may fluctuate within wide limits. These fluctuations in particle size become especially large near the critical point of a gas and cause pronounced light scattering, known as *critical opalescence*.

That nuclei of condensation or crystallization are in a state of metastable equilibrium is to be expected. A cluster of N primary particles has $3N - 6$ degrees of vibrational and internal rotational freedom (see Sec. 18). These allow the aggregate to assume various geometrical configurations and to assemble statistically enough energy in any one bond to break it. It is therefore a question of competition between the lifetime of a crystallization nucleus before it loses particles by dissociation and the time interval between the arrival of new partners from the outside that decides whether the nucleus shrinks or grows. Temperature *and* pressure are thus decisive. Since each partner contributes its share to the overall attraction of the nucleus, larger nuclei, as is well known, prove more stable than smaller ones. Macroscopically speaking, the smaller ones have a higher surface tension.

Once liquids or solids have formed, their structure and packing density must be ruled by the shape and size of the individual partners and the various modes in which they can be arranged. In a first approximation, the individual atoms can be characterized as spheres which, according to the type of bonding, act with different radii (van der Waals', covalent, ionic, or metallic) (see Sec. 13). The problem of structure, if this approximation suffices, can be discussed in terms of the *closest packing of spheres*.

Equivalent spheres, as Barlow first discovered,[8] can be arranged in two types of closest packing (Fig. 24.6). In a single layer, each sphere is always in contact with six others; a second layer B can be superposed on the

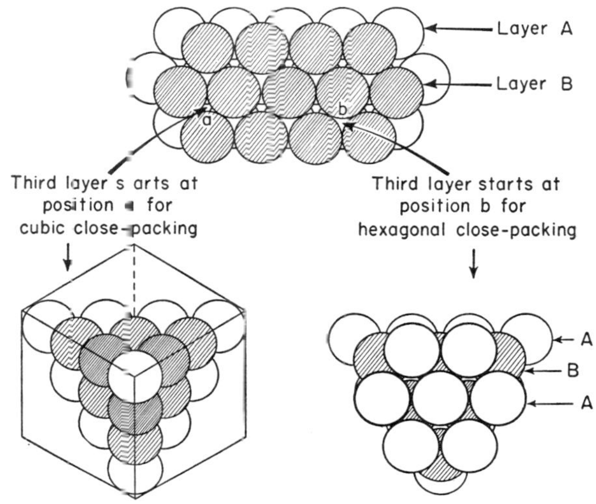

Fig. 24.6. Closest packing of spheres.

first layer A by having each sphere of layer A in contact with three spheres of the adjacent layer B. A third layer can be added in either one of two possible positions: (1) repeating the first layer A or (2) placed over the holes of the first layer not occupied by the second layer, producing layer C. In the first position we arrive at the closest packing of *hexagonal* symmetry and of the sequence $ABABAB\ldots$; in the second position the closest-packed *cubic* structure of the sequence $ABCABC\ldots$ results.

In each of these closest-packed structures every sphere is in contact with twelve others, with a hexagon of six in its own plane and a triangle of three in the two adjacent planes. The sphere is said to have the *co-ordination number* 12. In the hexagonal packing the triangles have identical orientation; in the cubic arrangement the upper is rotated against the lower by 60°.

Metal atoms and the atoms of the noble gases approach nearest the concept of spherical particles, and these elements are therefore found to crystallize in such closest-packed arrangements. In ionic crystals of

[8] W. Barlow and W. J. Pope, *J. Chem. Soc. 89*, 1675 (1906); *91*, 1150 (1907).

nonequivalent spheres, this simple situation can be realized only when one ion type is of a much larger size than the others and can form by itself a closest-packed structure which accommodates the smaller ions in its interstices. This situation is frequently realized in silicate minerals, as Bragg and his co-workers found,[9] when the oxygen ions dominate in size. Also numerous halides crystallize in structures characterized by densest-packed halogen-ion arrangements.

In building crystals from nonequivalent spheres, the *radius ratio* enters as a decisive parameter. Goldschmidt [10] was the first to study and emphasize the

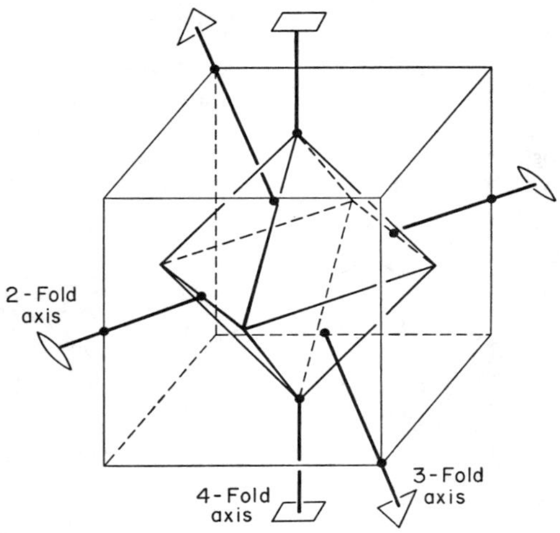

Fig. 24.7. Symmetry axes of NaCl structure.

importance of this factor in crystal chemistry. A simple example is offered by the alkali halides. The great majority crystallize in the *sodium chloride structure,* where each ion is surrounded by six oppositely charged ions in an octahedral arrangement. However, cesium chloride, bromide, and iodide form the body-centered *cesium chloride structure* (see Fig. 17.4), where each cesium in the center of the cube is surrounded by eight ions of the opposite kind. It is reasonable to assume that each ion will surround itself with the greatest possible number of oppositely charged partners. That the co-ordination number six is formed for most of the alkali halides, but eight for the cesium salts is obviously due to the large size of the Cs^+ ion which accommodates eight halogen neighbors.

By studying the radius-ratio limits inside which typical co-ordination arrangements are geometrically possible, we find in a first approximation the results given in Table 24.1 Near the stability limits of the various co-ordinations, peculiarities in the crystal properties will occur; for example, CsCl assumes the sodium chloride structure at elevated temperature, LiCl is very deliquescent, etc.

[9] W. L. Bragg, *Atomic Structure of Minerals,* Cornell University Press, Ithaca, N. Y., 1937, p. 30.

[10] V. M. Goldschmidt, *Geochemische Verteilungsgesetze der Elemente,* Oslo. Akad. Wiss. Vol. 8 (1927); see also the rules given by L. Pauling, *J. Am. Chem. Soc. 51,* 1010 (1929).

Table 24.1. Relation between co-ordination groups and radius ratio

(After Goldschmidt [10])

Co-ordination Group of Nearest Neighbors	Co-ordination Number	Radius Ratio Limits
Triangular	3	0.155 ↔ 0.225
Tetrahedral	4	0.225 ↔ 0.414
Octahedral	6	0.414 ↔ 0.732
Cubic (body-centered)	8	0.732 ↔ 1.000
Closest-packed cubic (cubo-octahedral)	12	1.000 → >1

The validity of the radius ratio concept, just as that of the radii themselves (see Sec. 13), should not be taken too seriously. These are useful first approximations which lost much of their stringency after it became obvious that mixed-bond types are frequently encountered.

Additional complications enter when structural units are not individual atoms or ions but composed of groups of atoms. Highly symmetrical composite building stones like NH_4^+ and $SO_4^=$ may still be approximated by spheres as far as their spatial requirements go. However, the potential variety of compounds that may form becomes apparent only when the actual shape of such group molecules is taken into account. Many silicates and titanates, for instance, can be visualized as frameworks of SiO_4 tetrahedra or TiO_6 octahedra, respectively, with cations interspersed as electrostatic binder. These nonspherical building stones may share corners, edges, or faces and thus produce a tremendous variety of inorganic compounds.

For less symmetrical building stones like the S_8 rings of Fig. 24.4 or for organic structures, where the directive properties of the covalent bonds dominate, the tendency for close packing may be submerged completely. The structures assume more and more individuality, and each new analysis becomes a scientific achievement.

An ideal crystal is composed of a periodic sequence of building stones that can be repeated indefinitely in space without leaving holes. Crystals possess, therefore, particular kinds of macroscopic symmetry elements (axes, planes, and centers of symmetry) consistent with the repetition pattern of the atoms or molecules. The actual faces a crystalline material develops, its *crystal habit,* depend on the conditions of crystal growth, but the symmetry elements remain the same. Sodium chloride, for example, may be grown as cube or octahedron, but in both cases we count three axes of fourfold, four axes of threefold, and six axes of twofold symmetry (Fig. 24.7). The possible combinations of symmetry elements lead to 32 *crystal classes* grouped according to the reference co-ordinates, the *crystallographic axes,* in seven *crystal systems* † (Fig. 24.8).[11]

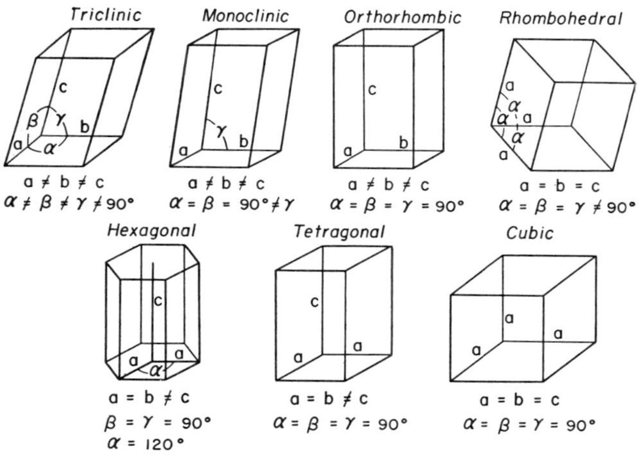

Fig. 24.8. Reference co-ordinates of the seven crystal systems. (After Barrett.[11])

A real crystal has molecular holes and other flaws which may alter its physical properties decisively, as Smekal [12] first stressed forcefully. Defects of special importance are: missing lattice points (*Schottky defects* [13]); atoms displaced to interstitial positions (*Frenkel defects* [14]); and dislocations,[15] where a whole row (or plane) of atoms may terminate in a dead end (Fig. 24.9). The lattice point defects are of great importance in conduction phenomena, but dislocations seem to play a decisive role for crystal growth and mechanical strength.

With increasing temperature the crystal defects become more numerous, until the crystal structure collapses at the melting point. The *long-range order* of the periodic structure is destroyed by melting, but a *short-range order* persists in liquids. Each Na^+ ion of a sodium chloride melt, for example, is still preferentially surrounded by Cl^- ions, and vice versa, because of the ordering influence of the Coulomb attraction. Glasses show short-range order and behave akin to supercooled liquids.

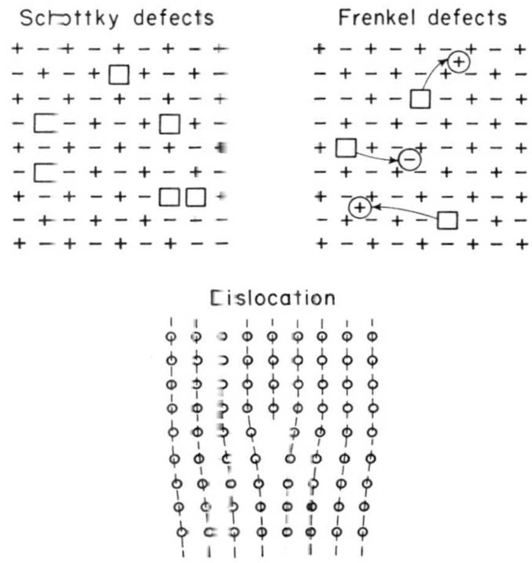

Fig. 24.9 Crystal defects.

† For a further discussion of crystal symmetry, see Sec. 26.

[11] See, for example, C. G. Barrett, *Structure of Metals,* McGraw-Hill Book Co., New York, 1952.

[12] A. Smekal, *Handbuch der Physik,* Vol. 24, Pt. 2, Chap. 5.

[13] C. Wagner and W. Schottky, *Z. physik. Chem. 11,* 163 (1930).

[14] J. Frenkel, *Z. Physik 35,* 652 (1926).

[15] See, for example, *Imperfections in Nearly Perfect Crystals,* W. Shockley Editor, John Wiley and Sons, New York, 1952.

25 · Various Models for the Discussion of Orientation Polarization in Liquids and Solids

When matter aggregates from atoms to small molecules, macromolecules, and, finally, to the condensed phases of liquids and solids, a bewildering variety of structures may arise (see Sec. 24). Accurate expressions for ϵ^* and μ^*, for the response of a specific material to electric and magnetic fields, must, therefore, be based on a detailed structure analysis of the material in question. Theory, while acknowledging this need for individual analysis, has to construct more generalized models which represent characteristics common to groups of materials and which contain parameters adjustable to the individual case.

The theory of polarization in its simplest form was based on the concept of the Mosotti cavity and led to the Clausius-Mosotti-Lorentz-Lorenz equation (see Sec. 2). It introduced the reference molecule as a mathematical point around which is scooped out an empty sphere of such a large radius that the molecules beyond it blend into a continuum. The electric field in this continuum is not supposed to notice the existence of the sphere. This picture has been incorporated without difficulty into the classical theory of resonance or relaxation spectra (Secs. 4 and 22) but led, when applied to polar molecules, to the Mosotti catastrophe of spontaneous polarization (see Sec. 23).

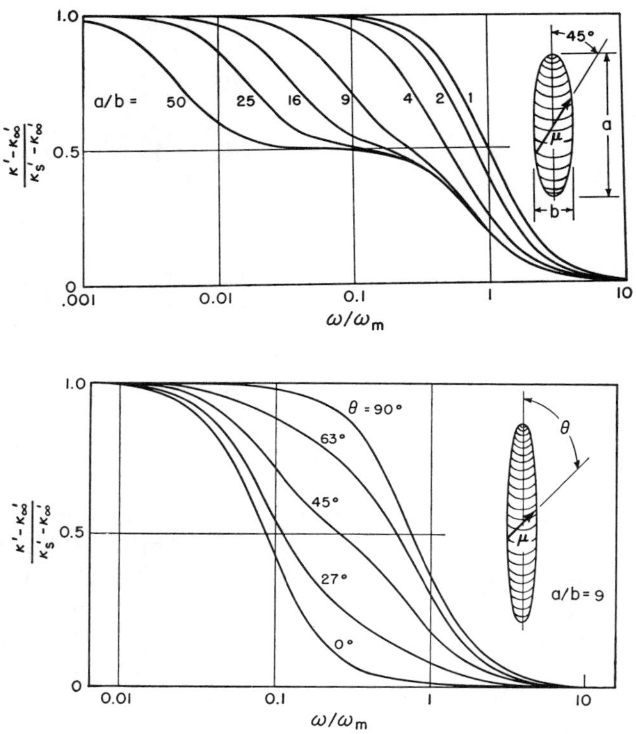

Fig. 25.1. Various shapes of relaxation characteristics for ellipsoids of revolution. (After Oncley.[1])

For a discussion of the polarization arising from the orientation of permanent moments, the reference molecule obviously should not be a mathematical point but a dipole carrier. In the refinement of the model by Onsager (see Sec. 23), the cavity is therefore visualized as a molecular sphere in which a mathematical dipole is centered. The sphere is thought to be filled with a medium possessing the deformation polarization of the dielectric, and is assumed to distort the field in the surrounding continuum like a real spherical inclusion.

This spherical model of a dipole molecule, reacting back upon its surroundings, avoids the Mosotti catastrophe. It provides a reasonable first-order model for the discussion of a dilute solution of spherical polar molecules in a nonpolar solvent and of nonassociated polar liquids (Fig. 23.2). Calculations based upon it have been carried through for static fields only, but we may assume that in alternating fields the dipole in Onsager's cavity behaves like Debye's spherical molecules in a viscous medium, that is, that a simple relaxation spectrum results.

We can add one more refinement to the continuum theory by replacing the spherical molecule with an ellipsoid possessing three principal moments of inertia (see Sec. 18). Whereas a spherical top rotating in a viscous liquid shows a relaxation spectrum with one time constant only (see Eq. 22.16), the asymmetric top has six main time constants which correspond to the various orientations of one principal axis pointing in the field direction and the other two parallel or perpendicular to the direction of motion. If the ellipsoid points obliquely to the field, for instance, because its dipole moment is not lined up with one of the principal axes, any intermediate time constant may be realized. Thus it is possible that even one single type of elliptical dipole molecule, by assuming a statistical distribution of orientations in the field, may cause a complex relaxation spectrum with a broad distribution of relaxation times.

The inherent possibility of deducing from relaxation spectra some information on the size and shape of polar molecules and on the orientation of the momentum axis has been exploited for the macromolecules of biology, especially by Oncley and his co-workers.[1] The perhaps dangerously simple assumption is that deviations of the dispersion and absorption characteristic from a Debye curve with one relaxation time can be interpreted as indicating that the molecule in question is not a sphere characterized by one time constant τ but an ellipsoid of revolution of an axial ratio a/b in which the electric momentum axis deviates by a dipole angle θ from the major axis a. Perrin,[2] and Budó, Fischer, and Miyamoto[3] have calculated the orientation polarization of such ellipsoidal particles in viscous media. By matching the measured dispersion characteristic with curves thus computed (Fig. 25.1), approximate values for a/b, θ, and τ may be obtained.

The hypothetical time constant τ of a spherical molecule rotating in the solvent of the viscosity η gives, according to Eq. 22.16, an idea what the size of the molecule and its molecular weight might be. This information can be checked against information obtained from diffusion, ultracentrifuge, viscosity measurements, and from light scattering.[4] Finally a dipole moment

1 J. L. Oncley, *J. Am. Chem. Soc. 60*, 1115 (1938); J. D. Ferry and J. L. Oncley, *ibid. 60*, 1123 (1938); *63*, 272 (1941); J. L. Oncley, *Chem. Rev. 30*, 433 (1942).

2 F. Perrin, *J. phys. radium 5*, 497 (1934).

3 A. Budó, E. Fischer, and S. Miyamoto, *Physik. Z. 40*, 337 (1939).

4 P. Debye, *J. Appl. Phys. 15*, 338 (1944).

can be determined from the difference between static and optical dielectric constant by assuming some kind of a local field. The Mosotti field leads to dipole moments which are much too small for solvents of high dielectric constant, as Wyman [5] and others have shown. The actual moments, calculated on the basis of Onsager's model, range for protein molecules in the order of 10^2 to 10^3 debyes. They are so large, not because the dipole charge is higher than in normal molecules, but because the dipole length is great, since the molecular weight of these molecules lies in the range of 100,000. Figure 25.2 shows the shapes of some macromolecules as inferred from dielectric and other methods.

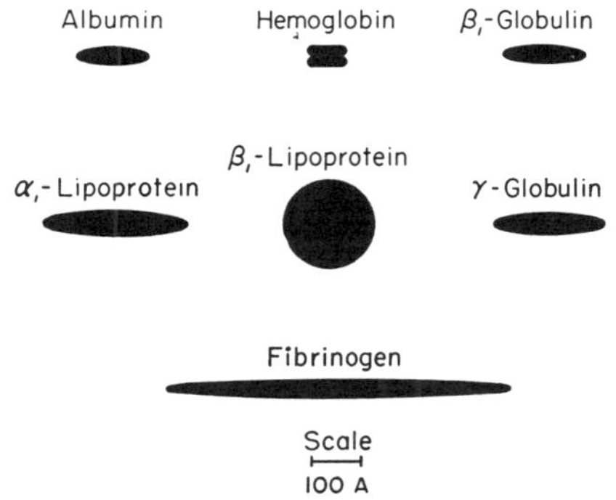

Fig. 25.2. Postulated shapes of some protein molecules in blood serum. (After Oncley.[1])

To treat the immediate neighborhood of a dipole as a continuum in which the molecule can rotate according to Stokes's law, with a friction factor given by the macroscopic viscosity of the solvent, may be an appropriate picture for these giant molecules of biology,† but is, in general, an unsatisfactory concept. Figure 25.3 makes this convincingly clear with the example of methyl palmitate dissolved in benzene, paraffin oil, polyisobutylene, and polyethylene.[6] The macroscopic viscosity of the solution varies in this sequence from fractions of a centipoise to near infinity, but the relaxation time of the polar molecule, as indicated by a specific absorption near 10^{10} cycles, remains practically unaltered.

We saw in Sec. 24 that spherical particles tend to assume regular geometrical arrangements—structures of densest packing consistent with their radius ratio. When a crystalline state cannot be realized, a short-range order still exists which betrays itself by diffuse diffraction rings in the X-ray patterns of liquids and glasses.[7] An improved model for the discussion of orientation polarization in polar media will therefore result if the dipole molecule with its first layer of neighbors is treated as a dynamic structural unit—a molecular island floating in a dielectric continuum in which

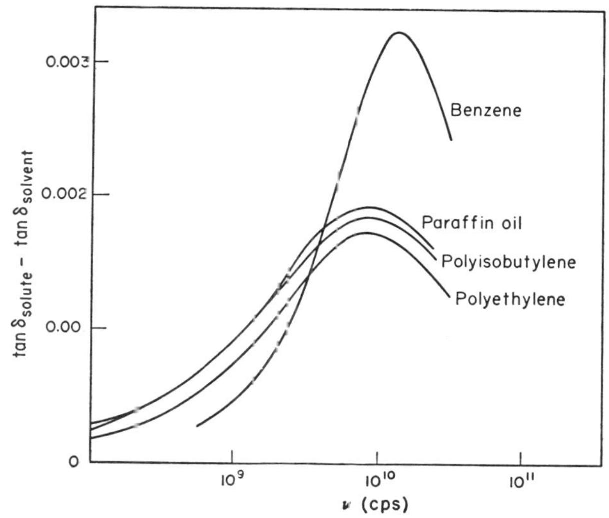

Fig. 25.3. Relaxation spectrum of methyl palmitate in various solvents. (After Srivastava.[6])

bonds are broken and reformed statistically. Kirkwood [8] has developed the molecular theory in this direction and found for water, for example, by considering the first shell of neighbors in tetrahedral co-ordination (Fig. 25.4), a static dielectric constant between 55 and

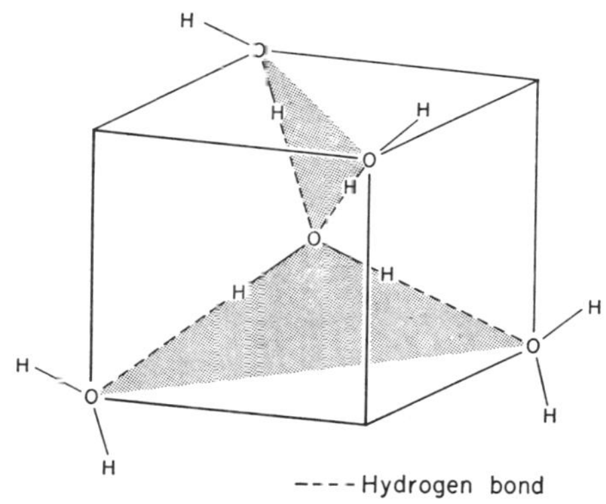

Fig. 25.4. Short-range order in water.

82, depending on the bond angle, whereas the unmodified Onsager model gives the value 31. Still better results are obtained if the molecular island is extended to

[5] J. Wyman, Jr., *J. Am. Chem. Soc. 58,* 1482 (1936).

† That the situation can be much more complex has been shown, for example, by R. M. Fuoss and others in recent studies on polyelectrolytes.

[6] S. S. Srivastava, *Thesis,* University of Lucknow, India, 1951.

[7] See, for example, W. H. Zachariasen, *J. Am. Chem. Soc. 54,* 3841 (1932); B. E. Warren, H. Krutter, and O. Morningstar, *J. Am. Ceram. Soc. 19,* 202 (1936).

[8] J. G. Kirkwood, *J. Chem. Phys. 7,* 911 (1939); G. Oster and J. G. Kirkwood, *ibid. 11,* 175 (1943). See also H. Fröhlich, *Theory of Dielectrics,* pp. 49 ff, 137 ff (1949).

include the second nearest neighbors, but the mathematical problem becomes increasingly formidable.

As soon as molecular structure is introduced in the surroundings of a dipole molecule, we have to realize that the dipole moment itself represents a structural element, likely to be anchored in its environment. Hence, an activation energy U will be required to dislodge it, and unimpeded rotation becomes feasible only at sufficiently high temperature ($U \ll kT$). In general, a polar molecule in solids assumes prescribed positions and switches orientation in discontinuous jumps.

Debye, already at an early date, became aware of this situation and tried to improve his model of a sphere immersed in a homogeneous liquid of stream-lined motion by adding the concept of *hindered rotation*.[9] The quasi-crystalline surroundings are thought to superpose an angular energy dependence $-U \sin \phi$ on the position of the dipole, in addition to the dependence arising from the external field. Such phenomena of *steric hindrance* are well known for the intramolecular rotation of gas molecules. A simple case in kind is the ethane molecule (Fig. 25.5), where the methyl radicals are not rotating freely around the C-C axis as the single bond might imply, but have to surmount an activation energy of about 3 kcal between consecutive staggered positions.

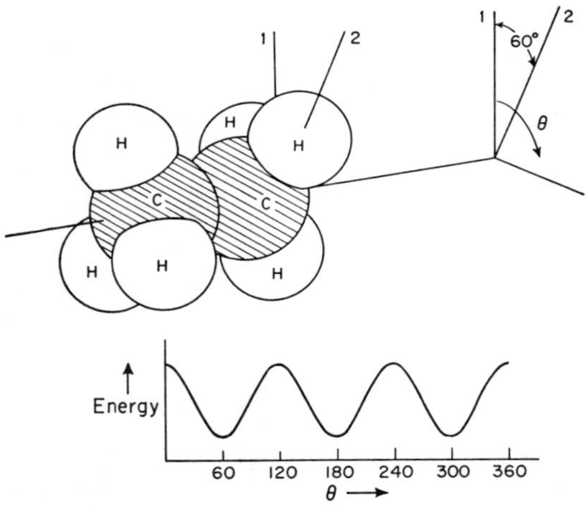

Fig. 25.5. Steric hindrance of ethane molecule.

The ethane molecule does not carry a dipole moment. Its potential barrier can therefore not be deduced directly from dielectric measurements, but heat capacity data, Raman effect, and vibration-rotation spectra give relatively conclusive quantitative evidence.[10] More interesting from our standpoint is the hindered rotation in the halogenated ethanes, especially a case such as symmetrical dichloroethane, where the mutual interaction of two dipole moments increases the complications. The two equivalent moments μ are set a distance d apart and inclined against the C-C axis by an angle θ (Fig. 25.6);[11,12] the angle ϕ designates the relative angular orientation of the two individual moments.

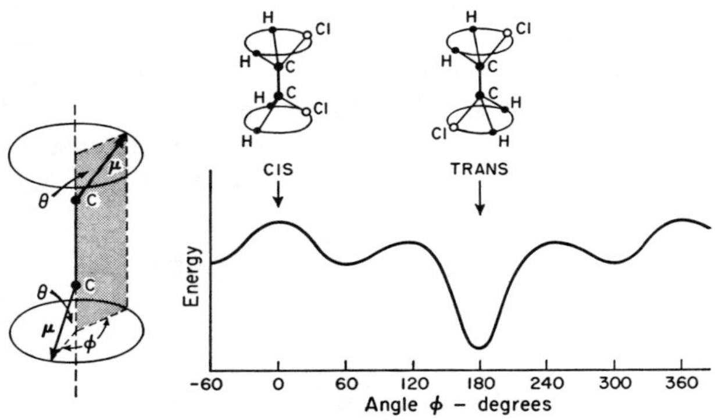

Fig. 25.6. Orientations of the ethylene dichloride molecule. (Adapted from Palmer [11] and Le Fèvre.[12])

The resultant moment μ' for any given position can be derived by vector addition as

$$\mu' = 2\mu \sin \theta \cos \frac{\phi}{2}. \tag{25.1}$$

The average moment per CH_2Cl-CH_2Cl molecule, $\bar{\mu}$, follows from the mean square moment,

$$\bar{\mu}^2 = \frac{1}{2\pi} \int_0^{2\pi} \left(2\mu \sin \theta \cos \frac{\phi}{2}\right)^2 d = 2\mu^2 \sin^2 \theta, \tag{25.2}$$

as

$$\bar{\mu} = \sqrt{2}\, \mu \sin \theta \tag{25.3}$$

if free rotation is realized.

Actually, the rotation is restricted by two kinds of interaction: (*a*) steric hindrance, as in the case of the ethane molecule, produces three potential minima, probably favoring the staggered positions (60°, 180°, 300°); (*b*) the electrostatic coupling between the two dipole moments makes the *trans configuration* (C_{2h}) more favorable than the *cis configuration* (C_{2v}), that is, it produces a potential curve with an energy minimum in the 180° position. Both effects together create a potential curve with 3 minima, one deep well at 180° stabilizing the trans configuration, and two equal, flatter minima producing a metastable orientation near 60° and 300°, respectively.

The symmetrical dichloroethane molecule exists therefore in the region of hindered rotation as two

[9] P. Debye, *Physik. Z. 36*, 100, 193 (1935); P. Debye and W. Ramm, *Ann. Physik 28*, 28 (1937).

[10] For a critical discussion of rotation in ethane and halogenated ethane, see G. Glockler, *Revs. Mod. Phys. 15*, 145 (1943).

[11] W. G. Palmer, *Valency*, Cambridge University Press, 1945.

[12] See R. J. W. Le Fèvre, *Dipole Moments*, Methuen and Co., Ltd., London, 1938, p. 82.

isomers. At low temperature the trans configuration dominates, and, since its dipole moment is zero, the average moment $\bar{\mu}$ tends towards zero as the temperature decreases (Fig. 25.7). For higher temperatures the metastable configuration begins to compete, and finally the moment should increase asymptotically to the value of about 2.5 debyes predicted for free rotation.[12]

At normal temperature, the molecule will linger in the potential trough of each isomeric form, executing

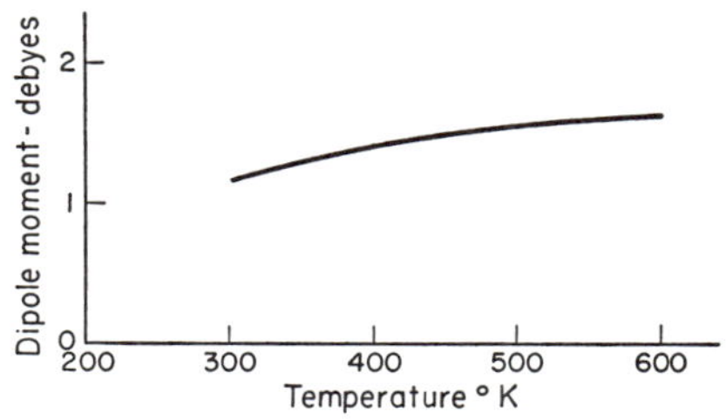

Fig. 25.7. Variation of the dipole moment of ethylene dichloride with temperature in the gaseous state. (After Le Fèvre.[12])

torsional oscillations until, after a statistical waiting period τ, it receives sufficient energy by thermal fluctuations to traverse the potential barrier U. The relaxation time τ, or its inverse, the *jump rate* k_0 (reaction rate), is given by Boltzmann statistics as

$$\tau \equiv \frac{1}{k_0} = \tau_0 e^{U/kT}. \tag{25.4}$$

The factor τ_0 represents the time of a single oscillation in the potential well. Since we deal in the dichloroethane molecule with two different wells and three modes of jumping, three different relaxation times should be distinguished. If we consider that the depth of the potential troughs will vary with the total vibration state of the molecule, we can foresee that, especially in the condensed states of solids and liquids, hindered rotation will also lead to complex relaxation spectra with more or less broad distributions of relaxation times.

Actually, the relaxation spectra of solids and liquids are, in general, broader and flatter than the simple Debye characteristic of Fig. 22.3 predicts. The characteristic of vulcanized rubber is a typical example in kind[13] (Fig. 25.8); plotted in a Cole-Cole diagram (see Fig. 22.4), the points fall on a semicircle which, in this case, has its center below the abscissa. Mathematically speaking, the complex permittivity of Eq. 22.11 has been replaced by the more general expression[14]

$$\kappa^* = \kappa_\infty' + \frac{\kappa_s' - \kappa_\infty'}{1 + (j\omega\tau_e)^{1-\alpha}}, \tag{25.5}$$

[13] W. Kauzmann, *Revs. Mod. Phys. 14,* 12 (1942).
[14] K. S. Cole and R. H. Cole, *J. Chem. Phys. 9,* 341 (1941).

where $\alpha \frac{\pi}{2}$ is the angle between the abscissa and the radius drawn from the point of intersection. The empirical constant α, which may vary between 0 and 1, describes the broadening of the relaxation region; for $\alpha = 0$ we return to the case of a single relaxation time.

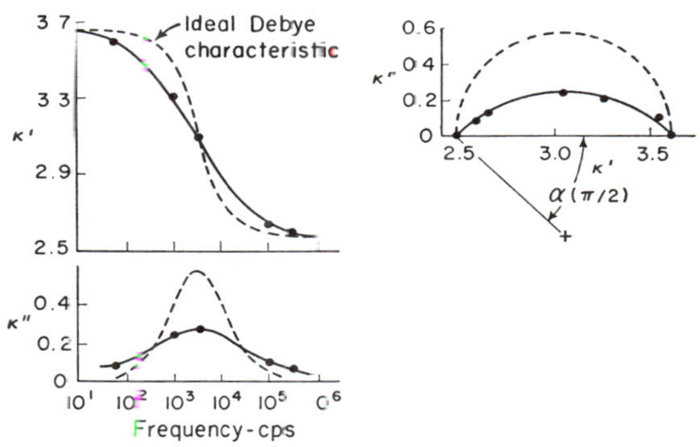

Fig. 25.8. Relaxation spectrum and Cole-Cole diagram of vulcanized rubber. (After Kauzmann.[13])

The circuit equivalent of Debye's dipole molecule rotating in a medium of dominating friction was an RC circuit shunted by a capacitance (see Fig. 22.1). The Cole-Cole expression of Eq. 25.5 corresponds to the more general network of Fig. 25.9, in which the resistance is replaced by a complex impedance. In the Debye case we dealt with only one time constant τ_e, which could be derived simply from the critical wavelength at which the loss factor reaches its maximum as (see Eq. 22.24)

$$\lambda_m = 2\pi c\tau_e. \tag{25.6}$$

In the case of the depressed circular-arc locus this equation remains valid, but τ_e becomes a mean time constant around which other time constants are spread in a symmetrical distribution when plotted on a log τ/τ_e scale. The spreading factor α can be determined from the plot of Fig. 25.9, for example, as tan $\alpha\pi/2$, or calculated from the ratio

$$v/u = (\omega\tau_e)^{1-\alpha}, \tag{25.7}$$

since τ_e is known from Eq. 25.5.

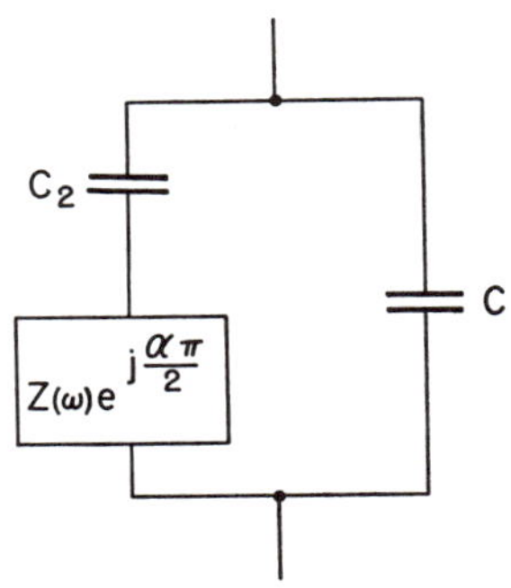

Fig. 25.9. Equivalent circuit of dipole relaxation spectrum with distribution of relaxation times.

The circular arc locus in the complex plane, well known to the electrical engineer as the Argand diagram and used, for example, in admittance and impedance representations (see I, Fig. 24.3), is valid when

κ^* is prescribed by a single relaxation time or by special distribution functions of relaxation times grouped symmetrically around a center frequency.[15] The factor α is relatively insensitive for the actual shape of the function, and the circular arc locus is an empirical assumption without basic molecular foundation. But the Cole-Cole plot gives a useful check on the reliability of experimental data and is a guide for a tentative extrapolation of κ^* into frequency regions which are hard to reach by direct measurements. It provides in addition a valuable method for resolving a complicated relaxation spectrum into its principal constituents, as R. H. Cole has shown recently for some hydrogen-bonded liquids [16] (Fig. 25.10).

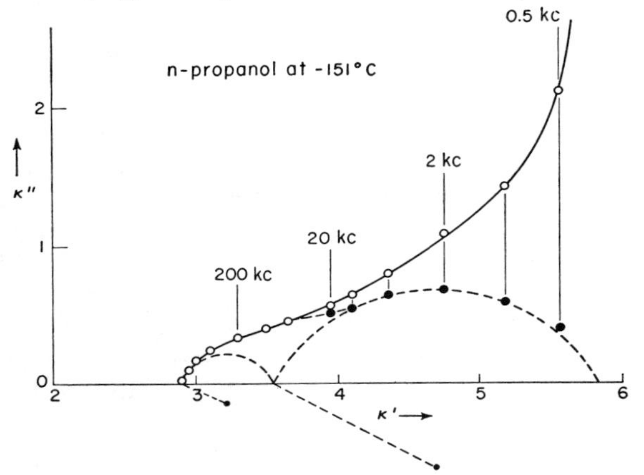

Fig. 25.10. Interpretation of relaxation spectrum of n-propanol by Cole-Cole plot of two circles. (After Davidson and Cole.[16])

By considering the structure of dielectrics, whether the shape of the dipole molecule itself or the fixation of the polar molecule in its surroundings, we arrived at a distribution of relaxation times. If activation energies hinder the orientation of the dipoles, the relaxation times are the inverse of reaction rate constants (see Eq. 25.4). Hence the theory of chemical reaction rates, as developed by Eyring [17] and his co-workers, can be applied with advantage to the treatment of dielectric relaxation phenomena, as Kauzmann [13] first showed in detail. Several other causes may produce broad relaxation spectra, as we shall see later. An unbiased analysis of all pertinent physical and chemical information is required before we decide in favor of a special model.

Models, as shown above, can frequently be based on the concept of short-range order, where only a very small section of the true structure need be considered and the rest of the dielectric may be lumped in a continuum. However, in other cases we may find that long-range interaction effects are decisive and that quite different approximations have to be made. An example in kind is the potential energy binding an Na^+ or Cl^- ion in the lattice of the rock-salt crystal. The Coulomb-force interaction with the neighbors falls with the distance r as $1/r^2$, but simultaneously the number of partners coupled to our reference ion increases proportional to r^2. Hence the interaction terms have no tendency to converge as the distance gets greater though they alternate in sign. The problem cannot be solved by lumping the more remote ions in a neutral continuum but only lumping lattice points into groups that provide rapidly converging series.[18] This procedure leads to the expression

$$U = -A\frac{(ze)^2}{r^2} \tag{25.8}$$

for the Coulomb energy binding one ion of the charge ze to all its neighbors in an ionic crystal. The *Madelung factor* A, named after the physicist who first computed some of the structures, has the value 1.742 for the sodium-chloride lattice [19] (see also Appendix A, II, Sec. 9).

Other types of long-range interaction arise. Just as the motion of a snake cannot be predicted by observing only one of its vertebrae, the squirming and coiling of long-chain molecules cannot be handled statistically by focusing attention on only the immediate surroundings of a polar group. Kirkwood and Fuoss [20] have treated such squirming motions of long-chain polymers and their effect on the distribution of relaxation times. This is only one illustration of a more general situation, where the mechanical coupling of a whole system influences decisively its electrical reaction.

[15] W. A. Yager, *Physics 7*, 434 (1936); R. M. Fuoss and J. G. Kirkwood, *J. Am. Chem. Soc. 63*, 385 (1941); W. Kauzmann, *Revs. Mod. Phys. 14*, 12 (1942).

[16] P. W. Davidson and R. H. Cole, *J. Chem. Phys. 19*, 1484 (1952).

[17] H. Eyring, *J. Chem. Phys. 4*, 283 (1936).

[18] For a discussion of the forces in crystals, see M. Born and M. Göppert-Mayer, "Dynamische Gittertheorie der Kristalle," *Handbuch der Physik*, Vol. 24, Pt. 2, Springer, Berlin, 1933.

[19] E. Madelung, *Physik. Z. 19*, 524 (1918); P. P. Ewald, *Ann. Physik 4*, 253 (1921).

[20] See the summary "Physical Chemistry of Polymers," by R. M. Fuoss, *Am. Scientist 36*, 258 (1948).

26 · Piezoelectricity

One of the most striking manifestations of the freezing-in of permanent dipole moments as structural elements and of their compulsory coupling to the mechanical distortions of a macroscopic system is provided by the *piezoelectric effect*, discovered in 1880 by the brothers Pierre and Jacques Curie.[1] They found that certain types of asymmetrical crystals, like quartz, tourmaline, and rochelle salt, when compressed in specific directions, develop a potential difference. The effect was called *piezoelectricity* (pressure-electricity) from the Greek word πιέζειν, to press. Obviously, when this *direct effect* exists (mechanical distortion creates a voltage), the *converse piezoelectric effect* must exist also (the application of an electric voltage creates a mechanical distortion). This was predicted by Lippmann in 1881 on thermodynamic grounds,[2] and verified by the Curies.[3]

Until now we have tended to restrict our discussion to *isotropic* materials. A piezoelectric effect can exist only in special *anisotropic* dielectrics. The application of an electric field **E** creates in anisotropic materials a polarization **P** that points, in general, not parallel to **E**. Thus the simple vector relations between the field strength **E**, the polarization **P**, and the electric displacement **D** (see I, Eqs. 2.6 and 2.7),

$$\mathbf{P} = \chi\epsilon_0\mathbf{E} \tag{26.1}$$

and

$$\mathbf{D} = \epsilon'\mathbf{E}, \tag{26.2}$$

have to be replaced by the more general expressions

$$\begin{aligned}
\mathbf{P}_1 &= \chi_{11}\epsilon_0\mathbf{E}_1 + \chi_{12}\epsilon_0\mathbf{E}_2 + \chi_{13}\epsilon_0\mathbf{E}_3,\\
\mathbf{P}_2 &= \chi_{21}\epsilon_0\mathbf{E}_1 + \chi_{22}\epsilon_0\mathbf{E}_2 + \chi_{23}\epsilon_0\mathbf{E}_3,\\
\mathbf{P}_3 &= \chi_{31}\epsilon_0\mathbf{E}_1 + \chi_{32}\epsilon_0\mathbf{E}_2 + \chi_{33}\epsilon_0\mathbf{E}_3,
\end{aligned} \tag{26.3}$$

and

$$\begin{aligned}
\mathbf{D}_1 &= \epsilon_{11}\mathbf{E}_1 + \epsilon_{12}\mathbf{E}_2 + \epsilon_{13}\mathbf{E}_3,\\
\mathbf{D}_2 &= \epsilon_{21}\mathbf{E}_1 + \epsilon_{22}\mathbf{E}_2 + \epsilon_{23}\mathbf{E}_3,\\
\mathbf{D}_3 &= \epsilon_{31}\mathbf{E}_1 + \epsilon_{32}\mathbf{E}_2 + \epsilon_{33}\mathbf{E}_3.
\end{aligned} \tag{26.4}$$

The subscripts 1, 2, 3 refer to orthogonal directions x, y, z. The relationship between the vectors is still assumed to be a linear one, but each component of **P** or **D** has now become a linear function of the three components of **E**. The set of linear equations for **P** or **D** is designated as a *linear transformation*. Their coefficients (the susceptibilities χ or permittivities ϵ) characterize the transformation and may be written in the rectangular array

$$\begin{bmatrix} \chi_{11} & \chi_{12} & \chi_{13} \\ \chi_{21} & \chi_{22} & \chi_{23} \\ \chi_{31} & \chi_{32} & \chi_{33} \end{bmatrix} \tag{26.5}$$

and

$$\begin{bmatrix} \epsilon_{11} & \epsilon_{12} & \epsilon_{13} \\ \epsilon_{21} & \epsilon_{22} & \epsilon_{23} \\ \epsilon_{31} & \epsilon_{32} & \epsilon_{33} \end{bmatrix}, \tag{26.6}$$

called the *matrix* of the transformation. The susceptibilities and permittivities are interrelated according to Eq. I, 2.8 as

$$\chi_{11} = \frac{\epsilon_{11}}{\epsilon_0} - 1 = \kappa_{11} - 1, \tag{26.7}$$

and so on.

Turning to the mechanical deformation of our dielectric, we specify that we will operate within the limits of Hooke's law, that is, the application of a mechanical stress results in a strain proportional to this stress. A mechanical stress, designated by the symbol $\overleftrightarrow{T}$, operates when equal and opposite forces put a material under tension, compression, or shear. The dimension of $\overleftrightarrow{T}$ is therefore that of a pressure,

$$\begin{aligned}
[\overleftrightarrow{T}] \equiv [\text{stress}] &= [\text{pressure}]\\
&= [\text{force per unit area}]\\
&= [\text{newton/m}^2],
\end{aligned} \tag{26.8}$$

but $\overleftrightarrow{T}$ is not a vector like **D**, **E**, or **P** pointing in one direction; it is a physical quantity, for which a direction and its opposite are equivalent. Voigt[4] was the first to investigate such quantities systematically in his classical studies on crystals and to introduce for them the name *tensor*. A compressional stress normal to some surface can be distinguished from an extensional one by giving the tensor a positive or a negative

[1] J. and P. Curie, *Compt. rend. 91*, 294 (1880).
[2] M. G. Lippmann, *Ann. chim. 24*, 145 (1881).
[3] J. and P. Curie, *Compt. rend. 93*, 1137 (1881).

[4] See W. Voigt, *Lehrbuch der Kristallphysik*, Teubner, Leipzig and Berlin, 1910.

sign. Graphically, it may be represented by an arrow with heads on both ends pointing in opposite directions, inwards for compressional and outwards for extensional stress. These types of stress are *polar* tensors acting parallel to an axis, in distinction from shear

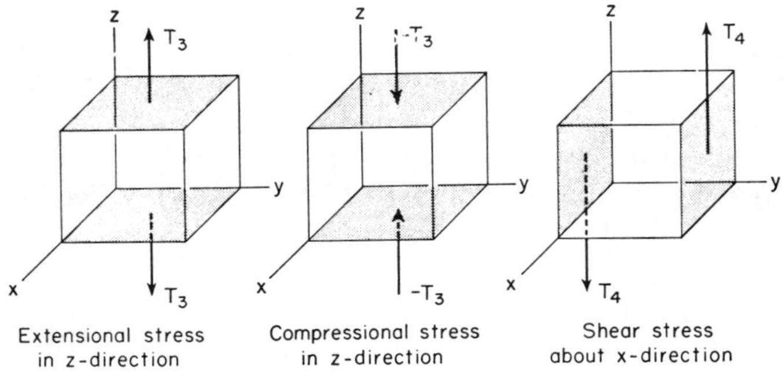

Fig. 26.1. Polar and axial stresses.

stresses, which are *axial* tensors and result from the balancing action of torque and countertorque around an axis (Fig. 26.1).

Stress applied to an elastic body produces a deformation. Its general description is that two points, originally located at P (0, 0, 0) and Q (x, y, z), have been shifted to P' (ξ_1, η_1, ζ_1) and Q' $(x + \xi_2, y + \eta_2, z + \zeta_2)$ (Fig. 26.2). The displacement must be a con-

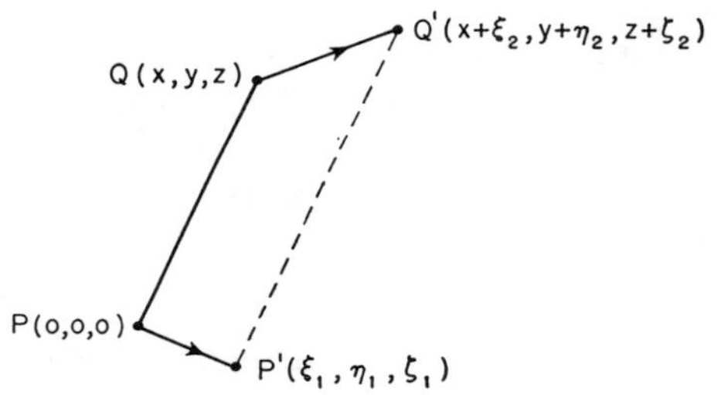

Fig. 26.2. General description of deformation.

tinuous function of the co-ordinates x, y, and z; hence

$$\xi_2 = \xi_1 + \frac{\partial \xi}{\partial x} x + \frac{\partial \xi}{\partial y} y + \frac{\partial \xi}{\partial z} z,$$

$$\eta_2 = \eta_1 + \frac{\partial \eta}{\partial x} x + \frac{\partial \eta}{\partial y} y + \frac{\partial \eta}{\partial z} z, \qquad (26.9)$$

$$\zeta_2 = \zeta_1 + \frac{\partial \zeta}{\partial x} x + \frac{\partial \zeta}{\partial y} y + \frac{\partial \zeta}{\partial z} z,$$

if higher derivatives are neglected.

An elongation per unit length in the x-, y-, and z-directions is defined as the *linear strain* in these directions and designated by the polar-tensor components

$$\underset{\leftrightarrow}{S_1} \equiv \frac{\partial \xi}{\partial x},$$

$$\underset{\leftrightarrow}{S_2} \equiv \frac{\partial \eta}{\partial y}, \qquad (26.10)$$

$$\underset{\leftrightarrow}{S_3} \equiv \frac{\partial \zeta}{\partial z}.$$

In addition, *shearing* strains around the x, y, and z axes occur, as represented by the three axial-tensor components

$$\underset{\leftrightarrow}{S_4} \equiv \frac{\partial \zeta}{\partial y} + \frac{\partial \eta}{\partial z},$$

$$\underset{\leftrightarrow}{S_5} \equiv \frac{\partial \xi}{\partial z} + \frac{\partial \zeta}{\partial x}, \qquad (26.11)$$

$$\underset{\leftrightarrow}{S_6} \equiv \frac{\partial \eta}{\partial x} + \frac{\partial \xi}{\partial y}.$$

Strain tensors, according to this definition, have no dimension ([m/m]). We will designate linear strains as positive for extension and shears as positive when clockwise.

According to Hooke's law, the stresses are proportional to the strains, that is,

$$\underset{\leftrightarrow}{T_i} = \sum_j c_{ij} \underset{\leftrightarrow}{S_j} \qquad \text{for } i, j = 1 \text{ to } 6, \qquad (26.12)$$

$$\underset{\leftrightarrow}{S_j} = \sum_i s_{ji} \underset{\leftrightarrow}{T_i}. \qquad (26.13)$$

The factors s_{ji} of the linear transformation, which indicates the strain produced by the application of stress, are called *elastic compliance coefficients* or simply *compliances*. Since Eq. 26.13 is the mechanical analogue of Eq. 26.3, the coefficients are also sometimes called *elastic susceptibilities*. Owing to the existence of the axial shear components in addition to the polar linear components, we have thirty-six instead of nine of these coefficients, which may be written in matrix form as

$$\begin{vmatrix} s_{11} & s_{12} & s_{13} & s_{14} & s_{15} & s_{16} \\ s_{21} & s_{22} & s_{23} & s_{24} & s_{25} & s_{26} \\ s_{31} & s_{32} & s_{33} & s_{34} & s_{35} & s_{36} \\ s_{41} & s_{42} & s_{43} & s_{44} & s_{45} & s_{46} \\ s_{51} & s_{52} & s_{53} & s_{54} & s_{55} & s_{56} \\ s_{61} & s_{62} & s_{63} & s_{64} & s_{65} & s_{66} \end{vmatrix}. \qquad (26.14)$$

Similarly, we have thirty-six *elastic constants* c_{ij} or *stiffness coefficients* forming a matrix that is the inverse of Eq. 26.14. Fortunately, only twenty-one of the coefficients are independent since it can be shown that the matrices are *symmetrical*, that is,

$$s_{ij} = s_{ji}$$

and $$c_{ij} = c_{ji}. \tag{26.15}$$

Also the matrices of the susceptibilities and permittivities (Eqs. 26.5 and 26.6, respectively) are symmetrical,

$$\chi_{ij} = \chi_{ji}, \qquad \epsilon_{ij} = \epsilon_{ji}; \tag{26.16}$$

thus the possible maximum number of independent electrical coefficients is reduced from nine to six.

The existence of the piezoelectric effect in certain crystals implies a coupling between the mechanical and the electrical parameters. The application of a mechanical stress $\underleftrightarrow{T}$ produces not only a mechanical strain $\underleftrightarrow{S}$ (Eq. 26.13) but also an electrical polarization **P** (direct effect),

$$\mathbf{P}_i = \sum_j d_{ij} \underleftrightarrow{T_j} \quad (i = 1, 2, 3; j = 1, \ldots 6). \tag{26.17}$$

Similarly, the application of an electrical stress **E** causes not only a polarization **P** (Eq. 26.3) but also a mechanical strain $\underleftrightarrow{S}$ (converse effect),

$$\underleftrightarrow{S_j} = \sum_i d_{ij}\mathbf{E}_i \quad (i = 1, 2, 3; j = 1, \ldots 6). \tag{26.18}$$

A material exhibiting piezoelectricity is therefore to be described not only in its mechanical behavior by a matrix of elastic coefficients and in its electric behavior by a matrix of electrical coefficients, but, in addition, by a matrix of electromechanical coupling coefficients, the *piezoelectric coefficients* d_{ij} †

$$\begin{vmatrix} d_{11} & d_{12} & d_{13} & d_{14} & d_{15} & d_{16} \\ d_{21} & d_{22} & d_{23} & d_{24} & d_{25} & d_{26} \\ d_{31} & d_{32} & d_{33} & d_{34} & d_{35} & d_{36} \end{vmatrix}. \tag{26.19}$$

This matrix is again symmetrical,

$$d_{ij} = d_{ji}, \tag{26.20}$$

hence of its eighteen coefficients only fifteen can be independent.

How many coefficients actually exist and are independent of each other depends on the *macroscopic crystal symmetry*. This symmetry may be dessribed in terms of three types of *symmetry elements:* (*a*) a *center of symmetry*, which requires that all crystal faces occur in parallel pairs on opposite sides of the crystal; (*b*) a *plane of symmetry*, which makes a crystal bilaterally symmetrical; (*c*) an n-fold *axis of symmetry* around which a crystal may be rotated by $360°/n$ to arrive in a congruent position; n may assume the values 1, 2, 3, 4, and 6, that is, there can be *diad, triad, tetrad*, and *hexad* axes.[5]

According to the symmetries of their external forms, crystals may be grouped into seven *crystal systems* (Fig. 24.9), which may be characterized in terms of axes of symmetry. Each of these systems is subdivided into a number of *crystal classes*. All classes of a system have in common the characteristic symmetry of the system, but may have additional symmetry elements in various combinations (see Fig. 24.8). There are thirty-two crystal classes in all, as mentioned previously (see Sec. 24).

The introduction of any symmetry element reflects in the matrices of the elastic, electric, and piezoelectric coefficients by reducing the number of independent coefficients and making some of the constants equal to zero. All piezoelectric coefficients disappear when a crystal has a center of symmetry. This eliminates eleven crystal classes; in addition, the coefficients become zero in class 29 because of holoaxial symmetry (432). Thus as Voigt[4] first showed, of the total thirty-two crystal classes only twenty *piezoelectric classes* remain. Whether or not a crystal belonging to one of these classes is actually piezoelectric has to be established by experiment, for instance, by the Giebe and Scheibe click method.[6]

After a piezoelectric crystal has been found, characteristic cuts of slabs or bars can be made in specific orientations to the crystal axes, as shown in Fig. 26.3 for quartz,[7] to adapt the material to special electromechanical transducer purposes. For example, various mechanical vibrational modes can be excited, such as lengthwise thickness, flexure, and shear (Fig. 26.4).[7] If a frequency standard of extreme accuracy is desired, like the famous *quartz clock* first realized by Giebe and

† Instead of the coefficients d_{ij} we can use the piezoelectric constants e_{ij}, introduced by the defining equation $\mathbf{P}_i = \sum_j e_{ij} \underleftrightarrow{S_j}$.

[5] In place of the *elementary symmetry elements a* and *b*, we usually use today a *compound symmetry element, axes of rotary inversion* (*inversion axes*), symbolized in terms of their degree as $\bar{1}, \bar{2}, \bar{3}, \bar{4}, \bar{6}$ An n-fold axis of rotary inversion combines a rotation through $360°/n$ with an inversion through the center of the crystal. Thus, as we can easily verify, $\bar{1}$ is equivalent to a center of symmetry, $\bar{2}$ to a plane of symmetry, and $\bar{6}$ to a simple threefold axis (triad) perpendicular to a plane of symmetry (see, for example, F. C. Phillips, *An Introduction to Crystallography*, Longmans, Green and Co., London, New York, 1949).

[6] E. Giebe and A. Scheibe, *Z. Physik 33*, 760 (1925).

[7] W. G. Cady, "Crystals and Electricity," *Scientific American*, December, 1949, p. 46.

Scheibe,[8] cuts are made in an orientation that assures a minimum temperature dependence of the vibrational modes. All transducer properties may be evaluated by measuring the mechanical, electric, and piezoelectric coefficients as function of temperature with due reference to the mechanical freedom of the material (*free* or *clamped*) and to its electrical status (*short-circuited* or *open-circuited*).

By the macroscopic theory certain crystal classes can be singled out as permitting piezoelectricity, and

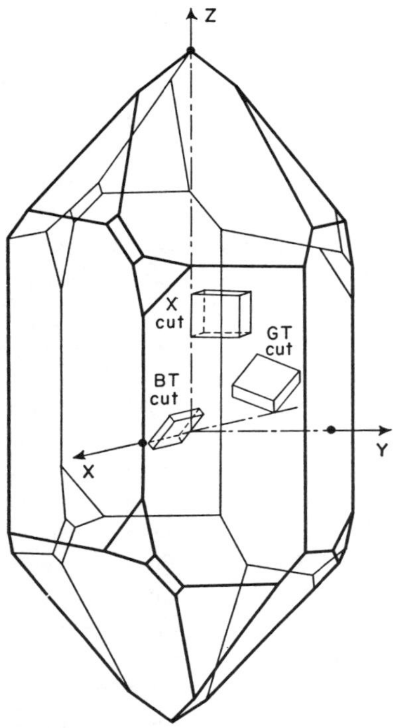

Fig. 26.3. Some characteristic cuts of quartz crystal. (After Cady.[7])

coefficients and methods of measurement can be prescribed which determine the behavior of a crystal transducer.[9] But which crystals are actually piezoelectric, a macroscopic theory cannot predict. Thus the hunt is on for better piezoelectrics as well as for a copious supply, because piezoelectric devices have become of outstanding technical importance. Especially well known are the applications of piezoelectrics as transducers in the sound and ultrasonic field initiated by Langevin[10] with his sonar experiments in World War I, and as frequency-control elements introduced by Cady[11] shortly afterwards.

The empirical search for piezoelectric crystals might be much more effective if guided by a molecular theory of piezoelectricity. In the preceding discussion we have confined ourselves to the external geometry of crystals. The thirty-two crystal classes resulted from symmetry operations which referred to the origin of the crystallographic axes. The center of symmetry was located in this origin, and the planes and axes of symmetry passed through this point. The thirty-two

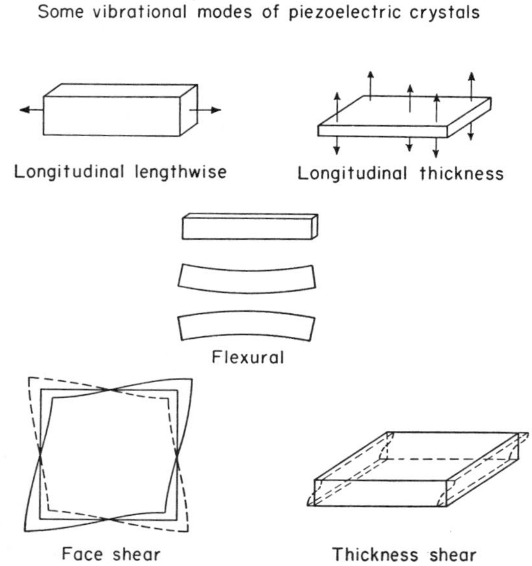

Fig. 26.4. Typical mechanical vibration modes. (After Cady.[7])

groups of macroscopic crystal symmetry are therefore also designated as *point groups*. For the formulation of a molecular theory we have to shift from this external contemplation of the habit of crystals to a consideration of their internal constitution.

Crystallography began this molecular approach in 1784 with the work of the abbé Haüy,[12] who formulated the idea that a crystal is an orderly assemblage obtained by the regular repetition of some unit of pattern. Haüy visualized this unit as a solid geometrical building block, for example, a parallelepiped. If, on the other hand, the unit is represented quite unspecifically by a point, its center of gravity, *space lattices* may be constructed by the parallel translation of such point in the x, y, and z directions over repetitive periodic distances. Thus a skeleton is formed of the internal

[8] E. Giebe and A. Scheibe, *Elekt. Nachr.-Technik 5*, 65 (1928); A. Scheibe, *Piezoelektrizität des Quarzes*, Theodor Steinkopf, Dresden and Leipzig, 1938.

[9] For detailed discussion of the macroscopic aspects of piezoelectricity and of the applications of piezoelectric materials, see the books of W. G. Cady, *Piezoelectricity*, McGraw-Hill Book Co., New York, 1946, p. 287, and W. P. Mason, *Piezoelectric Crystals and Their Applications to Ultrasonics*, D. Van Nostrand, New York, 1950.

[10] P. Langevin, French Patent 505,703 (appl. Sept. 17, 1918).

[11] W. G. Cady, U. S. Patent 1,450,246 (appl. Jan. 28, 1920).

[12] R. J. Haüy, *Traité de minéralogie*, Paris (1801).

structure of a crystal like the steel skeleton of a skyscraper. The condition is that each point finds itself in an identical environment. Bravais[13] found that fourteen different space lattices can be constructed on the basis of this requirement; Fig. 26.5 shows the *unit cells* of these *Bravais lattices*.[14]

These fourteen space lattices designate crystals of the highest possible symmetry (*holosymmetry*) in their respective crystal systems because the lattice points are

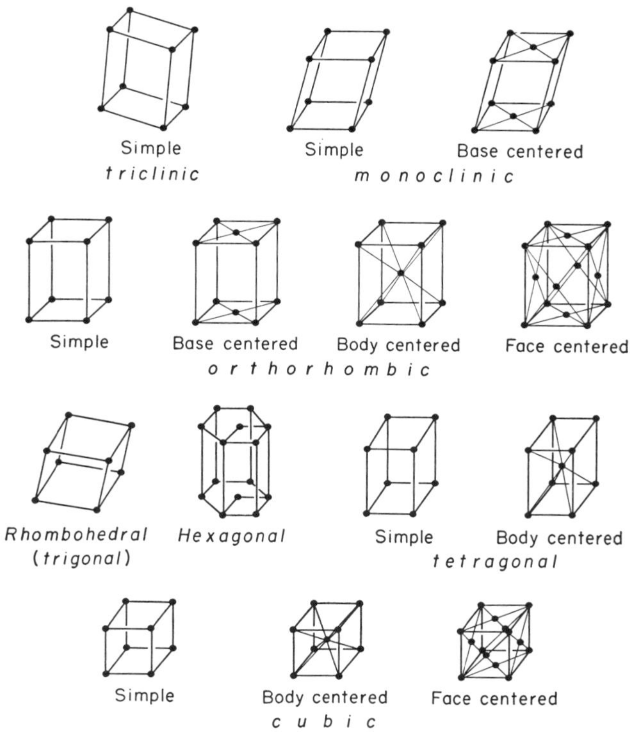

Fig. 26.5. The fourteen Bravais lattices. (After Barrett.[14])

visualized without any structure of their own. Bravais already realized that the lower symmetry exhibited by many crystals must result from the constitution of the molecular groups signified by the lattice points. The actual structure of such groups can be of a bewildering variety. However, for the construction of crystal lattices the *symmetry* of the arrangement of these molecular building stones around a lattice point counts and not the structural details.

Sohncke[15] realized that the axes of rotation introduced by morphological crystallography might represent, from the standpoint of molecular composition, a simple axis or a *screw axis* that combines with the rotation a translation of the molecular unit parallel to the axis (Fig. 26.6). He arrived thus at 65 point systems which satisfied the requirement of identical environment for each lattice point if screw axes are admitted.

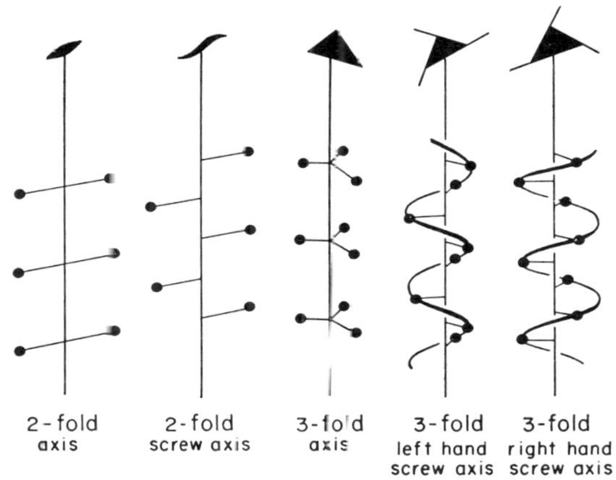

Fig. 26.6. Axes and screw axes. (Adapted from Phillips.[5])

Fedorow,[16] Schoenflies,[17] and Barlow,[18] by independent discovery, completed the theory of molecular crystal symmetry. They conceived that the symmetry planes of the macroscopic crystal geometry might signify, from the molecular standpoint, either a simple mirror reflection or a reflection combined with a translation of the building stones parallel to the mirror plane (Fig. 26.7) This introduction of *glide reflection planes*

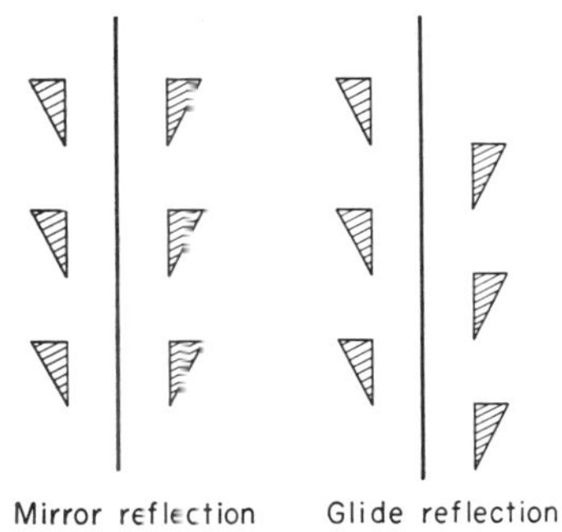

Fig. 26.7. Reflection and glide reflection planes.

in addition to screw axes expanded the possible number of point systems to 230.

The 230 different combinations of symmetry elements from which the space structure of crystals can be constructed, are called the *space groups* of crystallography. Von Laue's discovery of X-ray diffraction in 1912[19]

[13] A. Bravais, *J. Math. 14*, 141 (1849).

[14] C. G. Barrett, *Structure of Metals*, McGraw-Hill Book Co., 1952.

[15] L. Sohncke, *Entwicklung einer Theorie der Kristallstruktur*, Teubner, Leipzig, 1879.

[16] E. v. Fedorow, *Z. Krist. 38*, 321 (1904).

[17] A. M. Schoenflies, *Kristallsysteme und Kristallstruktur*, Teubner, Leipzig, 1891.

[18] W. Barlow, *Mineral. Mag. 11*, 119 (1895).

[19] M. von Laue, W. Friedrich, and P. Knipping, *Münchener Sitzungsberichte*, 1912, p. 303; *Ann. Physik 41*, 971 (1913).

opened the way for an experimental allocation of crystals to space groups by the analysis of their diffraction patterns.

The molecular theory of crystal symmetry, based on the arrangement of point *masses*, cannot provide a direct answer to our question: Which crystals are piezoelectric? What is required is a molecular re-interpretation of crystal symmetry on the basis of point *charges* and an analysis of the *charge* arrangements, which may lead to piezoelectricity. As a final step these charge patterns have to be identified with actual molecular groups occurring in crystal chemistry.

27 · Dipole Moments, Piezoelectricity and Crystal Structure[1]

An electric field polarizes a material by inducing new dipole moments or by changing the magnitude and orientation of pre-existing permanent moments. Both effects alter the mechanical dimensions of a solid by moving electrons and nuclei to new equilibrium positions. Mechanical stress changes also the dimensions of a material, but it acts on the mass points without discerning their charges. Therefore no dipole moments can be induced from the neutral state by such stress, nor will a dipole polarization arise by the mechanical distortion of a centrosymmetrical array of permanent moments. An electric field removes a center of charge symmetry by creating a polar axis; stress creates a bipolar axis. Hence, whereas polarization in such cases produces mechanical distortion, mechanical distortion produces no dipole polarization. This electromechanical effect, which has no inverse and is always present in dielectrics, is called *electrostriction.*

Piezoelectricity, in contrast, is characterized by a one-to-one correspondence of direct and inverse effect (see Sec. 26). Hence, it must be caused by pre-existing permanent dipole moments arranged without a center of symmetry. The absence of a center of charge symmetry does not imply that piezoelectric crystals must have a polar axis. An axis will exist only when the individual momentum vectors in the unit cell, instead of mutually canceling, add up to a permanent resultant moment. Since the magnitude of such resultant moment depends on the separation distance of the particles, that is, on the lattice dimensions, uniform heating or cooling of a permanently polarized crystal will change its inherent polarization. An outside observer, in consequence, will notice that a temperature change produces a voltage across the crystal, opposite in sign for heating and cooling. This effect is known as *pyroelectricity* (from the Greek πυρός = fire). Crystals that have a polar axis are pyroelectric in addition to being piezoelectric.

Macroscopic crystal symmetry prescribes that a polar axis can develop only in an acentric crystal that has neither a reflection plane nor a twofold axis perpendicular to the prospective direction. This restriction imparts to only 10 of the 20 piezoelectric classes the potentiality for pyroelectricity, as Voigt[2] pointed out. To arrive at more detailed conclusions we have to investigate the role played by permanent electric dipole moments in the formation of crystal structures.

In Sec. 26 we applied molecular symmetry considerations to the geometrical arrangement of the mass points in a unit cell; by admitting screw axes and glide reflection planes, 230 possible space groups resulted. This breaking down of the 32 classes of macroscopic symmetry (point groups) into 230 space groups does not help offhand to pin-point piezoelectric crystals more accurately. The molecular interpretation of crystal symmetry is concerned with the constellations of point masses without discerning their charges. By identifying the mass points as positive and negative ions, Born,[3] in his famous dynamics of crystal lattices, arrived at an electrical structure of crystals and could formulate the piezoelectric response in terms of the displacement of the individual ions under the action of mechanical stress and electric fields.[4] For the diatomic ionic lattice of the regular system, characterized by three elastic (c_{11}, c_{12}, c_{44}), one piezoelectric (e_{14}), and one dielectric constant (ϵ), an interrelating equation resulted

$$e_{14}{}^2 = \frac{\epsilon - \epsilon_0}{4\pi}(c_{12} - c_{44})\frac{c_{44}}{c_{12}}. \tag{27.1}$$

Checked on the sphalerite structure of ZnS, this prediction of the piezoelectric coefficient from elastic and dielectric constants proved a disappointment; the theo-

[1] See A. von Hippel, *Z. Physik 133,* 158 (1952).

[2] W. Voigt, *Lehrbuch der Kristallphysik,* Teubner, Leipzig and Berlin, 1910.

[3] See M. Born and M. Göppert-Mayer, "Dynamische Gittertheorie der Kristalle," *Handbuch der Physik,* J. Springer, Berlin, Vol. 24, Part 2, 1933.

[4] M. Born, *Physik. Z. 19,* 539 (1918); M. Born and E. Bormann, *Ann. Physik 62,* 218 (1920).

retical value was about ten times the measured one. Born suspected correctly that zinc blende was not completely ionic. Heckmann[5] repeated the calculation with inclusion of the polarizability of the ions, but the numerical agreement did not improve. The discrepancy obviously can be removed by postulating for ZnS a proper mixture of covalent and ionic binding, but this procedure is not very illuminating from the standpoint of obtaining a deeper insight as to why certain crystals are piezoelectric. The ionic point-lattice theory suffers quite generally from the fact that it arrives at its results by a cumbersome summation of the interactions of individual displacements and that a graphical picture of any concerted action of lattice points is lost by this pulverizing process.

True piezoelectricity, as stated above, is based on the existence of permanent dipole moments. A feasible alternative procedure is therefore to picture piezoelectric crystals as networks of permanent moments and to investigate the possible arrangements and reactions of such moments. Obviously, the constellations have to be compatible with crystal symmetry; hence, as a first step, we consider the dipole configurations resulting from elementary symmetry operations (Fig. 27.1).

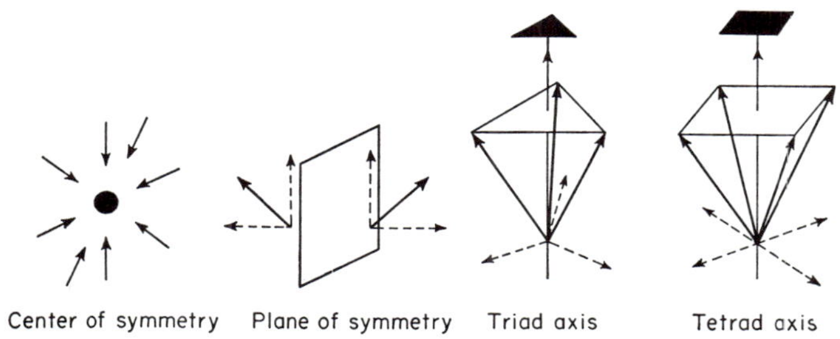

Fig. 27.1. Dipole configurations resulting from elementary symmetry operations.

Dipole moments facing a center of symmetry are converted into antiparallel pairs; components normal to a plane of symmetry or to a diad, tetrad, or hexad axis become also antiparallel pairs. The condition that piezoelectric crystals cannot have a center of symmetry, may be generalized into the statement that antiparallel pairs do not contribute to the first-order effects of piezoelectricity. Dipole components parallel to a symmetry plane become parallel pairs; components parallel to an n-fold axis of rotation are converted into parallel groups of n dipoles. Such components are piezoelectrically active and may produce a polar axis. A unique role is played by the triad axis due to the property of converting a normal component, not into antiparallel pairs, but into three dipole moments oriented at 120°. This constellation leads to piezoelectricity because pressure parallel to one moment of such star pattern will spread the angle between the other two moments and thus unbalance the arrangement.

Following up this approach, we can characterize the 20 piezoelectric classes by dipole configurations according to their symmetry elements (Fig. 27.2). These configurations do not represent the actual arrangements of the moments in the unit cell but their net constellation required by crystal symmetry. It becomes graphically clear how a polar axis arises either by a one-sided arrangement of dipole moments or the unbalance between geometrically similar but electrically unlike sets of moments. The arrow direction of the polar axis in piezoelectric crystals is prescribed by the asymmetry in the grouping of the moments and cannot be reversed.

The dipole symbols of Fig. 27.2 allow us to derive the matrices of the piezoelectric coefficients by visual inspection. The general matrix

$$\left.\begin{matrix} d_{11} & d_{12} & d_{13} & d_{14} & d_{15} & d_{16} \\ d_{21} & d_{22} & d_{23} & d_{24} & d_{25} & d_{26} \\ d_{31} & d_{32} & d_{33} & d_{34} & d_{35} & d_{36} \end{matrix}\right| \qquad (27.2)$$

symbolizes the electric polarization **P** produced in the three orthogonal directions 1, 2, 3, by the application of a mechanical stress $\overleftrightarrow{T}$, that is,

$$\mathbf{P}_i = \sum_j d_{ij} \overleftrightarrow{T}_j \qquad (i = 1, 2, 3; j = 1, \ldots, 6). \qquad (27.3)$$

Considering, for example, the monoclinic crystal class III with its two equal dipole moments oriented symmetrically to the diad axis 3 upwards from the 1,2-plane (Fig. 27.3), we observe that the 1,2-plane projection contains an antiparallel pair which cannot contribute; hence $d_{11} = d_{12} = d_{21} = d_{22} = 0$. However, application of a stress T_1 or T_2 stretches the circle in the 1,2-plane into an ellipse, the momentum vectors tip symmetrically to the polar axis; hence the components parallel to this axis and thus the moment in direction 3 changes (d_{31} and $d_{32} \neq 0$). A stress T_3 parallel to the polar axis changes also the angle of the vector symmetrically hence alters this moment ($d_{33} \neq 0$); but the components in the 1,2-plane remain an antiparallel pair ($d_{13} = d_{23} = 0$). A shearing stress around the axes 1 or 2 (T_4 or T_5) tips both vectors towards the 1 and 2 axis in the same direction; hence d_{14}, d_{15}, d_{24}, $d_{25} \neq 0$. This tilting shortens the one and lengthens the other vector in the direction of the polar axis by equal amounts; hence $d_{34} = d_{35} = 0$. Finally, a shear around the polar axis 3 tips the two vectors in opposite directions; hence again the antiparallel components balance ($d_{16} = d_{26} = 0$), while the parallel components

[5] G. Heckmann, *Z. Physik* *23*, 47 (1924); *33*, 646 (1925).

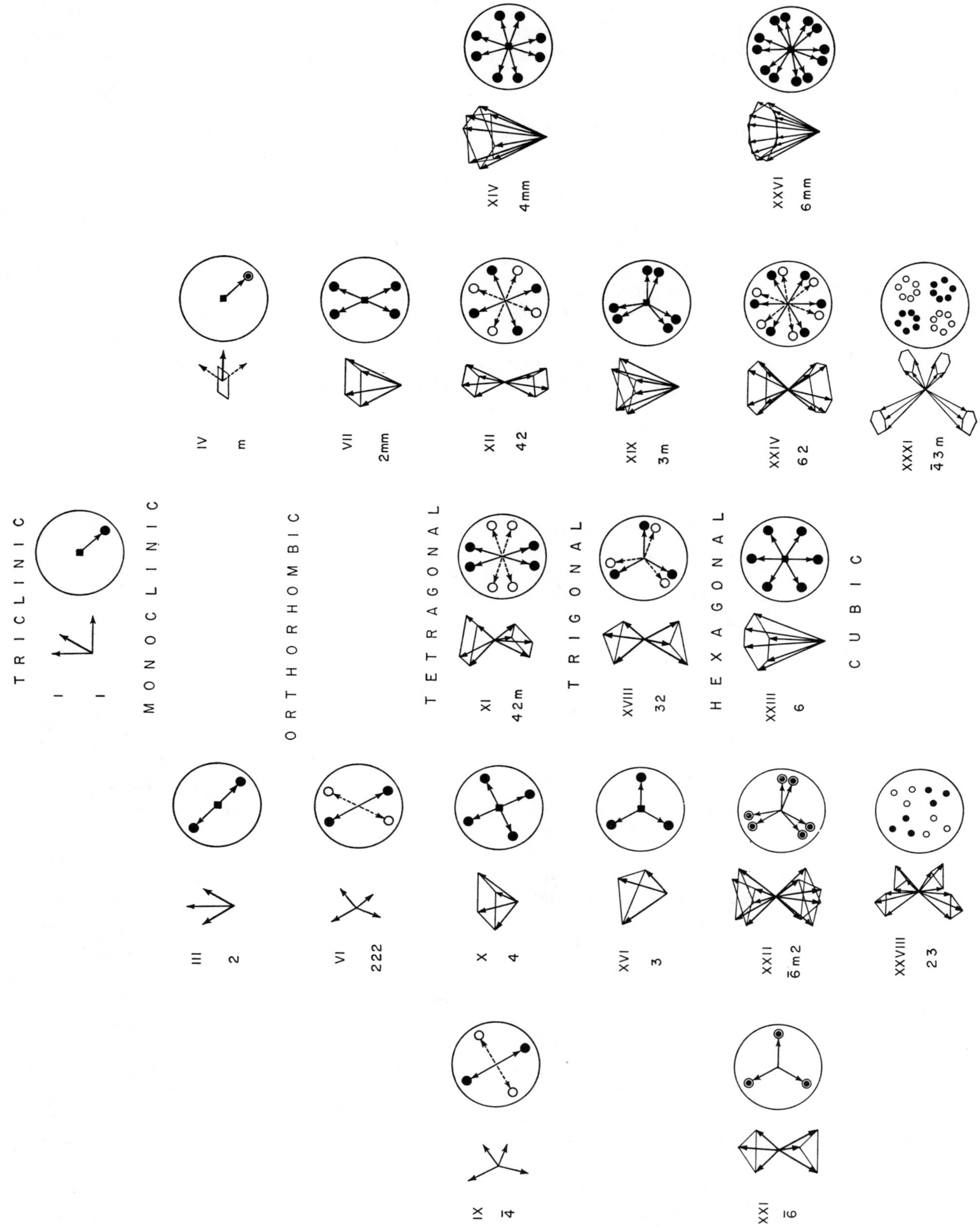

Fig. 27.2. Dipole configurations characterizing the 20 piezoelectric classes.

lengthen or shorten ($d_{36} \neq 0$). Thus we arrive at the matrix of class III:

$$\begin{vmatrix} 0 & 0 & 0 & d_{14} & d_{15} & 0 \\ 0 & 0 & 0 & d_{24} & d_{25} & 0 \\ d_{31} & d_{32} & d_{33} & 0 & 0 & d_{36} \end{vmatrix} . \qquad (27.4)$$

The actual planar or spatial arrangement of dipoles can be discussed in simple terms only when the ions are represented as spheres that assume neutralizing configurations of densest packing compatible with their radius ratio. Since only diad, triad, tetrad, and hexad axes of symmetry exist, an ion can be surrounded in a plane by 2, 3, 4, or 6 counterions, or in space by 3, 4, 6, 8, or 12 nearest neighbors of opposite polarity, as the co-ordination number indicates (see Table 24.1).

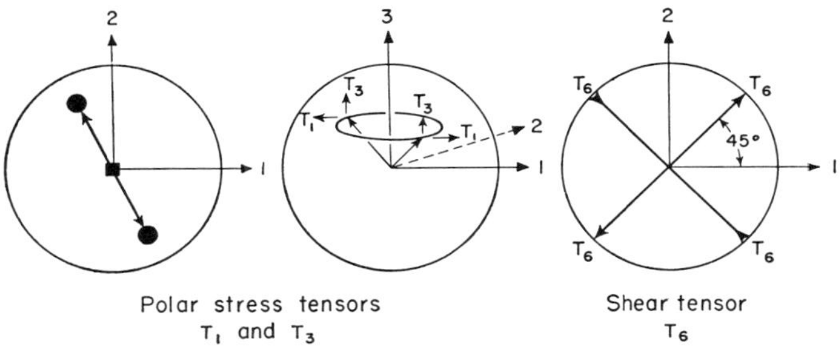

Fig. 27.3. Derivation of the piezoelectric matrix for class III.

Of these ionic groups of perfect geometrical regularity, only those with a triad axis produce dipole moments in an oblique mutual orientation. They correspond to a tetrahedral arrangement of the ions in space. Hence we may expect that tetrahedra are decisive piezoelectric building stones in crystals constructed of regular polyhedra. The tetrahedron is the only regular polyhedron involving nearest neighbors that has no center of symmetry. Hence, we would have come to the same conclusion by demanding that a piezoelectric crystal must not only be acentric in its macroscopic structure but also in the molecular structure of the groups responsible for piezoelectricity.

Let us illustrate the situation on Born's classical example of zinc sulfide. The material exists in two polymorphic forms as cubic zinc blende (*sphalerite*) and hexagonal *wurtzite*. Compared in their usual reference systems, the diamond structure of sphalerite looks very different from the crystal structure of wurtzite,[6] but by orienting the space diagonal of the former parallel to the *c*-axis of the latter, the correlation between the two arrangements becomes evident (Fig. 27.4). Both structures can be built up from zinc and sulfur atoms bonding mutually with tetrahedral *s,p*-hybrid bonds of partly ionic character. Thus each atom type can be characterized by a tetrahedral configuration of dipole moments (Fig. 27.5). In joining an

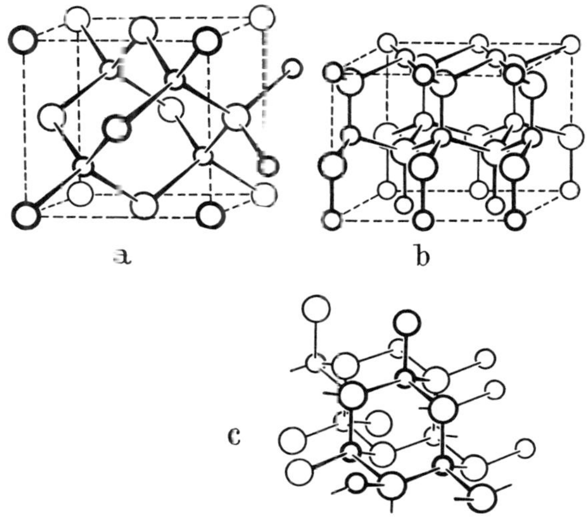

Fig. 27.4. Crystal structure of sphalerite (*a*), of wurtzite (*b*) (after Wells [6]), and of sphalerite (*c*) with [111] direction vertical.

atom of zinc and sulfur we face the same choice as in the formation of an ethane molecule from two methane radicals; we may arrange the dipole moments of the two partners in either an eclipsed or a staggered configuration. Thus the two structures of ZnS result, formed by diatomic ZnS molecules, orientated with their axis parallel to the *c*-direction and polymerized either in the staggered (sphalerite) or eclipsed (wurtzite) orientation of their moments.

It can be easily seen that both configurations will be piezoelectric. Pressure in the *c*-direction compresses

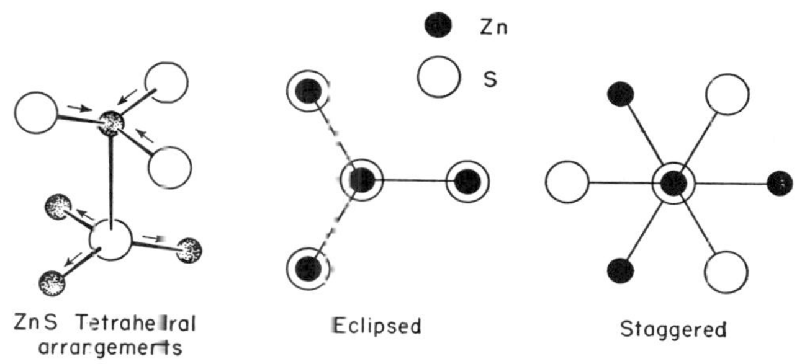

Fig. 27.5. Possible orientations of ZnS in tetrahedral bonding.

the originating diatomic ZnS molecules in their valence direction while applying to the other three bonds a bending stress which flattens their angle with the *c*-axis. This reduces the parallel components of these bonds which balanced the S $\rightarrow$ Zn moment in the *c*-direction; thus a net polarization in this direction appears.

[6] A. F. Wells, *Structural Inorganic Chemistry*, Oxford University Press, 1945, p. 83.

Although both forms are piezoelectric, only wurtzite is pyroelectric. This again is understandable on the basis of Figs. 27.4 and 27.5; the eclipsed orientation of the dipole moments in wurtzite places the zinc and sulfur atoms of the side groups above each other in the *c*-direction. The resulting electrostatic attraction acts like a compression, and a permanent polar *c*-axis is created.

Additional evidence for the usefulness of the dipole moment approach to the analysis of these structures can be found in the sequence of zinc compounds: ZnO → ZnS → ZnSe → ZnTe. X-ray analysis shows that ZnO forms only the wurtzite structure, ZnS and ZnSe are known in both modifications, whereas ZnTe occurs only in the sphalerite structure. This trend is easily understandable if we return once more to the diatomic ZnX molecule of Fig. 27.5 and its tetrahedral array of dipole moments. We know from organic chemistry that for ethane and similar molecules the staggered position is the stable one because the CH_3 groups in the eclipsed position are in steric contact and thus repel each other. In our case the contending groups are ZnX_3 and XZn_3 of opposite polarity; thus electrostatic attraction favors the eclipsed, and steric hindrance favors the staggered position. The polarity decreases and the size increases from oxygen to tellurium. Hence ZnO, the most polar of these compounds, assumes the eclipsed position and forms the wurtzite structure only; the attraction between the opposing groups is strong enough even to shorten the Zn-O distance in the *c*-direction by about 5 percent. In ZnTe, at the other extreme, the steric hindrance dominates so completely that only the staggered configuration of the sphalerite structure can be realized.

28 · Ferroelectricity

Permanent electric dipole moments, as the preceding discussion showed, are constructional elements in the formation of crystal structures and are therefore, in general, firmly anchored in place. There may exist several possible arrangements, as discussed for ZnS; and depending on pressure and temperature, the one or the other proves to be the more stable one. If the transformation can take place by some change in parameters without a fundamental rebuilding of the structure from crystal nuclei, we observe at some critical temperature or pressure a sudden *phase transition of the first kind* (also called *of first order*) in which energy, volume, and crystal structure change discontinuously.

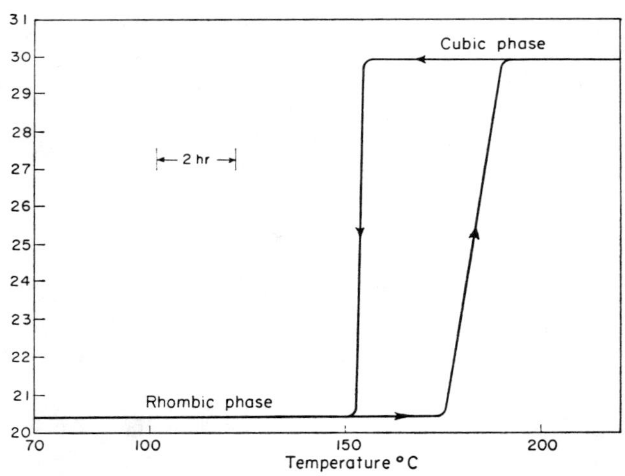

Fig. 28.1. Thermal hysteresis in first-order transition of thallous iodide (determined by dielectric constant measurement).

Actually, as dielectric analysis testifies, these transitions are frequently not as sudden as thermodynamics leads us to expect. Figure 28.1, for example, shows the transition of TlI from cubic to rhombic near 150°C, as observed by Smakula and Westphal of the Laboratory for Insulation Research. Approached from the cubic phase, the transition takes place at a lower temperature than from the rhombic phase. Similar cases of thermal hysteresis have been observed by Smyth and co-workers for organic and inorganic materials containing permanent dipoles.[1] Such hysteresis arises when the transition between the two states is delayed by high activation energies; in consequence, a sharp transition temperature will be observed only when the temperature is lowered at an excessively slow rate.

In contrast to first-order transformations, in which permanent dipoles may assume new orientations as in the sphalerite-wurtzite transition but stay "frozen in," the Debye theory of polar molecules based on Mosotti's local field foresees the occurrence of phase transitions by the spontaneous alignment of "free" electric dipoles at a critical Curie temperature (Mosotti catastrophe, Sec. 23). In this case, energy, volume, and structure might change more or less continuously, but the tem-

[1] See, for example, R. W. Crowe and C. P. Smyth, *J. Am. Chem. Soc. 72*, 1098 (1950).

perature derivatives of these quantities would have singularities. Such continuous phase transitions have been named by Ehrenfest [2] *transitions of the second kind.* Since at the Curie temperature the first derivative of the energy, the specific heat, jumps and describes a typical λ-shaped characteristic, this transition is *of the second order*, and the Curie point is also called a λ-point.

Ferroelectricity, the spontaneous alignment of electric dipoles by mutual interaction, was not observed until recently, and few materials are, as yet, known to be true ferroelectrics. Their outstanding representatives are rochelle salt, the tetrahydrate of potassium-sodium tartrate, recognized as a ferroelectric by Valasek [3] in 1921; potassium dihydrogen phosphate and arsenate by Busch and Scherrer [4] in 1935; and barium titanate, noticed for its unusual dielectric properties by Wainer and Salomon [5] in 1942–1943 and established as a new ferroelectric material [6] at the Massachusetts Institute of Technology in 1943–1944.[7]

Additional ferroelectrics, related to the titanates, have been found recently by Matthias.[8] They are $NaCbO_3$, $KCbO_3$, $NaTaO_3$ and $KTaO_3$ of the perovskite structure; WO_3, in which the WO_6 octohedra correspond in their arrangement to the TiO_6 octohedra; and $LiTaO_3$ and $LiCbO_3$ of the ilmenite structure.

The crystals of the first and second group are of relatively complicated structure; the ferroelectric range of rochelle salt is very narrow and that of the phosphates and arsenates limited to low temperatures (Fig. 28.2). The materials of both groups are piezoelectric above their Curie points and develop ferroelectricity in one axis direction only. Barium titanate, in contrast, crystallizes in the simple cubic perovskite structure (Fig. 28.3), hence has a center of symmetry and is not piezoelectric above its Curie point, at ca. 120°C. It may be employed as a single crystal or as a rugged ceramic material that can be formed in any shape desired. Thus this substance and its derivatives lend themselves much better to fundamental investigations and a variety of applications. Since also most of our own studies on ferroelectrics were concerned with $BaTiO_3$,[9] we shall illustrate the pertinent phenomena on this material, and then draw some general conclusions.

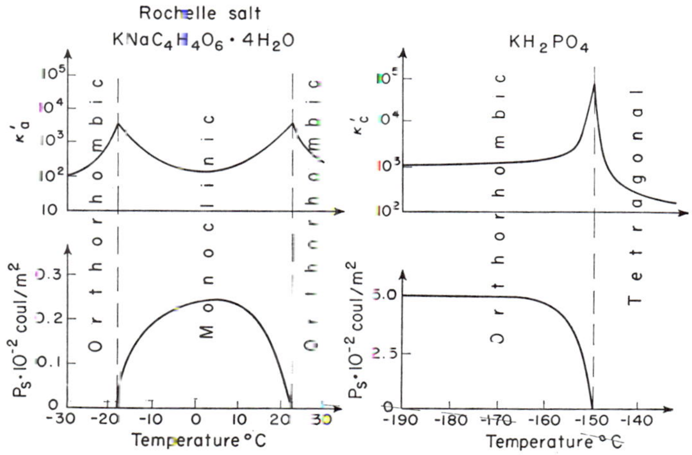

Fig. 28.2. Ferroelectric range and crystal symmetry of rochelle salt and potassium dihydrogen phosphate (initial dielectric constant and spontaneous polarization).

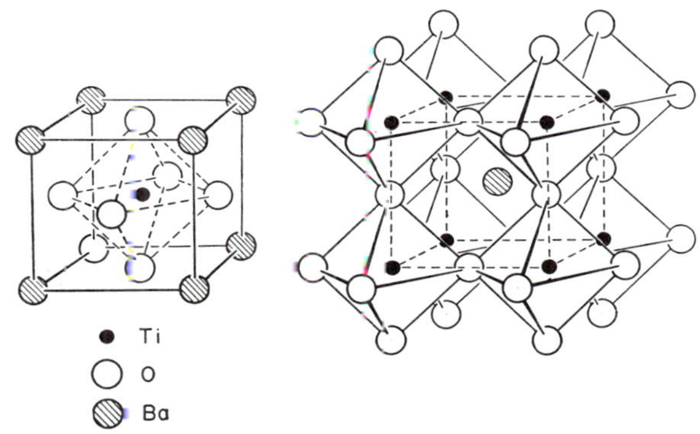

Fig. 28.3. Ideal perovskite structure.

Barium titanate ceramics

When a multicrystalline sample of $BaTiO_3$ cools down through the Curie region, a number of properties undergo rapid changes. The dielectric constant and loss tangent traverse a sharp maximum and minimum,

[2] P. Ehrenfest, *Comm. Phys. Lab. Leiden Univ. Suppl.* 75b, 1933; see also L. Tisza, in *Phase Transformations in Solids*, John Wiley and Sons, New York, 1951, pp. 1–37.

[3] J. Valasek, *Phys. Rev. 17*, 475 (1921).

[4] G. Busch and P. Scherrer, *Naturwiss. 23*, 737 (1935).

[5] E. Wainer and A. N. Salomon, *Titanium Alloy Manufacturing Co. Elec. Rep. 8* (1942); *9* and *10* (1943).

[6] A. von Hippel and co-workers, *N.D.R.C. Reps.* 300 (August, 1944) and 540 (October, 1945); A. von Hippel, R. G. Breckenridge, F. G. Chesley, and L. Tisza, *Ind. Eng. Chem. 38*, 1097 (1946).

[7] The existence of ferroelectricity in $BaTiO_3$ was initially denied by B. M. Vul and I. M. Goldman [*Compt. rend. acad. sci. U.R.S.S. 46*, 1939 (1945)] and then confirmed in the *U.S.S.R.* [B. M. Vul and F. L. Vereschagen, *Compt. rend. acad. sci. U.R.S.S. 48*, 634 (1946); B. M. Vul, *J. Phys. U.S.S.R. 10*, 95 (1946)] and in England, Holland, and Switzerland. For a more general literature survey see, for example, W. Jackson, *Proc. Inst. Elec. Engrs. (London)* Pt. 3, *97*, 285 (1950), and E. T. Jaynes, *Ferroelectricity*, Princeton University Press, 1953.

[8] B. T. Matthias, *Phys. Rev. 75*, 1771 (1949); *76*, 175, 430, 1886 (1949); J. K. Hulm, B. T. Matthias, and E. A. Long, *Phys. Rev. 79*, 885 (1950).

[9] See A. von Hippel, *Revs. Mod. Phys. 22*, 221 (1950); *Z. Physik 133*, 158 (1952).

respectively, the slope of the thermal expansion characteristic alters, and ferroelectric hysteresis loops appear (Fig. 28.4).[6] The X-ray diagram of the cubic

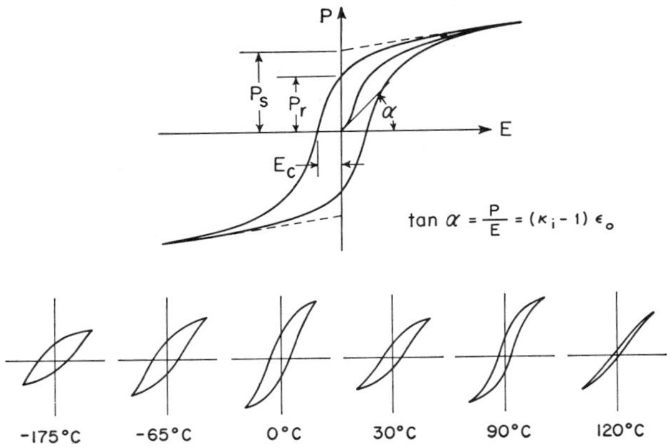

Fig. 28.4. Ferroelectric hysteresis loops.[9]

structure simultaneously becomes transformed progressively into that of a pseudo-cubic arrangement, expressed clearly in the multiplicity of the CuK_α doublet of the back-reflection lines.[6] Accurate measurements of Megaw [10] established this new phase as tetragonal (Fig. 28.5).

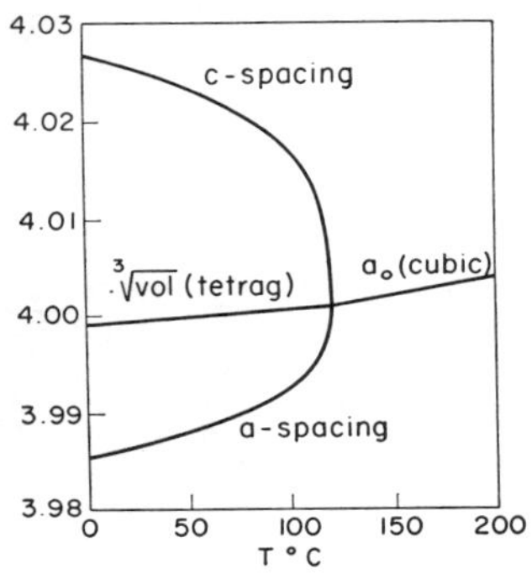

Fig. 28.5. Variation of all dimensions of barium titanate with temperature. (After Megaw.[10])

The loss curve and the thermal expansion characteristic make it apparent that some additional phase transition occurs near 0°C. When the initial permittivity derived from the hysteresis loops is plotted against temperature, two lower transition points, located near 0° and −70°C become clearly visible [6] (Fig. 28.6). They become more pronounced as the voltage increases, but the material remains ferroelectric throughout as the existence of the hysteresis loops indicates.

A successive replacement of barium by strontium ions lowers the Curie point systematically as Wainer and Salomon [5] already observed. Jackson and Reddish [11] followed this shifting down to −190°C, and Rushman and Strivens [12] established a linear dependence of the

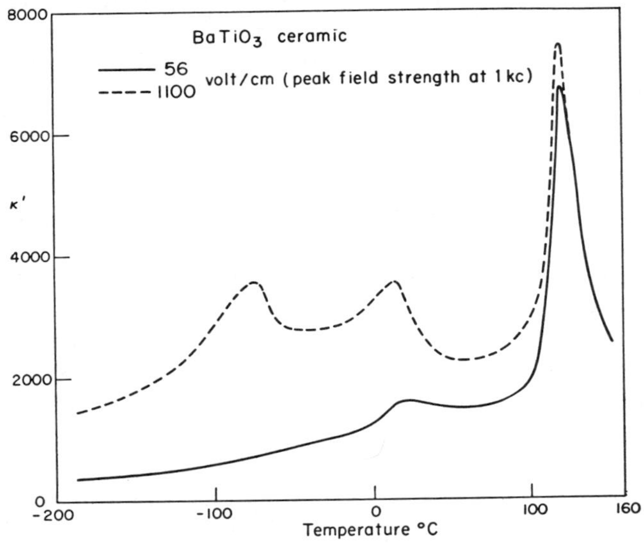

Fig. 28.6. Dielectric constant of barium titanate ceramic as function of temperature. (Measurements of W. B. Westphal, Laboratory for Insulation Research.)

Curie temperature on the lattice constant of the barium-strontium titanate mixed crystals.

If we observe the temperature dependence of the electric susceptibility above the Curie point, we find

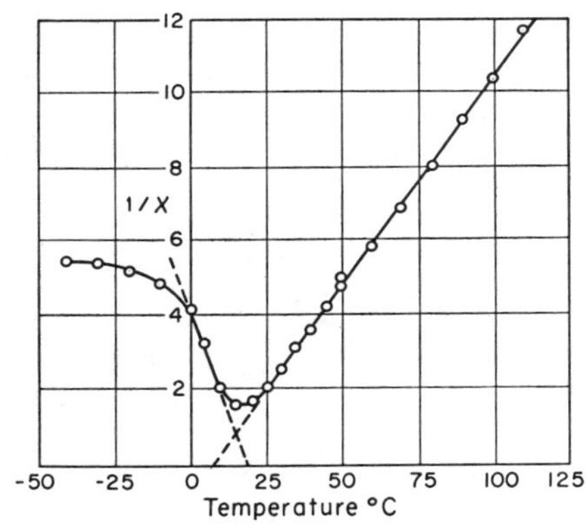

Fig. 28.7. Confirmation of Curie-Weiss law on barium-strontium titanate ceramic. (After Roberts.[13])

at low field strength the Curie-Weiss law fulfilled (Fig. 28.7), but at higher fields a strong dependence of χ on the applied field [13,14] (Fig. 28.8). The useful range

[10] H. D. Megaw, *Trans. Faraday Soc.* *42*A, 224 (1946); *Proc. Roy. Soc.* (*London*) A*189*, 261 (1947); see also R. G. Rhodes, *Acta Cryst.* *4*, 105 (1951).

[11] W. Jackson and W. Reddish, *Nature 156*, 717 (1945).

[12] D. F. Rushman and M. A. Strivens, *Trans. Faraday Soc.* *42*A, 231 (1946).

[13] S. Roberts, *Phys. Rev. 71*, 890 (1947).

[14] A consequence of this pulling-up of the Curie point by external fields is the "partial" hysteresis loops observed recently by W. J. Merz [*Phys. Rev. 90*, 375 (1953)].

of this nonlinearity of the polarization in high fields extends to about 40°C above the thermal Curie point and promises to be of special technical importance since the accompanying losses are much smaller than

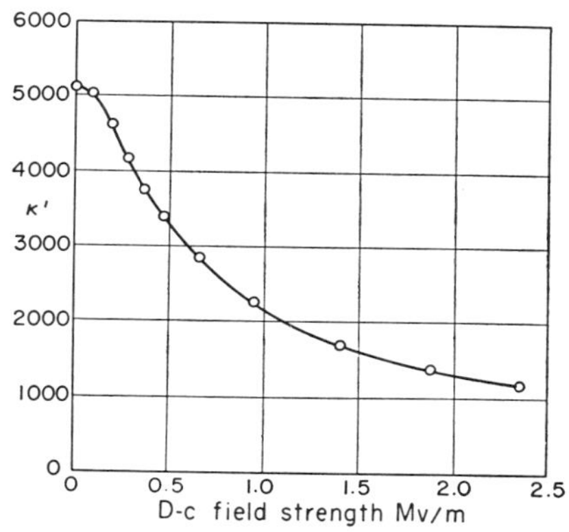

Fig. 28.8. Field-strength dependence of dielectric constant for barium-strontium titanate ceramic (above Curie point). (After Roberts.[13])

below the Curie point (Fig. 28.9). The reason for this lower loss is that in polarizing above the Curie point we do not have to overcome a pre-existing domain pattern (see below). The constant in the Curie-Weiss law is not

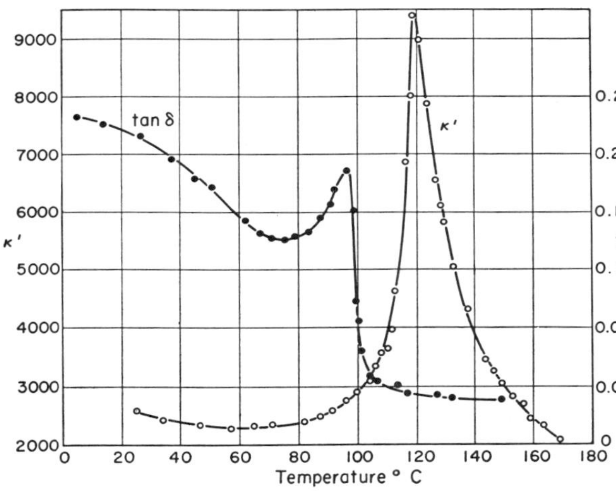

Fig. 28.9. Temperature dependence of dielectric constant and loss tangent of barium titanate ceramic. (Measurements of W. B. Westphal, Laboratory for Insulation Research.)

$3T_c$ as the simple theory predicts (see Eq. 23.7) but much larger; for the Ba-$SrTiO_3$ sample of Fig. 28.7 with $T_c = 281°K$, a value of 88,000 was obtained.

Since the field-strength dependence of the dielectric constant makes the ceramics useful as nonlinear dielectrics for a number of applications such as dielectric amplifiers, modulators, and memory devices, and for the tuning of circuits, their behavior under the action of a biasing field was investigated in the Laboratory for Insulation Research by Roberts[13] in more detail. As the measuring frequency approached the megacycle range, strong resonance effects occurred † (Fig. 28.10).

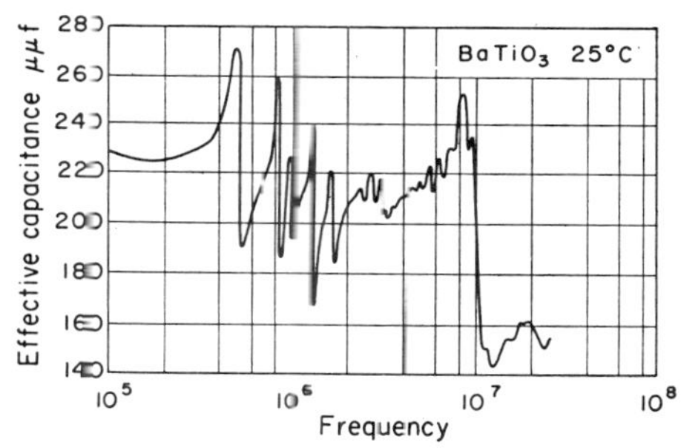

Fig. 28.10. Resonance spectrum of $BaTiO_3$ disk. (After Roberts.[13])

These resonances persisted after the biasing d-c voltage was disconnected, but could be wiped out by a bias of the opposite sign. When this resonance absorption of a disk was measured and then a piece of the sample broken off, the spectrum shifted to higher frequencies. The biasing field transformed the ceramic disk into a

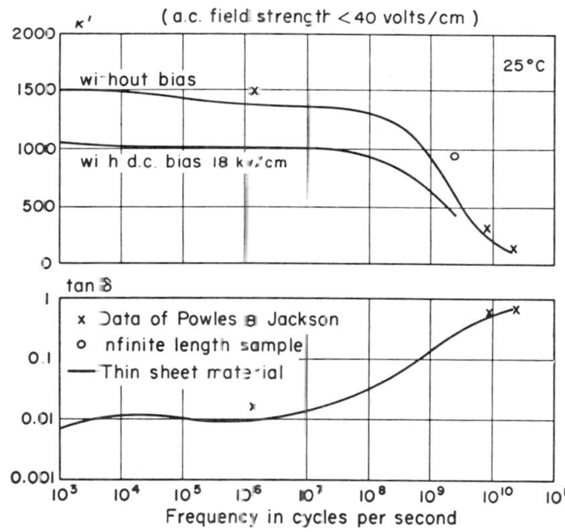

Fig. 28.11. Relaxation spectrum of the ferroelectric state in barium titanate ceramic at room temperature. (Measurements of W. B. Westphal, Laboratory for Insulation Research.)

piezoelectric resonator and its piezoelectric response persisted also without this field because of the remanence of the polarization.

† We expected in 1944 that $BaTiO_3$ might be piezoelectric below the Curie point, but the experiments with the Giebe-Scheibe click tester on powdered material, except in one experiment, gave a negative result. This was caused by the domain structure, as we realized later.

Systematic frequency response measurements have been in progress in our laboratory for some time;[15] the high dielectric constant and absorption make it very difficult to obtain accurate temperature characteristics in the microwave range. The essential phenomenon is a relaxation spectrum, as also Powles and Jackson[16] found, commencing at about 10^8 cycles and reducing the dielectric constant at about 3×10^{10} cycles to a

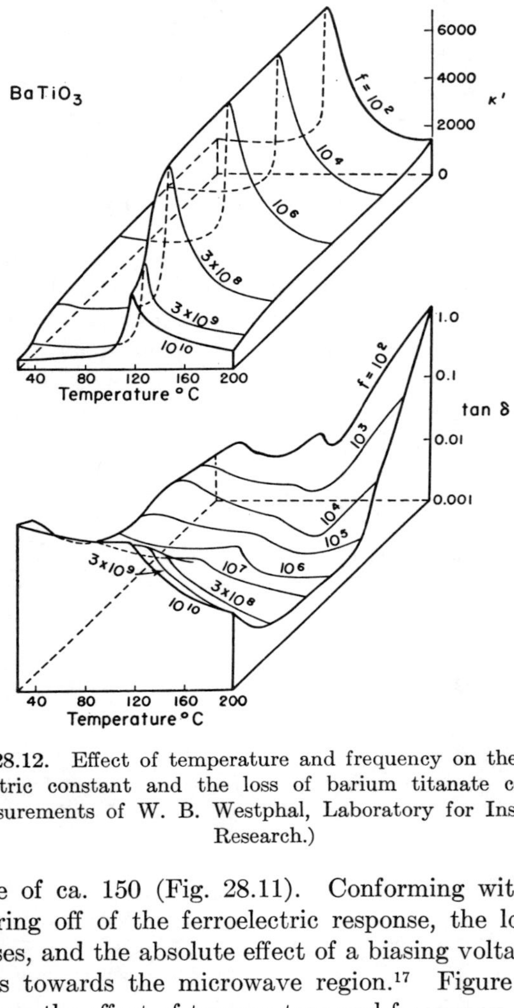

Fig. 28.12. Effect of temperature and frequency on the initial dielectric constant and the loss of barium titanate ceramic. (Measurements of W. B. Westphal, Laboratory for Insulation Research.)

value of ca. 150 (Fig. 28.11). Conforming with this tapering off of the ferroelectric response, the loss increases, and the absolute effect of a biasing voltage declines towards the microwave region.[17] Figure 28.12 surveys the effect of temperature and frequency on the initial dielectric constant and the loss of $BaTiO_3$ ceramic. The temperature dependence can be changed within wide limits by addition agents.

[15] A. von Hippel and W. B. Westphal, National Research Council Conference on Electrical Insulation, October, 1948.

[16] J. G. Powles and W. Jackson, *Proc. Inst. Elec. Engrs.* (*London*) Pt. 3, *96*, 383 (1949).

[17] Possible explanations for this frequency dependence are discussed by C. Kittel, *Phys. Rev. 83*, 458 (1951), and A. von Hippel, *Z. Physik 133*, 171 (1952); more experimental information is required.

Single crystals

We cannot expect to arrive at a real understanding of solid-state phenomena without a study of single crystals. $BaTiO_3$ crystals of hexagonal and of cubic symmetry were obtained in Switzerland by Blattner, Matthias, Merz, and Scherrer[18] from ternary melts, and this method was further improved by Matthias in our laboratory.

A microscopic inspection of the pseudo-cubic type revealed that the crystals, cooled to room temperature, contained a variety of shaded areas[19] (Fig. 28.13). In an electric field these areas were seen to grow or contract, sections of new shading appeared suddenly, and disconnecting of the voltage left a remanent state which required a countervoltage for its removal. Observed in an alternating field with stroboscopic illumination, the whole crystal appeared to be in violent agitation, and viewed in polarized light flickering birefringence colors appeared. The domain structure of the material, only indirectly deducible for the opaque ferromagnetics, was visible in all details in the transparent ferroelectrics.[19,20] Since that time, domain patterns have also been found in the other groups of ferroelectrics.[21]

A careful analysis of the patterns revealed their laws of formation in various ramifications.[19,20] As the cubic crystal cools through the Curie temperature, a polar axis develops and forces the crystal into tetragonal symmetry, with the axis forming the *c*-direction (Fig. 28.14). Since any one of the cube edges may develop into the polar axis, a twinning in the (110) planes tends to take place and laminae form, with the *c*-axis alternating its direction by 90°. Seen from the front these laminae produce a diagonal striation whereas in the top and side views striae parallel to the cube edges result. Since the optical index of refraction is smaller parallel to the polar axis than perpendicular to it ($n_c - n_a = -0.055$ for sodium light at room temperature), the orientation of the axis and the thickness of the lamellae can be measured by the birefringence colors in polarized light. A strong electric field may wipe out the domain pattern more or less completely; the crystallographer has thus to accept the unusual situation that the *c*-axis of the $BaTiO_3$ crystal can be turned around at will by the application of an electric field.

[18] H. Blattner, B. Matthias, W. Merz, and P. Scherrer, *Experientia 3*, 148 (1947).

[19] B. Matthias and A. von Hippel, *Phys. Rev. 73*, 1378 (1948); see also H. Blattner, W. Känzig, W. J. Merz, and H. Sutter, *Helv. Phys. Acta 21*, 207 (1948).

[20] P. W. Forsbergh, Jr., *Phys. Rev. 76*, 1187 (1949).

[21] See, for example, the study of T. Mitsui and J. Furuichi on rochelle salt [*Phys. Rev. 90*, 193 (1953)].

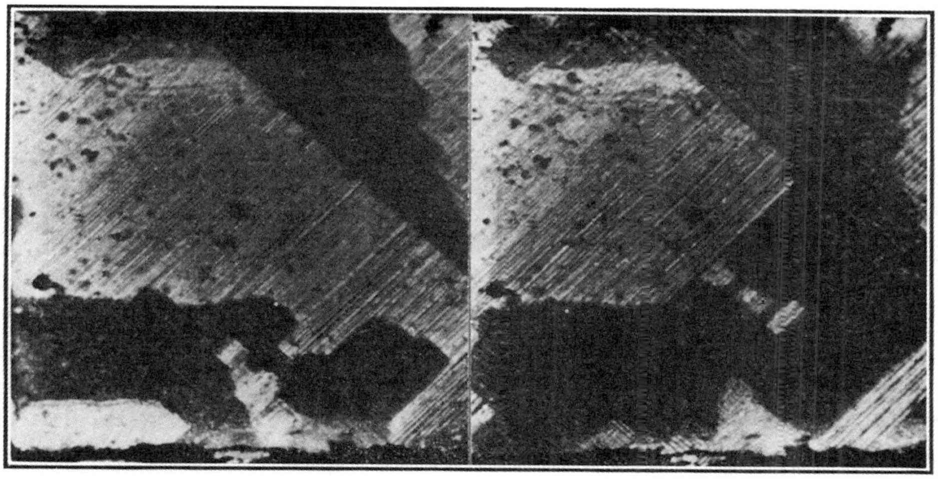

No field applied +2000 volts/cm

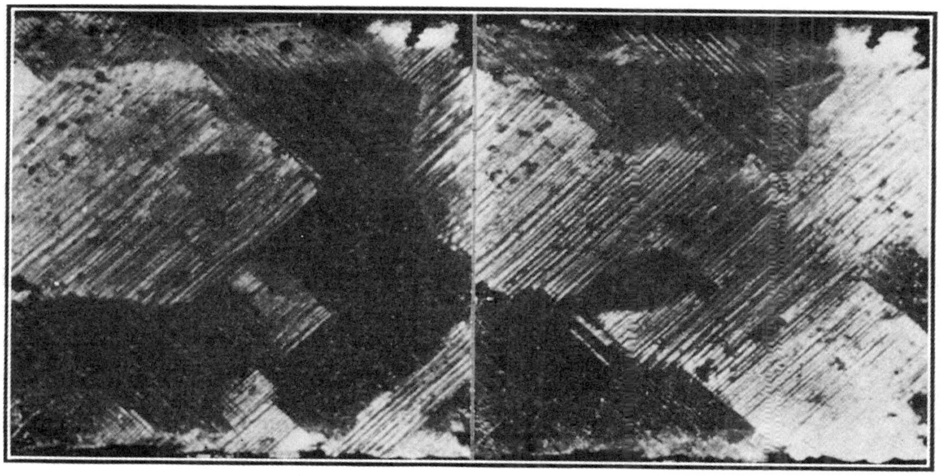

Field removed −2000 volts/cm

Fig. 28.13. Domain areas in $BaTiO_3$ crystal and effect of electric field.[19]

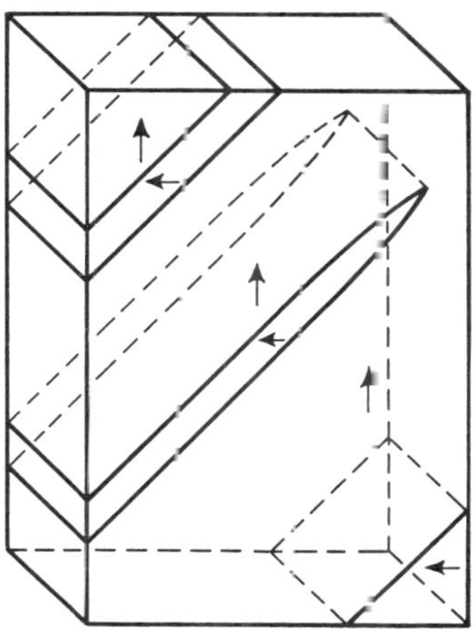

Fig. 28.14. Twinning of $BaTiO_3$ crystal.[19]

The build-up of the domains as Forsbergh[20] established in accurate detail begins with the formation of wedge-shaped laminae (Fig. 28.15). Mechanical pressure or electric fields can drive these wedges through the crystal or squeeze them out again. The final domain structure is therefore dependent on the strains originally contained in the material and on their modification by the domain of the ferroelectric state. If the first set of wedges is crossed by a second one at right angles, a system of intersecting laminae can be developed which leads finally to the beautiful square-net pattern of Fig. 28.16. Its building blocks have been identified in successive stages by birefringence measurements.

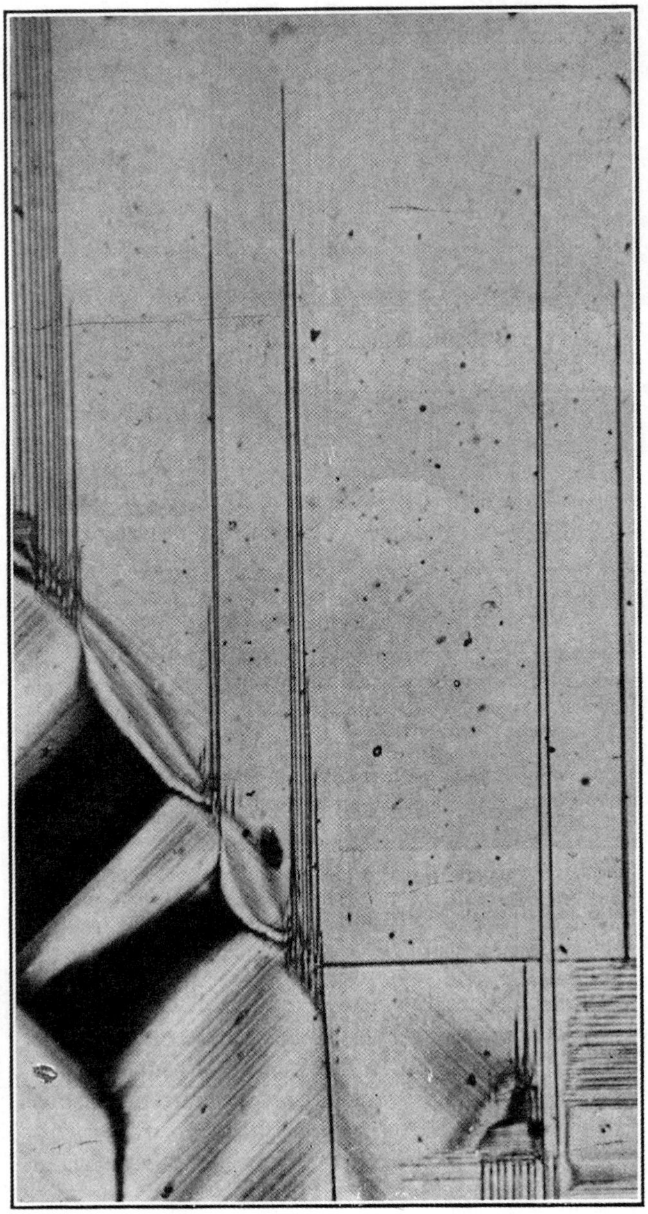

Fig. 28.15. Wedge-shaped laminar domains in $BaTiO_3$ crystal. (After Forsbergh.[20])

Fig. 28.16. Square-net domain pattern. (After Forsbergh[20]).

Above the Curie point the cubic crystal appears dark between crossed nicols; as it cools through the Curie region, the domain patterns develop as described. Further cooling through the transition regions near 0° and −70°C, produces sudden changes in the domain structure. Optical analysis established the fact that the polar axis, originally formed in the cube-edge direction, jumps abruptly, first in the face diagonal, and near −70°C into the space diagonal position (Fig. 28.17). The cubic $BaTiO_3$, changing at the Curie point into a tetragonal crystal, transforms by these two subsequent first-order transitions into an orthorhombic and finally a trigonal modification.[20] Electrical analy-

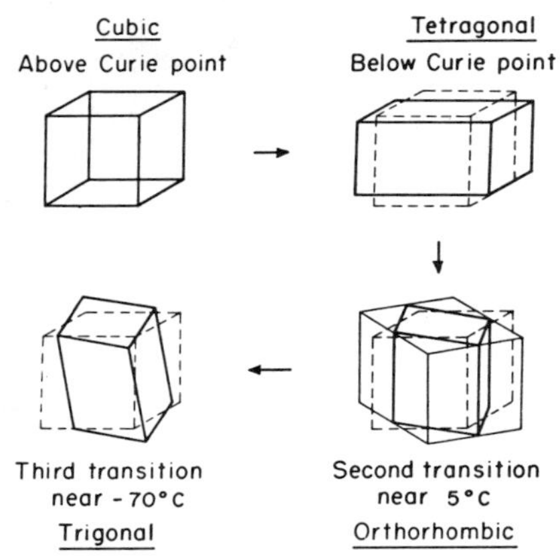

Fig. 28.17. Phase transitions of barium titanate. (After Forsbergh.[20])

sis by Merz[22] in the Laboratory for Insulation Research and a careful X-ray study carried through in England by Kay and Vousden[23] have led to the same conclusion.

A decisive factor in the formation of ferromagnetic domains is the demagnetizing field which originates

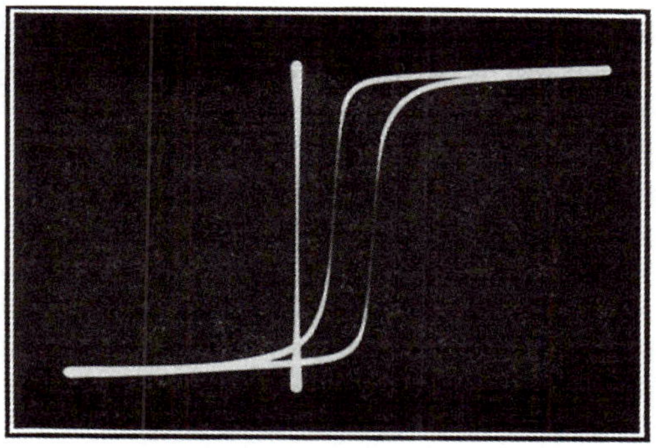

Fig. 28.18. Square hysteresis loop of barium titanate single-domain crystal. (After Merz.[22])

from the free ends of dipole chains at the boundaries of the material. By breaking up a single macroscopic block of parallel dipoles into a compensating array of micro-blocks, this field can be practically eliminated. In ferroelectrics the free surface charge of terminating dipoles may be compensated for by the attraction of countercharges through conduction. We were therefore not completely surprised to find crystals which give the optical appearance of a single domain. Since, for the theory of ferroelectricity, the properties of a single domain are decisive, these crystals were investigated in detail.[22]

Application of an alternating field produces initially distorted hysteresis loops, but the disturbances disappear after the crystal has been taken through several heating cycles with systematically increased field strength.

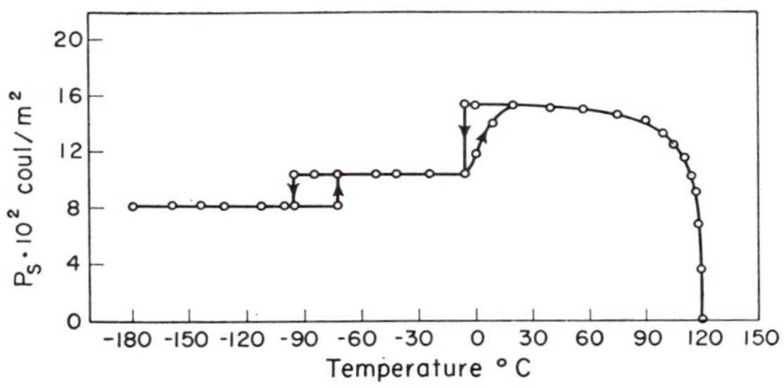

Fig. 28.19. Spontaneous polarization P_s of barium titanate crystal as function of temperature. (After Merz.[22])

It seems, therefore, that the *single domain crystals* consist initially of antiparallel domains; this has been confirmed optically in a study by Merz.[24] The final loops of these crystals are rectangular like those of properly oriented ferromagnetic crystals (Fig. 28.18). Any switching of domains is accompanied by electric pulses in analogy to the *Barkhausen noise* of ferromagnetics. Noise and a temperature hysteresis accompany the two lower transitions, where the reorientation of the polar axis creates new domain patterns. Figure 28.19 shows this hysteresis in the temperature characteristic of the spontaneous polarization.[22]

Origin of the ferroelectric state

The titanium ions of $BaTiO_3$ are surrounded by six oxygen ions in an octahedral configuration. This coordination is to be expected from the radius ratio of the partners when visualized as ionic spheres (see Tables 13.1 and 24.1); alternatively, the TiO_6 groups may be explained as resulting from covalent binding by octahedral *s,p,d*-hybrid bonds (see Sec. 17). Actually, a compromise between these two points of view is in order. The TiO_6 constellation leads to a high dielectric constant in all crystals containing it and reaches its high value through the dispersion stemming from the infrared vibrations. The temperature coefficient of the dielectric constant outside the ferroelectric range is strongly negative. We find ourselves, therefore, in a transition region between polar and nonpolar binding, where slight changes in internuclear separation produce large changes in the electric dipole moments.

The regular TiO_6 octahedron has a center of symmetry; its permanent moments cancel in antiparallel pairs. A mechanical distortion per se does not impair this mutual cancellation, as discussed above, and therefore leads neither to piezo- nor to ferroelectricity. This

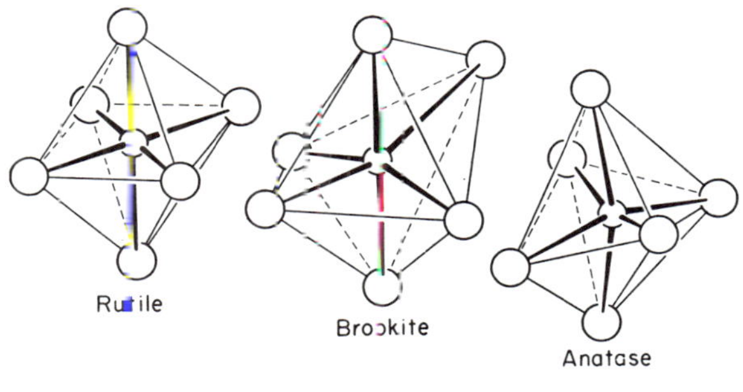

Fig. 28.20. The three modifications of TiO_2.[9]

is exemplified in the distortions of the octahedra occurring in the three TiO_2 modifications: rutile, brookite, and anatase (Fig. 28.20). A net permanent moment of the octahedron can result only by a unilateral displacement of the positive titanium ion against its negative

[22] W. J. Merz, *Phys. Rev. 76*, 1221 (1949).

[23] H. F. Kay and P. Vousden, *Phil. Mag.* [7] *40*, 1019 (1949).

[24] W. J. Merz, *Phys. Rev. 88*, 421 (1952).

oxygen surroundings; ferroelectricity, in addition, demands the proper coupling of such moments.

In rutile, brookite, and anatase, the octahedra are grouped in various compensating arrays by sharing two, three, and four edges, respectively, with their neighbors. These constellations are demanded by the composition formula TiO_2: each oxygen ion has to be coupled to three titanium ions if each titanium is surrounded by six oxygens. In $BaTiO_3$ each oxygen has to be coupled to only two titanium ions; the octahedra can be placed therefore in identical orientation, joined at their corners, and fixed in position by barium ions. Thus the stage is set in the perovskite structure for an effective additive coupling of the net moments.

Above the Curie point $BaTiO_3$ is isotropic, the titanium ions have their equilibrium position in the center of the octahedra, and the permanent dipole moments cancel mutually in antiparallel pairs. However, strong fluctuations of the net moments around the zero value occur by thermal agitation. An external field upsets, in addition, the average moment balance by displacing the titanium ions unilaterally. The octahedra and their moments are coupled by common oxygen ions, and any displacement of one titanium ion towards a specific oxygen ion unbalances strongly two of the neighboring titanium ions and more weakly the four remaining ones, tending to push all the titanium ions in the same direction. Figure 28.21 explains this situation.

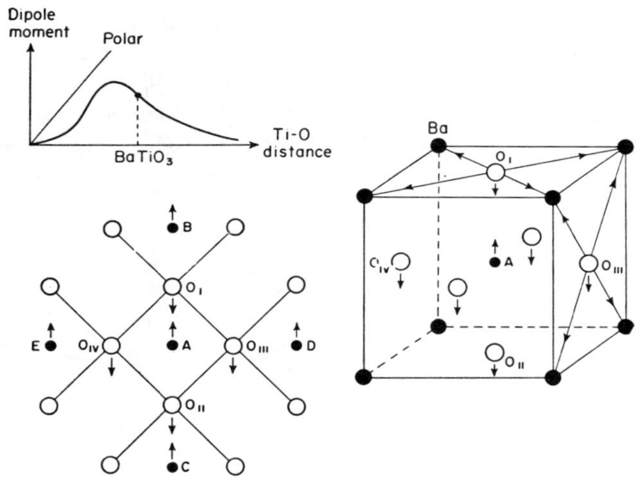

Fig. 28.21. Feedback coupling and displacement of ions in barium titanate.[9]

The dipole moment of a completely ionic compound, plotted as function of the separation distance of the partners, is represented by an ascending straight line; a nonpolar compound without dipole moment follows the abscissa. Compounds of mixed bonding, but separating into atoms, follow a characteristic traversing a maximum and then falling asymptotically to the zero line[25] (see Fig. 19.4). $BaTiO_3$ seems to correspond to this latter type and to have its equilibrium distance, according to the strongly negative temperature coefficient of the dielectric constant, on the far side of the maximum. If the titanium ion A moves towards the oxygen ion O_I, the dipole moment $O_I \rightarrow A$ becomes stronger, and the moment $O_{II} \rightarrow A$ weaker. Hence O_I moves towards A and O_{II} away from it, with the result that the titanium ions B and C follow the motion of the titanium ion A. The side titanium ions D, E, and their counterparts in front and back of the figure plane tend to follow suit because the coupling oxygen ions O_{III}, O_{IV}, etc., tend to move downwards, repelled by O_I. The whole action is akin to that of a Mosotti field. An applied field or thermal motion creates a net dipole moment in an octahedron by displacing the titanium ion against its oxygen surroundings; this displacement, in turn, produces through the oxygen coupling an increase of the displacement, that is, of the locally acting field, until at a critical temperature the thermal agitation can be overcome and the Mosotti catastrophe occurs.

The postulated displacement directions of the titanium and oxygen ions can be checked in part by X-ray analysis. Evans[26] measured at room temperature the diffraction intensities from small single crystals with refined methods and obtained two possible structures which fit the intensity patterns about equally well. The titanium ion A has moved up and the oxygen ions O_I and O_{II} down in the reference frame of the barium ions, as expected. Their thermal amplitudes around these new equilibrium positions are appreciably smaller in the vertical than in the horizontal direction. The coupling oxygen ions O_{III}, etc., either have moved down or their vertical thermal amplitudes have become very large.[27]

Thus far, the barium ions have been treated as inert spacers that permit the proper joining of the TiO_6 octahedra. Actually, the role of these divalent ions is much more important, as their influence on the Curie point indicates. By substituting strontium for barium, the

[25] See J. G. Kirkwood, *Physik. Z.* *33*, 259 (1932); A. Eucken and A. Büchner, *Z. physik. Chem.* B*27*, 321 (1934).

[26] H. T. Evans, Jr., Tech. Rep. 58, ONR Contract N5ori-07801, Laboratory for Insulation Research, Massachusetts Institute of Technology, January, 1953.

[27] The structure, in which O_{III} is not displaced, corresponds about to the results obtained by W. Känzig, *Helv. phys. Acta* *24*, 175 (1951). X-ray analysis fails to give an unambiguous final answer because the temperature motion of the oxygen ions is anisotropic and interferes with an accurate determination of the structure parameters. Neutron diffraction experiments are planned to overcome this difficulty.

Curie point can be lowered systematically from 120°C to a theoretical temperature beyond the absolute zero point. A replacement of Ba^{2+} by Pb^{2+}, on the other hand, raises the Curie point, until, for pure $PbTiO_3$, it reaches about 490°C.[28] This dependence is not explainable on the basis of ionic radii since Pb^{2+} at 1.21 A lies between Ba^{2+} (1.35) and Sr^{2+} (1.13); it becomes understandable when we draw the dipole moments from the oxygen ions to their four divalent metal-ion neighbors (Fig. 28.21). Obviously, this set of moments tends to hold the oxygen in place against the action of the titanium ions. Hence the divalent ions reduce the feedback effect that leads to the Mosotti catastrophe, the more so the higher their bond strength to oxygen. The bond distance Ba—O in barium titanate is practically that of barium oxide; the melting points of the oxides may therefore be considered as a measure of the bond strength of the divalent ions to oxygen. The melting points increase from PbO (888°C) to BaO (1933°C) and SrO (2430°C); hence the feedback action and Curie points decrease in this order. Also three-dimensional pressure reduces the displacement of the ions and with it the feedback effect and Curie temperature [29] (Fig. 28.22) just as clamping diminishes the piezoelectric response of a crystal.

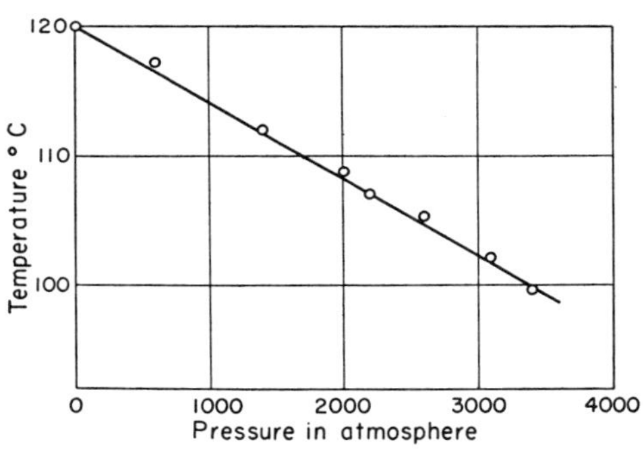

Fig. 28.22. Dependence of Curie temperature on hydrostatic pressure for barium titanate single crystal. (After Merz.[29])

In this consideration of the effect of dipole moments and ion replacement and displacement on the Curie temperature we have tacitly assumed that the cubic perovskite structure of Fig. 28.3 remains unaffected. This is not necessarily true, as the X-ray structure of $CaTiO_3$ shows, in which the neighboring TiO_6 groups are slightly tilted against each other (Fig. 28.23).[30] This may be the reason why the influence of Ca^{2+} in replacing Ba^{2+} appears to be small as far as the Curie point is concerned, while strongly affecting the lower transition point, as has been found recently.[31] Complicated effects of distortion, also of the octahedra themselves, may arise by substitution and have to be investigated in detail.

The preceding interpretation of the onset of spontaneous polarization and its dependence on structure parameters clarifies qualitatively the prerequisites for the formation of a ferroelectric state.[9] The old idea

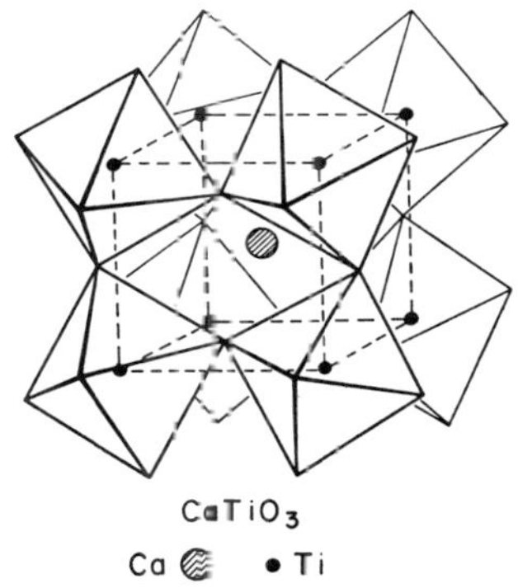

Fig. 28.23. Distorted perovskite structure of $CaTiO_3$. (After Náray-Szabó.[30])

that the rotation of permanent moments leads to a Mosotti catastrophe has to be discarded; such moments are built in and not available for free rotation. Ferroelectricity arises, not from rotation, but from vibration states; the displacement of certain ions from their equilibrium position strongly unbalances the equilibrium of the permanent moments. By a proper structural arrangement this upset induces a motion of the neighboring ions in a supporting sense; this increases the original displacement by feedback. The tendency to bring the vibrations of neighboring ions into ordered phase relations prevails at the Curie point against the random agitation, and the equilibrium position of the critical ions shifts to one side since the whole effect was made possible only by the displacement of these ions. The old balance of the permanent moments is destroyed and a polar axis created by the transformation of induced moments into additional permanent moments.

Since the ferroelectric state arises as a deviation from a previous equilibrium condition, the polar axis can develop in any one of the equivalent crystallographic directions; in $BaTiO_3$ it may point in any one of the six cube-edge directions. Thus, akin to crystallization, *nuclei of the ferroelectric state* may form and grow at random, as the Curie temperature is approached, until they divide the crystal into an array of domains determined by prehistory and energy balance. Cooling

[28] G. Shirane, S. Hoshino, and K. Suzuki, *Phys. Rev. 80*, 1105 (1950).

[29] W. J. Merz, *Phys. Rev. 78*, 52 (1950).

[30] St. v. Náray-Szabó, *Naturwiss. 31*, 202 (1943).

[31] D. A. Berlincourt and F. Kulcsar, *J. Acoust. Soc. Am. 24*, 709 (1952).

through the Curie point in an electric field would make one of the possible directions the preferred one; in analogy to the growing of single crystals, a single domain crystal might result.

Actually, there is an electric memory effect when we cool barium titanate through its Curie region in an electric field, but it may be overshadowed by the strain pattern that has been frozen in at high temperature during crystallization. In the extreme case of a very regular strain distribution by spherical warping of a crystal, the spectacular square-net domain pattern of Fig. 28.16 results.[20]

Even in the absence of pre-existing strain, and by starting with a single ferroelectric nucleus, we may still end up with a multi-domain configuration. In ferromagnetics, domains form in order to minimize the closing field originating from the free ends of dipole chains since no magnetic monopoles exist that could neutralize the free magnetism. In ferroelectrics, the free dipole ends can be neutralized by electric countercharges, but whether or not this happens fast enough to obviate domain formation depends on the competition between the growth of the ferroelectric phase and the availability of mobile charge carriers. A careful investigation of the dynamics of domain formation, of the various types of domain walls, their development and motion, is now in progress in this laboratory.

If Fig. 28.21 gives a qualitatively true picture of the situation, the onset of ferroelectricity is obviously a very precarious phenomenon. A change in the size of the octahedra may move the equilibrium point in the dipole moment characteristic to the left of the maximum and destroy the feedback action. A distortion of the oxygen lattice from its regular pattern may cause the coupling oxygen ions O_{III} and O_{IV} to move upward when O_I and O_{II} move downward. This would lead to antiparallel rows of dipole moments and thus to *antiferroelectricity*. The existence of antiferroelectricity has been demonstrated for $PbZrO_3$ by Shirane and coworkers [32] and peculiar transitions have been observed between ferro- and antiferroelectricity in solid solutions of Pb—$BaZrO_3$.[33]

A quantitative molecular theory of the ferroelectricity in the titanate group requires obviously that the polar charges and relative motions of the titanium, oxygen, and barium ions are known in detail and incorporated into a local field expression. We can calculate the dipole moment μ per elementary cube of the $BaTiO_3$ crystal from the saturation moment $P_s \simeq 16 \times 10^{-2}$ [coul/m^2] measured for single domain crystals (Fig. 28.19). Since the volume of such a cube is about $(4 \times 10^{-10})^3$ [m^3], that is, $N = \dfrac{1}{64 \times 10^{-30}} \simeq 1.6 \times 10^{28}$ [m^{-3}] are contained in the unit volume, this dipole moment is

$$\mu = \frac{P_s}{N} \simeq 10^{-29} \quad [\text{coul m}] \simeq 3 \quad [\text{debyes}]. \qquad (28.1)$$

If this moment is caused by the displacement of the Ti^{4+} from the center of the octahedra, a displacement of

$$d = \frac{\mu}{4e} \simeq 0.15 \text{ A} \qquad (28.2)$$

would be expected. The shortening of the distance between the titanium and the nearest oxygen ion is about of this magnitude,[26] but, instead of assigning formal charges, we have to measure the dipole moment distance characteristic and the feedback coupling schematized in Fig. 28.21. High-pressure experiments for this purpose are contemplated.

Useful information, on the other hand, can be obtained by the thermodynamic approach, which correlates the various macroscopic parameters by energy considerations. This latter type of treatment was first carried through for rochelle salt by Mueller [34] in his *interaction theory*, undertaken for $BaTiO_3$ by Devonshire,[35] and extended by Slater [36] by including the electronic polarization.

Whereas in $BaTiO_3$ the ferroelectricity seems to arise from a feedback coupling between the titanium and oxygen sublattices, in the other two groups hydrogen bonds may play an important part as suggested by Slater for KH_2PO_4.[37] If the bonding hydrogen atoms have a choice between various positions, as is well known for many crystal structures, the Curie point may correspond to an order-disorder transition at which a polar axis is created. However, it must be kept in mind that such axis need not by necessity lie in the direction of the hydrogen bonds; the ordering of the hydrogen bonds with parallel, antiparallel, or even oblique moments may create the ferroelectric acting moments, for example, by forcing the phosphorous cations into an eccentric-ordered position in their oxygen tetrahedra.

The preceding discussion of the formation of the ferroelectric state has clarified by implication the relation between ferro- and piezoelectricity. A piezoelectric

[32] G. Shirane, E. Sawaguchi, and Y. Takagi, *Phys. Rev. 84*, 476 (1951).

[33] G. Shirane, *Phys. Rev. 86*, 219 (1952).

[34] H. Mueller, *Phys. Rev. 47*, 175 (1935); *57*, 829, 842 (1940); *58*, 565, 805 (1940); *Ann. N. Y. Acad. Sci. 40*, 321 (1940).

[35] A. F. Devonshire, *Phil. Mag. 40*, 1040 (1940); *42*, 1065 (1951).

[36] J. C. Slater, *Phys. Rev. 78*, 748 (1950).

[37] J. C. Slater, *J. Chem. Phys. 9*, 16 (1941).

crystal, when pyroelectric, has a polar axis like a ferroelectric crystal, but the arrow direction of its axis is prescribed by the arrangement of the ions and cannot be reversed; this is clearly apparent in the wurtzite structure of Fig. 27.4. For the ferroelectric crystal, in contrast, the possibility of reversal is inherent because the axis evolves at the Curie point from a state of higher symmetry. Only in ferroelectric crystals, therefore, can domain structure appear and the moment be reversed by a sufficiently strong opposing field. However, the response to pressure and voltage of a ferroelectric crystal is a truly piezoelectric one, involving the change of permanent moments and therefore, in a first approximation, a linear dependence of polarization and expansion on the applied field strength.[9]

Superposed on this linear effect are changes caused by alterations of the domain pattern. For example, compression in the *c*-direction will tend to turn the polar axis by 90°; or two-dimensional pressure on the crystal edges will orientate the *c*-axis perpendicular to the free crystal surface. This mechanical production of a preferential *c*-axis direction by two-dimensional pressure raises the Curie point.[38]

[38] P. W. Forsbergh, Jr., Tech. Rep. 74, ONR Contract N5ori-07801, Laboratory for Insulation Research, Massachusetts Institute of Technology, October, 1953.

29 · Paramagnetism and Ferromagnetism

Permanent *electric* moments, with the exception of the electron cloud moments appearing in the linear Stark effect (see Sec. 9), are caused by ionic bonds *between* atoms. Such moments produce optically active resonance states, visible in the rotation spectra of gases (see Sec. 18); in liquids and solids these states are quenched by pressure broadening and hindered rotation, and only unquantized relaxation spectra remain. Permanent *magnetic* moments, in contrast, are located *in* atoms or molecules and originate from the circling of the electrons around the nuclei (*orbital moments*) and from the spin of the electrons themselves (*spin moments*) (see Sec. 10). These magnetic moments are causally related and proportional to angular mechanical momenta; the laws of quantum mechanics require that the angular momenta be quantized as integral or half-integral multiples of $\hbar$. Hence, whereas the torque of an external field on an electric or magnetic dipole can be treated classically in an identical manner, we have to expect deeper-lying differences in the actual behavior of electric and magnetic systems because the latter are always coupled to quantized gyroscopes.

The classical calculation of the average static moment $\bar{\boldsymbol{\mu}}$, which a gas molecule of the permanent electric dipole moment $\boldsymbol{\mu}$ contributes to the polarization $\mathbf{P}$ under the counteracting influences of electric torque and thermal agitation, has been given in Sec. 16. Derived by Debye,[1] in analogy to Langevin's calculation for paramagnetic gases,[2] it leads, when the field energy is small in comparison to the thermal energy ($\boldsymbol{\mu}\cdot\mathbf{E} \ll kT$), to the expression (Eq. 16.9)

$$\mathbf{P} = N\bar{\boldsymbol{\mu}} = \frac{N\boldsymbol{\mu}^2}{3kT}\mathbf{E}. \tag{29.1}$$

For the paramagnetic gas, since the torque is produced by the magnetic induction $\mathbf{B} = \mu'\mathbf{H}$ and not by the magnetic field $\mathbf{H}$ (see I, Eq. 2.16), the corresponding magnetization becomes

$$\mathbf{M} = N\bar{\mathbf{m}} = \frac{N\mathbf{m}^2}{3kT}\mu'\mathbf{H}. \tag{29.2}$$

The static susceptibilities of the paraelectric and the paramagnetic gas are consequently

$$\chi_e = \frac{\mathbf{P}}{\epsilon_0\mathbf{E}} = \frac{N\boldsymbol{\mu}^2}{\epsilon_0 3kT} \equiv \frac{C'}{T} \tag{29.3}$$

and

$$\chi_m = \frac{\mathbf{M}}{\mathbf{H}} = \frac{N\mathbf{m}^2\mu'}{3kT} \equiv \frac{C}{T}. \tag{29.4}$$

The fact that the susceptibility is inversely proportional to the absolute temperature was first discovered by P. Curie[3] in his studies on oxygen gas and is therefore known as *Curie's law*. Langevin's theory provides its theoretical foundation and leads to a first interpretation of the Curie constant C in Eq. 29.4. For oxygen, referring to the susceptibility per gram, Curie measured

$$\chi = \frac{33700 \times 10^{-6}}{T}. \tag{29.5}$$

[1] P. Debye, *Physik. Z. 13*, 97 (1912).

[2] M. P. Langevin, *J. physique 4*, 678 (1905); *Ann. chim. phys. 5*, 70 (1905).

[3] P. Curie, *Ann. chim. phys. 5*, 289 (1895).

Calculating the magnetic moment **m** per oxygen molecule from this value in mks units, we find

$$\mathbf{m} \simeq 7.7 \times 10^{-24} \quad [\text{joule/weber m}^{-2}]. \tag{29.6}$$

This is about 0.83 Bohr magnetons (see Eq. 10.7); hence the simple Langevin approximation gives a reasonable order of magnitude for the magnetic moment.

Deviations from Curie's law are expected by the classical theory as soon as the magnetic energy $|\mathbf{m}|\,|\mathbf{B}|$ is not very small in comparison to the thermal energy kT. In this case we have to return to the Langevin characteristic itself (see Fig. 16.2 and Eq. 16.8),

$$\frac{\overline{\mathbf{m}}}{\mathbf{m}} = \coth x - \frac{1}{x} \equiv L(x) \quad \text{with } x = \frac{|\mathbf{m}|\,|\mathbf{B}|}{kT}, \tag{29.7}$$

and expect saturation for sufficiently low temperatures or high field strength. Kammerlingh Onnes [4] was able to show that this situation is actually realized in gadolinium sulfate [$Gd_2(SO_4)_3$, $8H_2O$] (Fig. 29.1). The large magnetic moment of the gadolinium ion (**m** ca. 8 Bohr magnetons) makes the onset of saturation possible at relatively small values of $|\mathbf{B}|/T$, whereas the interaction of the moments, not considered in Langevin's theory, is held down by the diluting effect of the crystal water. Thus, at 1.31°K and 22,000 gausses, the average moment $\overline{\mathbf{m}}$ reaches 84 percent of the actual moment **m**.

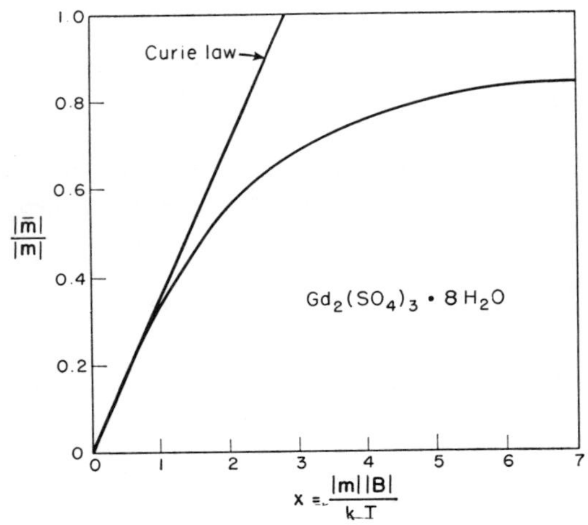

Fig. 29.1. Paramagnetic saturation characteristic of gadolinium sulfate. (After Woltjer and Kammerlingh Onnes.[4])

If the moments are not diluted, for example, in gadolinium metal,[5] we find frequently the simple Curie law of Eq. 29.4 replaced by the *Curie-Weiss law* [6] (see Eq. 23.7):

$$\chi_m = \frac{C}{T - T_c}. \tag{29.8}$$

The mutual interaction of the magnetic moments increases the locally acting field, until, at a critical Curie temperature T_c a spontaneous magnetization occurs; the material becomes *ferromagnetic*.

Weiss based his derivation of Eq. 29.8 on the assumption of a *molecular field* $w\mathbf{M}$ which adds its action to that of the applied magnetic field **H** to create a local field

$$\mathbf{H}' = \mathbf{H} + w\mathbf{M}, \tag{29.9}$$

just as the polarization **P** produces the local *Mosotti field* of Eq. 2.9,

$$\mathbf{E} = \mathbf{E} + \frac{\mathbf{P}}{3\epsilon_0}. \tag{29.10}$$

By assuming further that the magnetization **M** can be represented by the first term of the Langevin equation (see Eq. 29.2) he arrived at his susceptibility law. The derivation of the corresponding polarization law for ferroelectrics, given in Sec. 23, is patterned after his calculation. Weiss furthermore accounted for the absence of any appreciable permanent magnetization in pure iron and other soft ferromagnetics by the postulate that the material subdivides into microregions of compensating *domains* (Fig. 29.2). Such domains can actually be seen in the transparent ferroelectrics, as we have shown in Sec. 28.

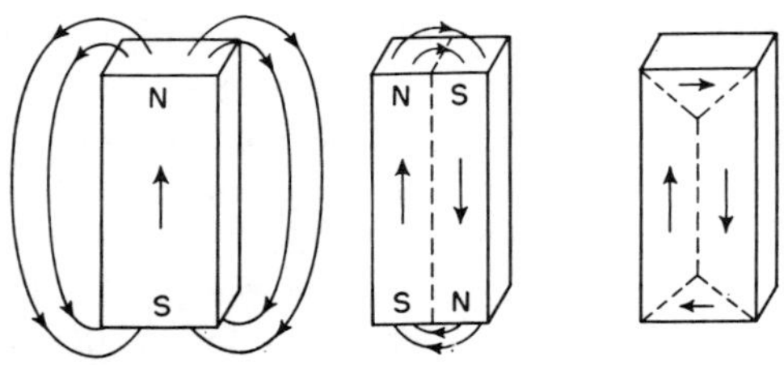

Fig. 29.2. Reduction of permanent magnetization by domain formation.

Ferromagnetic materials are characterized by a nonlinear magnetization characteristic (Fig. 29.3). A sample of soft iron, for example, exposed for the first time to a slowly increasing magnetic field **H**, follows in its magnetization the *virgin* curve $A \rightarrow B$, reaching a *saturation induction* $\mathbf{B}_s$ of ca. 21,000 gausses for a *saturation field* $\mathbf{H}_s$ of ca. 500 [Oe] at room temperature.

[4] H. R. Woltjer and H. Kammerlingh Onnes, *Comm. Phys. Lab. Univ. Leiden*, Vol. 15, No. 167[c]; *Verslag. Amst. Acad. 32*, 772 (1923).

[5] G. Urbain, P. Weiss and F. Trombe, *Compt. rend. 200*, 2132 (1935).

[6] P. Weiss, *J. phys. et radium 6*, 661 (1907); *Physik. Z. 9*, 358 (1908).

Reducing the field again to zero and then reversing it, we traverse a curve $B \rightarrow C \rightarrow D \rightarrow E$, and a return trace $E \rightarrow F \rightarrow G \rightarrow B$, that is, a *hysteresis loop*, characterized by a *remanent induction* (*remanence*) $\mathbf{B}_r$ $(A \rightarrow C)$ (for $H = 0$) and a *coercive field* $\mathbf{H}_c$

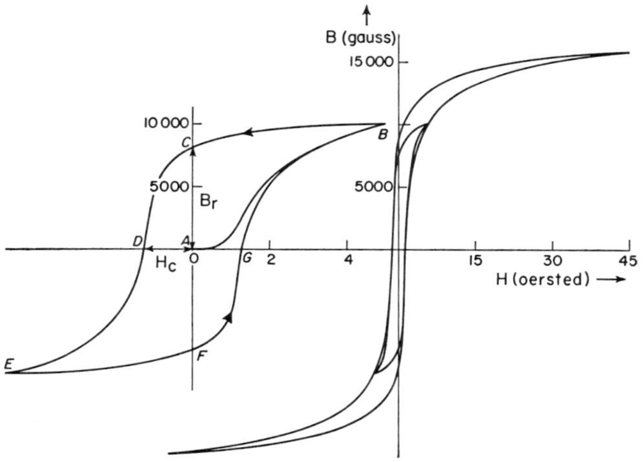

Fig. 29.3. Initial magnetization curve and hysteresis loops of soft iron.

$(A \rightarrow D)$, required for the wiping out of this remanence. $\mathbf{H}_c$ for soft iron at room temperature varies from ca. 1.5 [Oe] for the commercial product to 0.01 [Oe] for the purest material;[7] for hard steel alloys, on the other hand, it may be raised to several thousand oersteds.

The hysteresis characteristic is caused by domain effects and can therefore be changed within wide limits by the composition and pretreatment of the ferromagnetic. The saturation magnetization or *spontaneous magnetization* $\mathbf{M}_s$, on the other hand, is a characterizing constant of the material. As Weiss visualized: below the Curie temperature the elementary magnetic moments orient themselves parallel by mutual interaction; in each domain saturation magnetization prevails under the action of the molecular field; the external field $\mathbf{H}_s$ serves only for the alignment of the domains in the field direction. At the absolute zero point, the orientation of the elementary magnets is complete and absolute saturation achieved $(\mathbf{M}_\infty)$; with increasing temperature, the parallel orientation becomes disturbed by thermal agitation and finally destroyed at the Curie temperature T_c. If we plot the *relative spontaneous magnetization* $\mathbf{M}_s/\mathbf{M}_\infty$ as a function of the *normalized temperature* T/T_c, Fig. 29.4 is obtained.

This characteristic, according to the Weiss theory, is a universal one for all ferromagnetics and can be derived as follows. The saturation magnetization $\mathbf{M}_s$ is the magnetic moment per unit volume produced by the average magnetic moments $\overline{\mathbf{m}}$ (see Eq. 29.2), whereas the absolute saturation magnetization $\mathbf{M}_\infty$ corresponds to the additive action of the true moments,

$$\mathbf{M}_\infty = N\mathbf{m}. \tag{29.11}$$

The relative spontaneous magnetization is therefore given by the Langevin function (see Eq. 29.7)

$$\frac{\mathbf{M}_s}{\mathbf{M}_\infty} = L(x). \tag{29.12}$$

Below the Curie point, the Weiss molecular field dominates ($w\mathbf{M}_s \gg \mathbf{H}$), hence the local field of Eq. 29.9 can be approximated as

$$\mathbf{H}' \simeq w\mathbf{M}_s. \tag{29.13}$$

This local field enters into the parameter x of the Langevin function, the ratio of magnetic torque to thermal agitation, which may be written

$$x = \frac{|\,\mathbf{m}\,|\mu_0 H'}{kT} \simeq \frac{|\,\mathbf{m}\,|\mu_0 w|\,\mathbf{M}_s\,|}{kT}. \tag{29.14}$$

Dividing both sides by the absolute saturation magnetization and rearranging, we obtain a second equation for the relative spontaneous magnetization

$$\frac{\mathbf{M}_s}{\mathbf{M}_\infty} = \frac{kT}{\mu_0|\,\mathbf{m}\,|w|\,\mathbf{M}_\infty\,|}\,x. \tag{29.15}$$

Thus $\mathbf{M}_s/\mathbf{M}_\infty$ as $f(x)$ can be determined graphically as

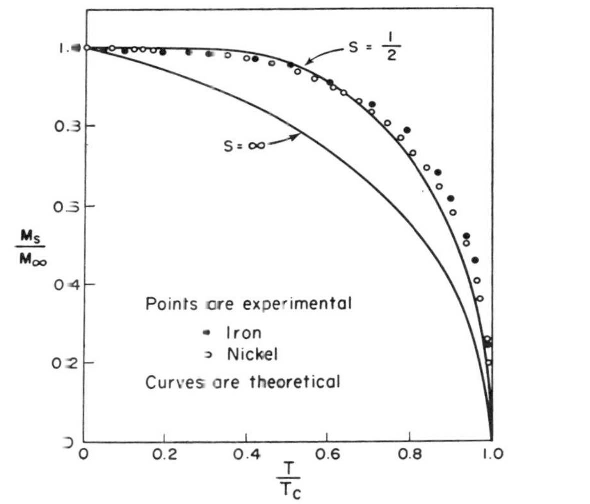

Fig. 29.4. Relative spontaneous magnetization as function of a normalized temperature ($S = \infty$, classical characteristic; $S = \frac{1}{2}$, characteristic based on spatial quantization). Experimental data: Nickel, P. Weiss and R. Forrer;[8] Iron, H. H. Potter.[9]

the intersection of the Langevin curve $L(x)$ with the straight line represented by Eq. 29.15 (Fig. 29.5).

[7] See R. M. Bozorth, *Ferromagnetism*, D. Van Nostrand, New York, 1951, p. 54.

At the Curie temperature T_c, the spontaneous magnetization disappears ($\mathbf{M}_s = 0$), that is, the straight line becomes tangent to the Langevin curve with a slope 1/3, since

$$L(x) = \frac{x}{3} - \frac{x^3}{45} + \cdots \qquad \text{for } x \ll 1. \tag{29.16}$$

Hence

$$\frac{kT_c}{\mu_0 \mathbf{m} w \mathbf{M}_\infty} = \frac{1}{3}$$

or

$$T_c = \frac{\mu_0 \mathbf{m} w \mathbf{M}_\infty}{3k}. \tag{29.17}$$

With this expression for the Curie temperature, the parameter x of Eq. 29.14 is dependent only on the nor-

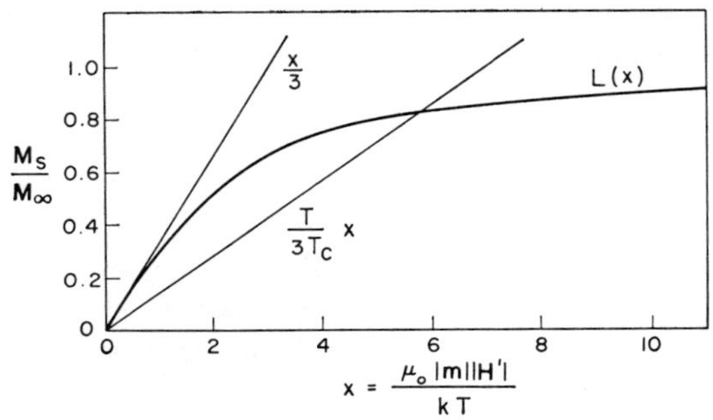

Fig. 29.5. Graphical derivation of the spontaneous magnetization characteristic.

malized temperature and relative spontaneous polarization,

$$x = \frac{3T_c}{T}\frac{\mathbf{M}_s}{\mathbf{M}_\infty}. \tag{29.18}$$

Hence the Langevin equation (29.12) becomes a *universal equation* for the relative magnetization when referring to the normalized temperature

$$\frac{\mathbf{M}_s}{\mathbf{M}_\infty} = L\left(\frac{3T_c}{T}\frac{\mathbf{M}_s}{\mathbf{M}_\infty}\right). \tag{29.19}$$

This universal saturation characteristic of the Langevin-Weiss theory is compared in Fig. 29.4 with the experimental characteristics for nickel [8] and iron.[9] The experimental curves approximately coincide as expected,[10] but rise much steeper than the classical theory predicts. Furthermore, if we calculate from the observed Curie temperatures of these metals ($T_{c\mathrm{Ni}} \simeq 630°\mathrm{K}$, $T_{c\mathrm{Fe}} \simeq 1060°\mathrm{K}$), the molecular field $w\mathbf{M}_\infty$ on the basis of Eq. 29.17, with $\mathbf{m}$ equal to one Bohr magneton, we obtain 2.5×10^7 and 4.2×10^7 oersteds, respectively. That such tremendous magnetic fields of the order of 10 million oersteds actually exist in ferromagnetics appears impossible. The mutual interaction of the magnetic dipole moments leads to much smaller values (see Eq. 11.1), and the magnetic moments of atoms can be uncoupled by fields of the order of 10^4 [Oe], as the Paschen-Back effect indicates (see Sec. 10). It is obvious that the Langevin-Weiss theory gives a valuable first approximation but requires modification and reinterpretation by quantum mechanics.

Beginning with Curie's law for paramagnetic gases, where the interaction between the magnetic moments can still be neglected (Eq. 29.4), we have obviously to modify Langevin's derivation since it allows any orientation on the basis of Boltzmann statistics. The Zeeman effect shows that the magnetic moments cannot assume an indefinite number of positions in respect to the field axis but only certain quantized orientations (see Sec. 10). A direct experimental confirmation of this *spatial quantization* was first furnished by the famous Stern-Gerlach experiment [11] on the deflection of an atomic silver beam in a nonhomogeneous magnetic field (Fig. 29.6). The silver atom carries one Bohr magneton; the beam splits into two discrete beams deflected upward and downward, respectively, instead of broadening in a continuous fashion according to classical statistics. Hence the magnetic moment of the atoms can assume only a parallel and an antiparallel orientation (cf. Fig. 8.9). Rabi's introduction of electrical resonance techniques [12] has made molecular

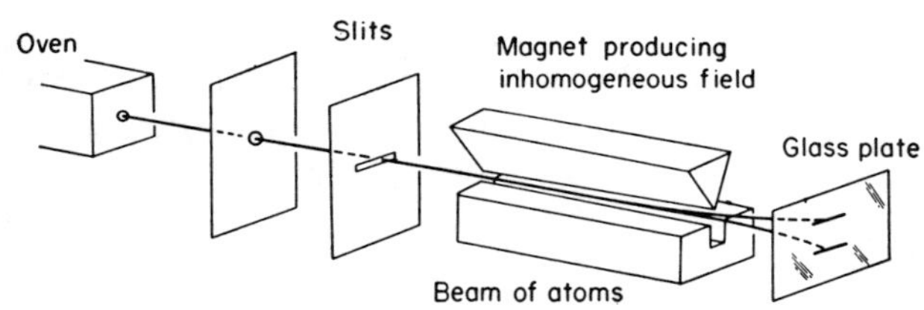

Fig. 29.6. Stern-Gerlach experiment.

beam experiments in nonhomogeneous fields a precision tool for the determination of elementary magnetic moments.

Contrary to the expectation of the old quantum theory, spatial quantization does not affect the paramagnetic (and electric) susceptibility, as Van Vleck has

[8] P. Weiss and R. Forrer, *Ann. phys. 5*, 153 (1926).
[9] H. H. Potter, *Proc. Roy. Soc. (London)*, A*146*, 362 (1934).
[10] See, however, the deviation recently found for Cu-Ni alloy (27.5 percent Cu) by W. Sucksmith, C. A. Clark, D. J. Oliver and J. E. Thompson, *Revs. Mod. Phys. 25*, 34 (1953).

[11] W. Gerlach and O. Stern, *Z. Physik 9*, 349 (1922).
[12] I. I. Rabi, J. R. Zacharias, S. Millman, and P. Kusch, *Phys. Rev. 53*, 318 (1938); *55*, 526 (1939).

shown (*principle of spectroscopic stability*).[13] A summation over the various quantized orientations leads to the same results as a classical integration over a random orientation of moments; the square of the direction cosine has the mean value 1/3. The only change in the formulation of Curie's law is a more detailed interpretation of the permanent magnetic moment **m**. The magnetic moment of an atom follows from its total angular momentum, characterized by the inner quantum number J (see Sec. 10) and quantized as

$$p' = \sqrt{J(J+1)}\hbar. \tag{29.20}$$

By introducing the Bohr magneton $\mathbf{m}_B$ and the g factor (see Eqs. 10.4 to 10.8), we obtain the magnetic moment as

$$\mathbf{m} = g\mathbf{m}_B\sqrt{J(J+1)}; \tag{29.21}$$

hence, in place of Eq. 29.4,

$$\chi_m = N\mu_0 \frac{g^2\mathbf{m}_B{}^2 J(J+1)}{3kT}. \tag{29.22}$$

This formulation of the Curie law holds not only for monatomic paramagnetic gases but represents well the susceptibility of salts of the rare earths. The trivalent rare-earth ions have, in addition to the completed rare-gas shells of the xenon core, an incomplete $4f$ shell far within the atom (see Table 8.1). These well-shielded electrons produce paramagnetism as in the corresponding free gaseous ions. That the terms of the $4f$ electrons are little perturbed by the surroundings, whether the ions are bound in crystals or placed in solution, is evidenced by their sharp spectral lines, comparable in width to those of gases.

Concluding by analogy we might expect that the paramagnetic salts of the iron group would obey Eq. 29.22 since these ions contain an incomplete $3d$ shell. However, here the situation is different: the $3d$ electrons are not well shielded and their orbits are greatly perturbed by neighboring ions. A measurement of the gyromagnetic ratio gives g factors of about 2; that is, in essence only the spins of the electrons are free to orient in relation to the magnetic field axis, while the orbitals are tied down in the lattice (*orbital quenching*). The strong interaction of the $3d$ electrons substitutes for the Curie law the Curie-Weiss law and makes the elements and compounds of the iron groups preferential candidates for ferromagnetism and antiferromagnetism.

Substances that obey Curie's law are said to exhibit *normal paramagnetism*. Deviations from this law must occur not only when we approach saturation (see Fig. 29.1) or when coupling effects between the magnetic moments enter, but also when we deal with several energy states of different magnetic properties so closely spaced that the thermal energy suffices for electronic transitions. In this case, realized especially in molecules, the population density of the various states has to be calculated, as discussed for the thermal occupation of rotation levels (Sec. 19). We deal with a mixture of magnetic moments that changes its composition with temperature, and should not be surprised to find paramagnetic characteristics completely at variance with Curie's law.

Paramagnetism originates from the orbital and spin moments of electrons in incomplete shells. Filled shells or subshells do not contribute because their moments cancel to zero (see Sec. 8). Covalent bonds, in addition, make the contribution of the binding electrons vanish because their spins orientate antiparallel and their orbital moments generally neutralize. Paramagnetic substances obey a Curie or Curie-Weiss law, that is, the susceptibility varies inversely proportional to the temperature, because the first term of the Langevin function (Eq. 29.16) is a sufficiently good approximation. Spatial quantization does not alter this term, as shown above. However, since spatial quantization of the moments leads to a different function from Langevin's, it will affect the magnetization characteristics when we approach saturation.

This can be shown in a simple way by prescribing $2J + 1$ possible orientations; the possible components of a moment **m** parallel to the field axis then vary as $\frac{J}{J}\mathbf{m}, \frac{J-1}{J}\mathbf{m}, \ldots, -\frac{J}{J}\mathbf{m}$. If these various orientations occur with Boltzmann probabilities, we would observe an average moment

$$\overline{\mathbf{m}} = \mathbf{m}\frac{\frac{J}{J}e^{Jx/J} + \frac{J-1}{J}e^{(J-1)x/J} + \cdots - \frac{J}{J}e^{-Jx/J}}{e^{Jx/J} + e^{(J-1)x/J} + \cdots - e^{-Jx/J}}. \tag{29.23}$$

When only the parallel and antiparallel orientations are allowed, as suggested for the iron group by the g factor of 2, the formula reduces to

$$\overline{\mathbf{m}}/\mathbf{m} = \tanh x \equiv Le(x), \tag{29.24}$$

a relation first proposed by Lenz.[14] An accurate calculation by Brillouin[15] leads for the case of orbital

[13] J. H. Van Vleck, *Electric and Magnetic Susceptibilities*, Oxford University Press, 1932, pp. 111 ff. and 152.

[14] W. Lenz, *Physik. Z.* *21*, 613 (1920).

[15] L. Brillouin, *J. phys* *8*, 74 (1927); see also F. Seitz, *The Modern Theory of Solids*, McGraw-Hill Book Co., New York, 1940, p. 581.

quenching to the more complicated expression

$$\mathbf{m} = \mathbf{m}_B \left[(2S+1) \coth \frac{(2S+1)|\,\mathbf{m}_B\,|\mu_0 H}{kT} - \coth \frac{|\,\mathbf{m}_B\,|\mu_0 H}{kT} \right], \quad (29.25)$$

where S is the total spin and $\mathbf{m}_B$ the Bohr magneton. This Brillouin function should obviously be used for the construction of saturation characteristics in place of the Langevin function (see Fig. 29.4), as also the recent magnetization studies of Henry [16] on the paramagnetic ions Cr^{3+}, Fe^{3+}, and Gd^{3+} in strong fields at low temperatures testify (Fig. 29.7).

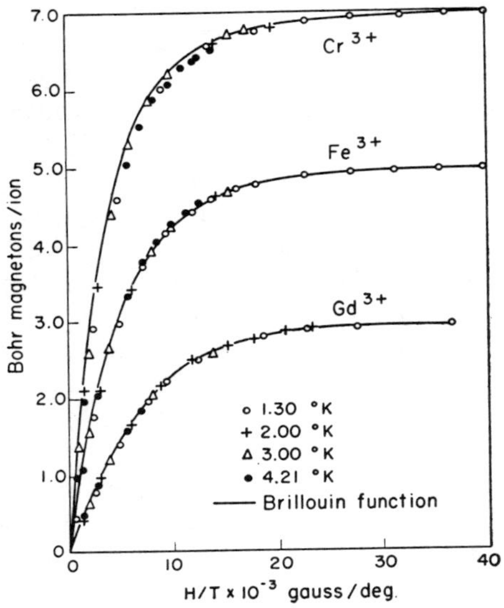

Fig. 29.7. Paramagnetic saturation magnetization of Cr^{3+}, Fe^{3+}, and Gd^{3+}. (After Henry.[16])

The discussion of the magnetization and its saturation on the basis of a Langevin-type characteristic, with inclusion of spatial quantization, suffices when the moments are of one type only and spaced sufficiently far apart, as in gases, aqueous solutions, or solids containing large amounts of crystal water as diluents. In metals and concentrated salts, the question of spatial quantization cannot be treated separately from the problem of mutual coupling.

The discussion of the interaction of magnetic moments in atoms (Sec. 11) and the formation of covalent bonds between atoms according to the Heitler-London approach (Sec. 14) led to the conclusion that the main contribution of the magnetic moments to the energy of stationary states is an indirect one. The electron spins codetermine, because of the Pauli exclusion principle, the standing-wave modes that may form. Parallel versus antiparallel spin orientation may therefore correspond to wave functions for which the electrostatic energy difference is several orders of magnitude larger than the difference in magnetic interaction energy. Quantum mechanics expresses this nonclassical interaction of two electrons according to spin orientation by their exchange energy. The exchange energy is consequently that part of the total energy of a system which depends on the relative spin orientations; it is calculated by exchange integrals (see Sec. 14). In the case of covalent binding the exchange energy causes the formation of electron pair bonds with antiparallel spins, that is, the exchange integral is negative. Heisenberg [17] suggested that in ferromagnetics the exchange integral between the 3*d* electrons of neighboring atoms is positive, hence that the molecular Weiss field of the order of 10^7 oerstedts is not a magnetic field at all, but expresses the exchange energy stabilizing the parallel spin orientation. This hypothesis of Heisenberg, that the exchange energies decide about the formation of the ferromagnetic state, has become the starting point of all subsequent theories of ferromagnetism.

That twenty-five years later we still have no universally accepted theory of ferromagnetism illustrates strikingly the dilemma of quantum mechanics. It can solve its problems only by approximations (see Sec. 7) and has therefore great difficulty in finding which exchange energies are decisive. Heisenberg assumed positive exchange integrals between an atom and its nearest neighbors in the lattice, but left unexplained why only a few of the transition elements are ferromagnetic. Slater [18] solved this dilemma by postulating that the ratio of internuclear distance to radius of the 3*d* shell is the important quantity which determines the sign of the exchange integral. With increasing nuclear charge the 3*d* shell shrinks, its overlap with neighboring atoms decreases, and the sign of the integral changes from negative to positive. Zener [19] recently proposed that this exchange integral is always negative, that is, that the coupling between the 3*d* electrons of neighboring atoms is antiferromagnetic, whereas the exchange between the 3*d* electrons and the conducting electrons leads to a ferromagnetic alignment. Bloch [20] tried to make the conducting electrons responsible for ferromagnetism by introducing exchange energy terms into Sommerfeld's description of a free electron gas (spin-wave theory); Slater [21] reconsidered the exchange energy between the 3*d* electrons on the basis of the elec-

[16] W. E. Henry, *Phys. Rev.* *85*, 487 (1952); *88*, 559 (1952).

[17] W. Heisenberg, *Z. Physik* *49*, 619 (1928).

[18] J. C. Slater, *Phys. Rev.* *36*, 57 (1930).

[19] C. Zener, *Phys. Rev.* *81*, 440 (1951); *83*, 299 (1951).

[20] F. Bloch, *Z. Physik* *57*, 545 (1929); *61*, 206 (1930).

[21] J. C. Slater, *Phys. Rev.* *49*, 537, 931 (1936); *52*, 198 (1937).

tron-band structure of solids; and Stoner[22] similarly based ferromagnetism on the holes in the 3*d* band of the metal (collective electron ferromagnetism). There is hope that the situation will clarify in the next few years, from the theoretical side by the extensive use of modern computing machines, and from the experimental side by more knowledge about exchange interactions through neutron diffraction experiments and work on ferromagnetic semiconductors.[23]

[22] E. C. Stoner, *Proc. Roy. Soc.* (*London*), A*165*, 372 (1938); *169*, 339 (1939).

[23] For the present status of the situation see the reports of the Washington Conference on Magnetism, *Revs. Mod. Phys. 25*, 1 (1953).

30 · Ferromagnetic Metals and Semiconductors

To understand in more detail what kind of information is available or can be provided, let us return to the elements of the iron group and follow their behavior in various states of bonding.

The designation *iron group* properly applies to the nine elements of the periodic system following Ca(20) (see Table 8.1). The argon core ($3s^2 3p^6$) has been completed, two 4*s* electrons ($4s^2$) have been added, and now, starting with Sc(21), a competition sets in between the 3*d* and 4*s* shells. The elements become multivalent because not only the 4*s* but the 3*d* electrons may act as valence electrons; the solutions of the salts become colored, because empty energy states are close enough above the filled ones to allow electron transfer by the absorption of visible light; the atoms become paramagnetic as long as the 3*d* shell is not complete.

As discussed in Sec. 8, the 3*d* shell can be subdivided into five 3*d* orbitals, each able to accommodate two electrons with antiparallel spins. The lowest energy state, according to Hund's theory of spectral terms,[1] will be the one that leads to the maximum number of multiplets, that is, the state with the largest inner quantum number J (see Eq. 11.5). We expect therefore that only after each 3*d* orbital has accepted one electron will the formation of electron pairs begin. The filling of the 3*d* shell terminates with the element Cu(29), but here the 4*s* orbital has been depleted by one electron ($4s^1$). In consequence, whereas cuprous salts are colorless and diamagnetic because the 3*d* shell of Cu^+ is still intact, the cupric salts are colored and paramagnetic, since the divalent Cu^{2+} has lost one 3*d* electron. Only with the next element, Zn(30), is the full stability of the 3*d* shell assured.

Table 30.1 compares[2] the measured magnetic moments of the paramagnetic ions with the values computed from Eq. 29.21 under the assumption that either the total angular momentum, represented by the inner quantum number J, is decisive or that only the spin quantum number S enters owing to orbital quenching. In the latter case with the gyromagnetic ratio $g = 2$, Eq. 29.21 simplifies to

$$\mathbf{m} = \mathbf{m}_B\sqrt{4S(S+1)}. \tag{30.1}$$

Obviously, the second assumption leads to a much better agreement with the experimental data. Hence the electron spins make the essential contribution and the magnetic moment reaches its maximum for five 3*d* electrons distributed over the the five orbitals in parallel alignment, as Hund's rule foresees.

A comparison of the magnetic moments of Cr^{3+} and Fe^{3+} in Table 30.1 with those of Fig. 29.10 shows that the tabulated values are appreciably larger than the saturation moments. This is not an experimental mistake but, on the contrary, can be interpreted as an impressive confirmation of spatial quantization. The tabulated values are derived from low field-strength measurements, where the first term of the Brillouin function (Eq. 29.25) applies. Here the total angular momentum $p' = \sqrt{S(S+1)}\,\hbar$ comes into play, whereas in the saturation measurements $\left(\frac{m_B\mu_0 H}{kT} \gg 1\right)$ the momentum in field direction is being measured as $p_z' = S\hbar$ (cf. Fig. 8.9).

Turning to the monatomic metals of the iron group, we find a change-over from paramagnetism (vanadium) to antiferromagnetism (chromium and manganese) to ferromagnetism (iron, cobalt, and nickel), and then an abrupt falling back to diamagnetism (copper) (Table 30.2). The approximately symmetrical rise and fall of the magnetic moments of the ions in Table 30.1 does not predict such a lopsided behavior. That ferromagnetism occurs only in the last three transition elements must therefore be explained by some special trend in

[1] F. Hund, *Linienspectren und periodisches System der Elemente*, Springer, Berlin, 1927, Chap. 5.

[2] Based on data in P. W. Selwood, *Magnetochemistry*, Interscience Publishers, Inc., New York, 1943, p. 99, and E. C. Stoner, *Magnetism and Matter*, Methuen and Co., London, 1934, p. 25.

Table 30.1. Magnetic moments of the ions of the iron group elements[2]

N_d Number of 3d Electrons	Ion Type	Term Symbol of Free Ion	J	S	Magnetic Moment **m** in Bohr Magnetons: Observed	Calculated: J	Calculated: S
0	K^+, Ca^{2+}, Sc^{3+}, Ti^{4+}, V^{5+}	$^1S_0$	0	0	0	0	0
1	Sc^{2+}, Ti^{3+}, V^{4+}	$^2D_{3/2}$	3/2	1/2	1.77–1.79	1.55	1.73
2	Ti^{2+}, V^{3+}	$^3F_2$	2	1	—, 2.76–2.85	1.63	2.83
3	V^{2+}, Cr^{3+}, Mn^{4+}	$^4F_{3/2}$	3/2	3/2	3.81–3.86, 3.68–3.86, 4.00	0.77	3.87
4	Cr^{2+}, Mn^{3+}	$^5D_0$	0	2	4.80, 5.0	0	4.90
5	Mn^{2+}, Fe^{3+}	$^6S_{5/2}$	5/2	5/2	5.2–5.96, 5.4–6.0	5.92	5.92
6	Fe^{2+}, Co^{3+}	$^5D_4$	4	2	5.0–5.5, (2.5)	6.70	4.90
7	Co^{2+}	$^4F_{9/2}$	9/2	3/2	4.4–5.2	6.64	3.87
8	Ni^{2+}	$^3F_4$	4	1	2.9–3.4	5.59	2.83
9	Cu^{2+}	$^2D_{5/2}$	5/2	1/2	1.8–2.2	3.55	1.73
10	Cu^+, Zn^{2+}	$^1S_0$	0	0	0, 0	0	0

the interaction between the atoms. Bethe[3] and Slater[4] emphasize the overlap of the 3*d* electrons of neighboring atoms. According to Slater, if the 3*d*-wave functions overlap much, it becomes more favorable energetically to have the electrons participate in conduction as a Fermi gas with antiparallel spins. With increasing order number the nuclear charge increases, the 3*d* orbitals shrink, and the exchange energy between the 3*d* electrons of the individual atom begins to enforce parallel spins (see Sec. 29). A measure of the overlap is the ratio of the interatomic distance *R* to the diameter of the 3*d* orbital 2*r*. This ratio, tabulated in Table 30.3, is in fact greatest for the ferromagnetic metals. The localized magnetic moment per atom in the ferromagnetic metals (Fig. 30.1) is appreciably smaller than in the paramagnetic salts (see Tables 30.1 and 30.2), as we would expect when 3*d* electrons are

[3] H. Bethe, *Handbuch der Physik 24*, Pt. I, 594 (1933).
[4] J. C. Slater, *Phys. Rev. 36*, 57 (1930).

Table 30.2. Saturation moments and magnetic state of metals of the iron group

	Saturation Moments μ_B	Magnetic State	Curie Temperature
Vanadium[5]	<0.1	para-	
Chromium[5]	0.4	antiferro-	ca. 473°K
Manganese[5]	0.5	antiferro	ca. 100°
Iron		ferro-	
paramagnetic	3.18		1093°
ferromagnetic	2.21		1043°
Cobalt		ferro-	
paramagnetic	3.13		1423°
ferromagnetic	1.72		1393°
Nickel		ferro-	
paramagnetic	1.60		650°
ferromagnetic	0.60		631°
Copper		dia-	
diamagnetic	0		

[5] C. G. Shull and M. K. Wilkinson, *Revs. Mod. Phys. 25*, 100 (1953).

Table 30.3. Ratio of interatomic distance to sum of radii of elements of the iron group

	Interatomic Distance	Sum of Radii	Ratio
Ti	2.93 A	1.10 A	2.65
V	2.63	0.98	2.69
Cr	2.51	0.90	2.79
Mn	2.52	0.84	3.00
Fe	2.50	0.78	3.20
Co	2.51	0.72	3.50
Ni	2.50	0.68	3.69

lost as conducting electrons by antiparallel pair formation.

In Table 30.2 two Curie temperatures have been given for the ferromagnetics. This is due to the fact that the straight line of $1/\chi$ versus T predicted by the Curie-Weiss law ends actually appreciably above the Curie point; near the Curie temperature the slope decreases (Fig. 30.2). Extrapolation of the straight line section leads to the *paramagnetic Curie temperature* T_{Cp}, extrapolation of the actually measured curve to the lower *ferromagnetic Curie temperature* T_{Cf}.[6]

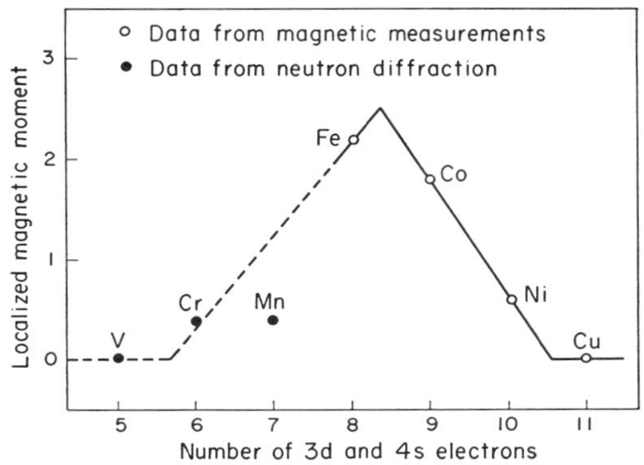

Fig. 30.1. Localized magnetic moment per atom of iron-group metals. (After Shull.[5])

Measurements on single crystals show that the magnetic moments tend to align themselves in preferential crystallographic orientations, that is, crystals are, in general, *magnetically anisotropic*. This is no great surprise when we recall that the orbital magnetic moments arise from eccentric electron cloud configurations which tend to produce bonds in special orientations (see Secs. 8 and 17). The supposition that the *crystalline field anisotropy* is primarily caused by the orbital moments, is supported by the fact that paramagnetic salts containing the magnetic ions in S states, such as Mn^{2+} and Fe^{3+} (see Table 30.1), are magnetically isotropic, whereas large anisotropies are found for salts containing Fe^{2-}, Co^{2+}, etc., where the spectroscopic states have orbital angular momenta.[7]

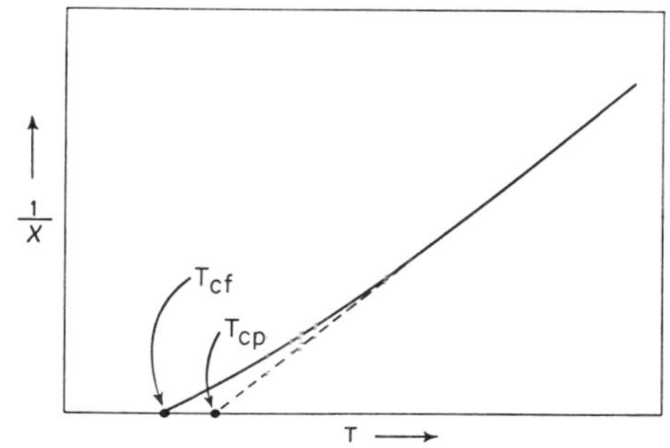

Fig. 30.2. Para- and ferromagnetic Curie temperature.

In ferromagnetics, the *direction of minimum free energy* for the orientation of the magnetic moments may be found by measuring the saturation characteristic of single crystals in various orientations (Fig. 30.3).[8] For the ferromagnetic monatomic metals at room temperature we find[9] that in iron the cube edge [100] is the

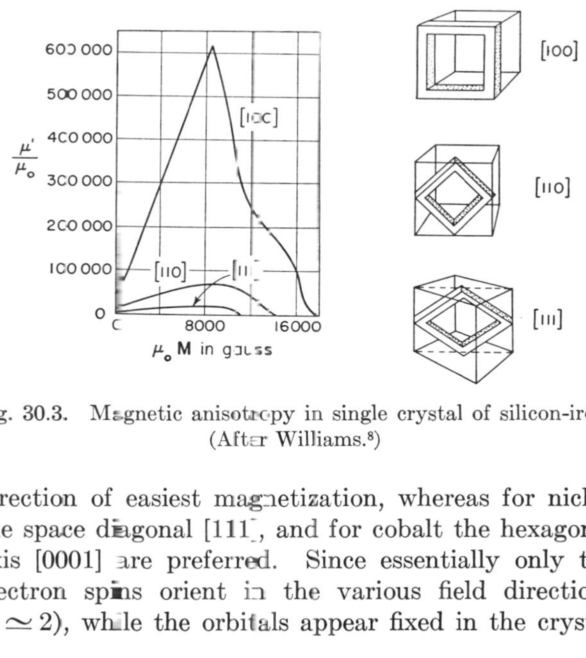

Fig. 30.3. Magnetic anisotropy in single crystal of silicon-iron. (After Williams.[8])

direction of easiest magnetization, whereas for nickel the space diagonal [111], and for cobalt the hexagonal axis [0001] are preferred. Since essentially only the electron spins orient in the various field directions ($g \simeq 2$), while the orbitals appear fixed in the crystal

[6] A similar uncertainty in the Curie point is observed when approached from the ferromagnetic side. With the magnetic balance one obtains a deviation of the spontaneous magnetization curve towards higher temperature, whereas by the magnetocaloric method a lower Curie temperature results [see W. Sucksmith, C. A. Clark, D. J. Oliver, and J. E. Thompson, *Revs. Mod. Phys.* *25*, 34 (1953)].

[7] For the extensive literature on magnetic anisotropy see R. Becker and W. Döring, *Ferromagnetismus*, Springer, Berlin, 1939, and R. M. Bozorth, *Ferromagnetism*, D. Van Nostrand Co., New York, 1951.

[8] H. J. Williams, *Phys. Rev.* *52*, 747 (1937).

[9] K. Honda and S. Kaya, *Science Rep. Tôhoku Imp. Univ.* *15*, 721 (1926); S. Kaya, *ibid.* *17*, 639, 1157 (1928).

lattice, we must assume[10] that the interaction between spin and orbital moment creates the directional variation of the free energy, the so-called *crystal* or *anisotropy energy.*

It is obvious that this anisotropy must lead to mechanical effects; the formation of a magnetic axis is accompanied by the development of a crystallographic axis, and any turning of this axis or changes in the magnetic moments must produce a mechanical distortion of the material. The situation is analogous to the piezoelectric response of ferroelectrics (see Sec. 28), but this analogy has not been recognized clearly. This becomes apparent in the unhappy choice of the name for the mechanical distortion accompanying magnetization changes as *magnetostriction.* The designation implies that the effect is akin to electrostriction, where we deal with induced moments (see Sec. 27), while actually we operate with permanent moments as in piezoelectrics.

The crystal energy U_c in its dependence on the direction of magnetization is obviously determined by the

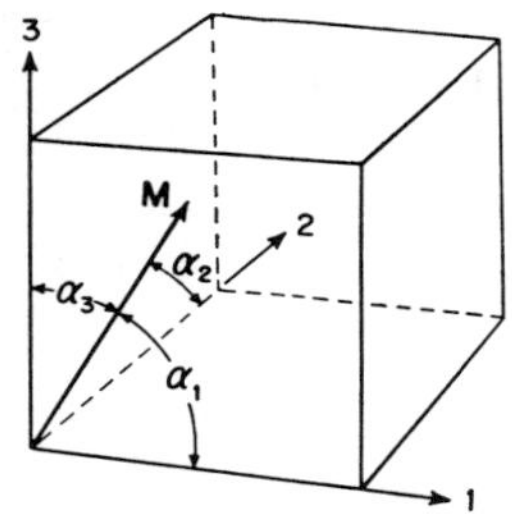

Fig. 30.4. Direction cosines of the magnetization for cubic crystals.

macroscopic crystal symmetry. For a cubic crystal, for example, the energy is given by the direction cosines α_1, α_2 and α_3 referring to axes in the cube-edge directions (Fig. 30.4), as

$$U_c = K_1(\alpha_1^2\alpha_2^2 + \alpha_2^2\alpha_3^2 + \alpha_3^2\alpha_1^2) + K_2(\alpha_1^2\alpha_2^2\alpha_3^2) + \cdots; \quad (30.2)$$

K_1 and K_2 are the *first-* and *second-order anisotropy constants.*[11]

In addition to this true *crystal anisotropy* we find, as in ferroelectrics, that external stress or residual strains in the material cause a *stress anisotropy.* How the magnetic moments orient in relation to such stress depends on the sign of the *magnetostriction constant* λ. The constant designates the relative change in length

$$\lambda \equiv \Delta l/l \quad (30.3)$$

occurring in the direction of an applied magnetic field under the influence of this field. It is negative in nickel, that is, a nickel bar shortens in field direction, whereas permalloy shows positive magnetostriction and iron, with its strong crystal anisotropy, gives λ values varying for different orientations, even in sign. Positive magnetostriction obviously leads to a positive stress anisotropy, that is, the moments line up preferentially *in* the direction of the stress on tension and perpendicular on compression.

A third anisotropy effect occurs in ferroelectrics as well as in ferromagnetics when nonspherical samples are used—a *shape anisotropy.* The free ends of the dipole chains produce a demagnetizing (closing) field; hence the material tends to magnetize in the direction in which this field is a minimum, that is, in the direction of its largest dimension. The effect can be calculated for rotational ellipsoids (see I, Sec. 10 and Appendix A, I, Sec. 2).

The phenomena discussed thus far dominate the behavior of a ferromagnetic near saturation when the moments are all lined up in the same direction. In the demagnetized state, the material is divided into Weiss domains (see Sec. 29), each of them magnetized to saturation and these microregions orientated in patterns that tend to minimize the closing field. Across the domain boundaries, called *Bloch walls* since Bloch[12] first treated their energy content and dimensions, the spin vectors gradually change their orientation from the one direction to the other over a distance of several hundred angstroms (Fig. 30.5). In cubic crystals the

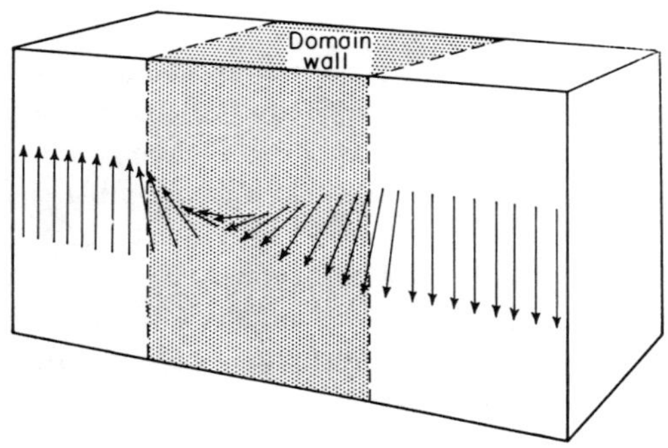

Fig. 30.5. Schematic representation of spin orientation across a 180° domain wall (Bloch wall).

domains are frequently mutually oriented at 90° or 180°; in consequence *90° walls* and *180° walls* play an important role in magnetization processes.

The intersection of the walls with the surface of the opaque ferromagnetics can be made visible by the deposition of a ferromagnetic powder, which is attracted by the strong local closing fields. By using colloidal Fe_3O_4 particles instead of coarse powders,

[10] J. H. Van Vleck, *Phys. Rev.* *52*, 1178 (1937).

[11] See, for example, R. Becker and W. Döring, *Ferromagnetismus,* Springer, Berlin, 1939, pp. 112 ff.

[12] F. Bloch, *Z. Physik* *74*, 295 (1932).

Bitter [13] could develop these surface domain patterns in beautiful detail. This method of the *Bitter stripes* has been used recently by scientists of the Bell Telephone Laboratories [14,15] and by Bates and his coworkers [16] with spectacular success for static and dynamic studies of the domain behavior.

The conventional method of mechanical polishing produces surface strains and powder patterns that cannot be used for the interpretation of the internal domain structure. Electrolytic polishing removes these disturbances [17] and leads, with properly oriented single crystals, to definite conclusions about the stable arrangement of domains and the motion of the Bloch walls under the influence of external fields. The minimum energy constellations, predicted for single crystals of various shapes and orientations by Néel,[18] could be verified, and the decisive influence of nonmagnetic inclusions or cavities on the wall motion and coercitive force, indirectly established especially by Kersten's work,[19] could be photographed in convincing detail.

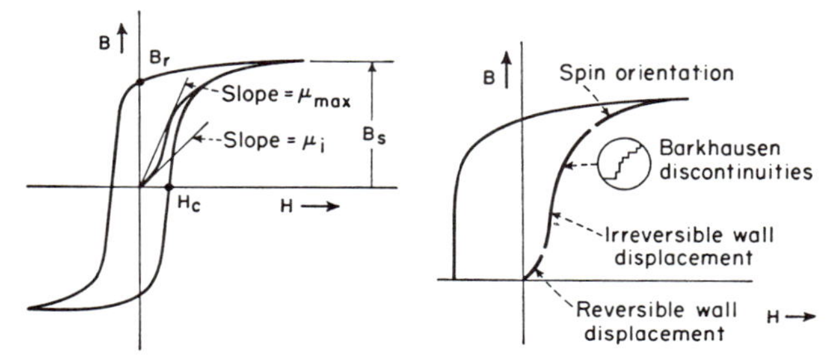

(a) Typical hysteresis loop for a ferromagnetic material (b) Interpretation of magnetization curve

Fig. 30.6. *B-H* characteristic and its possible interpretation (180°).

Let us visualize from the molecular standpoint the magnetization process, measured as a *B-H* curve and characterized by *initial permeability* μ_i, *maximum permeability* μ_{max}, *coercitive* field H_c, and *saturation induction* B_s (Fig. 30.6). In the region of the initial permeability μ_i reversible displacements of the Bloch walls take place. Domains orientated favorably towards the field direction grow at the expense of domains of opposite orientation. This motion of 180° walls under the influence of a slightly unbalancing external field has been made visible by Williams and Shockley [14] on a single crystal of silicon iron, cut in a hollow rectangle with edges and surfaces accurately parallel to the direction of easy magnetization [100]. The domain pattern in the demagnetized state consists of eight domains, four forming an inner rectangle magnetized in one direction, while the other four compose an outer rectangle of opposite magnetization (Fig. 30.7). The separating Bloch wall, lined out by a deposit of colloidal iron, halves the crystal as a white line. Changes of magnetization correspond to the growth of one set of domains at the expense of the other, that is, the line moves. When saturation is reached, the Bloch wall has moved completely to one side, leaving each leg of the rectangle one single domain. It should be noted that these 180° Bloch walls move perpendicularly to the external field. This strange behavior becomes understandable, when we recall that the magnetic moments are coupled to mechanical spins, hence act like gyroscopes producing by their precession a force component normal to the external force.[20]

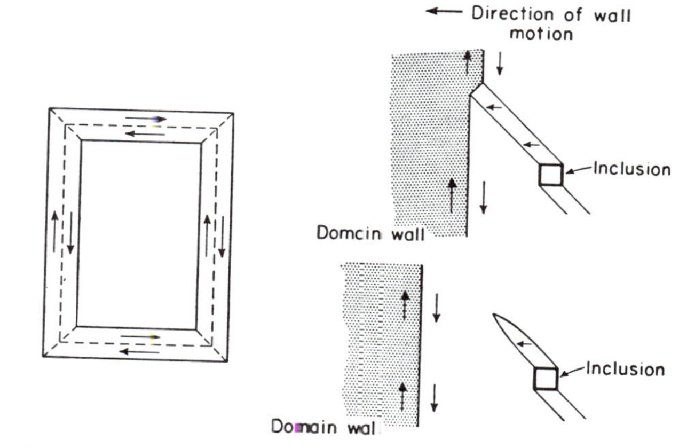

(a) Domain structure of hollow parallelogram (b) Effect of inclusion

Fig. 30.7. (*a*) Domain wall (180°) in single crystal core of silicon-iron. (*b*) Wall motion hindered by inclusion. (After Williams and Shockley.[14])

In an ideal single crystal the slightest unbalance should suffice to roll the Bloch wall to one side and thus to achieve immediate saturation in the direction of easiest magnetization. Actually, this does not happen because the wall entangles itself at inclusions and crystal imperfections. It becomes entrapped by secondary domain patterns forming around the irregularities; these patterns elongate into restraining spikes

[13] F. Bitter, *Phys. Rev. 38,* 1903 (1932); the same method was developed independently by L. von Hamos and P. A. Thiessen, *Z. Physik 71,* 442 (1931).

[14] H. J. Williams, R. M. Bozorth, and W. Shockley, *Phys. Rev. 75,* 155 (1949); H. J. Williams and W. Shockley, *ibid. 75,* 178 (1949); C. Kittel, *Revs. Mod. Phys. 21,* 541 (1949).

[15] J. K. Galt, J. Andrus, and H. G. Hopper, *Revs. Mod. Phys. 25,* 93 (1953).

[16] See the summarizing report by L. F. Bates, *Proc. Phys. Soc. (London)* A*165,* 577 (1952).

[17] C. Elmore, *Phys. Rev. 51,* 982 (1937); *53,* 757 (1938).

[18] L. Néel, *J. phys. et radium* [8] *5,* 241 (1944).

[19] M. Kersten, *Grundlagen einer Theorie der ferromagnetischen Hysterese und der Koerzitivkraft,* Hirzel, Leipzig, 1943.

[20] See, for example, R. Becker, *J. phys. et radium 12,* 332 (1951).

when the domain wall is being forced ahead by an external field (see Fig. 30.7). Finally, the holding spikes break and the wall rolls on, until caught again by new irregularities. This advancing of the domain walls in jerks produces the *Barkhausen noise*,[21] that is, the jumps in magnetization heard over the loudspeaker or registered on indicating instruments.

The region of the Barkhausen effect corresponds to irreversible wall displacements, that is, to the hysteresis loop part of the magnetization characteristic (Fig. 30.6). When the magnetization cannot increase further by such wall displacements, it may still be enhanced by turning the magnetization of the individual domains

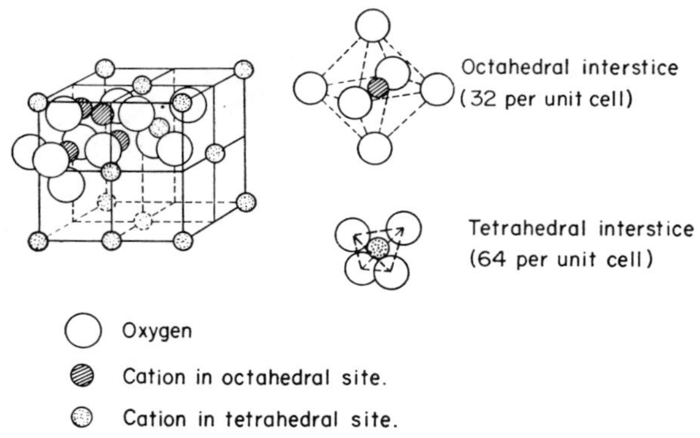

Fig. 30.8. The spinel structure.

from the direction of easiest magnetization into the field direction. By this rotation of spins, characterizing the upper part of the magnetization curve, even a multicrystalline material reaches finally the status of a single domain.[22]

In addition to paramagnetic salts and ferromagnetic metals, the elements of the iron group form ferromagnetic nonmetals, especially oxides. The mother substance of these materials is *magnetite* (Fe_3O_4), noticed as natural *lodestone* by the ancient Greeks for its power to attract small pieces of iron.[23] Magnetite and its derivatives, the *ferrites* or *ferrospinels*, have quite recently moved into the center of attention. This delay time of more than 2500 years may be considered somewhat extensive, but there is a good technological reason: the plastic metals are easier to handle than the brittle ceramics and have appreciably higher magnetic saturation. Hence, only as the high-frequency properties of the ferromagnetics became of importance, the low conductivity of the ceramics proved an advantage outweighing their shortcomings.[24]

The mineral *spinel* ($MgOAl_2O_3$) and the cubic ferrites ($MeOFe_2O_3$) crystallize in a structure characterized by a cubic, close-packed array of O^{2-} spheres (see Sec. 24) with the divalent and trivalent cations distributed in the interstices. There are two kinds of interstitial positions in the oxygen lattice, *tetrahedral* sites enclosed by four oxygen ions and *octahedral* sites formed by six oxygen ions (Fig. 30.8). In the spinel structure twice as many octahedral as tetrahedral sites are occupied. The unit cell, formed by 32 cubic close-packed oxygens (see Fig. 24.6), has 96 interstices, 64 of tetrahedral and 32 of octahedral configuration; hence, since it contains eight $MeOFe_2O_3$ groups, 24 cations have to be distributed. From the standpoint of charge compensation we might foresee that the divalent metal Me^{2+} ions will occupy the tetrahedral sites and the trivalent Fe^{3+} ions the octahedral sites. However, this *normal* spinel structure has been found thus far only for the zinc and cadmium ferrites; the other simple ferrites of only one kind of divalent metal ion are *inverted*.[25] The 8 divalent Me^{2+} ions occupy entirely (or preferentially) octahedral sites, whereas the 16 trivalent Fe^{3+} ions are placed half in tetrahedral and half in octahedral positions.

Magnetite ($FeOFe_2O_3$), with only one cation element, is the most interesting starting material for fundamental studies; it has been grown as single crystals in the Laboratory for Insulation Research [26] and is under intensive investigation.[27] The equal number of ferrous and ferric ions in octahedral lattice sites allows a simple electron exchange between Fe^{2+} and Fe^{3+}, as first pointed out by Verwey and his co-workers.[28] In consequence, magnetite has the highest conductivity of all the ferrites (ca. 10^4 [ohm^{-1} m^{-1}] at room temperature or about 10^{-4} times the conductivity of copper) and is therefore at present of minor technical importance.

The magnetic properties of magnetite near room tem-

[21] H. Barkhausen, *Physik. Z. 20*, 401 (1919).

[22] The detailed analysis of magnetization curves poses special problems for each individual material. See, for example, E. C. Stoner, *Revs. Mod. Phys. 25*, 2 (1953).

[23] The Latin name *magnes* is supposedly derived from the district of Magnesia, Asia Minor, where the ore is found in quantity (see Titus Lucretius Carus, *De Rerum Natura*—about 60 B.C.).

[24] Much pioneering work on the ferromagnetic nonmetals has been done at the Philip's Research Laboratories, for example, by E. J. W. Verwey and by J. L. Snoek and his co-workers; see J. L. Snoek, *New Developments in Ferromagnetic Materials*, Elsevier Publishing Co., New York and Amsterdam, 1947.

[25] T. F. W. Barth and E. Posnjak, *Z. Krist. 82*, 325 (1932); E. J. W. Verwey and E. L. Heilman, *J. Chem. Phys. 15*, 174 (1947).

[26] J. Smiltens, *J. Chem. Phys. 20*, 990 (1952).

[27] See L. R. Bickford, Jr., *Phys. Rev. 78*, 449 (1950); C. A. Domenicali, *Phys. Rev. 78*, 458 (1950); B. A. Calhoun, Tech. Rep. 68, ONR Contract N5ori-07801, Laboratory for Insulation Research, Massachusetts Institute of Technology, July, 1953.

[28] E. J. W. Verwey, P. W. Haayman, and F. C. Romeijn, *J. Chem. Phys. 15*, 181 (1947).

perature are similar to those of nickel. Its saturation magnetization $M_s \simeq 4.75 \times 10^5$ [amp/m] is about the same and its direction of easy magnetization the body diagonal [111]; the Curie temperature (ca. 585°C) lies ca. 200°C above that of nickel. The value of the saturation magnetization corresponds to 4 Bohr magnetons per Fe_3O_4 group. If the cations in magnetite can be represented as Fe^{2+} and Fe^{3+} ions with their four and five unpaired $3d$ electrons respectively cooperating in parallel orientation, we should expect $4 + 2 \times 5 = 14$ Bohr magnetons. This discrepancy was explained by Néel [29] as indicating that an antiferromagnetic coupling exists between the ferric ions in the tetrahedral and the ferric and ferrous ions at the octahedral sites, thus leaving only the ferromagnetic contribution of the Fe^{2+} magnetons. The existence of antiferromagnetism has been confirmed by Shull and co-workers [30] through neutron-scattering experiments.

To treat this more complicated situation, Néel expanded the Weiss concept of the molecular field (Eq. 29.9) by subdividing the crystal into ferromagnetic sublattices and deriving the local field by a superposition of their action. If the sublattices A and B (Fig. 30.9), for example, signify the Fe^{3+} ions in the tetrahedral and octahedral positions, the magnetizations $\mathbf{M}_A$ and $\mathbf{M}_B$ of these two groups of magnetic moments create two local fields, one acting on the A, the other on the B position:

$$\mathbf{H}_A' = \mathbf{H} + w_A'\mathbf{M}_A + w_A''\mathbf{M}_B$$
and
$$\mathbf{H}_B' = \mathbf{H} + w_B'\mathbf{M}_B + w_B''\mathbf{M}_A. \tag{30.4}$$

The primed coefficients refer to the action of the magnetizations on their own sites, the double primed factors, w_A'' and w_B'', to the interactions of sublattice B on site A and A on site B.

Above the Curie point the magnetization, just as the corresponding polarization of Eq. 23.1, may be written as

$$\mathbf{M} = N\alpha_m\mathbf{H}' \simeq \frac{C}{T}\mathbf{H}' \tag{30.5}$$

if the first term of the Langevin function (Eq. 29.16) suffices. Hence the two magnetizations of the sublattices become

$$\mathbf{M}_A = \frac{C_A}{T}\mathbf{H}_A'$$
and
$$\mathbf{M}_B = \frac{C_B}{T}\mathbf{H}_B'. \tag{30.6}$$

By assuming that $\mathbf{H}$, $\mathbf{M}_A$ and $\mathbf{M}_B$ are collinear, we find for the magnetic susceptibility above the Curie temperature (see I, Eq. 2.18)

$$\chi_m = \frac{M_A + M_B}{H}. \tag{30.7}$$

When

$$C_A = C_B, \qquad w_A' = w_B', \tag{30.8}$$

and since for reasons of symmetry

$$w_A'' = w_B'',$$

the susceptibility becomes simply

$$\chi_m = \frac{2C}{T - T_c}, \tag{30.9}$$

where the paramagnetic Curie temperature is

$$T_c = C(w_A' + w_A''). \tag{30.10}$$

Since antiferromagnetic coupling is assumed between the A and B sites, w_A'' is negative.

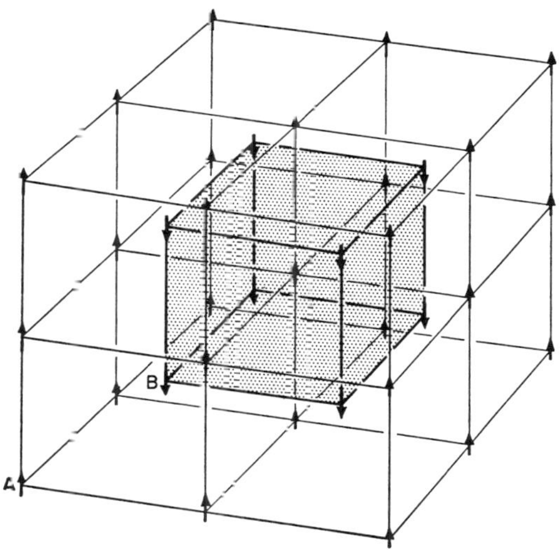

Fig. 30.9. Sublattices A and B, creating two local fields. (After Néel.[29])

In general, the simplifications of Eq. 30.8 will not apply; furthermore the ions may be distributed in different numbers over the A and B sites, additional sites C may enter, and the situation then becomes extremely complex. Néel has considered a number of cases and predicted various types of magnetization characteristics. Some of these curves have already been verified by experiment, and Néel's concepts have proved extremely valuable in clarifying the basic behavior of the ferrites. A summarizing treatment of the theory of antiferromagnetism has been given by Van Vleck.[31]

[29] L. Néel, *Ann. phys.* [12] *3*, 137 (1948); the first concepts of antiferromagnetism were developed by Néel in 1936, *Ann. phys.* *5*, 232 (1936), see also L. Néel, *Revs. Mod. Phys.* *25*, 58 (1953).

[30] C. G. Shull, E. O. Wollan, and W. C. Kochler, *Phys. Rev.* *84*, 912 (1951).

[31] J. H. Van Vleck, *J. phys. et radium* *12*, 262 (1951).

Ferrites of a wide variety of electric and magnetic properties have been derived by replacing the Fe^{2+} ions of magnetite with divalent ions of magnesium, manganese, cobalt, nickel, copper, zinc, or cadmium. The technically important materials are in general mixtures of two or more of these single ferrites. The mixed zinc-ferrites offer an especially interesting case because zinc occupies only tetrahedral sites, and $ZnFe_2O_4$ itself is nonferromagnetic. Thus, if x denotes the fraction of Zn^{2+} replacing Fe^{2+}, the distribution may be represented symbolically, as Gorter has investigated in detail,[32] by the formula

$$(Zn_x^{2+}Fe_{1-x}^{3+})_{\text{tetr.}}(Fe_{1-x}^{2+}Fe_{1+x}^{3+})_{\text{oct.}}O_4.$$

As long as the zinc content is small, the ferric ions at the tetrahedral sites, antiparallel to the spins at the octahedral sites, are still able to enforce parallel orientation of the ferric and ferrous ion on the octahedral sites antiparallel to the spins at the tetrahedral sites. Since the number of antiparallel spins has decreased and zinc does not contribute, the saturation magnetization rises (Fig. 30.10). For high zinc concentrations, the enforcing action from the tetrahedral sites begins to vanish and the iron ions at the octahedral sites tend to go into antiparallel or random orientation. The saturation moment decreases again and falls towards zero. A similar trend has been observed for the nickel-zinc and manganese-zinc ferrites.

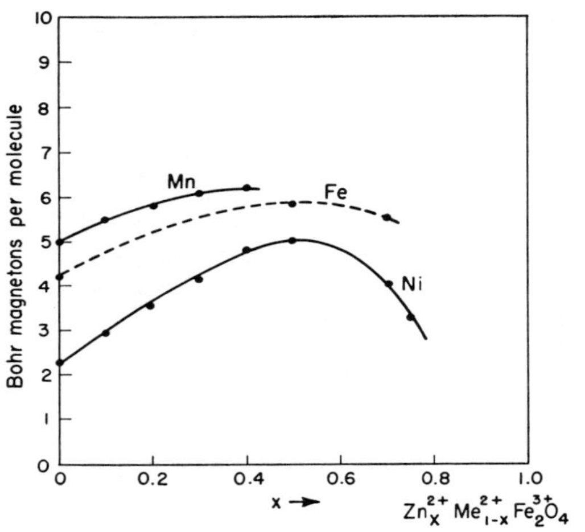

Fig. 30.10. Saturation magnetization of zinc ferrite as function of composition. (After Gorter.[32])

Obviously, ferromagnetic research has assumed a new breadth by this extension to semiconductors. Not only can new types of magnetic coupling effects be realized in wide variations, but the screening effect of the conducting electrons has been lifted, and the specific participation of the conducting electrons in magnetization and polarization may be observed. An example in kind is the transition of magnetite near $-160°C$. The λ-shaped specific heat anomaly[33] suggests a second-order transition. When a single crystal is cooled through the critical region, the conductivity drops by about two orders of magnitude; the magnetization in weak fields decreases sharply; and if a strong magnetic field is applied during cooling and then the crystal demagnetized below the transition, a most unusual effect is observed: a magnetic axis has been frozen in, oriented in the [100] direction closest to the external magnetization. In this direction the crystal can be remagnetized easily; in all other directions much larger fields are required; and this magnetic axis can be turned by a strong external magnetic field into other [100] directions[34] just as the polar axis in $BaTiO_3$ can be turned by an external electric field (see Sec. 28). Furthermore, the conductivity becomes anisotropic and dependent on the magnetic axis direction.

This surprising state of affairs becomes understandable if the transition, as first suggested by Verwey,[35] is interpreted as an electronic order-disorder transition. As mentioned previously, an equal number of Fe^{2+} and Fe^{3+} ions is distributed over the octahedral lattice sites. At room temperature, these ions exchange electrons freely and no distinction can be made between them; the lattice parameters are adjusted to an intermediate average charge distribution. With decreasing temperature the speed of the electron exchange slows down like any other process requiring activation energy. Near $-160°C$ the transfer has become so slow that the lattice vibrations can react to the momentary charge of the individual iron ions. The structure distorts selectively around the ferrous and ferric sites and freezes them into place. Since the ions are the carriers of ferro- and antiferromagnetism, the electronic ordering, when influenced by a strong magnetic field, causes also the fixation of a magnetic structure pattern.

Some other aspects should be mentioned. It was well established for metals that ferromagnetism vanishes in the microwave region;[36] shielding of the metal interior by eddy currents was blamed for this phenomenon. For the low-conducting ferrites, ranging in resistivity to 10^{10} [ohm m], this shielding by skin effect is eliminated, but still the ferromagnetism disappears as before. Closer inspection reveals wide variations in the onset and shape of the dispersion characteristics

[32] E. W. Gorter, *Nature 165,* 798 (1950).

[33] R. W. Millar, *J. Am. Chem. Soc. 51,* 215 (1929).

[34] See L. R. Bickford, Ref. 27, and B. A. Calhoun, Ref. 27.

[35] E. J. Verwey, P. W. Haayman, and F. C. Romeijn, *J. Chem. Phys. 15,* 181 (1947).

[36] See the collected data in R. M. Bozorth, *Ferromagnetism,* D. Van Nostrand Co., New York, 1951, p. 800.

for the various ferrites (Fig. 30.11). The curves frequently resemble not so much relaxation spectra but characteristics integrating over various resonance states.

Several mechanisms have been proposed to account for the ferromagnetic dispersion,[37] and each of them may have its range of application. One resonance phenomenon, not recognized by some early observers, is avoidable by selecting proper sample dimensions; the high permeability and permittivity of ferrites may make them act as cavity resonators at unexpectedly low frequencies.[38] Furthermore, in ferroelectric as well as in ferromagnetic ceramics, the individual grains, acting as electromechanical and magnetomechanical transducers, respectively, may operate at cross purposes and produce a dispersion and absorption as observed.[39]

A more intricate phenomenon is that domain walls are quasi-elastically bound to their equilibrium positions and, when moving, possess an apparent inertial mass due to the gyromagnetic spin precession; in consequence, domain wall resonances may appear.[40] Finally, the magnetic crystal anisotropy acts equivalent to an internal magnetic field that causes a Larmor precession of the magnetic spins and leads to a *ferromagnetic resonance* when an external frequency field approaches this Larmor frequency.[41] Quantum-mechanically expressed, a transition is induced between two Zeeman levels at this frequency.

This last effect is the most interesting one and belongs to the group of nuclear, para- and ferromagnetic resonance phenomena,[42, 43] which have become one of the most important tools for the study of solids and liquids since the famous nuclear resonance experiments by Bloch [44] and Purcell and co-workers.[45]

A phenomenon closely related to magnetic resonance

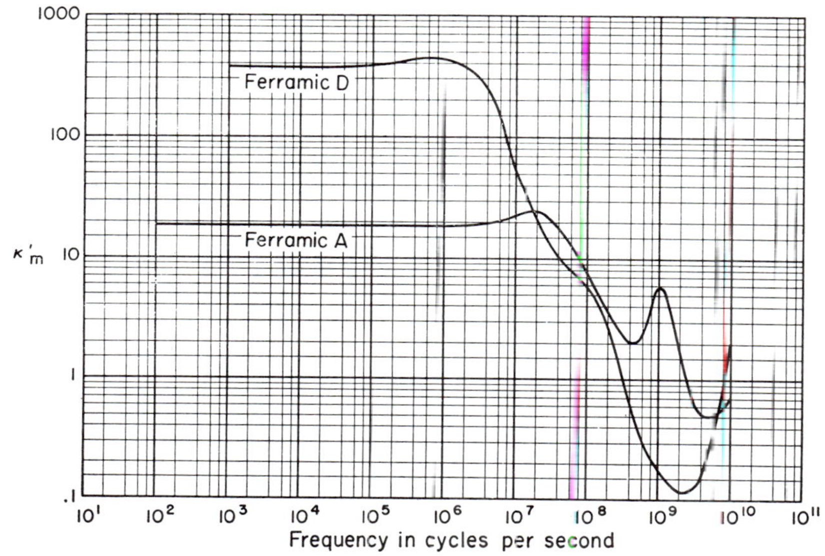

Fig. 30.11. Dispersion characteristics of two ferrites. (Samples from General Ceramics and Steatite Corp.; measurements by W. B. Westphal, Laboratory for Insulation Research.)

[37] See, for example, G. T. Rado, *Advances in Electronics*, Academic Press, Inc., New York, 1950, Vol. 2, pp. 251 ff.; *Revs. Mod. Phys. 25*, 81 (1953).

[38] F. G. Brockman, P. H. Dowling, and W. G. Steneck, *Phys. Rev. 77*, 85 (1950).

[39] A. von Hippel, *Z. Physik 133*, 158 (1952).

[40] W. Döring, *Z. Naturforsch. 3a*, 373 (1948); C. Kittel, *Phys. Rev. 80*, 918 (1950); R. Becker, *J. phys. et radium 12*, 332 (1951); G. T. Rado, R. W. Wright, and W. H. Emerson, *Phys. Rev. 80*, 273 (1950); G. T. Rado, *Phys. Rev. 83*, 821 (1951).

[41] L. Laudan and E. Lifshitz, *Phys. Z. Sovjetunion 8*, 153 (1935); J. L. Snoek, *Philips Tech. Rev. 8*, 353 (1946); *Physica 14*, 207 (1948); D. Polder and J. Smit, *Revs. Mod. Phys. 25*, 89 (1953).

[42] See, for example, the surveys by K. K. Darrow, "Magnetic Resonance," *Bell System Tech. J. 32*, 74, 384 (1953), and G. E. Pake, "Fundamentals of Nuclear Magnetic Resonance Absorption," *Am. J. Phys. 18*, 438, 473 (1950).

[43] Ferromagnetic resonance was first observed by Griffiths [J. H. E. Griffiths, *Nature 158*, 670 (1946)] in iron and the precession theory for ferromagnetic substances formulated by Kittel [C. Kittel, *Phys. Rev. 73*, 155 (1948); *J. phys. et radium 12*, 291 (1951)]. Resonance experiments on magnetic single crystals in the Laboratory for Insulation Research by Bickford [27] demonstrated the value of this new tool for the determination of g factors and crystal anisotropy in ferromagnetics.

[44] F. Bloch, *Phys. Rev. 70*, 460 (1946); F. Bloch, W. W. Hansen, and M. Packard, *Phys. Rev. 70*, 474 (1946).

[45] E. M. Purcell, H. C. Torrey, and R. V. Pound, *Phys. Rev. 69*, 37 (1946).

is the Faraday effect [46] (see Appendix AII, 5). If a magnetic axis is established in a material around which electron spins can precess, this axis acts on an electromagnetic wave like a dynamic screw axis. A linearly polarized wave propagated parallel to the axis can be thought of as composed of a right- and a left-hand circular component (see I, Sec. 12) which encounter different permeabilities, hence travel with different velocities. In consequence, a rotation of the plane of polarization results. In addition, when the frequency of the wave corresponds to the precession frequency of the electrons, the circular component rotating in the precession direction is strongly absorbed and thus highly attenuated. The ferromagnetic Faraday effect in the ferrites has become recently of technical importance, especially through the work of Hogan,[47] for the development of new types of switching components in the microwave range, called *gyrators* (Fig. 30.12). These components do not obey the reciprocity law; the plane of polarization of a reflected wave is rotated in the same direction as that of the incident wave.

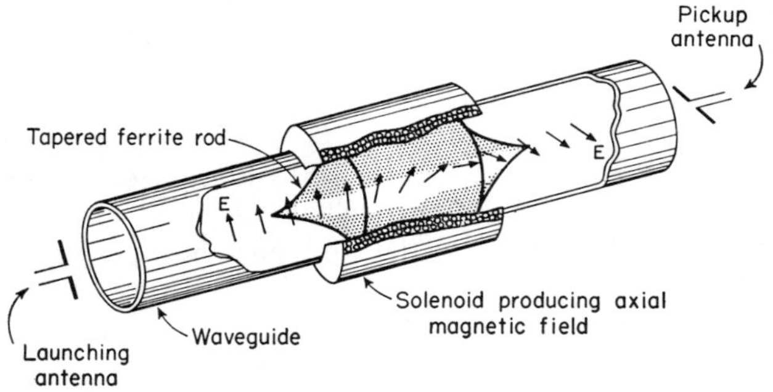

Fig. 30.12. Gyrator operating by ferromagnetic Faraday effect.

[46] M. Faraday, *Phil. Trans. 1* (1846); *Pogg. Ann. 68*, 105 (1846).

[47] C. L. Hogan, *Bell System Tech. J. 31*, 1 (1952); *Revs. Mod. Phys. 25*, 253 (1953).

31 · Interfacial and Space-Charge Polarization

At the outset of the molecular discussion (Sec. 1), four mechanisms of polarization were introduced, but only three have been discussed thus far: the electronic, the atomic, and the dipole orientation polarization. These three effects have in common that they are caused by the displacement or orientation of bound charge carriers. The remaining process, the space-charge or interfacial polarization, is produced by traveling charge carriers and confronts us with a completely different situation. Previously, the atoms and molecules found themselves under the influence of a local field consisting essentially of the applied field, modulated by the polarization of the surroundings. Now, large-scale field distortions enter, caused by the piling up of space charges in the volume or of surface charges at the interfaces of dielectrics.

The classical example of interfacial polarization is the Maxwell-Wagner two-layer condenser [1] (Fig. 31.1). The dielectric consists of two parallel sheets of materials (1) and (2), characterized by their dielectric constant, conductivity, and thickness (ϵ_1', σ_1, d_1) and (ϵ_2', σ_2, d_2), respectively. When a d-c field is suddenly applied, the initial field distribution corresponds to the electrostatic requirement of constant flux density

$$D_1 = D_2 \quad \text{or} \quad E_1/E_2 = \epsilon_2'/\epsilon_1', \tag{31.1}$$

whereas the final distribution follows from the condition of current continuity

$$J_1 = J_2 \quad \text{or} \quad E_1/E_2 = \sigma_2/\sigma_1. \tag{31.2}$$

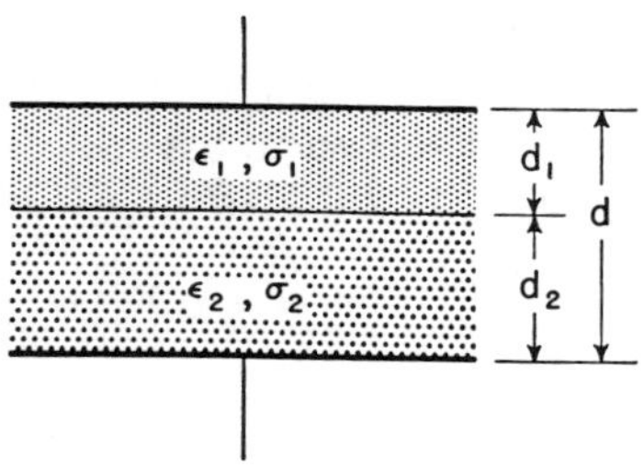

Fig. 31.1. Maxwell-Wagner two-layer condenser.

The transient which links the initial and final state may be derived from the equivalent circuit of Fig. 31.2 with

[1] J. C. Maxwell, *Electricity and Magnetism*, Clarendon Press, Oxford, Vol. 1, 452 (1892); K. W. Wagner, *Die Isolierstoffe der Elektrotechnik*, edited by H. Schering, Berlin, Springer, 1924, pp. 1 ff.

$$\mathcal{V} = \mathcal{V}_1 + \mathcal{V}_2$$

$$I = C_1 \frac{d\mathcal{V}_1}{dt} + \frac{\mathcal{V}_1}{R_1} = C_2 \frac{d\mathcal{V}_2}{dt} + \frac{\mathcal{V}_2}{R_2}, \tag{31.3}$$

$$C_1 = \frac{A}{d_1}\epsilon_1' \qquad C_2 = \frac{A}{d_2}\epsilon_2'$$

$$R_1 = \frac{d_1}{A\sigma_1} \qquad R_2 = \frac{d_2}{A\sigma_2}, \tag{31.4}$$

where A is the surface area of the plate capacitor.

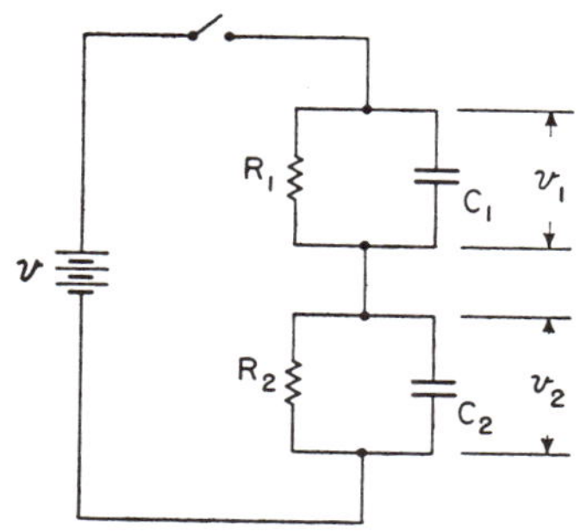

Fig. 31.2. Equivalent circuit of two-layer condenser.

Solving Eqs. 31.3 for $\mathcal{V}_1$ and $\mathcal{V}_2$, we obtain

$$\mathcal{V}_1 = \mathcal{V}\frac{R_1}{R_1 + R_2}\left\{1 - \left(1 - \frac{C_2 R_2}{\tau}\right)e^{-t/\tau}\right\},$$

$$\mathcal{V}_2 = \mathcal{V}\frac{R_2}{R_1 + R_2}\left\{1 - \left(1 - \frac{C_1 R_1}{\tau}\right)e^{-t/\tau}\right\}, \tag{31.5}$$

where the relaxation time

$$\tau = \frac{R_1 R_2 (C_1 + C_2)}{R_1 + R_2} = \frac{\epsilon_1' d_2 + \epsilon_2' d_1}{\sigma_1 d_2 + \sigma_2 d_1}. \tag{31.6}$$

The field changes exponentially from initial to final distribution.

The steady-state solution, when an a-c voltage $\mathcal{V} = \mathcal{V}_0 e^{j\omega t}$ is applied, can be derived from the admittance of the circuit

$$Y = \frac{I}{\mathcal{V}} = \frac{Y_1 Y_2}{Y_1 + Y_2}, \tag{31.7}$$

where

$$Y_1 = \frac{1}{R_1} + j\omega C_1 = \frac{1 + j\omega\tau_1}{R_1},$$

$$Y_2 = \frac{1}{R_2} + j\omega C_2 = \frac{1 + j\omega\tau_2}{R_2}. \tag{31.8}$$

The time constant τ of Eq. 31.6, expressed in terms of τ_1 and τ_2 of the two individual RC circuits, is

$$\tau = \frac{R_1\tau_2 + R_2\tau_1}{R_1 + R_2}; \tag{31.9}$$

hence the admittance may be rewritten

$$Y = \frac{1}{R_1 + R_2}\,\frac{(1 + j\omega\tau_1)(1 + j\omega\tau_2)}{1 + j\omega\tau}. \tag{31.10}$$

According to I, Eq. 1.10, the admittance determines the complex permittivity of the capacitor as

$$Y = j\omega\kappa^* C_0,$$

where

$$C_0 = \frac{A\epsilon_0}{d}, \qquad d = d_1 + d_2. \tag{31.11}$$

Hence the two-layer condenser appears to the outside observer as a dielectric of the dielectric constant

$$\kappa' = \frac{\tau_1 + \tau_2 - \tau + \omega^2\tau_1\tau_2}{C_0(R_1 + R_2)(1 + \omega^2\tau^2)}, \tag{31.12}$$

which changes from its static value ($\omega = 0$)

$$\kappa_s' = \frac{\tau_1 + \tau_2 - \tau}{C_0(R_1 + R_2)}, \tag{31.13}$$

to its optical value ($\omega \to \infty$)

$$\kappa_\infty' = \frac{\tau_1\tau_2}{C_0(R_1 + R_2)\tau} = \frac{1}{C_0}\,\frac{1}{\left(\dfrac{1}{C_1} + \dfrac{1}{C_2}\right)} \tag{31.14}$$

as the frequency ω increases. Introducing κ_s' and κ_∞' into Eq. 31.12, we obtain

$$\kappa' = \kappa_\infty\left\{1 + \frac{k}{1 + \omega^2\tau^2}\right\}, \tag{31.15}$$

where

$$k = \frac{(\tau_1 + \tau_2 - \tau)\tau - \tau_1\tau_2}{\tau_1\tau_2} = \frac{\kappa_s' - \kappa_\infty'}{\kappa_\infty'}. \tag{31.16}$$

The dissipation factor of the two-layer dielectric becomes

$$\kappa'' = \kappa_\infty'\left(\frac{\tau}{\omega\tau_1\tau_2} + \frac{k\omega\tau}{1 + \omega^2\tau^2}\right). \tag{31.17}$$

A comparison of these expressions with Eq. 22.17 shows that the two-layer condenser gives a relaxation spectrum indistinguishable from the simple orientation polarization of the Debye theory as far as κ' is concerned (Fig. 31.3). The κ'' characteristic contains, in addition, the ohmic conductivity term

$$\sigma = \frac{\kappa_\infty'\tau}{\tau_1\tau_2} = \frac{1}{C_0(R_1 + R_2)} \tag{31.18}$$

caused by the series resistor $(R_1 + R_2)$. The complex permittivity may thus be written

$$\kappa^* = \kappa_\infty' + \frac{\kappa_s' - \kappa_\infty'}{1 + j\omega\tau} - j\frac{\sigma}{\omega}. \tag{31.19}$$

From the molecular point of view, the optical dielectric constant of the two-layer condenser (Eq. 31.14)

$$\kappa_\infty' = \frac{d/\epsilon_0}{\left(\frac{d_1}{\epsilon_1'} + \frac{d_2}{\epsilon_2'}\right)} \tag{31.20}$$

is determined by the real permittivities of the two media, composed of the static contributions of the elec-

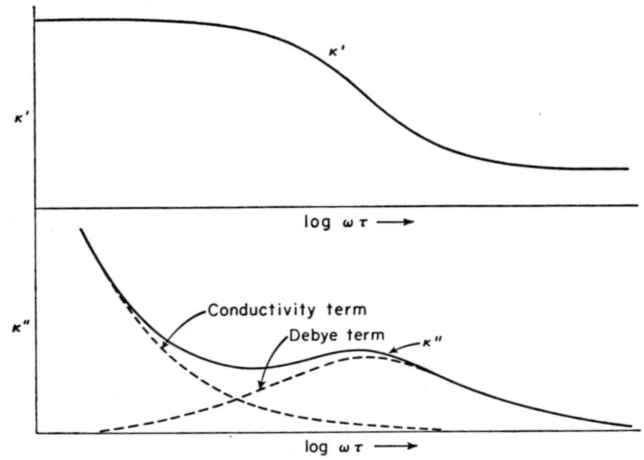

Fig. 31.3. Relaxation spectrum of the two-layer condenser.

tronic, atomic, and orientation polarizations. The static dielectric constant (see Eq. 31.13)

$$\kappa_s' = \kappa_\infty'\left\{1 + d_1 d_2\left(\frac{\frac{1}{\sigma_1}\sqrt{\frac{\epsilon_1'}{\epsilon_2'}} - \frac{1}{\sigma_2}\sqrt{\frac{\epsilon_2'}{\epsilon_1'}}}{\frac{d_1}{\sigma_1} + \frac{d_2}{\sigma_2}}\right)^2\right\} \tag{31.21}$$

is larger than the optical one because media (1) and (2) contain mobile charge carriers of the densities N_1 and N_2 [m^{-3}], transporting charges e_1 and e_2 with the mobilities b_1 and b_2. Because of these conductivities

$$\sigma_1 = N_1 e_1 b_1, \quad \sigma_2 = N_2 e_2 b_2, \tag{31.22}$$

charges pile up at the interface between media (1) and (2) until constant current transfer is established and the static conductivity results (see Eq. 31.18),

$$\sigma = \frac{d/\epsilon_0}{\left(\frac{d_1}{\sigma_1} + \frac{d_2}{\sigma_2}\right)}. \tag{31.23}$$

As the frequency increases, the interfacial polarization begins to lag. The dielectric constant κ' decreases to the midpoint between the static and optical value when $\omega = 1/\tau$; the relaxation time τ (Eq. 31.6) may be written in analogy to the preceding equations

$$\tau = \frac{\epsilon_1' d_2 + \epsilon_2' d_1}{\sigma_1 d_2 + \sigma_2 d_1} = \frac{\epsilon_1'\epsilon_2'\left(\frac{d_1}{\epsilon_1'} + \frac{d_2}{\epsilon_2'}\right)}{\sigma_1\sigma_2\left(\frac{d_1}{\sigma_1} + \frac{d_2}{\sigma_2}\right)}. \tag{31.24}$$

By measuring κ_∞', κ_s', σ, and τ we can calculate the dielectric parameters ϵ_1', ϵ_2', σ_1, and σ_2 of the two layers if the thickness ratio d_1/d_2 is given.

Since the polarization $\mathbf{P}$ of the two-layer condenser is proportional to the applied field, it is possible to define a space charge polarizability α_s of the overall dielectric in analogy to Eq. 1.5 by writing

$$\mathbf{P} = N\alpha_s\mathbf{E} = (\kappa^* - \kappa_\infty')\epsilon_0\mathbf{E}; \tag{31.25}$$

hence from Eq. 31.19

$$\alpha_s = \left(\frac{\kappa_s' - \kappa_\infty'}{1 + j\omega\tau} - j\frac{\sigma}{\omega}\right)\frac{\epsilon_0}{N}. \tag{31.26}$$

The only merit of this expression is that it allows a discussion of the total polarization as the sum of the four terms indicated in Eq. 1.6.

By dividing the layers (1) and (2) into n_1 and n_2 equal sublayers of the thicknesses Δ_1 and Δ_2, obviously nothing is changed; we replace in our equations

$$d_1 = n_1\Delta_1, \quad d_2 = n_2\Delta_2. \tag{31.27}$$

Next we might stack the sublayers (1) and (2) alternately or in any other sequence (Fig. 31.4); since the

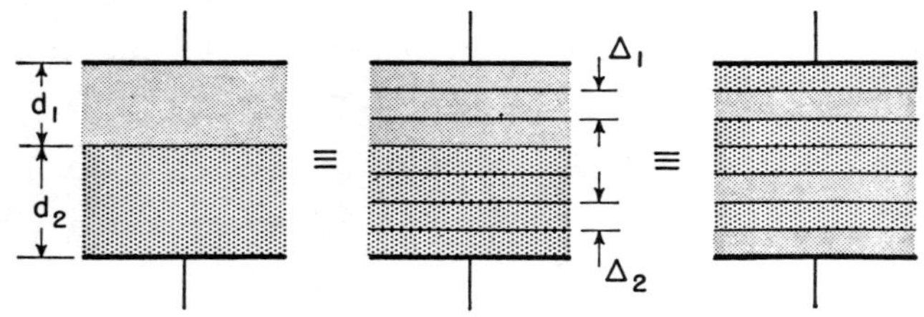

Fig. 31.4. Electrical identity of two-layer and multi-layer condenser (relative amount and shape of the two media maintained).

total impedance is not altered, the dielectric response still remains unchanged. By subdividing to molecular thicknesses, we could actually make a continuous transition from interfacial to atomic polarization if the latter would correspond to a relaxation and not to a resonance phenomenon. As long as the geometrical shape and orientation in the field remains the same

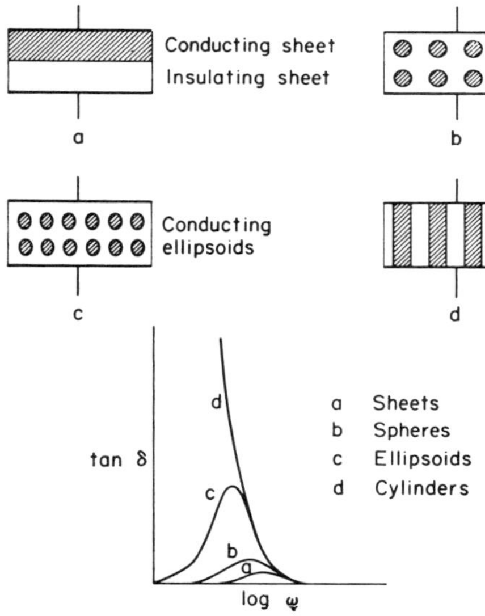

Fig. 31.5. Dependence of dielectric response on shape and orientation of particles of medium 2. (After Sillars.[3])

(layers stacked in series) and the relative amount of medium (1) to (2) unaltered (expressed by the ratio of the total thicknesses, d_1/d_2), the complex permittivity of the composite dielectric remains unaffected.

We might change the geometry of medium (2), while preserving its mass, and disperse it in medium (1) in the shape of spheres [2] or ellipsoids or rods [3] so diluted that the interaction between the particles can be neglected. Since the field distortion caused by a polarized particle depends on its shape and orientation relative to the applied field, κ^* will now vary with the shape and orientation of the medium (2) particles, as Figs. 31.5 and 25.1 indicate. Thus, for example, a distribution of relaxation times may be produced by varying the orientation of ellipsoidal particles. The absolute particle size, just as above the absolute thickness of the layers, does not enter into the calculations but only the mass ratio of the two media.

Medium (2), distributed in base medium (1), highly diluted and in prescribed shape and orientation, gives a simple relaxation spectrum of only one relaxation time. When n media are dispersed in medium (1), n relaxation times occur; correspondingly, a capacitor of $(n + 1)$ different dielectric layers has n relaxation times. Hence, broad-band absorbers may be realized, for example, by stacks of properly selected dielectric disks.

When the concentration c of material (2) in the base medium (1) is increased, the distorted field areas around each particle begin to overlap and affect each other. Obviously, the dielectric characteristics, plotted as $f(c)$, must start with the properties of medium (1) and terminate with those of medium (2). The shape of the characteristics between these two end points has been a problem for discussion for many years,[4] and frequently a *logarithmic mixing rule* given by Lichtenecker [5] is used to determine the permittivity κ_m' of a mixture from κ_1', κ_2' and the volume ratios θ_1, θ_2 of the components:

$$\log \kappa_m' = \theta_1 \log \kappa_1' + \theta_2 \log \kappa_2'. \tag{31.28}$$

This rule has been well-confirmed by Büchner [6] for mixed dielectrics containing TiO_2 (rutile) as the one of their two phases (Fig. 31.6). However, the validity of any rule is obviously limited, because the shape, size, and distribution of the particles are empirical parameters affecting the polarization.

Interfacial polarization results when abutting dielectrics differ in conductivity and therefore require different voltage gradients to transconduct a current of constant density. The migration of the charge carriers

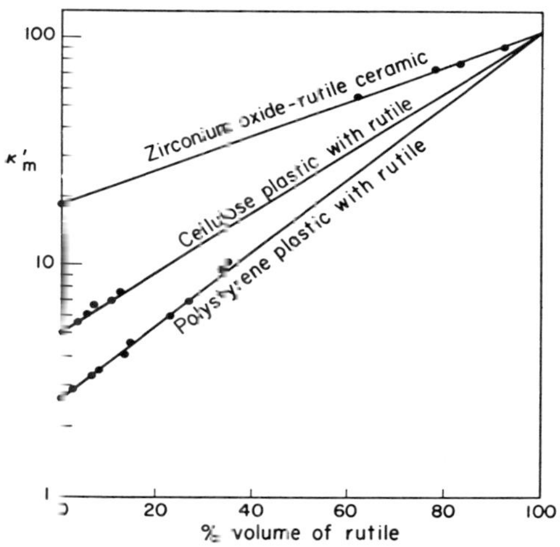

Fig. 31.6. The logarithmic mixing rule applied to TiO_2 compositions. (After Büchner.[6])

through each individual dielectric and their transfer to the electrodes is thought to be unimpeded. This tacit assumption is actually very stringent and not at all fulfilled for most materials, as anyone knows who has

[2] K. W. Wagner, *Arch. Elektrotech. 2,* 371 (1914).
[3] R. W. Sillars, *J. Inst. Elec. Engrs. (London) 80,* 378 (1937).

[4] See K. Lichtenecker, *Physik. Z. 27,* 115 (1926); *37,* 906 (1936); D. A. G. Bruggeman, *Ann. Physik* [5] *24,* 636 (1935).
[5] K. Lichtenecker, *Physik. Z. 10,* 1005 (1909).
[6] A. Büchner, *Wiss. Veröffentl. Siemens-Werken 18,* 84 (1939).

been shocked by the *dielectric after-effect* of a homogeneous capacitor dielectric. The phenomenon, as described by many investigators, is as follows: a constant voltage $\mathcal{V}$ applied at $t = 0$ to a uniform material (for example, a crystal plate) causes, after the initial surge of charge ($C\mathcal{V}$) has passed, a small charging current I_c that decays over minutes, hours, days, or even months. Shorting of the electrodes after a time T produces an opposite discharging current $-I_d$ that decreases according to the identical law. The discharge current $-I_d$ may be represented at any time $t_1 \geq T$ as the difference of a current $-I_c$, starting at $t = T$, and the primary positive current $+I_c$, extrapolated to $t = t_1$. This *law of superposition* (Fig. 31.7),

$$-I_d(t_1) = -I_c(t_1 - T) + I_c(t_1), \qquad (31.29)$$

expresses, in a somewhat involved fashion, the simple statement that the total residual charge stored by the dielectric is redelivered in the discharge process.

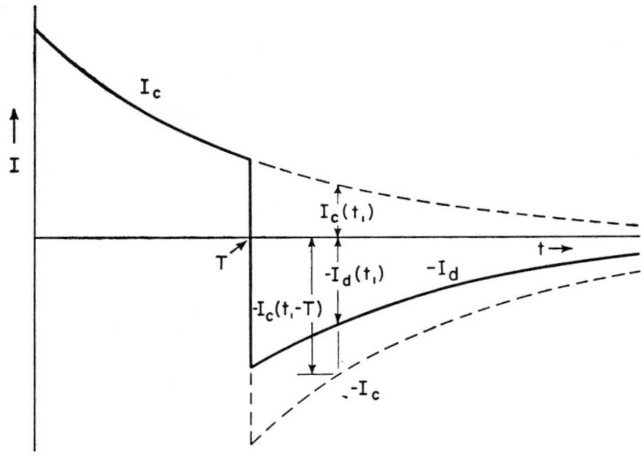

Fig. 31.7. The law of superposition.

This dielectric after-effect is, in general, the sign of space-charge build-up in the dielectric material. The charge carriers are required to migrate through the volume, to discharge freely at the one electrode, and to be replaced spontaneously at the other electrode, if field distortion is to be avoided. This is a most unlikely combination of circumstances in nonmetals. Thus space-charge polarization plays a most important and frequently not clearly recognized role in electric polarization phenomena and is, for example, the working principle of most *electrets*. Depending on the relative mobility of cations and anions and the electrode situation, a variety of field distribution characteristics can be realized; some typical curves, obtained by Joffé and his co-workers in their early work,[7] are shown in Fig. 31.8.

[7] See A. F. Joffé, *The Physics of Crystals*, McGraw-Hill Book Co., New York, 1928.

A very clear-cut case has been investigated recently:[8] the space-charge build-up by photoeffect in additively colored alkali halide crystals. Here exists, in a first approximation, the ideal situation of an electron cloud

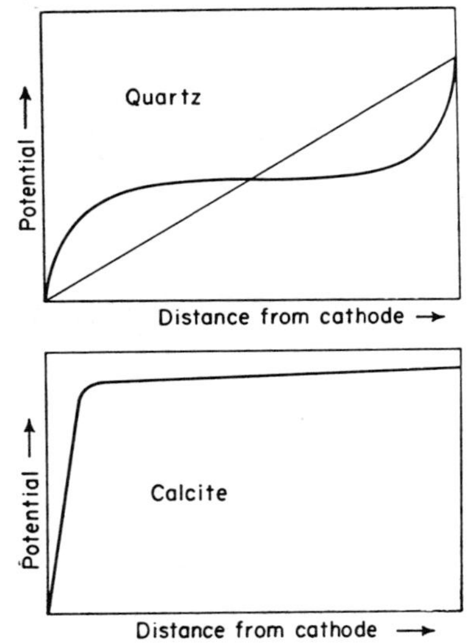

Fig. 31.8. Polarization in quartz and calcite. (After Joffé.[7])

of uniform density, frozen into a compensating positive matrix. By heating the crystal in alkali vapor, we can replace a number of the halogen ions by electrons attached to the anion vacancies, and the F-center absorption results (Fig. 31.9). These electrons can be

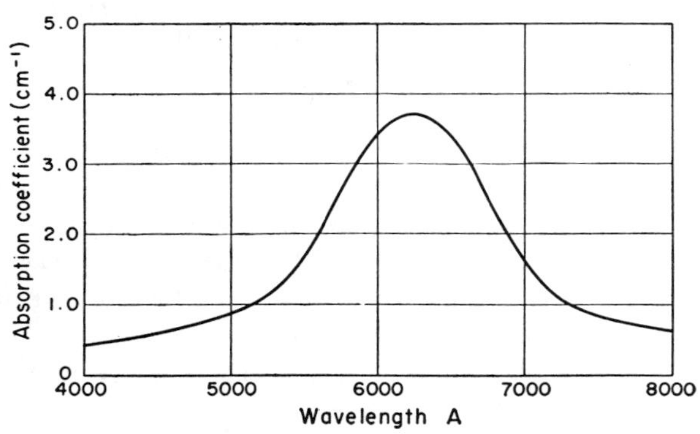

Fig. 31.9. F-center absorption in KBr.

mobilized by light absorption in the F band, in number equal to the quanta absorbed times the quantum yield, and will then drift towards the anode until they are discharged or retrapped by anion vacancies. This motion of the electron cloud towards the anode leaves a bleached region of positive space charge in front of the cathode, as long as no electrons can re-enter the crystal. If the initial concentration of color centers is suffi-

[8] A. von Hippel, E. P. Gross, J. G. Jelatis, and M. Geller, *Phys. Rev. 91*, 568 (1953).

ciently small and the applied voltage sufficiently high, the final state will be a completely bleached crystal. For higher concentrations or lower voltages, the crystal will consist of two distinct sections: a completely bleached region I in front of the cathode taking the total voltage, and a field-free unbleached region II with the original color-center concentration (Fig. 31.10).

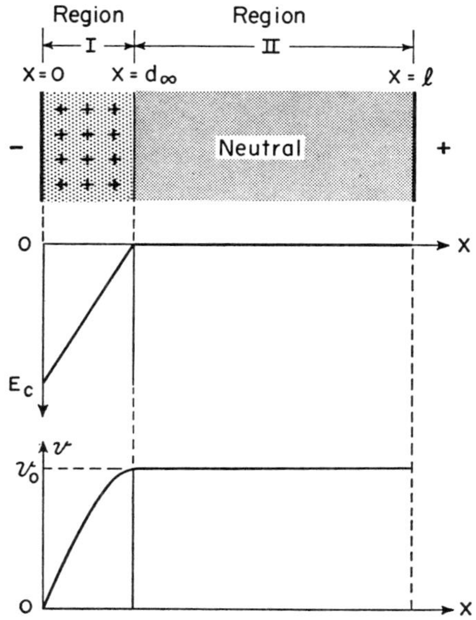

Fig. 31.10. Space charge, field strength, and voltage distribution in colored alkali halide crystal after charging.

The progress of the *shock front* ($x = d(t)$) separating regions I and II can be calculated as it advances from the cathode and, slowing down, comes to a final standstill at the distance $x = d_\infty$. The plunger-like motion of the shock front pushing the electron cloud out is reflected in the external circuit by a current decreasing as a function of time.

We find that this current is represented by the somewhat complicated expression

$$I(t) = I(0)e^{-t/\tau}\left\{\frac{(1 - e^{-t/\tau})\dfrac{d_\infty}{l} + (1 + e^{-t/\tau})}{(1 + e^{-t/\tau})^3}\right\}, \tag{31.30}$$

where the relaxation time τ, given by the positive space-charge density ρ_0, the mobility of the electrons b, the dielectric constant of the crystal ϵ', the length of the crystal l, and the applied voltage $\mathcal{V}_0$, is

$$\tau = \left(\frac{\epsilon' l^2}{2\rho_0 b^2 \mathcal{V}_0}\right)^{1/2}. \tag{31.31}$$

In case the final cathode fall d_∞ is very much smaller than the crystal length ($d_\infty/l \ll 1$), Eq. 31.30 simplifies to

$$I(t) \simeq \frac{I(0)}{\cosh^2\left(\dfrac{t}{2\tau}\right)} \tag{31.32}$$

(see Fig. 31.11).

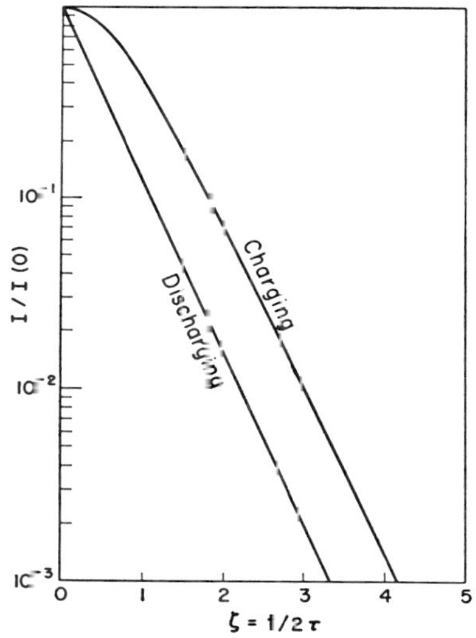

Fig. 31.11. Time dependence of photocurrent through colored alkali halide crystal (theoretical).

The initial deviation of the current from a simple exponential drop is caused by the charge transconducted to the anode. Once the final state is established, discharging and recharging at the same voltage simply

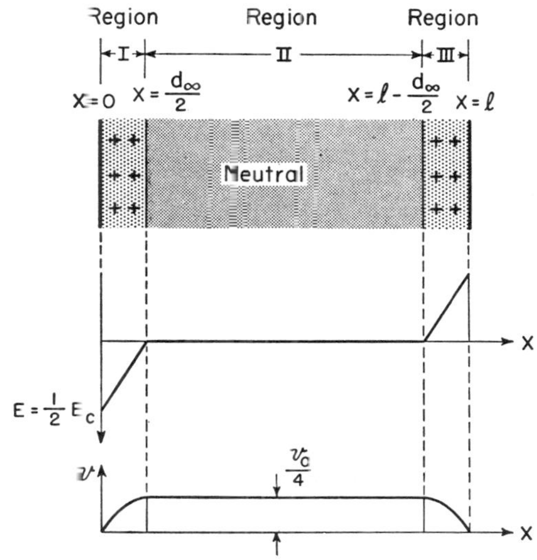

Fig. 31.12. Space charge, field strength and voltage distribution in shorted crystal (theoretical).

pushes the electron cloud forward and back without further loss of electrons. The field distortion after discharging is shown in Fig. 31.12. Recharging is accompanied by a simple exponential current drop

$$I(t) = I(0)\left(1 - \frac{d_\infty}{l}\right)e^{-t/\tau}. \qquad (31.33)$$

For this latter case, the law of superposition (Eq. 31.29) holds, while it will not strictly apply to the initial transient because electrons are removed from the crystal (Eq. 31.30). The initial polarization created in a dielectric cannot be wiped out completely by a subsequent discharge as long as the charge carriers transferred to the electrodes cannot be replaced.

The cathode fall built up in this experiment has the length

$$d_\infty = \left(\frac{2\epsilon'\mathcal{V}_0}{\rho_0}\right)^{1/2} \quad [\mathrm{m}], \qquad (31.34)$$

and produces directly in front of the cathode the maximum field strength

$$E_c = -\frac{\rho_0}{\epsilon'}d_\infty = -\left(\frac{2\mathcal{V}_0\rho_0}{\epsilon'}\right)^{1/2} \quad [\mathrm{v\ m^{-1}}]. \qquad (31.35)$$

The field strength at the cathode can thus be built up to extreme magnitudes by the simple device of using a highly colored crystal. The number of F centers per unit volume, N_0, determines, together with the elementary charge e, the space-charge density as

$$\rho_0 = N_0 e. \qquad (31.36)$$

For $N_0 = 10^{24}$ [$\mathrm{m^{-3}}$], a coloration achieved without great difficulty, the application of 10 v should produce a cathode fall

$$\begin{aligned} d_\infty &\simeq 8 \times 10^{-6} \quad [\mathrm{m}], \\ E_c &\simeq 2.5 \times 10^{8} \quad [\mathrm{v\ m^{-1}}]. \end{aligned} \qquad (31.37)$$

This field strength exceeds the breakdown strength of most alkali halide crystals.[9] Before it is reached, field emission sets in and stabilizes a lower gradient. In this way we have successfully produced field emission into alkali halide crystals with voltages $\leq$ 1 volt.

[9] See A. von Hippel, *J. Appl. Phys. 8*, 815 (1937).

In these experiments the mobility of the charge carriers and the space-charge build-up can be controlled from the outside by the wavelength and intensity I_0 of the illumination. This extreme case illustrates well that the existence of a mobility and the validity of Ohm's law does not presuppose a simple model of free-charge transfer through a medium of high friction. On the contrary, the photoelectrons in the alkali halides travel only a very short distance after their release, as Pohl and his co-workers established.[10] Ohm's law is valid because the drift distance $\bar{w}$ in the field direction increases proportionally to the driving field strength. Let $\bar{w}_0$ designate the average drift distance per unit field, and t_b and t_f the average time intervals, respectively, during which an electron is bound or travels freely. The drift velocity may then be written

$$\bar{v} = bE = \frac{\bar{w}_0}{t_b + t_f}E. \qquad (31.38)$$

Obviously, the time an electron stays bound varies in our experiments inversely proportionately to the intensity of the releasing light,

$$t_b = \frac{1}{\beta I_0}. \qquad (31.39)$$

When this time is long compared with that of free travel ($t_b \gg t_f$), the mobility

$$b \simeq \frac{\bar{w}_0}{t_b} \simeq \bar{w}_0\beta I_0 \qquad (31.40)$$

is proportional to the intensity of illumination.

The controlled build-up of space charges is obviously a powerful tool for learning more about the motion of charge carriers and about their release at the electrodes under the influence of very intense fields.

[10] W. Flechsig, *Physik. Z. 32*, 843 (1931); K. Hecht, *Z. Physik 77*, 235 (1932).

32 · Conduction and Breakdown

The preceding discussion of interfacial and space-charge polarization has introduced certain aspects of conduction phenomena. The existence of an ohmic conductivity in dielectrics was considered and a simple molecular interpretation of such conductivity given as

$$\sigma = Neb \qquad (32.1)$$

by visualizing mobile charge carriers of the density N carrying individual charges e and drifting with a mobility b in the field direction. Next, it was shown by the example of the colored alkali halide crystals that the actual molecular situation may be much more complex; the charge carriers may be trapped in general and released only at statistical intervals. This raises the general question why polar charge carriers can exist in dielectrics without being neutralized, what their laws of motion are in various surroundings, and how their behavior changes as a function of temperature and field strength. We will try to give answers by proceeding, as previously in the treatment of polarization, from gases to liquids and solids. In this way one starts with elementary, well-isolated phenomena and acquires some guiding principles for the handling of the complex situations encountered in the condensed phases.

Gases

If our plate capacitor of I, Sec. 1, is filled with a gas of low pressure and connected, not to an alternating voltage of increasing frequency as previously, but to a d-c voltage of increasing magnitude, a very sensitive current recorder traces a characteristic, as shown in Fig. 32.1. This type of curve, first analyzed by Town-

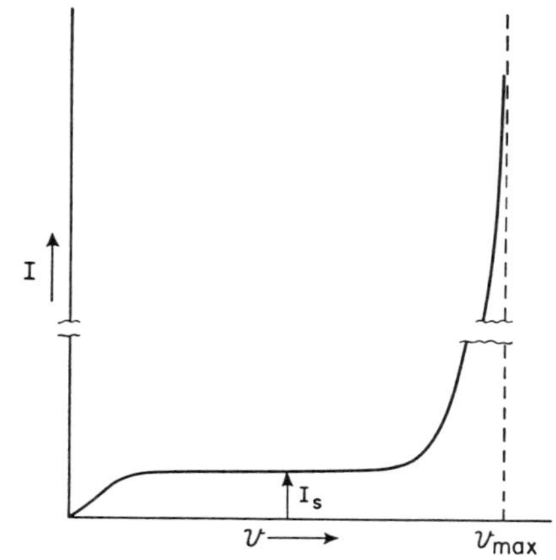

Fig. 32.1. Current-voltage characteristic at low gas pressure (schematic).

send,[1] contains two main items of information: (1) After an initial rise, the characteristic levels off to a *saturation current,* indicating that charge carriers are generated in the gas at a constant rate. (2) Increase of the voltage beyond this region leads to a second rapid current rise, until, at a critical *breakdown voltage* $V_{\max}$, the gas is transformed suddenly from a relatively good insulator into an extremely good conductor.

To explain this rapid increase of charge carriers, Townsend introduced the avalanche concept, the prototype of the neutron avalanches in atomic bomb explosions. An electron, falling in field direction a distance dx, liberates a new electron by *impact ionization* with a probability $\alpha\, dx$; hence n electrons at x increase to $n + dn$, where

$$dn = n\alpha\, dx. \qquad (32.2)$$

Thus n_0 electrons starting at the cathode ($x = 0$) have augmented to

$$N = n_0 e^{\alpha d}\,\dagger \qquad (32.3)$$

when arriving at the anode ($x = d$); that is, each starting electron produces an *electron avalanche,* and, in consequence, leaves behind a *positive ion avalanche* of the height

$$H = e^{\alpha d} - 1. \qquad (32.4)$$

Each positive ion, migrating in the opposite direction, may liberate an electron with the probability γ at the cathode. Hence, when the ion avalanche grows so large that the regeneration of a starting electron becomes certainty,

$$\gamma(e^{\alpha d} - 1) = 1, \qquad (32.5)$$

the discharge becomes self-supporting and breakdown must occur.

This *breakdown condition* of Townsend has as its molecular parameters an *ionization probability* α representing the number of ionizing impacts per electron and unit distance and a *regeneration probability* γ; both have to be interpreted in detail. In addition, we have to investigate how the layer thickness d of the gas dielectric actually affects the breakdown strength; finally, the origin of the starting electrons n_0 has to be established.

The probability that an electron ionizes by impact depends obviously on the kinetic energy of such electron, on its trajectory, and on the electronic structure of the atom or molecule with which it collides. Since the electron cloud configurations of atoms and molecules have many excited states (see Secs. 11 and 14), a strong probability exists that the kinetic energy, accumulated by the striking electron in the external field, is squandered prematurely in electronic excitation leading to light emission or other secondary processes. The electron is thus slowed down suddenly and has to start all over again, as graphically illustrated in the current-voltage characteristic of the famous Franck-Hertz experiment (Fig. 6.4). Furthermore, in electronegative gases the electrons may be trapped while slow (Fig. 32.2) and thus inactivated as negative ions.[2] Ions are not effective as ionizers by impact, until they reach much higher kinetic energies than electrons. To predict the build-up of an electron avalanche requires

[1] J. S. Townsend, *Electricity in Gases,* Oxford University Press, Oxford, 1914.

† More accurately, the exponential term in this and the consecutive equations should be written $e^{\int_0^d \alpha\, dx}$, since the field may be distorted by space charges as discussed below.

[2] See H. S. W. Massey and E. H. S. Burhop, *Electronic and Ionic Impact Phenomena,* Clarendon Press, Oxford, 1952; the measurements shown in Fig. 32.2 are by N. E. Bradbury, *Phys. Rev. 44,* 883 (1933), and H. L. Brose, *Phil. Mag. 50,* 536 (1925).

therefore a detailed discussion of the motion of an electron as a function of its velocity. Ohm's law is obviously not valid when the electrons become accelerated by the external field. It has to be replaced by a statistical calculation of the probability with which electrons

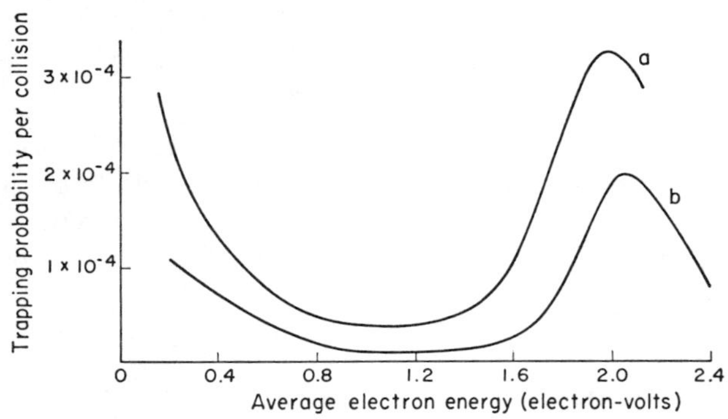

Fig. 32.2. Probability of electron trapping in O_2 as a function of the average electron energy. [(*a*) Bradbury,[2] electron-filter method; (*b*) Brose,[2] diffusion method.]

of various kinetic energies may penetrate the *friction barrier of electronic excitation states.* As found experimentally, the ionization probability characteristics traverse relatively flat maxima for electron velocities in the range of 10^2 ev (Fig. 32.3).[3] Townsend's ionization

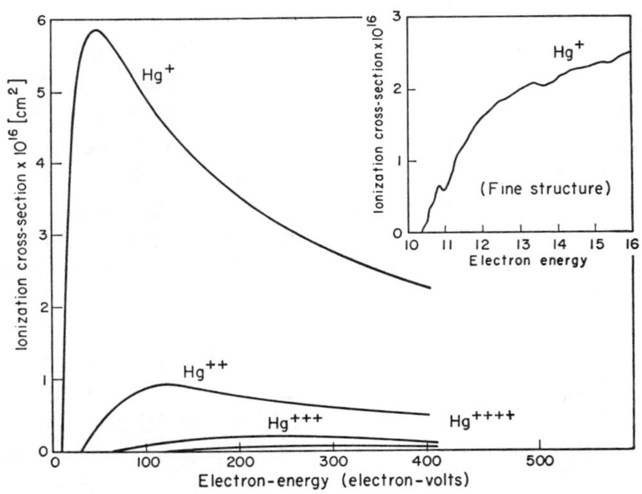

Fig. 32.3. Ionization cross sections of mercury for different degrees of ionization as a function of the electron velocity. (After Bleakney and Smith; [3] fine structure after Lawrence.[3])

probability α is an experimental term integrating over a wide and ill-defined velocity range of these characteristics.

[3] See Massey and Burhop; [2] the measurements shown in Fig. 32.3 in mercury are due to W. Bleakney and L. G. Smith, *Phys. Rev. 49*, 402 (1936), and E. O. Lawrence, *Phys. Rev. 28*, 947 (1926).

A similarly complex situation faces us in the evaluation of the regeneration probability γ.[4] The probability that a positive ion liberates an electron at the cathode is a function of the ion type, its kinetic energy, and of the metal and its surface treatment. Furthermore, a new starting electron may be liberated at the cathode, not only by ion impact, but also by photoeffect, or thermal or field emission. Finally, electrons may originate in the gas instead of at the cathode, whether by photoeffect or through the ionization by excited atoms or molecules (*collisions of the second kind* [5]). These effects frequently may enter in succession while the breakdown develops.

The thickness d of the gas dielectric enters exponentially into the height of the Townsend avalanche (see Eq. 32.4); hence H reaches its decisive magnitude during the last ionizing impacts in front of the anode. A slight increase in the gap distance d should cause a very large increase in H and a drastic lowering of the breakdown voltage. This is observed in the low-pressure range, where the number of impacts of an electron traversing the gap is small. The number of impacts across the gap is proportional to the product (pd) because the gas pressure p (or, more accurately, the gas density ρ) is inversely proportional to the free path λ of an electron. Paschen [6] found that the breakdown voltage of a given gas is a function of only this product, that is, that pressure can be traded for distance since only the number of collisions matters. This statement represents a typical *similarity law:* cutting the thickness of the dielectric by some factor, while increasing the gas density by the same factor, leaves the decisive parameter, the voltage drop per free path, unaltered.

In measuring *Paschen curves* (Fig. 32.4),[7] we observe that the trend reverses for higher pressures or longer gap distances: the breakdown voltage rises again. The minimum of the Paschen curve indicates that here the number of collisions is so adjusted that the kinetic energy of the electrons accumulates most efficiently for ionization. To the left of the minimum, the number of collisions is too small to build up the required height H of the avalanche across the gap with the previous voltage, whereas at the right too many collisions take place and an excessive amount of energy is squandered

[4] See, for example, F. Llewellyn Jones, "Electrical Discharges," Physical Society of London, *Reports on Progress in Physics 16*, 216 (1953).

[5] O. Klein and S. Rosseland, *Z. Physik 4*, 46 (1921); J. Franck, *ibid.*, *9*, 259 (1922).

[6] F. Paschen, *Ann. Physik 37*, 69 (1889).

[7] H. Fricke, *Z. Physik 86*, 464 (1933); B. Frey, *Ann. Physik 85*, 381 (1928); F. Ehrenkranz, *Phys. Rev. 55*, 219 (1939); A. A. Kruithof, *Physica 7*, 519 (1940).

in excitation processes. The Paschen curve itself may be visualized as equilibrium line ($1 \rightarrow 1$) separating a lower region in which the avalanche remains too small for regenerating one electron per starting electron ($1 \rightarrow <1$), from an upper region in which more than 1 electron is reliberated at the cathode ($1 \rightarrow >1$) (Fig. 32.5).

The picture thus far seems to imply that the development of the breakdown process requires at least the time interval needed for the positive ions to traverse the gap from anode to cathode. The first cathode-ray oscillograms of voltage-time characteristics, in consequence, were taken in the pioneering work of Rogowski and his co-workers [8] with the expectation of finding breakdown times in the order of 10^{-4} second at atmospheric pressure, but times of 10^{-6} to 10^{-7} second were actually observed. This outcome appeared to endanger the whole basis of the Townsend theory until it was recognized that one essential aspect was missing in the theory: the influence of space charge on the original field distribution. When the electrons disappear into the anode, the space charge of the positive ion avalanche remains behind. The space charge of a few successive avalanches may suffice to contract the field, without any motion of the ions, into a steep cathode fall.[9] The positive ions have, therefore, only to cross a diminutive gap in front of the cathode, aided by very intense fields, to produce a final breakdown. This may be accomplished in the times observed.

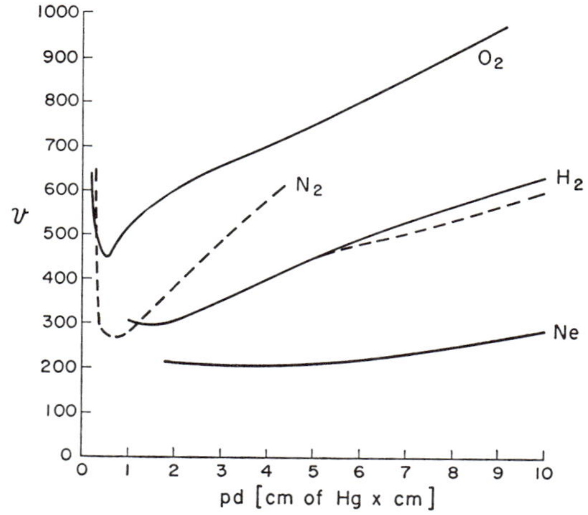

Fig. 32.4. Paschen curves for various gases. O_2 by Fricke [7] (Fe-cathode); N_2 by Frey [7] (brass-cathode); H_2 by Ehrenkranz [7] (Na-cathode) and (broken line) by Fricke [7] (Fe-cathode); Ne by Kruithof [7] (Cu-cathode).

The underlying concept, that the electrons ionize and the velocity difference between these fast electrons and the slow, positive ions polarizes the dielectric, has many useful applications. For example, returning to the Paschen curve and its interpretation as a boundary line characterized by a regeneration ratio 1:1, we observe that breakdown initiated at the right of the minimum must lead to instability [10] (see Fig. 32.5). When breakdown starts at a point A, the positive space charge left in front of the anode acts like a gap-shortener. The field towards the cathode increases, and the virtual operating point moves towards the left into a higher yield region.† The regeneration ratio, and with it the field contraction, increases at an accelerated pace, but slows down again when the left branch of the curve is approached. Final stabilization takes place near B if the voltage across the gap has been kept constant; the current has increased by many orders of magnitude.

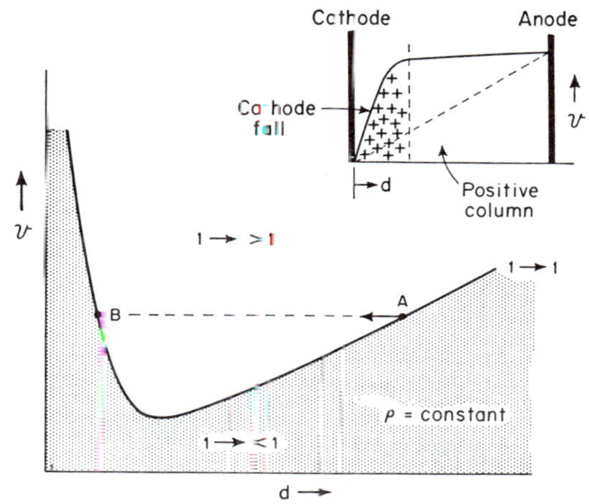

Fig. 32.5. Contraction into a glow discharge after breakdown voltage is reached at A.

The main field is now concentrated in a narrow cathode fall, the *dark space* of a *glow discharge*, through which the ions speed towards the cathode for electron regeneration. The remaining length of the discharge

[8] W. Rogowski, *Arch. Elektrotech. 16*, 496 (1926); *Der elektrische Durchschlag, Probleme der modernen Physik*, S. Hirzel, Leipzig, 1928, pp. 189 ff.

[9] A. von Hippel and J. Franck, *Z. Physik 57*, 696 (1929). That field distortion by space charge might be able to save Townsend's theory was already suspected by L. B. Loeb, *J. Franklin Inst. 205*, 305 (1928), who suggested in 1929 a step-breakdown mechanism [*Science 69*, 509 (1929)]. See also L. B. Loeb, *Fundamental Processes of Electrical Discharge in Gases*, Wiley, New York, 1939, for cases where the breakdown is so fast that an electron can cross only a fraction of the gap distance.

[10] M. Steenbeck, *Z. Physik 53*, 192 (1929); W. Rogowski, *Arch. Elektrotech. 25*, 551 (1931); *26*, 643 (1932); *27*, 743 (1933); A. von Hippel, *Z. Physik 97*, 455 (1935).

† This field contraction by space charge may not set in immediately at point A, but only after the applied voltage has risen into the instability region, as Llewellyn Jones [4] has shown in recent experiments.

space is bridged by the *positive column*, a well-conducting mixture of electrons and positive ions, that requires only a small voltage gradient to maintain the current flow. At its new operating point near B, the discharge has *longitudinal stability*, since a contraction of the cathode fall by a statistical increase of the ionization leads to a lower ionization yield and therefore to expansion, and vice versa. The positive column is the light source of a glow discharge, since here the electrons are slow and dissipate the major part of their energy in electronic excitation processes.

Since electrons and ions play such different roles in breakdown and the consecutive build-up of gas discharges (the electrons causing excitation and ionization; the ions, space-charge formation and regeneration), the breakdown in inhomogeneous fields becomes typically *polarity dependent*.[11] Positive and negative

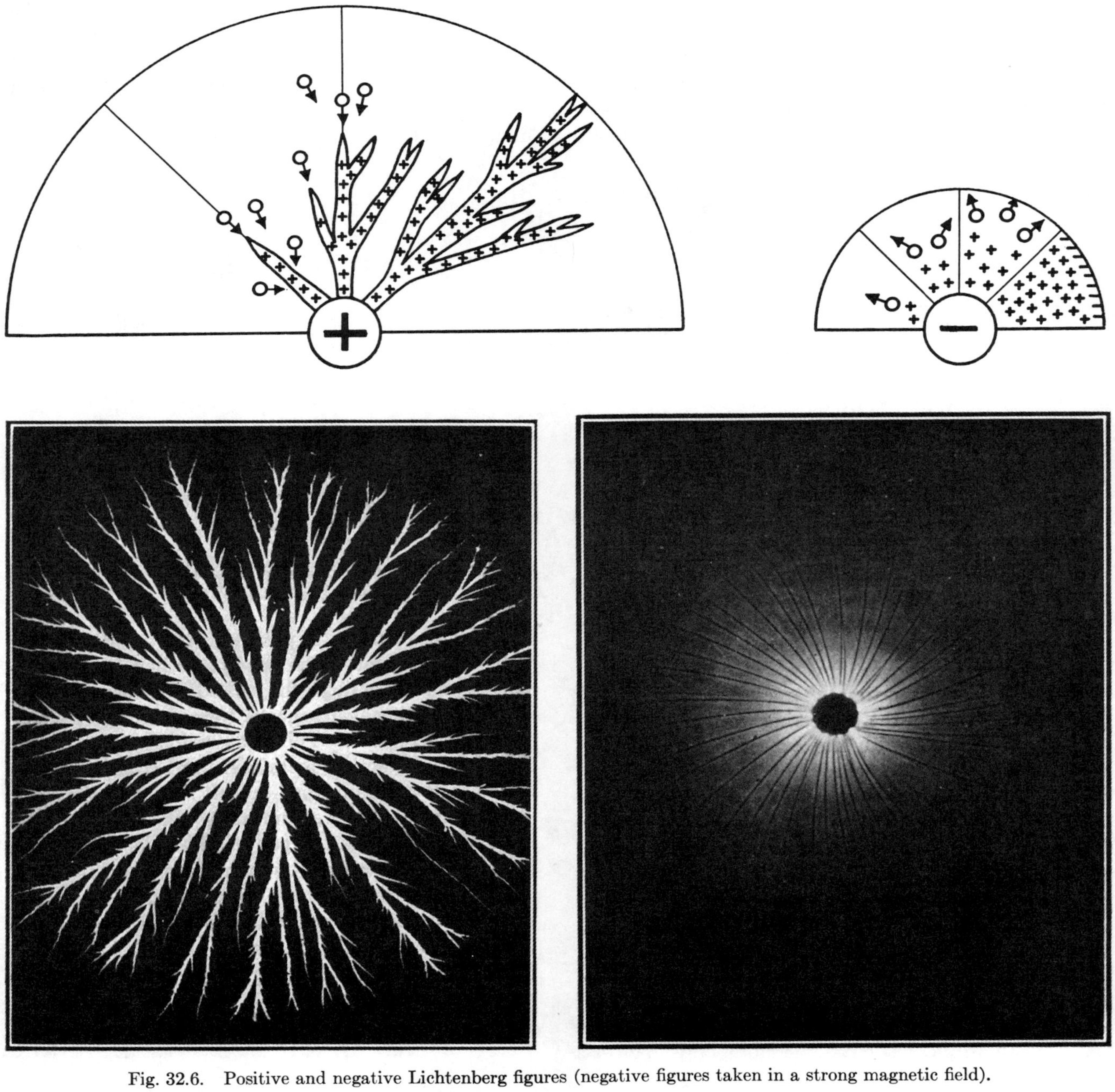

Fig. 32.6. Positive and negative Lichtenberg figures (negative figures taken in a strong magnetic field).

[11] See A. von Hippel, *Z. Physik 80*, 19 (1933); *97*, 455 (1935).

discharge patterns form and are of importance for a variety of phenomena from the operation of Geiger counters and Wilson cloud chambers[12] to the interpretation of lightning flashes.[13] These polarity effects are revealed in most beautiful detail in Lichtenberg figures,[14] where the developing discharge photographs itself in consecutive stages.[15]

Near a positive point, electrons are accelerated inwards into the steep field gradient of the anode. The disappearing electron avalanches leave positive space-charge branches behind, which reach out as auxiliary electrodes. They branch further by drawing in electron avalanches from the surroundings, and the brush-like *positive Lichtenberg figure* results (Fig. 32.6). The negative point, in contrast, ejects electrons into regions of diminishing field strength. As the impact ionization proceeds, the avalanches reduce their own driving field further by leaving their positive space charge behind and thus steepening the field directly in front of the cathode. This screening of the radial field in the rear of the avalanche produces tangential field components, and the space-charge pattern broadens into the well-known sector type of the *negative Lichtenberg figure.* Both figures reach their final expansion when the field intensity at the outer edge of the positive space charge falls below the limit of effective ionization. The *radius* of the figures indicates therefore the peak voltage of a transient, their *structure* reveals the polarity. These tell-tale properties of the primary Lichtenberg figures have been exploited in the *Klydonograph* for the recording of lightning strokes.[16]

When overvoltage, field or thermionic emission, or a long duration of the discharge provides an extremely copious electron supply, the broad space-charge patterns of the primary figures tend to contract into a narrow path, filled with a nearly metallic-conducting mixture of electrons and ions, the *plasma* of a *spark.* Such sparks still disclose in their wavering or smooth design their descent from a positive or negative primary discharge (Fig. 32.7). When a spark reaches the counterelectrode, the breakdown of the gas dielectric completes itself in a short circuit. It may continue as an *arc* if backed up by a powerful current source, with its electron supply for the gas discharge assured by thermionic or field emission.

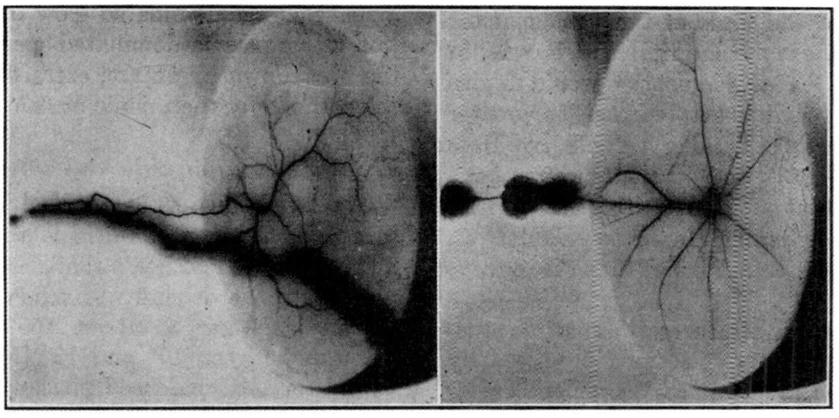

(*a*) (*b*)

Fig. 32.7. Positive (*a*) and negative (*b*) spark from point electrode continuing as gliding sparks on soapstone plate backed by metal electrode.[11]

This complex sequence of events was set in motion by a few starting electrons n_0 (Eq. 32.3). It remains to ascertain how these are provided, if spontaneous emission from the cathode is excluded. If we inquire quite generally as to how charge carriers can be created in gases, it might seem possible, at first glance, that collisions between electropositive and electronegative partners could produce at least ions by electron trans-

[12] See the work of H. Raether, *Z. Physik 107,* 91 (1937) and *Ergeb. exakt. Naturw. 22,* 73 (1949), in which the pre-breakdown stages of discharges are clarified with the help of cloud-chamber photographs.

[13] See A. von Hippel, *Naturwiss. 22,* 701 (1934); J. E. McDonald, *Sci. American 188,* April, 32 (1953).

[14] G. C. Lichtenberg, *Novi. Comment. Göttingen 8,* 168 (1777).

[15] Pioneering work on the macroscopic laws of the development of sparks and Lichtenberg figures was done by Max Toepler from 1897 to 1939, and by P. O. Pedersen, *Kgl. Danske Videnskab. Selskab.* 1919, 1922, and 1929. Compare the survey of K. Przibram, *Handb. Physik 14,* 391 (1927). The first extensive attempt at a molecular interpretation of the figures was made by F. H. Merrill and A. von Hippel, *J. Appl. Phys. 10,* 873 (1939).

[16] J. F. Peters, *Electrical World 83,* 769 (1924).

fer. However, the energy required for such process,

$$\mathcal{E} = \text{ionization energy } \mathcal{V}_i - \text{electron affinity } E + \text{electrostatic separation energy } \frac{e^2}{\epsilon_0 4\pi r}, \quad (32.6)$$

is prohibitive at normal temperatures. Even in a most favorable case such as the collision between a cesium and a fluorine atom, where E exceeds $\mathcal{V}_i$ by about 1/3 ev, the energy gain is much too small for a separation of the charged partners; they are bound together by about 4.7 ev of Coulomb attraction.

Detailed studies of the initial current in gases established about fifty years ago that extraneous ionizing agents are responsible. Spurious radioactive matter in the earth, the air, and the walls of the gas capacitor were originally held the only culprits; and it was, in consequence, expected that this ionization would diminish when a sealed capacitor was carried to greater heights above the earth. Balloon ascents by Hess (1911), Kohlhoerster (1913), and others, however, gave the surprising result that after an initial decrease an increase occurred, a fact leading Hess [17] to postulate an ionizing radiation coming from above; thus *cosmic rays* were discovered.

Systematic investigation of the recombination process [18] established that the free electrons produced by this primary ionization are trapped, in general, after an extremely short time to form *small negative ions* of molecular size. The small positive and negative ions in turn may be captured by dust particles, pollen, etc., to form *large ions*. Finally, the positive and negative charge carriers disappear again by recombination.

An external electric field can interfere with this sequence of events, as the Townsend characteristic of Fig. 32.1 certifies. First, recombination can be prevented and the ions formed drawn off as a saturation current. As the field increases further, the electrons can be kept free and made available for impact ionization, as discussed. The number n_0 of starting electrons is thus a statistical factor, depending on the ionizing conditions, the gas, and the applied field. It expresses itself in a scatter of the breakdown strength values; a gap in impulse tests may reach appreciable overvoltage before breaking down if no starting electrons are available at the decisive moment.[19] This factor is customarily brought under control by providing a copious supply of photoelectrons by strong ultraviolet illumination of the cathode.[20]

Summarizing, the discussion thus far has established that conduction and breakdown in gases at d-c voltages may develop in the following sequence of events: ionization by external sources; electron trapping and ion recombination; prevention of these two processes by external fields; acceleration of electrons; slowdown by the friction barrier of electronic excitations; impact ionization and avalanche formation; regeneration of electrons at the cathode by ion impact, photoeffect, etc. (possibly also in the gas by photoeffect and collisions of the second kind); field distortion by positive (sometimes, also negative) space charge; contraction of the field in a cathode fall; formation of a space-charge-dominated glow discharge; transformation into a plasma-dominated spark; and completion of the breakdown by an arc, extracting its electrons from the cathode by thermionic or field emission.[21]

From gases to liquids and solids

The breakdown strength of air at atmospheric pressure in a homogeneous field is about 30 kv per cm or 3 [Mv/m]. If Paschen's similarity law could be extrapolated to the densities of solids and liquids, that is, for an ideal gas to about 1000 atm, a breakdown strength between 10^3 and 10^4 [Mv/m] should be observed for the condensed phases. The actually measured strength of most insulators is about ten to one hundred times smaller than this extrapolated value.

One reason for this discrepancy becomes apparent when we observe the scatter of the breakdown values as a function of the gas density while illuminating the cathode with ultraviolet light (Fig. 32.8). Initially, as shown for CO_2 by Young,[22] the scatter without illumination is very large; it decreases, however, rapidly for high pressures and becomes independent of illumination when the gas density reaches about one-tenth of the density of liquid CO_2. Simultaneously, a marked dependence of the breakdown strength on the cathode material sets in, whereas at lower pressures the nature of the metal makes no essential difference (Fig. 32.9). This has also been found by Trump and his co-workers [23]

[17] V. F. Hess, *Physik. Z. 13*, 1084 (1912); *Wien. Ber. 121*, 2001 (1912).

[18] See L. B. Loeb (ref. 9); B. Gänger, *Der elektrische Durchschlag von Gasen*, Springer, Berlin, 1953, pp. 60 ff.

[19] See, for example, R. Strigel, *Elektrische Stossfestigkeit*, Springer, Berlin, 1939.

[20] Explanations for the remaining formative lag in the mechanism of spark breakdown have been given, for example, by L. B. Loeb and J. M. Meek, *The Mechanism of the Electric Spark*, Stanford University Press, 1941; H. Raether, *Elektrotech. Z. 63*, 301 (1942); and R. C. Fletcher, *Phys. Rev. 76*, 1501 (1949).

[21] High-frequency discharges may develop quite differently; see, for example, W. P. Allis and S. C. Brown, *Phys. Rev. 87*, 419 (1952) and F. Llewellyn Jones (ref. 4).

[22] D. R. Young, *J. Appl. Phys. 21*, 222 (1950).

[23] J. G. Trump, R. W. Cloud, J. G. Mann, and E. P. Hanson, *Elect. Eng. 69*, 961 (1950).

for much higher voltages and corresponding larger gap distances. Obviously, the voltage gradient before breakdown has become so steep that the ensuing discharge provides its own starting electrons by field emission. This is in agreement with our observations on the alkali halides,[24] where field emission sets in at comparable field strengths.

In Fig. 32.1 the current increase beyond saturation images the growth of the individual avalanches by electron impact ionization under the influence of a rising field. Since the initial starting electrons are produced

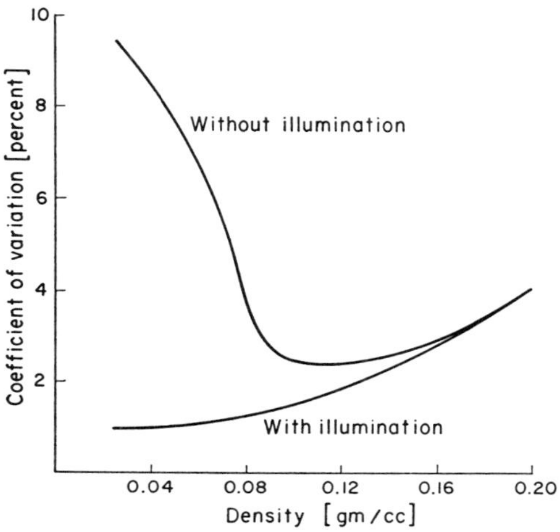

Fig. 32.8. Effect of illumination on the scatter of the breakdown values in CO_2 as function of gas density. (After Young.[22])

by extraneous ionization, the number of avalanches no more than doubles until the breakdown voltage is reached according to the Townsend condition, Eq. 32.5. Onset of field emission alters this situation radically. Townsend's breakdown condition and the interpretation of the Paschen characteristic as a yield curve for the regeneration of starting electrons loses its validity, since the necessity for regeneration has disappeared.

Field emission from a metal point into vacuum has been shown to follow the law

$$J = aE_c^2 e^{-b/E_c}. \tag{32.7}$$

This equation was first derived by Fowler and Nordheim [25] on the assumption that, as the field gradient E_c at the cathode steepens, electrons in rapidly increasing quantity escape from the Fermi lake of metal electrons by *tunnel effect* (see Sec. 14) through the potential barrier represented by the work function of the metal. The current density J is independent of temperature as long as the thermal energy kT is small in comparison to the energy required to traverse this barrier.

The field-strength dependence of the pre-breakdown current now observed should therefore reflect the increase of the field emission current as well as its multiplication by impact ionization. If no additional effects enter, we expect a current density

$$J = aE_c^2 e^{-b/E_c} e^{\alpha d}. \tag{32.8}$$

Young's measurements [22] on CO_2 at high pressures confirmed an exponential dependence on d and $1/E$ when

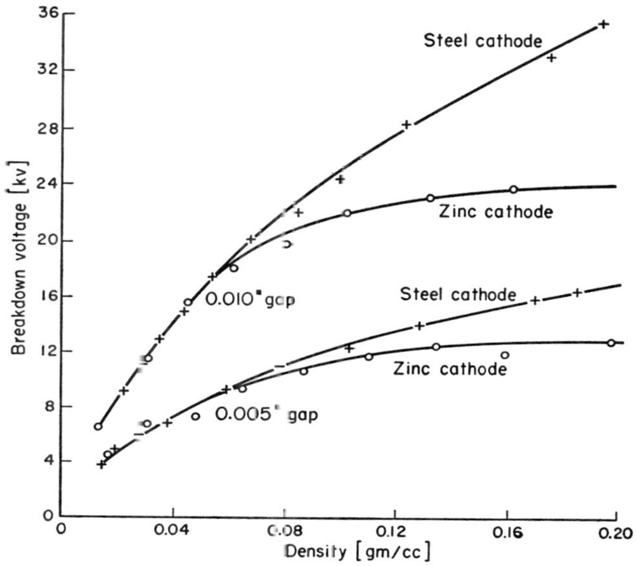

Fig. 32.9. Effect of cathode material on breakdown strength of CO_2 gas. (After Young.[22])

the gap is not too small ($d > 0.01$ in.); for narrower gaps the current density proved larger than predicted. The current density, furthermore, increased about exponentially with the density of the gas, an effect traced to a density dependence of the factor a. Equation 32.8 obviously contains some of the truth, but oversimplifies the situation. This becomes apparent when we return from the problem of pre-breakdown currents to that of breakdown itself.

Impact ionization, as discussed above, produces a positive space charge, hence an increase of the field gradient at the cathode. The field emission rises, this rise in turn increases the ionization, and thus, by a positive feedback effect, the initially homogeneous field must rapidly contract into a cathode fall; and the discharge transforms into a "field arc" whenever field emission and impact ionization co-operate unrestrictedly. Equation 32.8, therefore, corresponds to an unstable state, until the final contraction into the arc stage is reached (see Fig. 32.5). We must conclude that, in

[24] A. von Hippel and R. S. Alger, *Phys. Rev. 76*, 127 (1949); A. von Hippel, E. P. Gross, J. G. Jelatis, and M. Geller, *Phys. Rev. 91*, 568 (1953).

[25] R. H. Fowler and L. Nordheim, *Proc. Roy. Soc.* (*London*) A*119*, 173 (1928).

the presence of impact ionization, the onset of field emission causes breakdown; and, vice versa, in the presence of field emission the onset of impact ionization leads to breakdown, if no stabilizing phenomenon is introduced.

In searching for a stabilizing mechanism operative in the volume of the gas, we may not invoke the motion of the positive ions. The breakdown times in d-c fields at high pressures are so short that the ions stay practically fixed in place. The current is essentially an electron current, and, to achieve its stabilization, we have, macroscopically speaking, to increase the resistance of the anodic part of the gap until no more voltage can be transferred to the cathodic region. This increase in resistance might be produced by cutting down the density of the free electrons by recombination and trapping. Furthermore, the mobility of the electrons may decrease as the field strength is lowered. How far these effects exist and suffice for a stabilization of the pre-breakdown current, when field emission and ionization are present simultaneously, must be investigated. This presupposes some general knowledge of the existence and behavior of charge carriers in liquids and solids.

Charge carriers in liquids and solids

In gases, charge carriers have to be created by external means, and electric fields are required to prevent their recombination. In liquids and solids such carriers may have a stable existence because of the higher relative dielectric constant κ' of their surroundings. The electrostatic field energy of a charged particle is reduced to $1/\kappa'$ of its vacuum value. By transferring, for example, a mole of spherical particles of the radius r and charge Ze from vacuum into an aqueous solution, the *hydration energy* is released:

$$W_h = N_0 \frac{Z^2e^2}{\epsilon_0 8\pi r}\left(1 - \frac{1}{\kappa'}\right), \tag{32.9}$$

provided we may neglect the influence of the molecular structure of the solution. Since for water $\kappa' \simeq 80$, the electrostatic field is neutralized almost completely and the interdiction of Eq. 32.6 against the separation of oppositely charged partners loses its validity.

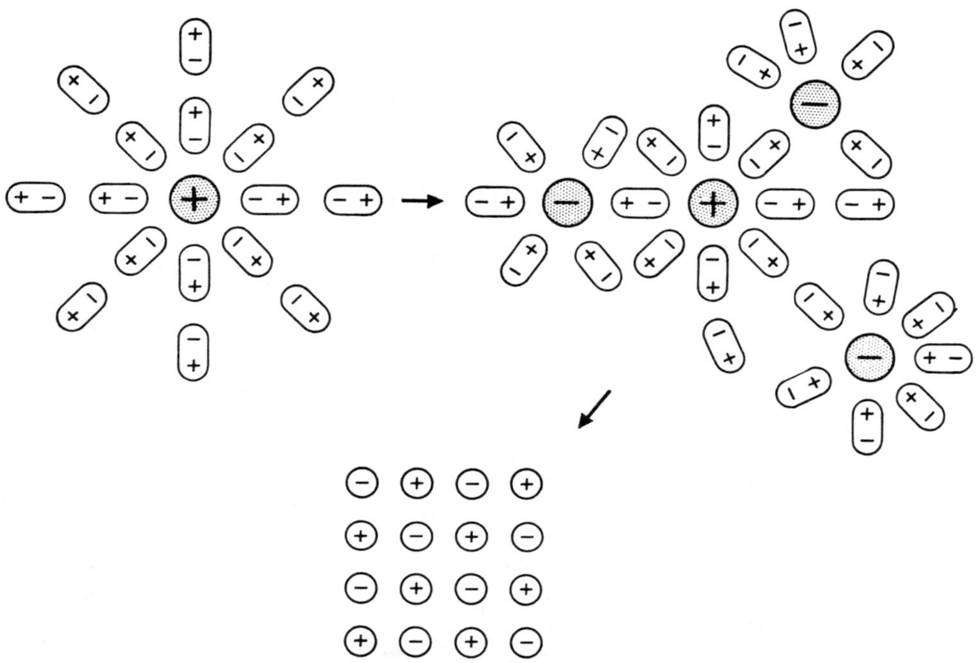

Fig. 32.10. From dipole cloud to ionic atmosphere to crystallization (schematic).

Seen from the molecular point of view, the polarization energy in very dilute aqueous solutions is provided by the clustering of water dipoles around the individual ions, as indicated schematically in Fig. 32.10. As the concentration of ions increases, a competition for water molecules sets in and the electrostatic field of the ions remains partly unshielded. In consequence, each ion tends to attract ions of the opposite polarity. With increasing ion concentration the *dipole clouds* give way to *ionic clouds* that surround each ion with a halo of countercharges. These *ionic atmospheres*, introduced by Debye into the theory of electrolytes,[26] decrease the mobility of the ions in an electric field. In sufficiently

[26] See P. Debye and E. Hückel, *Physik. Z. 24*, 185 (1923).

intense fields they can be partly stripped, and the conductivity increases, as M. Wien first showed in his impulse experiments (*Wien effect*).[27]

Each ion has a tendency to surround itself with the maximum number of counterions (see Sec. 24). In dilute solutions, where the binding energy for the halo is of the order kT, the ions of the atmosphere drift in and out statistically. With increasing concentration, the approach distance shortens and the binding energy increases, until, at a critical temperature and concentration, the cloud becomes transformed into a regular array of counterions. The *solubility limit* of the solution has been reached: ionic crystals form. Initially stabilized by water dipoles and mobile, because electrically shielded, the ions find themselves now locally bound in the giant molecule of a periodic ionic crystal structure.

To move ions in liquids or solids requires the existence of holes into which the ions may jump when the activation energy becomes available. Liquids, in general, are full of holes, and also in solids the problem is minor if the ions fit into interstices of the lattice structure. Since Frenkel [28] first invoked the transfer of ions to such lattice sites to make the mechanism of ionic conduction in solids understandable, the holes created by this kind of disorder in the lattice array are called *Frenkel defects* (Fig. 24.9). Alternatively, the required vacant lattice sites may be created by an activation process in which ion pairs are moved from the volume to the crystal surface (*Schottky defects*).[29]

If N_0 is the number of ions per unit volume of a given type and U the activation energy required for its mobilization, the density of potentially mobile charge carriers, according to Boltzmann statistics, is $N_0 e^{-U/2kT}$ and their contribution to the ionic conduction, according to Ohm's law,

$$J = \sigma E = N_0 e^{-U/2kT} bE. \tag{32.10}$$

If the drift distance in the field direction per unit field strength, the *mobility* b, is temperature independent, the conductivity increases exponentially with temperature as long as the majority of the ions are trapped in their surroundings.

Obviously, since ions can be stabilized in the condensed phases by the polarization of the surrounding medium, it will be possible in favorable cases to ionize atoms without anchoring the removed electrons firmly in counterions. Thus *metals* and *semiconductors* result. The classical prerequisite is that the electron affinity of any prospective negative ion is negligible for metals and small enough in semiconductors to allow thermal dissociation. However, here classical physics ends, since the wavelength of the electrons in the condensed phases is comparable to the separation distance of the particles. Hence, the stationary states as well as the motion of electrons have to be described by quantum mechanics.

[27] M. Wien, *Ann. Physik 83*, 327 (1927); *85*, 795 (1928); *11*, 429 (1931).

[28] J. Frenkel, *Z. Physik 35*, 652 (1926).

[29] C. Wagner and W. Schottky, *Z. physik. Chem.* B*11*, 163 (1930).

Electrons in periodic lattices

An electron in free space has the energy and momentum

$$\mathcal{E} = \frac{1}{2} mv^2 = \frac{p^2}{2m}, \qquad \mathbf{p} = \frac{\partial \mathcal{E}_{kin}}{\partial v} = m\mathbf{v}. \tag{32.11}$$

According to de Broglie's conditions (Eq. 7.6), the total energy determines the frequency, and the momentum the wavelength of a particle as

$$\mathcal{E} = h\nu, \qquad |\mathbf{p}| = \frac{h}{\lambda}. \tag{32.12}$$

To bring out the vector property of $\mathbf{p}$, it proves convenient to introduce a *wave vector* k of the magnitude

$$|\mathbf{k}| = \frac{2\pi}{\lambda} \tag{32.13}$$

that points in the direction of propagation; hence

$$\mathbf{p} = \hbar\mathbf{k}. \tag{32.14}$$

Plotted as $f(k)$, the energy of the free electron is a parabola (Fig. 32.11),

$$\mathcal{E} = \frac{\hbar^2}{2m}\mathbf{k}^2. \tag{32.15}$$

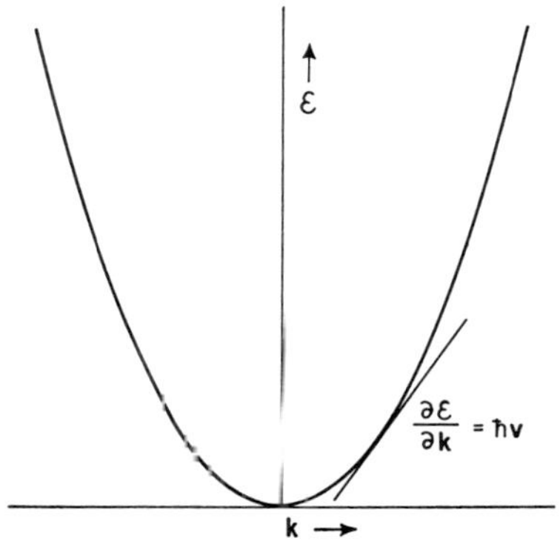

Fig. 32.11. Energy parabola for free electrons.

The tangent at any point

$$\frac{\partial \mathcal{E}}{\partial k} = \frac{\hbar^2 \mathbf{k}}{m} = \mathrm{v}\hbar \qquad (32.16)$$

determines the velocity of the electron. This particle velocity, given by

$$\frac{\partial \mathcal{E}}{\partial k} = \hbar\, \partial \nu\, \partial \lambda,$$

is, according to I, Eq. 11.17, the *group velocity* v_g of the wave phenomenon. Its *phase velocity* is, according to definition,

$$\nu\lambda = \frac{\mathcal{E}}{\mathbf{p}} = \frac{\hbar \mathbf{k}}{2m}, \qquad (32.17)$$

that is, for the free electron, equal to half the group velocity. The curvature of the energy characteristic

$$\frac{\partial^2 \mathcal{E}}{\partial k^2} = \frac{\hbar^2}{m} \qquad (32.18)$$

determines the mass of the particle.

We have encountered the concept of the wave vector previously, while discussing the propagation of electromagnetic waves through boundaries (see I, Sec. 14). An electric plane wave in space, traveling in the $+x$-direction without attenuation, was described at a given moment t_1 as (cf. I, Eq. 9.5)

$$E(x) = E_1 e^{-j\frac{2\pi}{\lambda}x}. \qquad (32.19)$$

To free ourselves from the fixed co-ordinate system, we introduced a unit vector $\mathbf{k}^0$ pointing in the direction of propagation and a position vector $\mathbf{r}$ drawn from some origin O to a point P in space (I, Fig. 14.3). The distance of a plane of constant phase from the origin is then given as $\mathbf{k}^0 \cdot \mathbf{r}$, and Eq. 32.19 in this more general formulation becomes

$$E = E_1 e^{j\frac{2\pi}{\lambda}\mathbf{k}^0 \cdot \mathbf{r}}. \qquad (32.20)$$

The wave vector is obviously connected to the unit vector $\mathbf{k}^0$ as

$$\mathbf{k} = \frac{2\pi}{\lambda}\mathbf{k}^0; \qquad (32.21)$$

hence the wave function of a plane electron wave in free space may be written as

$$\psi = A e^{j\mathbf{k}\cdot\mathbf{r}}. \qquad (32.22)$$

How will an electron plane wave fare in traversing an ideal periodic crystal structure? Obviously interference phenomena will occur as in the electromagnetic case (see I, Sec. 18), for which ratio of wavelength to periodicity distance is the decisive parameter.

Wave interference in space lattices was first encountered in X-ray diffraction. The elementary problem is the reflection of a parallel X-ray beam as a function of the angle of incidence from a set of parallel planes of uniform spacing d (Fig. 32.12). The index of refrac-

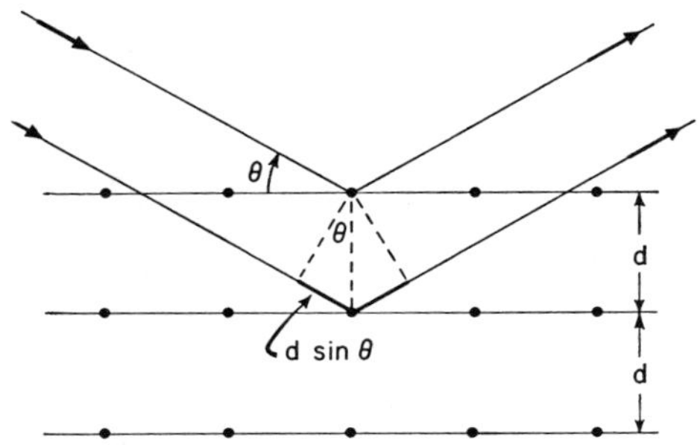

Fig. 32.12. Bragg's reflection condition.

tion of matter is taken as unity; hence the angle of incidence and refraction are the same and no phase jump takes place at the front or back surface. It is customary, furthermore, in X-ray diffraction to refer to the glancing angle θ instead of the angle ϕ between the normal and the beam as in optics. Since $\cos\phi = \cos\psi = \sin\theta$, the equation of interference optics (I, Eq. 18.25) for the phase difference Δ between two partial beams simplifies to

$$\Delta = \frac{4\pi d}{\lambda}\sin\theta. \qquad (32.23)$$

When the path between the partial beams differs by a multiple of the wavelength

$$\Delta = n2\pi \qquad \text{with } n = 1,2,3,\ldots, \qquad (32.24)$$

all partial amplitudes add and total reflection occurs. This statement,

$$2d\sin\theta = n\lambda, \qquad (32.25)$$

is *Braggs'* famous *reflection condition for X-rays*.[30] By introducing the wave vector $\mathbf{k}$ we may rewrite it as

$$|\mathbf{k}| = \frac{n\pi}{d\sin\theta} \qquad \text{for } n = 1, 2, 3, \ldots. \qquad (32.26)$$

This obviously is the answer also for an electron plane wave traversing a periodic lattice. At the critical $\mathbf{k}$ values prescribed by Eq. 32.26, total reflection will occur. The energy curve as $f(k)$ will flatten away from the parabola as these wavelengths are approached and assume a horizontal tangent, since the group velocity indicated by this tangent (see Eq. 32.16) becomes zero. Between the critical regions the parabolic law is re-

[30] W. H. and W. L. Bragg, *Proc. Roy. Soc. (London)* A*88*, 428 (1913).

sumed; hence the energy characteristic must have discontinuities at the critical values k_1, k_2, . . ., as shown (Fig. 32.13). An energy spectrum of allowed and forbidden zones results, corresponding to the transmission

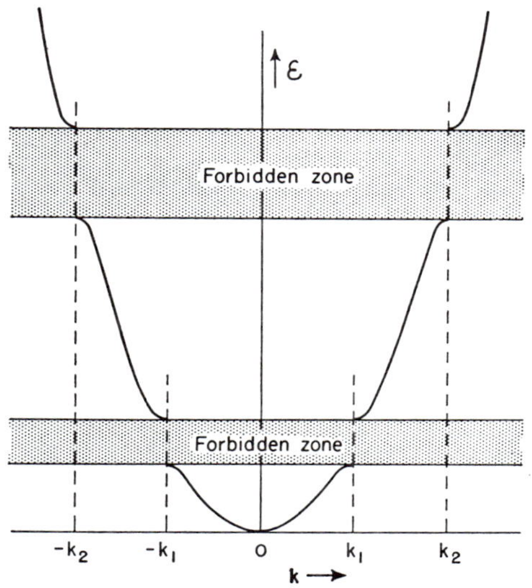

Fig. 32.13. The energy characteristic for electron plane waves in a periodic lattice.

spectrum of a periodically loaded transmission line or of a wave guide acting as a band-pass filter.

In the reciprocal wavelength space, the k space, the allowed zones are bounded by surfaces of total reflection. To traverse these surfaces, electrons have to leap to the next allowed energy value. These allowed zones

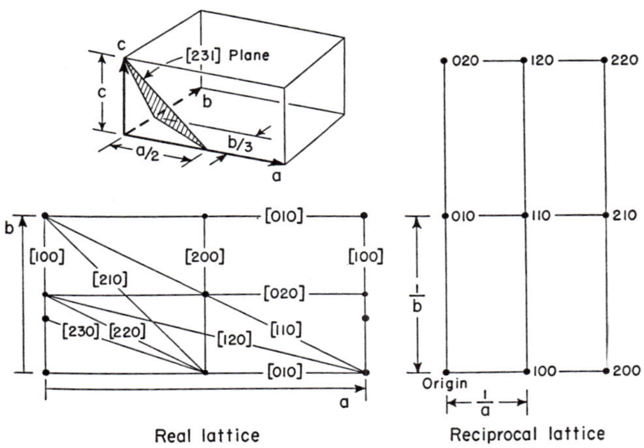

Fig. 32.14. Miller indices, real and reciprocal lattice.

in k space are called *Brillouin zones*, because Brillouin first investigated some of their properties.[31] They are well known to the crystallographer through the concept

[31] See L. Brillouin, *Wave Propagation in Periodic Structures*, McGraw-Hill Book Co., New York, 1946.

of the *reciprocal lattice*. In the real crystal lattice the lattice points indicate particles and the distances their molecular spacings. To construct the reciprocal lattice, lines are drawn perpendicular to the planes of the real lattice and points are marked at distances inversely proportional to the spacings of these planes. Each point of the reciprocal lattice corresponds thus to a set of planes of the real lattice and is designated by the *index numbers* of this set of planes (Fig. 32.14).† The k space, according to the reflection condition Eq. 32.26, is equivalent to the reciprocal lattice space.

Figure 32.15 illustrates the first three Brillouin zones of the simple cubic lattice. The first, bounded by reflecting surfaces corresponding to the set of (100) planes of widest lattice spacing (a), is a cube. The next closest

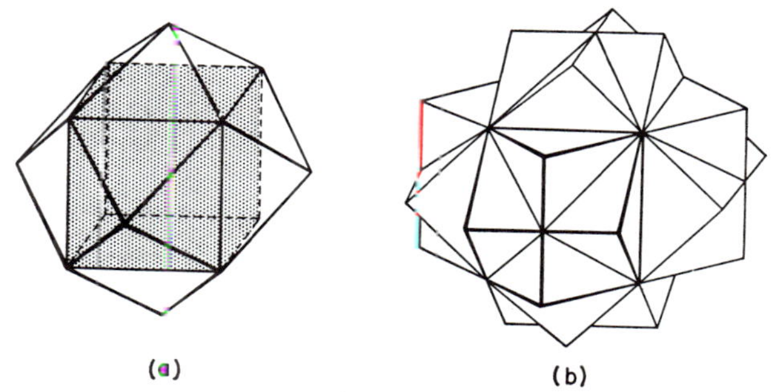

Fig. 32.15. The first three Brillouin zones of a cubic lattice: (*a*) first and second zone; (*b*) third zone added.

lattice spacing ($a/\sqrt{2}$) characterizes the face-diagonal planes (110); the second Brillouin zone, bounded by these planes, is a dodecahedron enclosing the cube. The third zone, bounded by the reflections on the (111) planes of spacing $a/\sqrt{3}$, is a polyhedron enclosing the dodecahedron, and so on.

These Brillouin zones and the geometrical considerations of wave interferences in space lattices are valid for any kind of plane waves, whether electromagnetic, acoustic, or electron waves. To understand the situation for electron waves in more detail, we have to return to Schrödinger's equation and solve it for plane waves modulated by a periodic lattice potential. This was first done by Bloch.[32] Assuming that the periodic potential acts as a small perturbation, he could show that the correct wave function is a product of the original wave function (Eq. 32.22) for free space and of a periodic function of the lattice period. From these

† These *Miller indices* characterizing a plane are derived, as the upper drawing in Fig. 32.14 shows, by designating on the crystallographic axes unit intercepts a, b, and c. The intercepts made by any other plane can then be expressed as a/h, b/k, c/l, where h, k, l are simple integers or zero. The set of these numbers [hkl] identifies the crystal plane.

[32] F. Bloch, *Z. Physik* *52*, 555 (1928).

Bloch functions, running waves can be built up and standing waves in analogy to the electromagnetic case discussed in I, Secs. 17 ff. Again, we arrive at a system of allowed bands separated by forbidden energy regions as in the geometrical interference discussion.

This *band scheme* lends itself to an intriguingly simple explanation of electronic conduction (Fig. 32.16). If the allowed bands overlap or are only partially filled, *metals* result, because electrons can be speeded up by external fields. If a lower band is completely filled, but a higher, unfilled conduction band is accessible by thermal excitation, an *intrinsic semiconductor* is obtained.

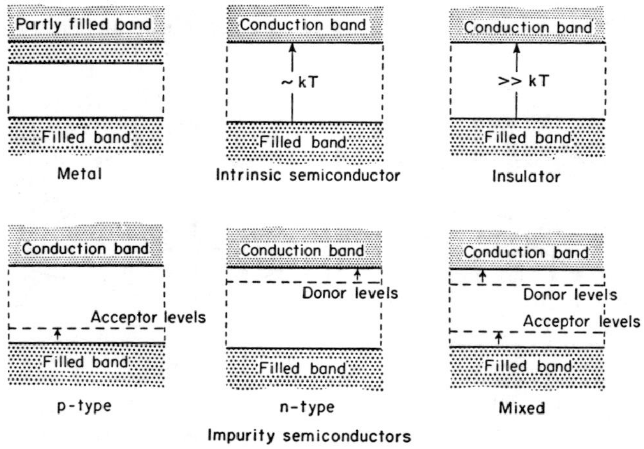

Fig. 32.16. Insulators, semiconductors, and metals according to band scheme.

It may be an *n-type semiconductor* (*n* for negative) in case only the electrons in the excited state contribute to the conduction; a *p-type semiconductor* (*p* for positive), if the *positive holes* left in the previously filled band act as the current carriers; or a *mixed semiconductor,* when both electrons and holes participate as migrating charge carriers. If the forbidden zone is too wide for bridging by thermal excitation, the ideal crystal will be an electronic *insulator.* However, in real dielectrics impurities or lattice disturbances are encountered, which tend to create intermediate localized energy levels and may change the prospective insulator into an *impurity semiconductor.* Electronegative addition agents, located above the filled band, may act as electron *acceptors* and produce hole conduction. Electropositive impurities may form localized states near enough below the conduction band to act as *donors* of electrons, thus causing *n*-type conduction. The electron supply in both cases may be easily exhausted, but there is also the possibility that the inserted levels act like the rungs of a ladder and permit an excitation in steps from the filled to the conduction band.

This treatment of semiconductors, initiated by Wilson,[33] is convincingly simple and has been invoked somewhat indiscriminately for a first explanation of rectification, optical absorption, photoeffect, temperature dependence of conduction currents, phosphorescence, and the effect of addition agents. In applying this band concept, its prerequisites are frequently forgotten: the electrons must be fast so that they can be treated as free and the crystal lattice considered as a rigid reflector of these electron waves. This is a reasonable first approximation for metals and for some very good semiconductors in which the conducting electrons abound.

To clarify this point we have to consider the occupation density of the various energy levels of a band.[34] In classical statistics, the number of electrons occupying an energy state $\mathcal{E}_i$ of a degeneracy g_i at a temperature T is (see Eq. 19.4)

$$N_i = g_i A e^{-\mathcal{E}_i/kT}. \tag{32.27}$$

For electrons we do not have this unrestricted occupancy of quantum states, because the Pauli exclusion principle prescribes that any orbital can accommodate two electrons with opposite spins (see Sec. 8). The band picture implies that each energy level of a band belongs to the crystal as a whole and is in this sense an orbital of the crystal.

The statistics embodying the Pauli principle is the *Fermi-Dirac statistics;*[35] it prescribes for the electrons an occupation density

$$N_i = \frac{g_i}{e^{\frac{\mathcal{E}_i - \mathcal{E}'}{kT}} + 1}. \tag{32.28}$$

The classical factor A and the parameter $\mathcal{E}'$ in electron statistics are normalizing factors, so adjusted that the summation over all levels leads to the total number of electrons in a band.

Obviously, at the absolute zero of temperature all quantum states below a certain energy level $\mathcal{E}_0'$ must be filled and those above that energy, empty. $\mathcal{E}_0'$ is the value of $\mathcal{E}'$ at 0°K. If we plot this situation for a metal with the energy $\mathcal{E}$ as the ordinate and a distance parameter x as the abscissa, where $x = 0$ corresponds to the interior of the metal and $x \to \infty$ leads through the surface into vacuum, a picture of the "Fermi lake" of metal electrons is obtained. The depth $-W_a$ measures the potential of the bottom of the lake against

[33] A. H. Wilson, *Proc. Roy. Soc.* (*London*) A*133*, 458 (1931); A*134*, 277 (1931).

[34] See, for example, F. Seitz, *Modern Theory of Solids,* McGraw-Hill Book Co., 1940.

[35] E. Fermi, *Z. Physik 36*, 902 (1926); P. A. M. Dirac, *Proc. Roy. Soc.* (*London*) A*112*, 661 (1926); see A. Sommerfeld and H. Bethe, *Handbuch der Physik* XXIV, Pt. 2 (1933), pp. 333 ff.

vacuum, the work function $P = e\phi$ that of the top of the lake (Fig. 32.17), whereas $\mathcal{E}_0'$ represents its depth. If we assume $\mathcal{E}' = \mathcal{E}_0'$ independent of temperature and

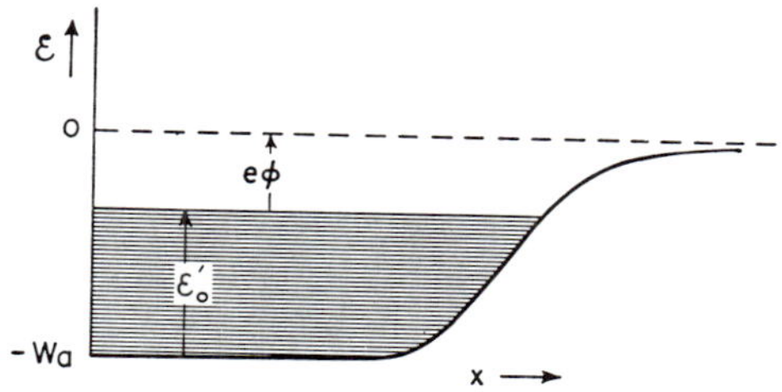

Fig. 32.17. The Fermi lake of metal electrons.

$g_i = 2$, we obtain the Fermi distribution function for the occupancy of the various energy states (Fig. 32.18). The energy level $\mathcal{E}_0'$, where the occupancy for $T > 0$ falls to one half the possible value, that is, to unity in our case, is called the *Fermi level.*[36]

The depth of the Fermi lake depends obviously on the number and distribution of the stationary states in a conduction band and on the number of electrons available for their occupation. If the length of the crystal

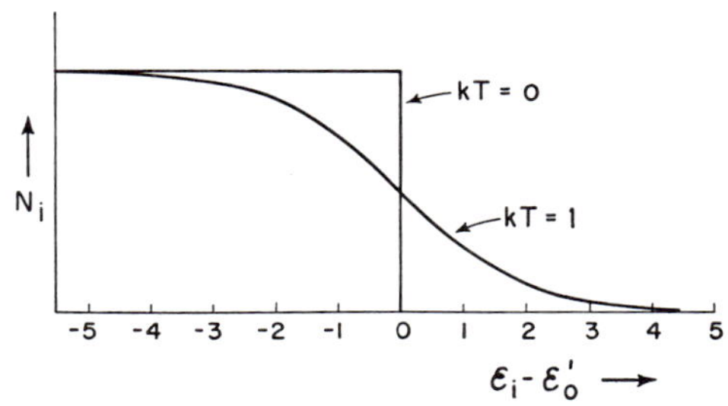

Fig. 32.18. The Fermi distribution function.

in the x-direction is L_x, stationary states in this direction can form only if this length corresponds to an integral number of wavelengths (see Eq. 7.10), that is,

$$L_x = n_x\lambda$$

or

$$n_x = \frac{L_x}{\lambda} = \frac{L_x}{h}p_x. \tag{32.29}$$

Hence the number of states in a momentum interval dp_x is

$$dn_x = \frac{L_x}{h}dp_x. \tag{32.30}$$

The corresponding equations are valid for the y and z components; hence the number of states in a momentum interval $d\mathbf{p}$ composed of the three components is

$$dn = \frac{L_xL_yL_z}{h^3}dp_x\,dp_y\,dp_z = \frac{V}{h^3}\,|\,d\mathbf{p}\,|, \tag{32.31}$$

where V is the volume of the crystal.

To relate the momentum interval $d\mathbf{p}$ to the energy parameter $\mathcal{E}$ of the preceding figures, we assume that the kinetic energy of an electron in a periodic lattice has the same form as in free space (see Eq. 32.11), but that the mass m has to be replaced by an effective electron mass m^*. Thus the total energy of an electron in the interior of a metal, the sum of the potential and kinetic energy, is

$$\mathcal{E}_t = -W_a + \frac{p^2}{2m^*}. \tag{32.32}$$

The energy parameter $\mathcal{E}$ of our figures is chosen as zero at the bottom of the lake, hence corresponds to the kinetic energy of the electron

$$\mathcal{E} = \mathcal{E}_t - W_a = \frac{p^2}{2m^*}. \tag{32.33}$$

A differential change of this energy, $d\mathcal{E}$, leads to a momentum change

$$|\,d\mathbf{p}\,| = 2\pi(2m^*)^{3/2}\sqrt{\mathcal{E}}\,d\mathcal{E}. \tag{32.34}$$

The number of states between bottom and top of the Fermi lake is, therefore,

$$n = \frac{V}{h^3}2\pi(2m^*)^{3/2}\int_0^{\mathcal{E}_0'}\sqrt{\mathcal{E}}\,d\mathcal{E}$$

$$= \frac{4\pi}{3}\frac{V}{h^3}(2m^*)^{3/2}\mathcal{E}_0'^{3/2}. \tag{32.35}$$

Each state can be occupied by two electrons; hence

$$n_0 = \frac{2n}{V} \tag{32.36}$$

designates the number of metal electrons per unit volume and

$$\mathcal{E}_0' = \frac{h^2}{2m^*}\left(\frac{3n_0}{8\pi}\right)^{2/3} \tag{32.37}$$

the depth of the Fermi lake, that is, the kinetic energy of the electrons at its top surface. Since in metals approximately every atom contributes a "free" electron, the Fermi lake is filled so high that the top electrons speed along with kinetic energies of several electron volts.

The stationary states of the Fermi lake correspond to running waves which traverse the crystal in various directions and are, therefore, in principle, able to carry current. However, because each direction has its coun-

[36] For the meaning of the Fermi level in semiconductors see, for instance, R. A. Hutner, E. S. Rittner, and F. K. Du Pré, *Philips Research Reports 5*, 188 (1950).

terpart in the interior of the metal, the current transfer cancels for all filled states below the top level and only the uppermost electrons can make a contribution. These electrons have a high kinetic energy, as just established, when the upper edge of the band is still far away. The band edges themselves, as shown above, correspond to standing waves, that is, here the group velocity and energy transfer become zero.

The "free-electron theory" of quantum mechanics is a good first approximation for metals, because the conducting electrons are fast and their electrostatic field well shielded by the high dielectric constant of the surroundings. The lattice acts like a rigid reflector because the time of local interaction is short and the coupling weak. For most semiconductors these prerequisites are not true; there are relatively few electrons in the conduction band and holes in the filled band, and these electronic charge carriers move in a medium of much lower dielectric constant. Hence the interaction between electrons and lattice, previously idealized as a periodic modulation of free waves, assumes new aspects. The electrons are so slow that they excite, with great probability, lattice vibrations (*phonons*); the *friction barrier* of these *vibrations* becomes the essential obstacle for the acceleration of electrons when external fields are applied.[37] Simultaneously the electrons are not well shielded and act as space-charge centers distorting the surroundings; this polarizing effect depends on the speed of the electrons as well as on the polar character of the crystal lattice.

Graphically speaking, an electron acting as a space charge surrounds itself in liquids and solids with a halo of countercharges. When the electron moves so fast that the nuclei cannot respond by displacement, this halo consists of electronic polarization only. For an electron at rest, electronic, atomic, and dipole polarization contribute fully. Thus the extent and composition of the halo changes with the speed of the electron. Simultaneously, it varies with distance, especially in ionic materials; for far-away lattice points the field pull exerted by the electron may alter direction so slowly that they still can respond, whereas neighboring ions may not be able to follow.

These concepts, introduced by the author,[37,38] are in a way a logical extension of the theory of electrolytes.[39] For ions, we deal, in general, with halos corresponding to the local static dielectric constant, with the finally formed dipole and ion clouds. Only the Wien effects introduce aspects of formation of these clouds. For electrons, we run the full gamut of halo formation and phonon exchange from the full development of the halo in the trapped state to its complete disappearance for free electrons in the Bloch band theory.

An electron in the intermediate state between free and bound is frequently designated as a *polaron*. The drag exercised by the halo can be expressed as an increase in the effective mass. Numerous calculations have been attempted concerning the magnitude of this effect and its dependence on the electron velocity, with quite contradictory results.[40] The theoretical difficulty is that the interaction between electron and lattice should not be handled as a simple perturbation problem, but ways must be found to calculate it reliably as a first-order effect.

Breakdown in liquids and solids

The diverse origin of the charge carriers in the condensed phases and the variety of their interactions with the surrounding medium makes us suspect that there will not be only one type of breakdown but a number of mechanisms may come into play, depending on circumstances. Electric breakdown, as the name implies, is initiated by some action of electrons or ions and terminates in the formation of a highly conductive path through the dielectric. A post-mortem examination of a solid may reveal some characteristic damage to its texture: a channel of molten material;[41] a jagged hole, as if a stone had been thrown through a window; a tree-like decomposition pattern of carbonized or metallic matter (Fig. 32.19). Such a visible evidence may be a significant clue or characterize only a subsequent phase of destruction after the actual demise.

Obviously, clear-cut mechanisms can be named that would cause breakdown, if nothing else happens. Earliest recognized was the *thermal breakdown*, based by Wagner[42] on the simple macroscopic concept of a competition between the Joule heat Q_1 generated by electric conduction and the heat loss Q_2 produced by thermal conduction (Fig. 32.20). As long as Q_1 rises more slowly than Q_2 with increasing temperature, stabilization at some point A occurs. When the heat begins to develop faster than it can be removed (point B or C), breakdown by melting or decomposition is imminent. Characteristic for this type of breakdown is its relatively slow development, and dependence on the cooling conditions, but these are not unique criteria.

[37] A. von Hippel, *Z. Physik 75*, 145 (1932); *Z. Elektrochem. 39*, 506 (1933).

[38] A. von Hippel, *Ergeb. exakt. Naturw. 14*, 79 (1935); *Z. Physik 101*, 680 (1936).

[39] A. von Hippel, *J. Chem. Phys. 8*, 605 (1940).

[40] See E. P. Gross, "Behavior of Slow Electrons in Polar Crystals," *Technical Report 55*, Laboratory for Insulation Research, Massachusetts Institute of Technology, December, 1952.

[41] L. Inge and A. Walther, *Z. Physik 34*, 15 (1925).

[42] K. W. Wagner, *Elec. Eng. 41*, 1034 (1922).

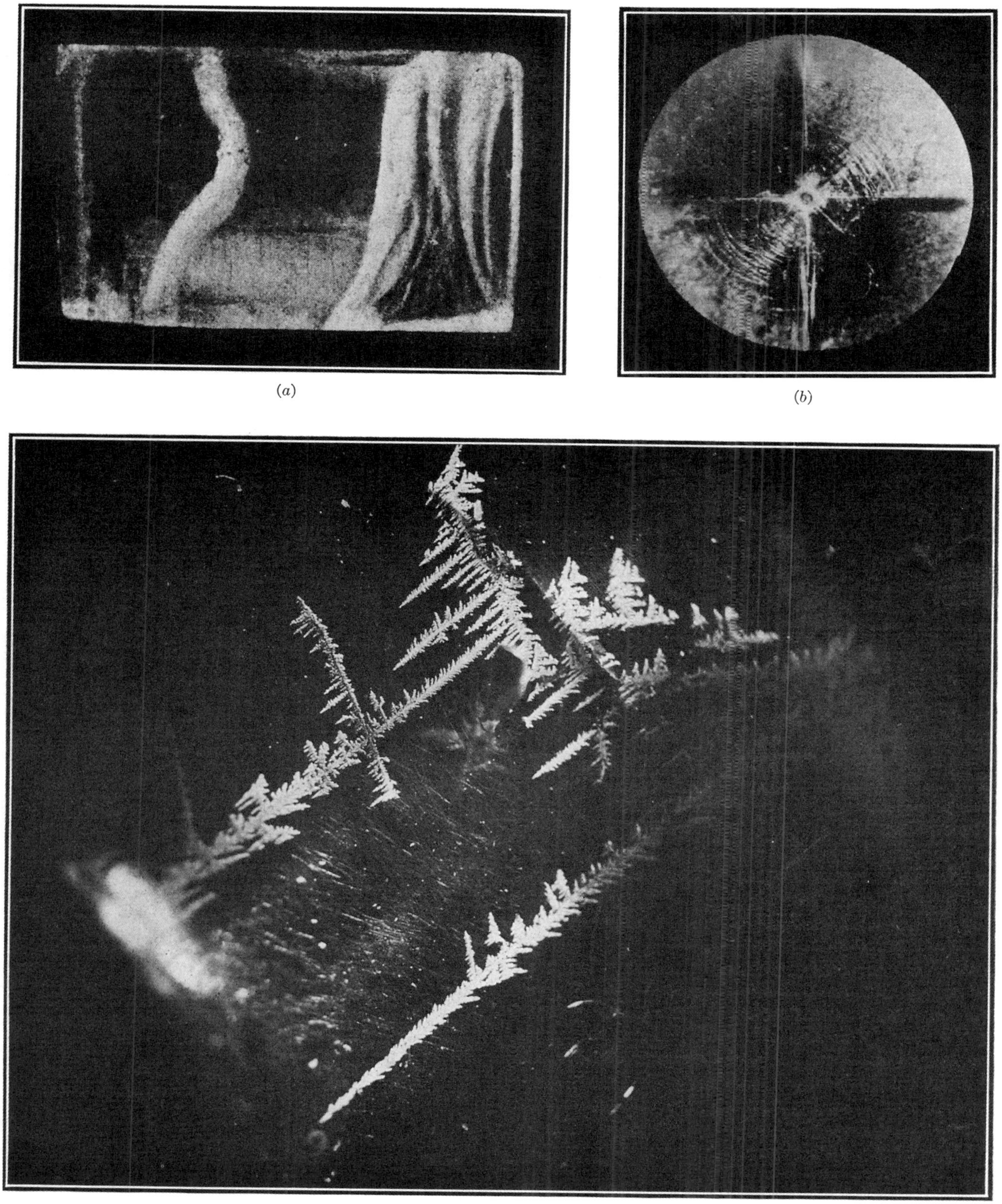

(a) (b) (c)

Fig. 32.19. Various types of breakdown: (a) breakdown channels produced by melting (after Inge and Walther [41]); (b) jagged hole; (c) metal dendrites bridging electrodes.

Seen from the molecular point of view, a thermal breakdown may be caused by electrons or ions requiring activation energy for their mobilization (see Eqs. 32.10 and 32.28). With increasing temperature, the number of charge carriers increases rapidly, and, if their supply is not exhausted or their mobility drastically reduced by thermal agitation, only a ballast resistor or the rapid increase in heat dissipation by radiation (see Eq. 5.22) may stabilize the situation. In this sense, a "Nernst glower" represents a semiconductor in an arrested state of thermal breakdown. Obviously, thermal breakdown may also take place in electrolytes with their abundant supply of ions, or becomes possible

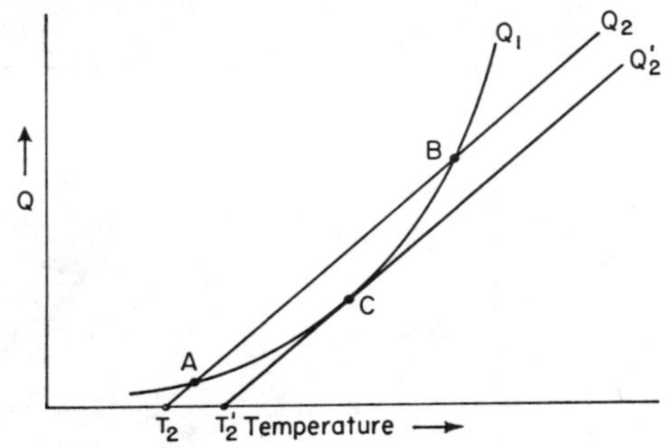

Fig. 32.20. Thermal breakdown. (After Wagner.[42]) (The straight lines represent heat conduction for two different electrode temperatures, T_2 and T_2'.)

when field emission provides electron currents that rapidly increase with field strength (see Eq. 32.7).

In these last-mentioned cases, various other destructive effects may accompany the flow of current and compete with the thermal mechanism as the primary cause of breakdown. Gas bubbles may form at the electrodes in the liquid because the electric field overcomes the surface tension or ions are discharged; thus a gas breakdown may trigger the liquid breakdown. Metal ions, discharged in the form of dendrites, may bridge the gap between the electrodes by metallic paths. Outstanding examples are the dendrites sprouting from the cathode in lead acetate solution or growing through crystals of the alkali [43] and silver halides [44] (Fig. 32.21). A closer observation of these dendrites in crystals shows that they do not represent a simple electrolytic deposition but are coupled with electronic phenomena. The dendrites in the silver halides seem to grow only along interfaces and with a speed incompatible with purely ionic conduction. In the alkali halides the dendrites grow through the volume itself in preformed crystallographic directions which change with temperature. It seems likely that the alkali ions are discharged mainly by field emission and not so much by cathodic deposition in forming the alkali-metal trees.

In the case of dendrites, the current deposits highly conducting decomposition products which ruin the insulation. Similar in appearance, but of different origin,

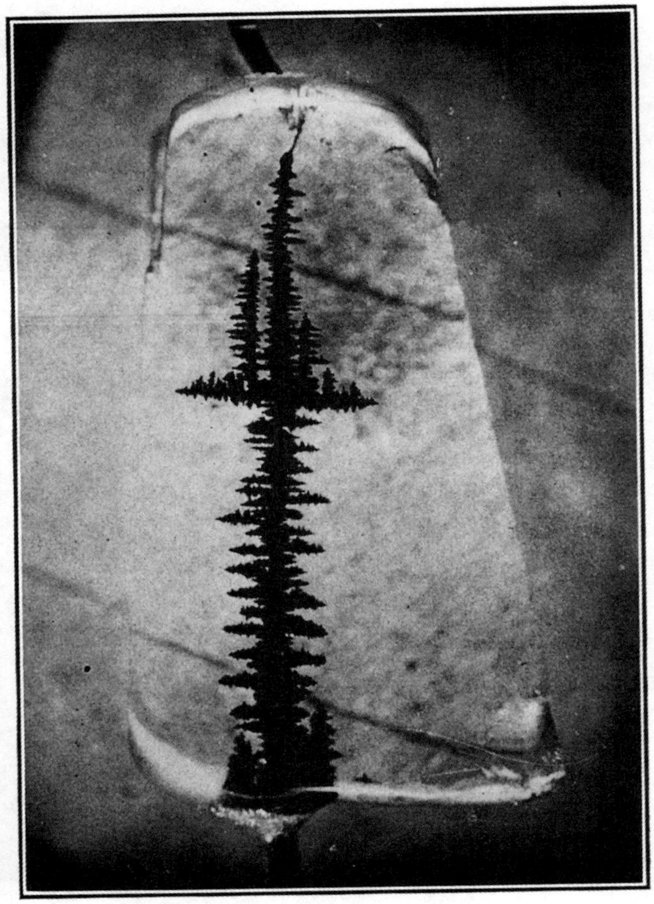

Fig. 32.21. Dendrite in NaCl.[43]

are the carbonized paths which often initiate the breakdown of cables. These are the result of discharges that originate in voids and correspond to a fixation of Lichtenberg figures by the chemical decomposition of the material: that is, electronic excitation and impact ionization in the gas phase along internal surfaces are responsible, with all their complicating consequences.

Although in these cases electrons act as a powerful chemical in producing conducting material, they may ruin a dielectric by a much more innocuous process. In recombining with ions they may produce atoms that are too big for the structure in question. Thus the insulator may be shattered by the build-up of mechanical strain (Fig. 32.22).[45]

[43] A. von Hippel, *Z. Physik 98*, 580 (1936).

[44] C. Tuband, *Handb. exper. Physik* XII, 1, Akademische Verlagsgesellschaft, Leipzig (1932).

[45] A. von Hippel, *J. Appl. Phys. 8*, 815 (1937).

We can mention here only the chemical action of discharge products, which may cause breakdown of capacitors if not suppressed by *inhibitors;* the action of impurities in liquids that may be pulled by the electric field into the gap and cause premature breakdown; the *self-healing* of dielectrics during breakdown by the formation of an insulating phase, as in electrolytic capacitors, selenium rectifiers, or layer capacitors employing evaporated zinc electrodes.[46] Unfortunately, no

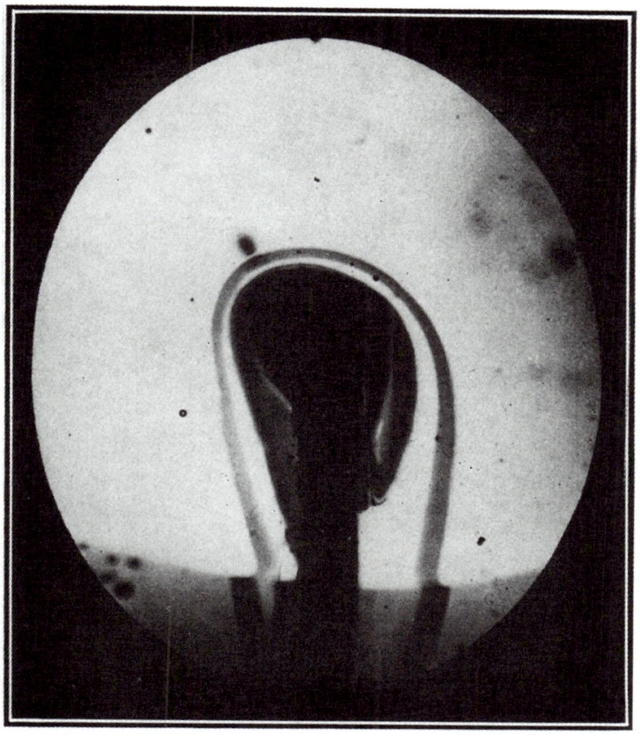

Fig. 32.22. Cracking of glass after electron emission.[45]

space remains in this short survey for more than a quick glance at the counterpart of thermal breakdown, the *intrinsic electric breakdown by impact ionization.*

We saw that, in gases at high pressures, the occurrence of field emission invalidates Townsend's breakdown condition based on the regeneration of charge carriers. Since the field at the cathode increases when positive space charge develops, we had to conclude that now even the onset of impact ionization should lead to breakdown if no new stabilizing mechanism can be invoked. In contrast to the situation in gases, where the electronic excitation of atoms and molecules saps the energy of the electrons before they reach ionizing velocities, in solids and liquids the excitation of vibrations seems to form the main friction barrier, preventing cumulative ionization until the breakdown strength is reached.[37] The stopping action of this barrier may be amplified by the polaron state of the electron; in many materials slow electrons will probably be trapped below a critical field strength.

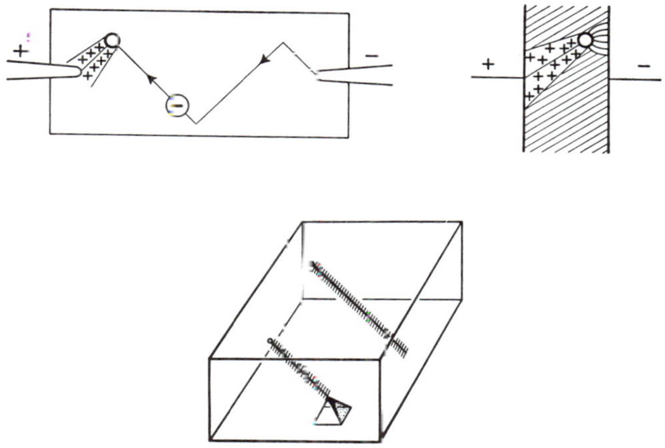

Fig. 32.23. Direction breakdown in NaCl at room temperature. (Breakdown direction is the face diagonal [110]; near cathode the direction may change to [111] and characteristic pyramids form by cracking of the crystal between the [111] edges.)

We discussed the fact that electrons, because of their wave nature, are subject, in periodic lattices, to interference effects which lead to the electronic band structure of crystals. We may expect that the breakdown of crystals reveals peculiarities connected with this quantum-mechanical feature. Such effects apparently have been found by the author and his co-workers in the *direction breakdown of crystals* [47] (Fig. 32.23) † and

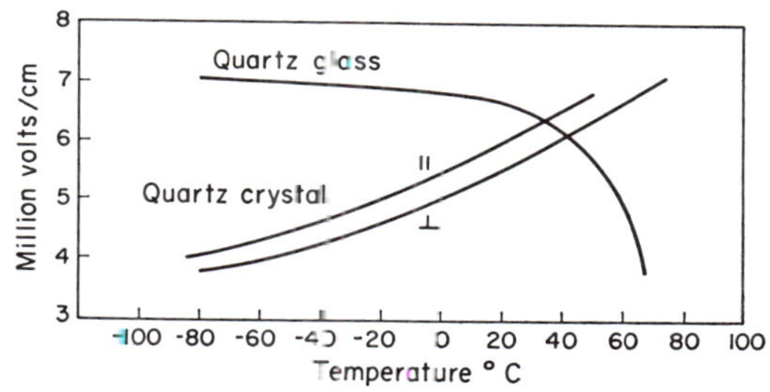

Fig. 32.24. Temperature dependence of the breakdown strength of quartz glass versus quartz crystal.[48]

in the *temperature dependence* of the breakdown strength of crystals versus glasses [48] (Fig. 32.24). There may be others, like the breakdown mechanism suggested by

[46] Some information on these subjects will be found in the companion volume, *Dielectric Materials and Applications.*

[47] A. von Hippel, *Z. Physik 67,* 707 (1931); *68,* 309 (1931); *88,* 358 (1934); J. W. Davisson, *Phys. Rev. 70,* 685 (1946).

† In all breakdown discussions, as we again emphasize in Fig. 32.23, we have to keep in mind that the original field becomes distorted by space charges and destruction paths as the breakdown develops. This dynamic field, in contrast to the applied field, determines the further course of events.

[48] A. von Hippel and R. J. Maurer, *Phys. Rev. 59,* 820 (1941).

Zener,[49] but here we reach the boundary of present-day research. Many interesting observations have been made and ingenious theoretical explanations proposed,[50] but the situation is still in doubt. The breakdown-strength measurements made until now scatter too widely; they indicate that in the repetition of the experiments some vital parameters, for example, field emission, are not under control.[24] Only in facing the challenging complexity of the situation with improved experimental and theoretical methods will we produce the better dielectrics of tomorrow.

[49] C. Zener, *Proc. Roy. Soc.* (*London*) A*145*, 523 (1934).

[50] See H. Fröhlich and J. H. Simpson, *Intrinsic Dielectric Breakdown in Solids, Advances in Electronics,* Vol. 2, Academic Press, 1950; S. Whitehead, *Dielectric Breakdown in Solids,* Clarendon Press, Oxford, 1951; W. Franz, "Theorie des reinen elektrischen Durchschlags fester Isolatoren," *Ergeb. exakt. Naturw. 27,* 1 (1953).

APPENDIX

A · Problems and Illustrative Examples

This supplementary part parallels the sections of the main text with the objective of amplifying important concepts and imparting some facility in the handling of dielectric problems.

I · Macroscopic Approach

1 · Interrelation between the Real and the Imaginary Part of the Complex Permittivity or Permeability

The parameters ϵ^* and μ^* allow a simple description of a dielectric if they are independent of field strength. In this case, the contributions of various frequency components superpose linearly and if either the dispersion phenomenon (ϵ' and μ') or the absorption phenomenon (ϵ'' and μ'') is known over the entire frequency scale ($\omega = 0$ to $\omega = \infty$), the accompanying absorption or dispersion characteristic is completely defined. This is evident to the physicist, because dispersion and absorption are two different aspects of the same phenomenon (see II, Fig. 19.1). The fact is obvious to the electrical engineer from his circuit analogues: the frequency behavior of the real part of an input impedance function determines the frequency dependence of the imaginary part of the function. The statement can be proved mathematically by Fourier analysis, as we shall show for the dipole relaxation spectrum (see I, Eq. 26.14, and II, Sec. 22). Here

$$\kappa^* = \kappa_\infty' + \frac{S}{1 + j\omega\tau},$$

that is,

$$\kappa'(\omega) - \kappa_\infty' = \frac{S}{1 + \omega^2\tau^2},$$

$$\kappa''(\omega) = \frac{S\omega\tau}{1 + \omega^2\tau^2}.$$

According to Fourier analysis,

$$\kappa'(\omega_1) - \kappa_\infty' = +\frac{2}{\pi} P \int_0^\infty \frac{\omega\kappa''(\omega)\,d\omega}{\omega^2 - \omega_1^2},$$

$$\kappa''(\omega_1) = -\frac{2}{\pi} P \int_0^\infty \frac{\omega\kappa'(\omega)\,d\omega}{\omega^2 - \omega_1^2}.$$

Hence the dispersion or the absorption at some frequency ω_1 is given by the frequency spectrum of the complementary component as indicated; P designates the principal value of the Fourier integral.[1] Inserting the expression for $\kappa''(\omega)$ into the first Fourier integral,

$$\kappa'(\omega_1) - \kappa_\infty' = \frac{2}{\pi} P \int_0^\infty \frac{\omega}{\omega^2 - \omega_1^2} \frac{S\omega\tau}{1 + \omega^2\tau^2}\,d\omega,$$

we note that the integral is even in ω. Integration from $-\infty$ to $+\infty$ therefore doubles the value of the integral or

$$\kappa'(\omega_1) - \kappa_\infty' = \frac{1}{\pi} P \int_{-\infty}^{+\infty} \frac{\omega^2}{\omega^2 - \omega_1^2} \frac{S\tau}{1 + \omega^2\tau^2}\,d\omega.$$

The integration may be carried through in the complex frequency plane (Fig. 1.1). The integral has poles at $-\omega_1$ and ω_1 on the real-frequency axis, where it becomes infinite; similarly, on the imaginary-frequency axis,

[1] See E. A. Guillemin, *The Mathematics of Circuit Analysis*, Wiley, New York, 1949, p. 335; J. H. Van Vleck, *Propagation of Short Radio Waves*, Massachusetts Institute of Technology Radiation Laboratory Series, No. 13, D. E. Kerr, Ed., McGraw-Hill, New York, 1951.

poles lie at $\omega = j/\tau$ and $-j/\tau$. To evaluate the principal part, we examine the integral over the closed contour of Fig. 1.1. By Cauchy's theorem this integral is $2\pi j$ times the residue at j/τ or $1/(1 + \omega_1^2\tau^2)$. The principal part may be expressed as

$$\frac{\pi}{1 + \omega_1^2\tau^2} - \left\{\int_{C_1} + \int_{C_2} + \int_{C_3}\right\}.$$

The integral along C_3 vanishes; the integrals along C_1 and C_2 cancel. We find

$$\kappa'(\omega) - \kappa_\infty' = \frac{S}{1 + \omega^2\tau^2}.$$

The ohmic conductivity of a conductor is an exception to the preceding consideration when represented

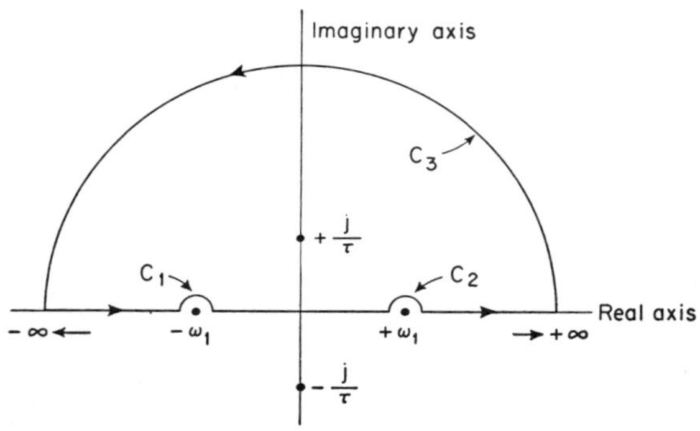

Fig. 1.1. Integration in the complex frequency plane.

by a resistance only, because no storage phenomenon accompanies the energy dissipation. Actually, however, the ohmic conductivity has a dispersion in the infrared or optical frequency range and should be represented in the Fourier integration by an RL circuit (see Appendix A, II, Sec. 2). *Show that with this proviso the exception disappears and the conductivity causes a negative contribution to the dielectric constant.*

2 · Depolarization and Demagnetization

In I, Fig. 2.1,† we assumed that the free ends of the dipole chains, produced by the polarization of the dielectric, are neutralized by bound countercharges at the electrodes. In this case the macroscopic electric field inside the dielectric is identical with the field $\mathbf{E}$ measured at the outside. If, however, there are no electrodes on the dielectric surface, this neutralization of the free terminal charges cannot take place and a depolarizing closing field extends from the positive to the negative ends of the dipole chains. Now we have to distinguish between the external field $\mathbf{E}_e$ and an internal field $\mathbf{E}_i$ that is smaller than $\mathbf{E}_e$ by an amount increasing proportionally to the polarization $\mathbf{P}$:

$$\mathbf{E}_i = \mathbf{E}_e - \frac{w}{\epsilon_0}\mathbf{P}.$$

Identical considerations hold for the magnetization and are here of even greater import, because no magnetic monopoles exist to neutralize the free ends of magnetic dipole chains. Since in the magnetic field theory $\mathbf{H}$ and $\mathbf{M}$ have the same dimensions, the equation becomes

$$\mathbf{H}_i = \mathbf{H}_e - w\mathbf{M}.$$

The factor w is determined by the geometry of the polarized or magnetized body. It can be calculated for spheres, cylinders, and ellipsoids because the field is homogeneous inside such bodies. A sample calculation for a polarized sphere is given in I, Sec. 10 (see also Appendix A, II, Sec. 7). The values of w for various shapes [1] are shown in Table 2.1. *Confirm the value of w for the case of an infinite cylinder, when the field is orientated perpendicularly to the axis.*

† References to figures and equations preceded by I or II refer to the main text.

[1] See, for example, E. C. Stoner, *Phil. Mag.* [7] *36*, 803 (1945), and J. A. Stratton, *Electromagnetic Theory*, McGraw-Hill, New York, 1941, pp. 207 ff.

3 · Image Dipole, Image Force, and Electron Emission

An electric charge $+Q$ at a distance x from a conducting plane produces an electric field terminating perpendicularly on this equipotential surface. The plane can be replaced by an image charge $-Q$ at a distance $-x$

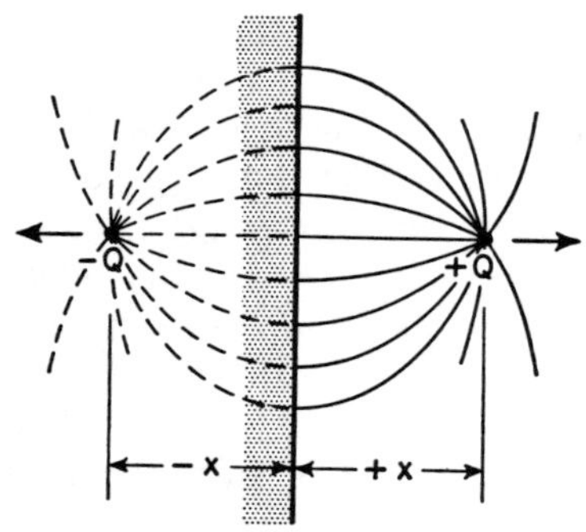

Fig. 3.1. Image dipole.

behind the plane in so far as this field is concerned (Fig. 3.1).† The charge and its mirror image form an image dipole, the special property of which is that a

† The solution of our electrostatic problem has to satisfy the boundary condition and Laplace's equation (see I, Eq. 4.13). The field of a point charge fulfills implicitly Laplace's equation, and the image dipole constellation satisfies the boundary condition.

Table 2.1. Depolarization or demagnetization factors [1]

Shape		w
Sphere		$1/3$
Ellipsoid of revolution	*(Oblate spheroid)*	
(*a*) Field $\parallel c$	$a = b > c$ $c = a\sqrt{1-\epsilon^2}$ $\epsilon = \sqrt{1-\frac{c^2}{a^2}}$ (ϵ = numerical eccentricity; see II, Fig. 6.2)	$\frac{b^2c}{2} \times \frac{2}{(a^2-c^2)^{3/2}} \left\{ \frac{\sqrt{a^2-c^2}}{c} - \tan^{-1}\frac{\sqrt{a^2-c^2}}{c} \right\}$ $= \frac{1}{\epsilon^2} - \frac{\sqrt{1-\epsilon^2}}{\epsilon^3} \arcsin \epsilon;$ $\doteq 1$ for $c \to 0$; i.e., $\epsilon \to 1$
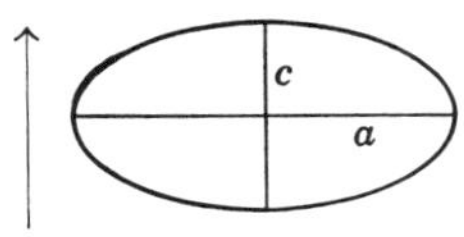		
	(Prolate spheroid)	
(*b*) Field $\parallel a$	$b = c < a$	$\frac{1-\epsilon^2}{\epsilon^2}\left\{\frac{1}{2\epsilon}\ln\frac{1+\epsilon}{1-\epsilon} - 1\right\};$ $\doteq 0$ for $\epsilon \to 1$; i.e., $a \to \infty$
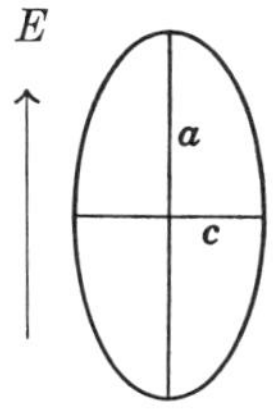		
General ellipsoid	$b \neq c \neq a$ Field $\parallel a$	$\frac{abc}{2}A$ A = an elliptic integral to be tabulated
Long cylindrical rod	$E \parallel$ to axis	0
Short cylindrical rod	$E \parallel$ to axis	1
Long cylindrical rod	$E \perp$ to axis	1/2

displacement of Q normal to the plane produces a displacement of the image charge $-Q$ by an equal amount in the opposite direction.

The image force between the two charges is, according to Coulomb's law (I, Eq. 3.8),

$$F_i = -\frac{Q^2}{\epsilon' 4\pi (2x)^2}.$$

To remove the charge Q from x to infinity against this force requires the work

$$W = \int_x^\infty F_i\,dx = \frac{Q^2}{\epsilon' 4\pi}\frac{1}{4x}.$$

The work would become infinite for $x = 0$, which makes no physical sense. We know that only the finite energy of the work function has to be overcome (see II, Fig. 32.17). The image-force law therefore holds only to a critical molecular approach distance x_c. That the image force actually plays a role in pulling electrons back to the cathode is established by the fact that the saturation current of a thermionic emitter increases with the applied field (*Schottky effect*).[1]

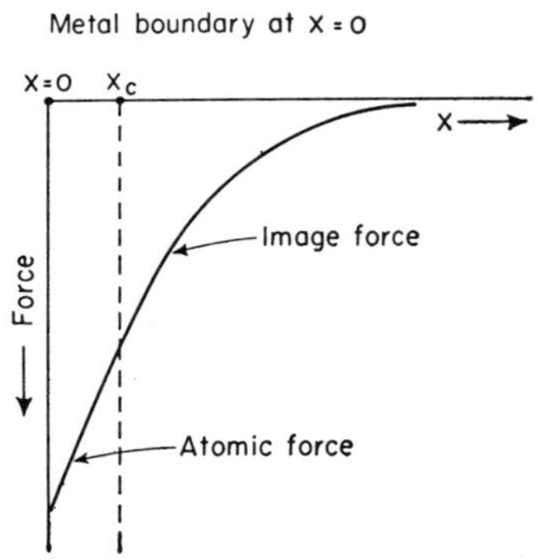

Fig. 3.2. Image force and atomic force curves.[2]

Calculate the external field strength required to produce field emission by assuming that the electron is held back by the image force beyond x_c and, from $x = 0$ to $x = x_c$, by an atomic force given by the tangent of the image-force curve at $x = x_c$ (Fig. 3.2).[2]

[1] W. Schottky, *Z. Physik 14*, 63 (1923).
[2] F. Ollendorff, *Potentialfelder der Elektrotechnik*, Springer, Berlin, 1932.

4 · Wilson's Bipolar Thundercloud

The earth is negatively charged with respect to the surrounding atmosphere; in fair weather a field gradient of about -130 volts per meter is observed near the earth's surface. Wilson[1] proposed that the negative charge of the earth is maintained by thunderstorms, and that the fair-weather field of the earth produces bipolar thunderclouds according to the following mechanism (Fig. 4.1): Raindrops are weakly polarized with their positive poles pointing downwards. The falling drops, therefore, catch preferentially negative ions from the atmosphere, because these are attracted towards the approaching drop whereas the positive ions, because of the initial repulsion, have less chance to reach the upper negative part of the raindrop as it passes. Thus a negative space charge is carried downward and increases the fair-weather field gradient in the cloud while reversing the field gradient near the earth. By this self-excitation process, the dipole charge in the cloud and the field between cloud and ground increase until lightning occurs.

The thundercloud represents a space-charge dipole above the conducting surface of the earth. *How can direction of this dipole, its charge and its height above ground be determined by electrostatic measurements on the earth?*

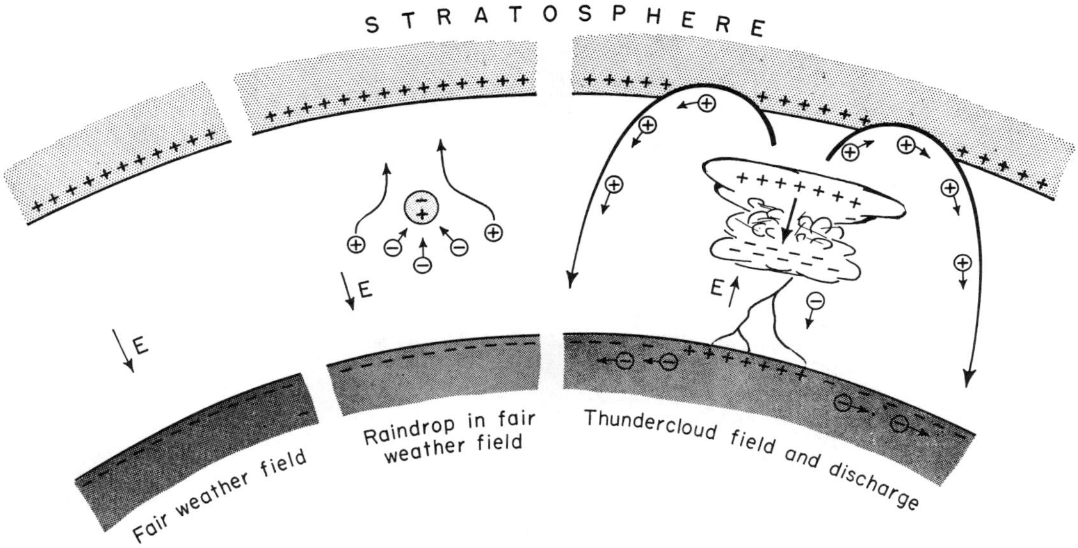

Fig. 4.1. Formation of bipolar thundercloud (raindrops polarized in the fair-weather field of the earth trap preferentially negative charges).

[1] C. T. R. Wilson, *J. Franklin Inst. 208*, 1 (1929).

5 · Cyclotron and Betatron

The cyclotron, invented by Lawrence and developed with Livingston,[1] makes use of the fact that a charged

[1] E. O. Lawrence and M. S. Livingston, *Phys. Rev. 45*, 608 (1934).

particle in a uniform magnetic field of the induction B revolves with an angular velocity

$$\frac{d\phi}{dt} \equiv \omega = \frac{eB}{m},$$

independently of the speed of the particle and the radius of the circle. Hence, such particles can be accelerated by an oscillator operating with this cyclotron frequency (Fig. 5.1).

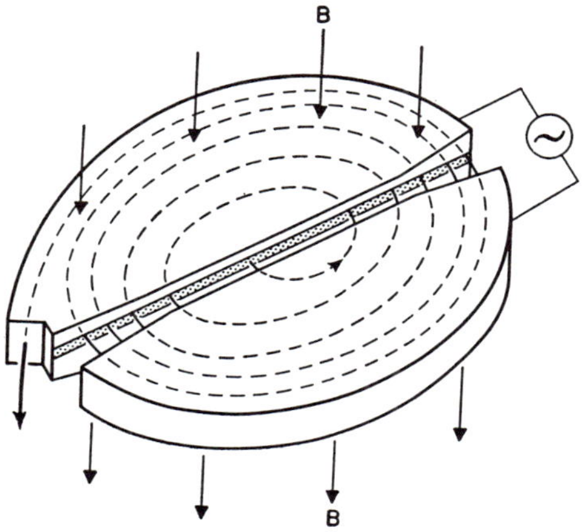

Fig. 5.1. Cyclotron principle.

Faraday's induction law suggests that if free charged particles could circle a varying magnetic flux, forming in effect the secondary winding of a transformer, they might acquire extremely high energies in the induced electric field. Kerst [2] was the first to realize a practical particle accelerator on this basis, the betatron (Fig. 5.2). To keep an electron on a circle of a fixed radius

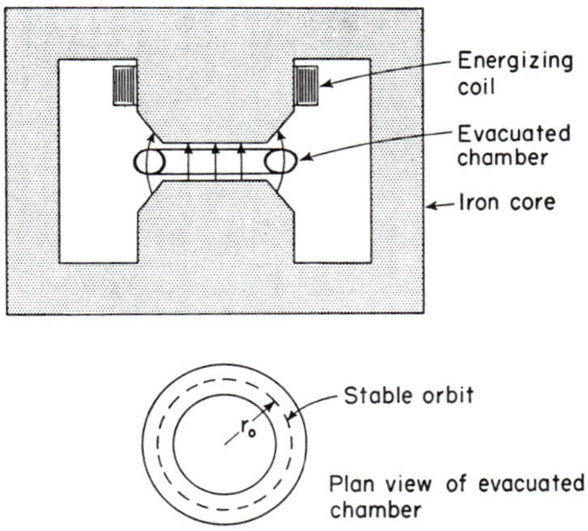

Fig. 5.2. Betatron principle.

r_0 in the betatron, one has, in addition to the cyclotron condition to fulfill Wideröe's condition,[3] that the total flux Φ through this orbit must be equal to twice the orbit area times the magnetic induction B at the orbit,

$$\Phi = 2\pi r_0^2 B.$$

Derive both conditions by assuming that an electron moves in a circular path in the x-y plane under the influence of a changing magnetic flux $\Phi(r,t)$ orientated in the z-direction. The changing flux produces, according to I, Eqs. 5.30 and 5.31, a tangential voltage

$$\int \mathbf{E}_\phi(r,t) \cdot dl = -\int \mathbf{B}(r,t) \cdot \mathbf{n}\, dA.$$

In addition to the electric field $E_\phi(r,t)$, the magnetic induction B acts on the electron, according to I, Eq. 10.21. The electric field and the angular component of the magnetic induction cause a change in the angular momentum of the electron, and the centripetal force and the radial component of the induction affect the radial momentum. By setting up these torque and force equations, the result is readily obtained.

[2] D. W. Kerst, *Phys. Rev. 60*, 47 (1941); *Am. Scientist 35*, 57 (1947).

[3] R. Wideröe, *Arch. Elektrotech. 21*, 387 (1928).

6 · Maxwell's Equations and the Conservation of Charge

Maxwell's theory is based on the two field equations (I, Eqs. 6.6 and 6.10) and the two scalar relations (I, Eqs. 5.3 and 5.4) defining the true electric charge density and postulating the nonexistence of true magnetic monopoles:

$$\nabla \times \mathbf{H} = J + \frac{\partial \mathbf{D}}{\partial t}$$

$$\nabla \times \mathbf{E} = -\frac{\partial \mathbf{B}}{\partial t}$$

$$\nabla \cdot \mathbf{D} = \rho$$

$$\nabla \cdot \mathbf{B} = 0.$$

The divergence of a curl is zero; hence by forming the divergence of the first equation we find

$$\nabla \cdot (\nabla \times \mathbf{H}) = 0 = \nabla \cdot \mathbf{J} + \frac{\partial}{\partial t}(\nabla \cdot \mathbf{D})$$

or

$$\nabla \cdot \mathbf{J} = -\frac{\partial \rho}{\partial t}.$$

This is the *equation of current continuity* or *charge conservation*, which states that a divergence of the conduction current is causally connected to a change in the true charge density. Without introducing the displacement current, Maxwell's theory would have violated this law.

Show, by assuming that Ohm's law (I, Eq. 6.7) *is valid, that a space charge in a conducting medium decays expo-*

nentially to zero. Calculate the time constants for copper, sea water, and fused quartz and demonstrate that the charge will flow to the surface of a conductor (Faraday's pail experiment).

7 · Shrinkage of the Wavelength of an Electromagnetic Wave in a Semiconductor

The wavelength of an electromagnetic wave depends not only on the dielectric constant and permeability but also on the conductivity of a dielectric medium (see I, Eqs. 7.15 and 9.23). Assuming a dielectric constant κ' of 10 for selenium, *how large would its conductivity have to be in order to reduce to 5 cm a wavelength of 10 cm in the lossfree material?*

8 · Dimensional Analysis

Frequently the accurate form of physical relations can be derived, or at least surmised, by equalizing the dimensions of the two sides of an equation. Let us illustrate this procedure by deriving the time constant of an RC and of an LRC circuit.

According to the transient equation of an RC circuit

$$R\frac{dQ}{dt} + \frac{1}{C}Q = 0,$$

its time constant τ can only depend on R and C, hence

$$[\tau] = [R]^{\alpha}[C]^{\beta}.$$

To determine the exponents α and β in this dimensional equation, we introduce for $[R]$ and $[C]$ the dimensions given in I, Table 8.1, and rewrite

$$[\text{sec}] = [\text{kg}^{\alpha}\ \text{m}^{2\alpha}/\text{sec}^{\alpha}\ \text{coul}^{2\alpha}][\text{sec}^{2\beta}\ \text{coul}^{2\beta}/\text{kg}^{\beta}\ \text{m}^{2\beta}].$$

With $\alpha = \beta$, the equation balances, if $\beta = 1$; hence $\tau \simeq RC$. $\tau = RC$ is the result of the standard calculation (see I, Eq. 26.7).

If we follow the same procedure for the LRC circuit by writing

$$[\tau] = [L]^{\epsilon}[R]^{\alpha}[C]^{\beta},$$

that is,

$$\begin{aligned}[\text{sec}] &= [\text{kg}^{\epsilon}\ \text{m}^{2\epsilon}/\text{coul}^{2\epsilon}][\text{kg}^{\alpha}\ \text{m}^{2\alpha}/\text{sec}^{\alpha}\ \text{coul}^{2\alpha}] \\ &\quad \times [\text{sec}^{2\beta}\ \text{coul}^{2\beta}/\text{kg}^{\beta}\ \text{m}^{2\beta}] \\ &= [\text{kg}^{\epsilon}\ \text{m}^{2\epsilon}/\text{coul}^{2\epsilon}] \\ &\quad \times [\text{kg}^{\alpha-\beta}\ \text{m}^{2(\alpha-\beta)}/\text{coul}^{2(\alpha-\beta)}\ \text{sec}^{\alpha-2\beta}],\end{aligned}$$

we eliminate all dimensions, besides the time dependence, for $\epsilon = \beta - \alpha$ and balance the equation with $2\beta - \alpha = 1$. For $\epsilon = 0$ we return to the solution of the RC circuit. With $\epsilon = 1$ the equation balances, when $\beta = 0$, $\alpha = -1$, i.e., $\tau \simeq \dfrac{L}{R}$. According to II, Eq. 4.14, and I, Eq. 26.23, the result is $\tau = \dfrac{L}{R}$. However, here the limitations of dimensional analysis become apparent. If we would choose $\epsilon = 2$ the equation would balance with $\beta = -1$ and $\alpha = -3$, and the time constant would become $\tau \simeq \left(\dfrac{L}{RC^2}\right)\dfrac{L}{R}$. To derive the general case, we eliminate ϵ by substitution and find $\tau \simeq \left(\dfrac{L}{RC^2}\right)^{-\beta}\dfrac{L}{R}$, i.e., dimensional analysis leaves the existence of a factor $f\left(\dfrac{L}{RC^2}\right)$ in the realm of possibility.

Use dimensional analysis to derive the energy radiated per second from a dipole. This radiation loss of a dipole antenna will obviously depend on the frequency of the dipole, the dipole moment, and the phase velocity of the electromagnetic wave, i.e., on the dielectric constant and permeability of free space. For the actual radiation law see I, Eq. 13.18.

Dimensional analysis is obviously much simplified by the use of four dimensions (m, kg, sec, coul), because the exponents must be integers.

9 · Macroscopic Analysis of the Frequency-Response Characteristic of a Ferrite

Figure 9.1 shows the dielectric frequency-response characteristic of a ferromagnetic semiconductor (see

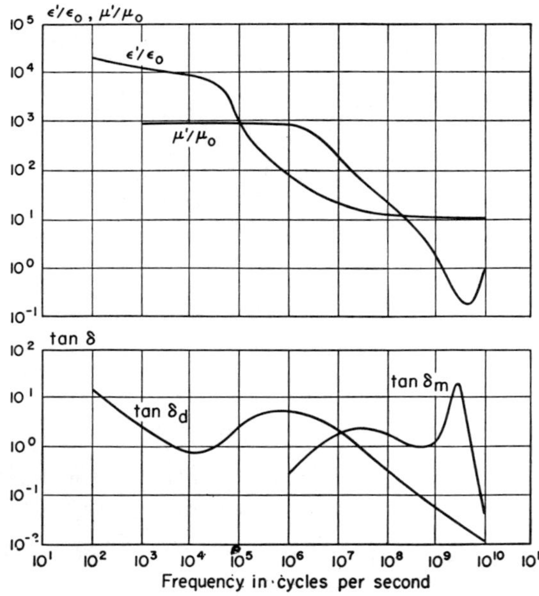

Fig. 9.1. Frequency-response characteristic of a ferrite.

II, Sec. 30) as represented by the real part of the permittivity and permeability and the electric and magnetic loss tangent (dissipation factor). *Answer the following questions by using the conversion formulas and charts of I, Sec. 9*: (1) *At what frequency is the magnitude of the intrinsic impedance most nearly that of free space?*

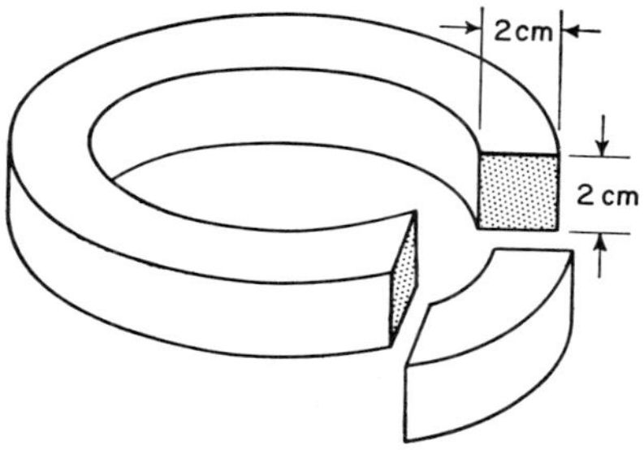

Fig. 9.2. Ferrite toroid.

(2) *Is the material suitable for a 20-db attenuator section in a coaxial line at 3000 Mc, if the length is not to exceed 2 cm?* (3) *What would the reflection coefficient of such an attenuator be if inserted without matching section* (cf. for this determination I, Secs. 17 and 18)?

Since a ferromagnetic dielectric allows the storage of electric and magnetic energy, the material itself can play the role of an *LRC* circuit or a space resonator. *If you make a toroid of this material with a cross section of 2 by 2 cm*^{2} (Fig. 9.2), *at which frequency will it tend to resonate regardless of the coil wound on this ferromagnetic core* (cf. I, Sec. 22)?

10 · Electric and Magnetic Balances

By measuring the force of attraction between the plates of a capacitor, for example, in a balance arrangement[1] (Fig. 10.1), either the voltage between the

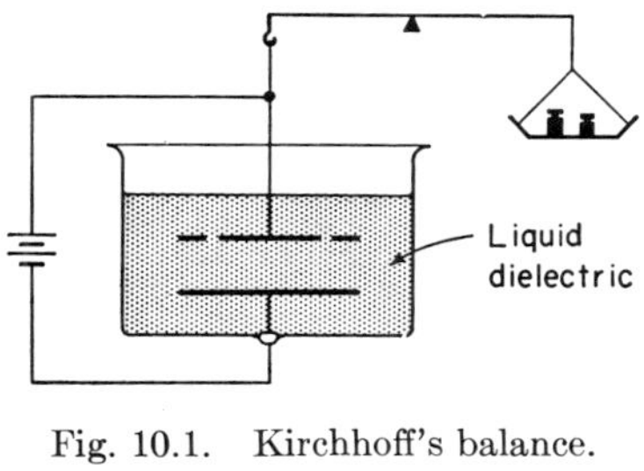

Fig. 10.1. Kirchhoff's balance.

plates or the dielectric constant of the filling medium can be determined. *Derive the operating equation by differentiation of the electrostatic energy* (cf. I, Eq. 3.25 and Sec. 11). *Is the true or free charge decisive for the force action? How well will such a balance operate at alternating current?*

The dielectric constant of liquids can be measured by employing a U tube with one branch of narrow, the other of wide bore (Fig. 10.2). The narrow branch is

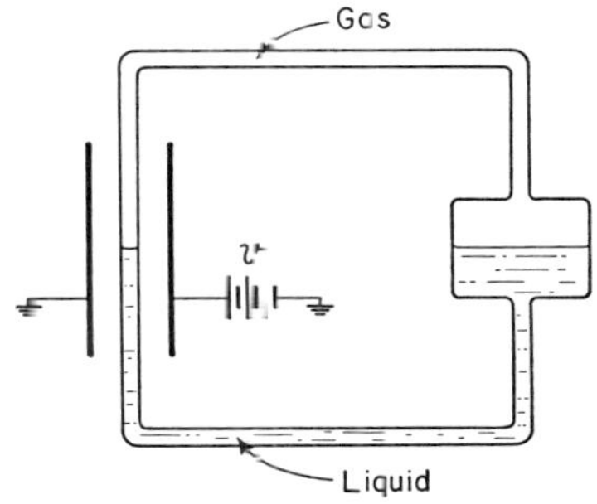

Fig. 10.2. Quincke's method.

placed in a constant electrostatic field, and the rise of the meniscus determines the difference in dielectric constant between liquid and gas. This method of Quincke[2] can also serve for the measurement of magnetic susceptibilities if the electrostatic field is replaced by a magnetic field. *Derive the electric and magnetic susceptibility expressions and calculate the rise of the meniscus for water in a field of 5000 v/cm or 5000 gausses, respectively.*

The dielectric constant or permeability of a solid may be measured by suspending a uniform cylinder of the material with one end in a homogeneous field and compensating the force, for example, by weights on the other balance arm (Fig. 10.3). This method of Gouy[3]

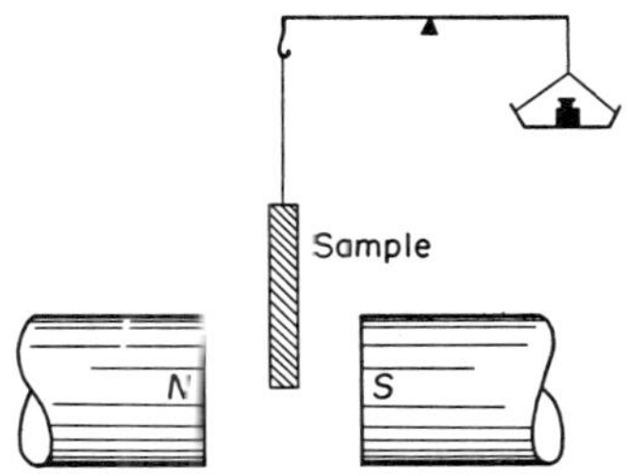

Fig. 10.3. Gouy balance.

is most commonly used for magnetic measurements. It can be applied also to liquids and gases enclosed in a cylindrical container. *Compare the merits and shortcomings of the three methods.*

[1] For a description of this arrangement due to Kirchhoff, see F. Kottler, *Handbuch der Physik*, Springer, Berlin, 1927, Vol. 16, p. 225, and Vol. 12, pp. 349 ff.

[2] G. Quincke, *Ann. Physik 24*, 347 (1885); see also E. C. Stoner, *Magnetism and Matter*, Methuen, London, 1934, pp. 81 ff; and R. M. Bozorth, *Ferromagnetism*, Van Nostrand, New York, 1951, pp. 859 ff.

[3] L. G. Gouy, *Compt. rend. 109*, 935 (1889); see also E. C. Stoner, Ref. 2.

11 · Mass Spectrographs

The classical mass spectrographs analyze ions by the combined action of electric and magnetic fields which separates the ions according to their ratio of charge to mass (e/m determinations).

The first method of this type was developed by J. J. Thomson [1] (Fig. 11.1), who passed the positive rays

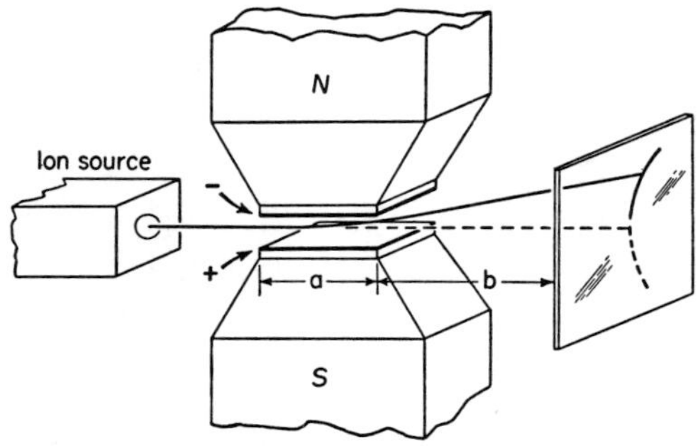

Fig. 11.1. Thomson's parabola method.

emerging from a long narrow channel in the cathode of a discharge tube (*canal rays*) through parallel and overlapping electric and magnetic fields. After traversing a field-free space of a length $b \gg a$, where a is the length of the field space, the ions strike a photographic plate. *Show with the help of I, Eq. 10.21, that the ions of the same e/m for all velocities are focused on a parabola.*

The next great advance was the mass spectrograph of Aston,[2] who arranged the electric and magnetic fields in tandem and perpendicular to each other (Fig. 11.2).

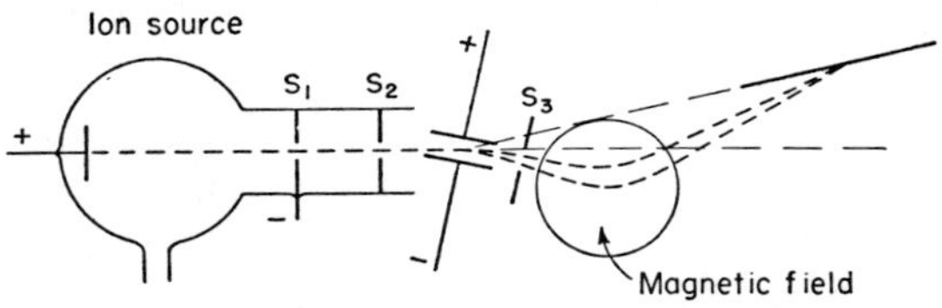

Fig. 11.2. Aston's mass spectrograph (magnetic field normal to plane of paper).

Show that in this case ions of a velocity range limited by aperture S_3, instead of being spread out along a parabola for the same e/m, *are focused in a single point.*

Ions of constant velocity emitted, for example, from a hot filament F (impregnated with a salt) and accelerated in an electric field, can be investigated in a mass spectrograph of the Dempster type [3] (Fig. 11.3). Here the particles describe a semicircle in the magnetic field, and it can be shown that ions of the same e/m entering the slit S_1 under various angles are brought to a focus at the exit slit S_3 by the proper adjustment of the accelerating field $\mathcal{V}$ or the magnetic field H. *Derive the expression for the resolving power dm/m of this spectrograph.*

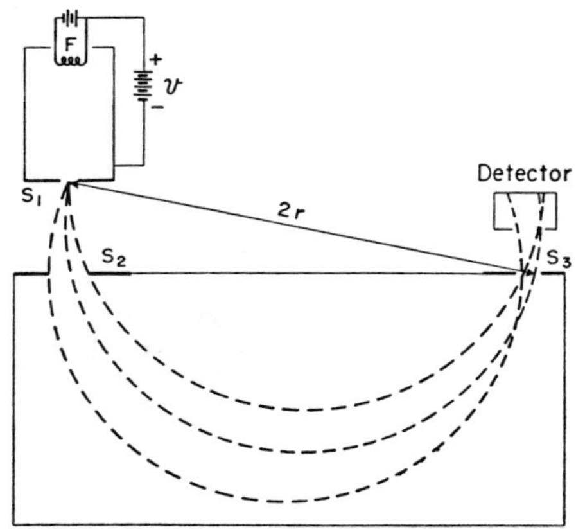

Fig. 11.3. Dempster's focusing method.

[1] J. J. Thomson, *Phil. Mag. 13*, 561 (1907); *Rays of Positive Electricity*, Longmans Green, London, 1921.

[2] F. W. Aston, *Phil. Mag. 38*, 707 (1919); *Isotopes*, Arnold, London, 1924.

[3] A. J. Dempster, *Phys. Rev. 11*, 316 (1918); *21*, 209 (1923).

12 · The Electromagnetic Field Energy and Its Flow According to Maxwell's Equations

Starting with the set of equations of Sec. 6,

$$\nabla \times \mathbf{H} = \mathbf{J} + \frac{\partial \mathbf{D}}{\partial t},$$

$$\nabla \times \mathbf{E} = -\frac{\partial \mathbf{B}}{\partial t},$$

$$\nabla \cdot \mathbf{D} = \rho,$$

$$\nabla \cdot \mathbf{B} = 0,$$

we form the scalar product of the first equation with **E** and of the second with **H**. Subtracting, applying the vector identity

$$\nabla \cdot (\mathbf{E} \times \mathbf{H}) = \mathbf{H} \cdot (\nabla \times \mathbf{E}) - \mathbf{E} \cdot (\nabla \times \mathbf{H}),$$

and integrating over the volume V, we obtain the change of the electromagnetic energy stored in this volume. *Show that the result is the Poynting theorem*, I, Eq. 11.13.

13 · Boundary Condition for the Electric and Magnetic Flux Densities

In I, Sec. 14, we obtained the refraction law for the static electric and magnetic field by introducing, in addition to the continuity of the tangential component of **E** or **H**, the continuity of the normal component of **D** or **B**. In deriving the reflection and refraction laws for the electromagnetic field we used as boundary conditions the continuity of the tangential components of **E** and **H** (see I, Eq. 14.9) without any reference to the normal components of **D** and **B**. *Show that also in the electromagnetic case the normal components of* **D** *and* **B** *are continuous, but that this additional boundary condition does not lead to any further information due to the coupling of the field vectors by Maxwell's equations.*

14 · Interference Filters

The dielectric layer of I, Fig. 18.3, shows maxima of transmission or "passbands" at those wavelengths at which its optical thickness corresponds to a multiple of $\lambda/2$. Such a layer can therefore serve as a filter for the transmission of a prescribed wavelength region. The properties of interest are: peak transmission; ratio of maximum to minimum transmission; bandwidth (see I, Fig. 24.7); wavelength separation between passbands; and the angular field of view. This last quantity is defined as the angle through which the filter must be tilted from normal incidence to shift the wavelength of peak transmission by a distance equal to the bandwidth.

Figure 14.1 shows three types of interference filter: (*a*) a single layer of ZnS of the thickness $n\frac{\lambda_2}{2}$ with $n = 1, 2, 3, \ldots$; (*b*) the same layer of ZnS backed on both sides by a highly reflecting (92 percent) layer of semitransparent silver (*solid Fabry-Perot filter* [1]); and (*c*) a *multi-layer Fabry-Perot filter*,[2] in which the silver reflector in front and back of the half-wavelength plate is replaced by a dielectric reflector consisting of seven $\lambda/4$ layers each of alternating high and low dielectric constant. The purpose of this last construction is to avoid the absorption loss in the silver layer. All three filters are assumed to be adjusted with their peak transmission to the wavelength of the green mercury line.

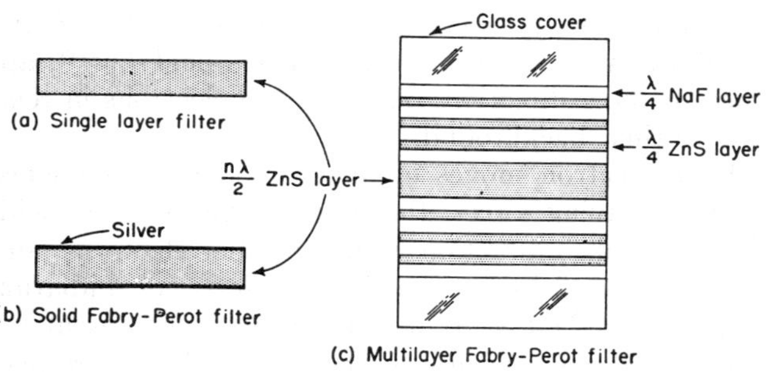

Fig. 14.1. Three types of interference filters.

[1] O. Struve, *Sky and Telescope*, January, 1951.

[2] H. D. Polster, *J. Opt. Soc. Am. 39*, 1054A (1949); see also B. H. Billings, *Photographic Engineering 2*, 45 (1951).

The indices of reflection of ZnS and NaF are $n \simeq 2.3$ and $\simeq 1.3$, respectively; the silver layers reduce the transmitted energy at the peak to about 35 percent.

Derive, by using the impedance concepts, the general equation for the transmission coefficient t and the energy coefficient of transmission T (I, Eqs. 15.17 and 15.22) *for oblique incidence* (cf. I, Sec. 18). *Compare the properties of the three filters.*

15 · Microwave Optics

Light optics operates with lenses, mirrors, and plates, that is, it uses the dielectric properties of materials and the shapes of their reflecting and refracting boundaries. Microwave optics has an additional degree of freedom because the beam width is of the order of magnitude of the wavelength. Thus it can use the lateral boundaries to produce optical effects since wave guides behave like highly dispersive dielectrics (see I, Fig. 22.3).

A TEM wave traveling in the x-direction can be polarized by giving its field components E_y and E_z a phase shift δ against each other (see I, Eq. 12.1). Depending on the magnitude and direction of this temporal phase shift, various types of polarized radiation result (I, Fig. 12.4). A simple optical polarizer consists

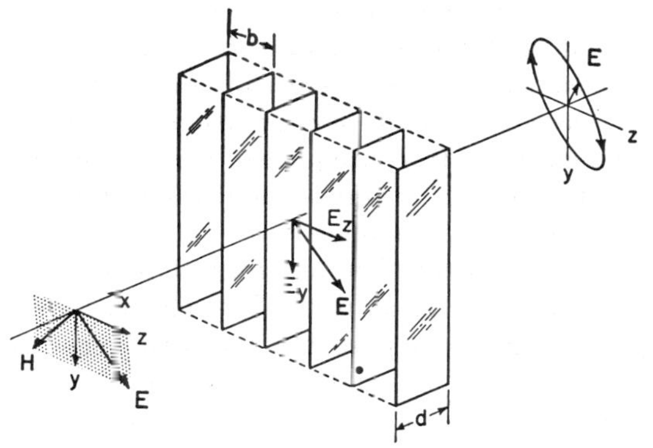

Fig. 15.1. Microwave polarizer.

of a layer of an anisotropic material (mono-axial crystal, stretched plastic), orientated with its axis, for example, in the y-direction. The index of refraction, and with it the propagation velocity, differs for the y and z component, and desired phase shifts δ are obtained by adjusting the magnitude of the anisotropy and the thickness of the layer.

In microwave optics the same purpose can be achieved by placing in the way of the TEM wave an array of parallel-plate wave guides, orientated with its plate surfaces, for example, parallel to the x,y-plane (Fig. 15.1). If we assume that the plates have zero thickness and infinite conductivity, and that fringing effects can be neglected, we can treat this plate polarizer very simply. The E_z-component of our TEM wave passes through the array completely undisturbed, since the electric field stands normal to the boundaries and cannot create any current. The E_y-component, on the other hand, traverses each wave guide section as a TE wave (cf. I, Sec. 22) with a phase velocity larger than that of free space (cf. I, Eqs. 21.13 and 22.23). Normally, the plate spacing is maintained as $\lambda > b > \lambda/2$, so that only the dominant TE_{10} mode will propagate. Thus the *microwave plate polarizer* acts like an *artificial dielectric*, an anisotropic parallel plate of the refractive index 1 in the z-direction and a refractive index $n < 1$ in the y-direction.

Calculate on the basis of this analogue the action of a parallel-plate array ($b = 8.33$ cm, $d = 15$ cm) *on a 10-cm wave entering at normal incidence and linearly polarized at 45°. What is the state of polarization of the emerging wave?* In solving this problem, higher-order modes can be disregarded, but standing waves in the artificial dielectric have to be taken into account (cf. I, Sec. 18).

II · Molecular Approach

1 · The Compton Effect

The particle nature of radiation is most directly observed in the scattering of a monochromatic X-ray beam by some light element. As Compton[1] first showed, there appears, in addition to the primary beam, a scattered secondary beam of longer wavelength, deflected by an angle ϕ (see II, Fig. 7.1). The phenomenon can be explained quantitatively by assuming that an incident photon $h\nu$ of the mass $h\nu/c^2$ (see II, Eq. 7.4) collides elastically with an electron and transfers some energy according to the conservation laws of energy and momentum. In consequence, the scattered quantum has the smaller energy $h\nu'$ and the electron recoils with a velocity v at an angle θ.

The simple calculation assumes that the electron is originally free and at rest. Its velocity afterwards is high enough to require a relativistic mass correction (see II, Eqs. 7.1 and 7.2). If the rest mass of the electron is m_0, its kinetic energy is accordingly $mc^2 - m_0c^2 = m_0c^2\left(\dfrac{1}{\sqrt{1-\dfrac{v^2}{c^2}}} - 1\right)$ and its momentum $mv = \dfrac{m_0 v}{\sqrt{1-\dfrac{v^2}{c^2}}}$.

Confirm the general equation for the wavelength shift

$$\Delta\lambda = \lambda' - \lambda = \frac{h}{m_0 c}(1 - \cos\phi)$$

observed in the Compton effect and calculate the kinetic energy and angle θ *of the recoil electron. Give numerical results, assuming that the incident radiation is molybdenum* Kα *radiation* ($\lambda = 0.71$ A).

[1] A. H. Compton, *Phys. Rev. 21*, 207, 483 (1923).

2 · Plasma Resonance and Dispersion of an Electron Gas

Assume a plasma of electrons and ions, overall neutral and suspended in free space as a spherical space-charge cloud. The positive ions are assumed to be smeared out into a stationary continuum. We wish to determine the polarization and permittivity of the cloud when an external homogeneous field $E_0e^{j\omega t}$, applied in the x-direction, displaces the electrons in relation to their background.

Each electron moves obviously under the influence of three driving forces: the positive space-charge field, the Coulomb interaction with the other electrons, and the external field. (The friction force of radiation damping [see II, Sec. 4] may be neglected at present.)

If the cloud contains S electrons and ions, respectively, the potential $\phi_i(r)$ of the positive space-charge sphere of radius r_0 is (see I, Eq. 4.20)

$$\phi_i(r) = \frac{Se(3r_0^2 - r^2)}{\epsilon_0 8\pi r_0^3} \qquad \text{for } r \leq r_0,$$

hence the space-charge force in the x-direction for jth electron at a distance x_j from the center is $-\dfrac{Se^2x_j}{\epsilon' 4\pi r_0^3}$.

The Coulomb interaction is given by the x derivative of the Coulomb potential as

$$-\frac{\partial}{\partial x_j}\sum_{k\neq j}\frac{e^2}{|\mathbf{r}_j - \mathbf{r}_k|},$$

and the applied-field amplitude $\mathbf{E}_0$ is modified to a cavity field (I, Eq. 10.13)

$$\mathbf{E} = \frac{3}{\kappa' + 2}\mathbf{E}_0.$$

Thus the equation of motion of the jth electron becomes

$$m\ddot{x} = -\frac{Se^2}{\epsilon_0 4\pi r_0^3}x_j - \frac{\partial}{\partial x_j}\sum_{k\neq j}\frac{e^2}{|r_j - r_k|} + e\frac{\kappa' + 2}{3}\mathbf{E}e^{j\omega t}.$$

The total dipole moment of the space-charge sphere is found by the summation over the contribution of all electrons, $\sum_{j=1}^{S} ex_j$. Summing over j in the equation of motion, we find that the Coulomb interaction vanishes because each force term is balanced by a counter-force term. Hence collisions between the electrons of the plasma gas do not affect the result.

From the overall force equation

$$m\sum_{j=1}^{S}\ddot{x}_j = -\frac{Se^2}{\epsilon_0 4\pi r_0^3}\sum_{j=1}^{S}x_j + Se\frac{\kappa' + 2}{3}Ee^{j\omega t},$$

we find, by introducing the number of electrons per unit volume,

$$N = \frac{3S}{4\pi r_0^3},$$

the polarization $\mathbf{P}$ (the dipole moment per unit volume) as

$$\mathbf{P} = \frac{3e\sum_{j=1}^{S}x_j}{4\pi r_0^3} = \frac{(N_e^2/m)(\kappa' + 2)Ee^{j\omega t}}{\omega_p^2 - 3\omega^2},$$

and the plasma resonance frequency as

$$\omega_p = \sqrt{\frac{Ne^2}{\epsilon_0 m}}.$$

Since the polarization by definition (I, Eq. 2.7)

$$\mathbf{P} = (\kappa' - 1)\epsilon_0 Ee^{j\omega t},$$

we obtain the equation for the determination of the dielectric constant of this plasma

$$\frac{\kappa' - 1}{\kappa' + 2} = \frac{\omega_p^2}{\omega_p^2 - 3\omega^2},$$

or

$$\kappa' = 1 - \frac{\omega_p^2}{\omega^2}.$$

Our electron gas behaves like a system of undamped electron oscillators (cf. II, Sec. 4) on account of the restoring force introduced by the positive ions. It makes a negative contribution to the dielectric constant because, somewhat unjustifiably, the static cavity field has been invoked. In plasmas extending without definite boundaries, the restoring force is missing, if no longitudinal, compressing and expanding waves are introduced. Examples are the Heaviside layers in the highest atmosphere and the electron gas in metals, which normally are treated equivalent to RL circuits. *Calculate the behavior of an electron gas that suffers friction by collisions and radiation but has no restoring force.*

Plasma oscillations have become recently of new interest for investigations of the stratosphere as well as for the generation of microwaves.

3 · Wave Functions of the Harmonic Oscillator in an Electric Field

The classical behavior of a harmonic oscillator and its quantum-mechanical counterpart has been discussed in II, Sec. 18. If the oscillator vibrates in the x-direction, its total energy $\mathcal{E}$ may be written, according to II, Eqs. 18.7 and 18.17,

$$\mathcal{E} = \frac{p^2}{2m} - \frac{1}{2}m\omega_0^2x^2.$$

We obtain the wave equation of the oscillator by introducing the total and the potential energy into Schrödinger's equation (II, Eq. 7.18) as

$$-\frac{\hbar^2}{2m}\frac{d^2\psi}{dx^2} + \frac{1}{2}m\omega_0^2x^2\psi = \mathcal{E}\psi.$$

Specifying the solution

$$\psi = Ae^{-\alpha x^2},$$

we find that the parameter α is given by the equation

$$-\frac{\hbar^2}{2m}(-2\alpha)(1 - 2\alpha x^2) + \frac{1}{2}m\omega_0^2x^2 = \mathcal{E}.$$

In quantum physics, the energy of the oscillator does not depend on the amplitude x but only on the frequency (see II, Eq. 18.27); hence the two terms of the equation depending on x^2 must cancel,

$$(2\alpha)^2\frac{\hbar^2}{2m} = \frac{1}{2}m\omega_0^2.$$

This determines α as

$$\alpha = \frac{m\omega_0}{2\hbar},$$

and with it the energy as

$$\mathcal{E} = \frac{\hbar\omega_0}{2}.$$

This is the zero-point energy of the oscillator, since our solution represents the ground state (see II, Fig. 18.2).

The factor A serves to normalize the wave function according to II, Eq. 7.20; from $\int_{-\infty}^{+\infty} \psi\tilde{\psi}\, dx = 1$ we obtain $A = \left(\frac{m\omega_0}{\pi\hbar}\right)^{1/4}$.

Application of an electric field E in the $+x$-direction adds to the potential energy $\frac{1}{2}m\omega_0{}^2x^2$ a dipole energy term $-eEx$. *Show that the wave function becomes now*

$$\psi' = Ae^{-\alpha\left(x - \frac{eE}{m\omega_0{}^2}\right)^2},$$

the energy

$$\mathcal{E}' = \mathcal{E} + \left(\frac{eE}{m\omega_0{}^2}\right)^2,$$

and that the matrix element of the dipole moment (see II, Eq. 12.14) *is*

$$\int_{-\infty}^{+\infty} \psi'\tilde{\psi}ex\, dx = \frac{e^2E}{m\omega_0{}^2}$$

in agreement with the static polarization contribution of the classical oscillator (see II, Eq. 19.2). This result confirms the contention made in II, Sec. 19, that the static dielectric constant can be obtained from the optical spectrum.

4 · The Kerr Effect

Kerr observed that glass [1] or liquids [2] become doubly refracting in an electrical field, that is, they behave optically anisotropic like a mono-axial crystal, orientated with its axis in the field direction. With more sensitive methods this *electro-optical Kerr effect* has later also been found in gases.[3] If n_p and n_s designate the indices of refraction for light of the wavelength λ, oscillating with its electric vector parallel or perpendicular to the electric field, the phase shift between the two components is

$$\Delta_\lambda = \frac{l}{\lambda}(n_p - n_s).$$

Here l designates the length of the dielectric exposed to the field and Δ_λ the shift measured in units of wavelength. Kerr found that

$$\Delta_\lambda = BlE^2,$$

where E is the field strength and B the Kerr constant. Depending on whether n_p is greater or smaller than n_s, the material is called positive or negative doubly refracting.

The effect can obviously stem from several causes. (*a*) Molecules, originally isotropic, may become anisotropic in their polarizability owing to pre-polarization by the electric field. This would be an action stemming from the anharmonicity of the oscillators responsible for the electronic and atomic polarization. The effect obviously will be small and temperature insensitive. (*b*) The molecules are a priori anisotropic in their polarizabilities, but carry no dipole moment. (*c*) The molecules are anisotropic and carry dipole moments. In cases *b* and *c*, the electric field E causes orientation polarization; hence the Langevin function with its large temperature dependence applies (II, Sec. 16). The effect *c*, produced by the orientation of permanent dipoles, will be the strongest one.

To derive an expression for the anisotropy $(n_p - n_s)$ and the Kerr constant caused by the orientation of permanent dipoles, we assume the following simple model. The permanent field $\mathbf{E}_0$ produces a preferential orientation of the dipoles in its direction according to Boltzmann statistics (cf. II, Eq. 16.2). The probability of finding a dipole pointing at a space angle inclined between θ and $\theta + d\theta$ against the field axis is

$$w(\theta)\, d\theta = Ce^{(|\mu|\,|\mathbf{E}_0|\cos\theta)/kT} \sin\theta\, d\theta\, d\phi.$$

Let the dipoles be polarizable by the electric field of the light wave only along their polar axis. If the polarizability is α and the electric field vector $\mathbf{E}$ of the light

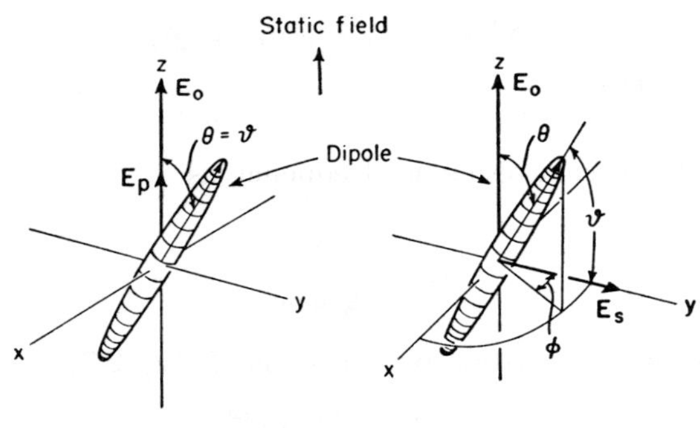

Fig. 4.1. Model for calculation of the Kerr effect caused by orientation of permanent dipoles.

wave makes an angle ϑ with the dipole axis, the component $E\cos\vartheta$ parallel to the axis induces a moment $\alpha E\cos\vartheta$ along the axis or a component $\alpha E\cos^2\vartheta$ in the $\mathbf{E}$-field direction (Fig. 4.1). Hence the polarization induced by the light wave in the E-direction is

$$\mathbf{P} = N\alpha\mathbf{E}\,\overline{\cos^2}\,\vartheta,$$

where $\overline{\cos^2}\,\vartheta$ describes the average statistical distribution of the dipole axes in relation to the $\mathbf{E}$ vector of the light wave.

[1] J. Kerr, *Phil. Mag.* [4] *50*, 337, 446 (1875).
[2] J. Kerr, *Phil. Mag.* [5] *8*, 85, 229 (1879).
[3] R. Leiser, *Physik. Z. 12*, 955 (1911); G. Szivessy, *Z. Physik 26*, 323 (1924).

The average statistical distribution of the dipole axes in relation to the applied static field $\mathbf{E}_0$ is given as

$$\overline{\cos^2\theta} = \frac{\iint w(\theta)\cos^2\theta\, d\theta\, d\phi}{\iint w(\theta)\, d\theta\, d\phi}.$$

For the parallel (p) component of the light wave, $\vartheta = \theta$; for the light component perpendicular to the static field (s component), $\cos\vartheta = \cos\phi\sin\theta$, or, after averaging over ϕ,

$$\overline{\cos^2\vartheta} = \tfrac{1}{2}(1 - \overline{\cos^2\theta}).$$

Calculate $\mathbf{P}_p$ *and* $\mathbf{P}_s$ *for our simple model, determine* $n_p - n_s$ (cf. I, Sec. 12), *and show that the Kerr constant has the value*

$$B = \frac{1}{\lambda}\left(\frac{n_0 - 1}{n_0}\right)\frac{1}{5}\left(\frac{\mu}{kT}\right)^2,$$

where n_0 is the isotropic index of refraction without static field.

If a Kerr cell is placed between crossed Nicols, light will pass in general through the analyzer when $\mathbf{E}_0$ is applied. *Can this light be cut off by turning the analyzer?*

5 · The Faraday Effect

Faraday [1] placed a transparent dielectric (lead glass) between the poles of a strong electromagnet and sent linearly polarized light, through a bore in the pole

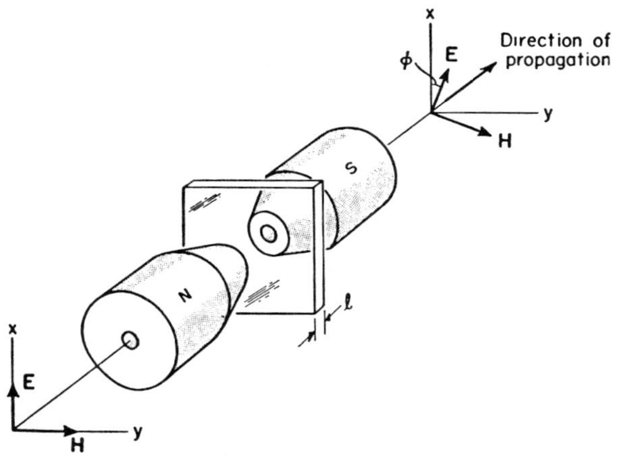

Fig. 5.1. Faraday effect.

pieces, parallel to the field through the material (Fig. 5.1). He observed a rotation of the plane of polarization by an angle

$$\phi = RlH,$$

[1] M. Faraday, *Experimental Researches in Electricity*, Quaritch, London, 1839.

where l is the length of the dielectric, H the magnetic field strength, and R the so-called Verdet [2] constant, which depends on the dielectric, its temperature, and on the frequency of the light.

That a rotation of the plane of polarization will occur can be derived with the simple model of electronic harmonic oscillators used for the discussion of anomalous dispersion and resonance absorption in II, Sec. 4. If the electrons oscillate in the y-z plane under the influence of the local electric field E' and the magnetic field H, we obtain, in place of II, Eq. 4.20, by referring to I, Sec. 10,

$$\frac{d\dot{\mathbf{r}}}{dt} + 2\alpha\frac{d\mathbf{r}}{dt} + \omega_0^2\mathbf{r} = \frac{e}{m}\left(E' + \frac{d\mathbf{r}}{dt}\times B\right).$$

By introducing the polarization $\mathbf{P}$ and the local Mosotti field (see II, Eqs. 4.22 and 4.23), the differential equation for the polarization results:

$$\ddot{\mathbf{P}} + 2\alpha\dot{\mathbf{P}} + \omega_0''^2\mathbf{P} = \frac{Ne^2}{m}\mathbf{E} + \frac{\mu'}{Nm}(\dot{\mathbf{P}}\times\mathbf{H}).$$

Since P is periodic with time (see I, Eq. 4.27), the components of the polarization in the y- and z -directions follow as

$$P_y(\omega_0''^2 - \omega^2 + j\omega 2\alpha) = \frac{Ne^2}{m}E_y + j\omega\frac{\mu' H}{Nm}P_z,$$

$$P_z(\omega_0''^2 - \omega^2 + j\omega 2\alpha) = \frac{Ne^2}{m}E_z - j\omega\frac{\mu' H}{Nm}P_y.$$

Using the relation I, Eq. 2.7, between $\mathbf{P}$ and $\mathbf{E}$ we substitute the electric field for the polarization and find

$$E_y = \pm jE_z,$$

$$\epsilon' - \epsilon_0 = \frac{Ne^2/m}{\left(\omega_0''^2 \mp \dfrac{\omega\mu' H}{Nm} - \omega^2\right) + j\omega 2\alpha}.$$

The first of these equations states that our linearly polarized light can be considered as composed of a right- and left-hand circularly polarized component (see I, Sec 12). The second equation shows that the dielectric constant, that is, the propagation velocity for these two components differs, the former being retarded, the latter advanced by equal amounts. *Investigate this situation in detail and calculate the Verdet constant of atomic hydrogen at atmospheric pressure according to this simple model for the light of the green mercury line* (cf. II, Eq. 20.3).

A special application of the Faraday effect in ferrites, the gyrator, is shown in II, Fig. 30.12.

[2] E. Verdet, *Ann. chim. phys.* *41*, 570 (1854).

6 · The Ionic Atmosphere in the Debye-Hückel Theory of Strong Electrolytes

The ions of one sign in electrolytes have the tendency, due to their Coulomb fields, to surround themselves preferentially with ions of the opposite sign. This concept of the ionic atmosphere (see II, Fig. 32.10), first visualized by Milner,[1] was used by Debye and his co-workers (especially Debye and Hückel [2]), to formulate a theory of strong electrolytes. The extension and potential of such an ion cloud can be calculated by using Maxwell-Boltzmann statistics in connection with the Poisson equation (I, Eq. 4.12).

Let us use a negative ion of the charge $-e$ as the reference ion. Its Coulomb potential at a distance r (see I, Eq. 3.7),

$$\phi = -\frac{e}{\epsilon' 4\pi r},$$

produces a statistical increase ΔN_+ of positive and a decrease ΔN_- of negative ions

$$\Delta N_+ = N_0 e^{-e\phi/kT},$$
$$\Delta N_- = N_0 e^{+e\phi/kT},$$

when N_0 represents the average density of the positive and negative ions, respectively, in the neutral mixture. The actual electrostatic potential has to obey Poisson's equation

$$\nabla^2\phi = -\frac{(\Delta N_+ - \Delta N_-)e}{\epsilon'}.$$

At large distances the field energy of the negative ion is small in comparison with the energy of thermal agitation ($e\phi/kT \ll 1$). By expanding the exponentials and limiting ourselves to the first term, the differential equation simplifies to

$$\nabla^2\phi = \frac{2e^2N_0\phi}{\epsilon' kT} \equiv \frac{\phi}{\lambda^2}$$

near the edge of the cloud. Close to the ion, on the other hand, the potential must approach the Coulomb potential; hence we have to fulfill the boundary condition

$$\phi \to -\frac{e}{\epsilon' 4\pi r} \quad \text{for } r \to 0.$$

Confirm, by checking dimensions (see I, Table 8.1) *that*

$$\lambda = \sqrt{\frac{\epsilon' kT}{2e^2N_0}}$$

represents a length, the "radius" of the ion cloud. Since our problem is a purely radial one without angular dependence, the Laplacian in spherical co-ordinates (I, Eq. 4.16) becomes

$$\nabla^2\phi = \frac{1}{r^2}\frac{\partial}{\partial r}\left(r^2\frac{\partial\phi}{\partial r}\right) = \frac{\phi}{\lambda^2}.$$

Verify that the solution of this differential equation is

$$\phi = -\frac{e}{\epsilon' 4\pi r}e^{-r/\lambda},$$

and calculate the radius of the atmosphere for a 1 *M* NaCl *solution at 18°C.*

[1] S. R. Milner, *Phil. Mag.* *23*, 551 (1912).
[2] P. Debye and E. Hückel, *Physik. Z.* *24*, 185, 305 (1923).

7 · Dipole Moments in Polar Liquids According to Onsager's Theory

In II, Sec. 23, we have discussed the essential aspects of the Onsager model which avoids the Mosotti catastrophe. Since Onsager's theory provides an important method of evaluating dipole moments of molecules in polar liquids (see II, Fig. 23.2), we return to these calculations in more detail.

Onsager's model is a point dipole of the moment $\boldsymbol{\mu}$ placed in the center of a sphere of the molecular radius R. The sphere is thought to be filled with a medium of the optical dielectric constant (κ_∞') and surrounded with a continuum of the static dielectric constant (κ_s'). An homogeneous field $\mathbf{E}_0$ is applied and acts as such at large distance from the sphere. Near the sphere it is distorted by the existence of the cavity; hence, instead of $\mathbf{E}_0$, the cavity field $\mathbf{E}_c$ exercises its torque on the dipole. The dipole in the sphere furthermore polarizes the continuum. This polarization of the surroundings in turn creates, by its free charges, a reaction field $\mathbf{E}_{re}$ parallel to the dipole that exercises a polarizing action on the dipole itself and the medium in its cavity. This effect increases the dipole moment in the cavity, but the reaction field turns with the dipole and cannot contribute to the torque. Thus, while the Mosotti-Lorentz-Debye approach introduces *one* local field $\mathbf{E}'$ (see II, Eq. 23.2) which both turns and polarizes, the Onsager theory introduces *two* fields: the cavity field $\mathbf{E}_c$ as the directing field, and the vector sum of the cavity field $\mathbf{E}_c$ and the reaction field $\mathbf{E}_r$, designated as the internal field $\mathbf{E}_i$, as the field that induces polarization.

(*a*) To calculate the cavity field E_c we have to solve the Laplace equation (see I, Eq. 4.13)

$$\nabla^2\phi = 0$$

under the following conditions:

(1) At large distances the potential must be that of the applied field, or $\phi \to E_0 r \cos\vartheta$ for $r \to \infty$.

(2) In the center of the cavity the potential must be finite, or ϕ finite for $r \to 0$.

(3) The electrostatic refraction law (I, Eq. 14.4) prescribes continuity of the normal component of **D** and of the tangential component of **E** at the cavity wall, or D_r and E_ϑ continuous at $r = R$.

Designating the media outside and inside the sphere as 1 and 2, respectively, we have shown in I, Sec. 10, that the solution of Laplace's equation for the potential ϕ_a outside and ϕ_i inside the sphere is (I, Eq. 10.12)

$$\phi_a = \left(\frac{\epsilon_2' - \epsilon_1'}{\epsilon_2' + 2\epsilon_1'}\frac{R^3}{r^3} - 1\right) E_0 z$$

$$\phi_i = -\frac{3\epsilon_1'}{2\epsilon_1' + \epsilon_2'} E_0 z.$$

The same potential ϕ_a would be created by superposing on the applied field the field of a dipole, located parallel to the applied field in the center of the sphere and of the moment (I, Eq. 10.16),

$$\boldsymbol{\mu}_i = 4\pi R^3 \epsilon_1' \frac{\epsilon_2' - \epsilon_1'}{\epsilon_2' + 2\epsilon_1'} \mathbf{E}_0.$$

The cavity field inside the sphere is homogeneous and has the value (I, Eq. 10.13)

$$\mathbf{E}_c = -\frac{\partial \phi_i}{\partial z} = \frac{3\epsilon_1'}{2\epsilon_1' + \epsilon_2'} \mathbf{E}_0.$$

(*b*) To calculate the reaction field we consider that a dipole of the moment $\boldsymbol{\mu}$ placed in the center of the sphere creates a dipole field of the potential $\frac{|\boldsymbol{\mu}|}{\epsilon' 4\pi r^2} \cos\theta$ (see I, Eq. 3.10). This dipole field polarizes the continuum (medium 1) outside the sphere. According to to the preceding calculation, this polarization field of dipole symmetry creates a homogeneous reaction field of the field strength R in medium 2 inside the sphere and this field, as seen from the outside, can be replaced by a dipole moment in the center of the sphere. Thus we can immediately formulate the potentials inside and outside; the former a superposition of the original dipole field and the homogeneous reaction field,

$$\phi_2 = \frac{|\boldsymbol{\mu}|}{\epsilon_2' 4\pi r^2} \cos\theta - E_{re} r \cos\theta \text{ for } r \lesseqgtr R;$$

the latter a dipole field produced by the original dipole and the superposed induced dipole,

$$\phi_1 = \frac{|\boldsymbol{\mu}'|}{\epsilon_1' 4\pi r^2} \cos\theta \text{ for } r \geq R.$$

The same transformation factor $\frac{3\epsilon_1'}{2\epsilon_1' + \epsilon_2'}$, which changes the applied field $\mathbf{E}_0$ in the preceding calculation to the cavity field $\mathbf{E}_c$, relates the actual dipole moment placed into the cavity to the effective dipole moment $\boldsymbol{\mu}'$, that is,

$$\boldsymbol{\mu}' = \frac{3\epsilon_1'}{2\epsilon_1' + \epsilon_2'} \boldsymbol{\mu}.$$

Thus the reaction field becomes, since $\phi_1 = \phi_2$ for $r = R$,

$$\mathbf{E}_{re} = \frac{2(\epsilon_1' - \epsilon_2')}{\epsilon_2'(2\epsilon_1' + \epsilon_2')} \frac{\boldsymbol{\mu}}{4\pi R^3}$$

(see II, Eq. 23.11).

The inner field $\mathbf{E}_i = \mathbf{E}_c + \mathbf{E}_{re}$ induces polarization (electronic and atomic) whereas the cavity field creates the orientation polarization according to the Langevin formula (see II, Sec. 16) *Calculate these polarizations and show that the Onsager model is transformed into the Clausius-Mosotti model, when the dipole in the cavity stands parallel to the cavity field and a special value for the radius R of the cavity is assumed. Calculate from the dielectric constant of nitrobenzene* ($\kappa_s' \simeq 35$, $\kappa_\infty' \simeq 1.5$) *its dipole moment according to the Onsager and the Debye formula and find the magnitude of the reaction field.*

The excessive magnitude of this reaction field and the doubt as to how it responds to changes in dipole position make the Onsager model suspect like any continuum model.[1]

[1] See the discussion of F. C. Frank, *Trans. Faraday Soc. 42A*, 19 (1946).

8 · The Franck-Condon Principle

Franck's basic idea,[1] later developed in wave-mechanical language by Condon,[2] was that electron transitions in molecules take place so rapidly in comparison to the vibrational motions that the nuclei, immediately after the jump, have still about the same relative position and velocity as before the jump. The result is that, depending on the mutual position of the potential curves of electronic states, more or less vibrational energy will be created in electronic transitions for which the light quantum has to pay (Fig. 8.1). This principle, later also extended to crystals,[3] proves of quite general importance for the electronic spectra of molecules and the condensed phases. We will therefore consider the situation more closely.

According to the old quantum theory we would visualize an electron in its vibration state as moving most rapidly through the middle position where the total energy has been converted into kinetic energy

[1] J. Franck *Trans. Faraday Soc. 21*, 536 (1925).
[2] E. U. Condon, *Phys. Rev. 32*, 858 (1928); *41*, 759 (1932).
[3] A. von Hippel, *Z. Physik 101*, 680 (1936).

and spending most of its time at the intersection points, the turning points, where only potential energy remains (see II, Sec. 18). Hence, the electron will be found with greatest probability at the turning points, and transitions vertically upwards or downwards from these points should correspond to the most intense bands spectroscopically observed.

Quantum mechanics formulates this situation more precisely by taking the probability distribution of the electron density into account for the two intercommunicating states. Let us demonstrate the procedure

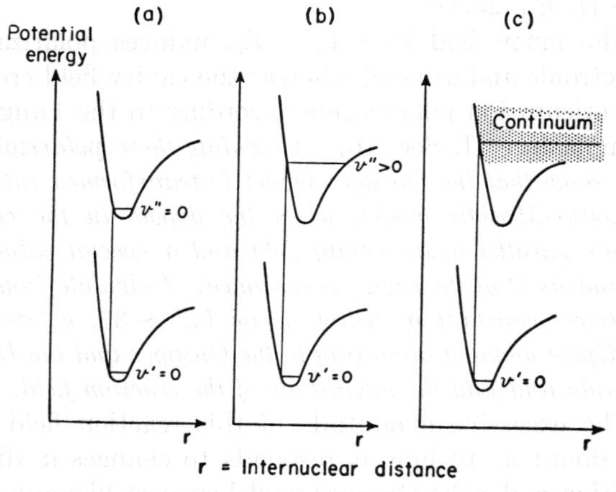

Fig. 8.1. Classical explanation of the Franck-Condon principle:[4] (*a*) no change of vibration energy; (*b*) appreciable excitation of vibration energy; (*c*) vibration excitation leading to dissociation.

by considering the transition from the unexcited ground state ($v' = 0$) of a diatomic molecule to some vibration states of the next higher potential characteristic (Fig. 8.2). In the lowest vibrational state the molecule has a zero-point energy of one-half a vibrational quantum, characterized by the wave function (see Appendix A, II, Sec. 3).

$$\psi'_{v'=0} = (m\omega_i/\pi\hbar)^{1/4} e^{-\frac{1}{2}(m\omega_i/\hbar)x^2}.$$

In the first two vibrational states of the upper potential characteristic, the wave functions are

$$\psi''_{v''=0} = (m\omega_j/\pi\hbar)^{1/4} e^{-\frac{1}{2}(m\omega_j/\hbar)x^2}$$

$$\psi''_{v''=1} = (m\omega_j/\pi\hbar)^{3/4} \sqrt{2\pi}\, x e^{-\frac{1}{2}(m\omega_j/\hbar)x^2}.$$

The probability of transition to one of these states is described by the matrix element of the dipole moment (II, Eq. 12.14):

$$\mathbf{\mu}_{ij} = \int_{\text{all space}} \psi'(e\mathbf{r})\tilde{\psi}''\, dV.$$

This matrix element takes all possible mutual constellations of the electron and of the nuclei in the two states into account. The dipole moment of these constellations determines the coupling strength to the electromagnetic field (dipole radiation). By integrating over all these moments the matrix element represents the transition moment which, after squaring, determines the transition probability (II, Eq. 12.20).

The main contribution to the dipole moment comes from the electrons. The $\mathbf{\mu}_{ij}$ contain two essential terms: a nearly constant contribution characterizing the two interacting electronic states and a contribution determined by the overlap of the electron functions in the

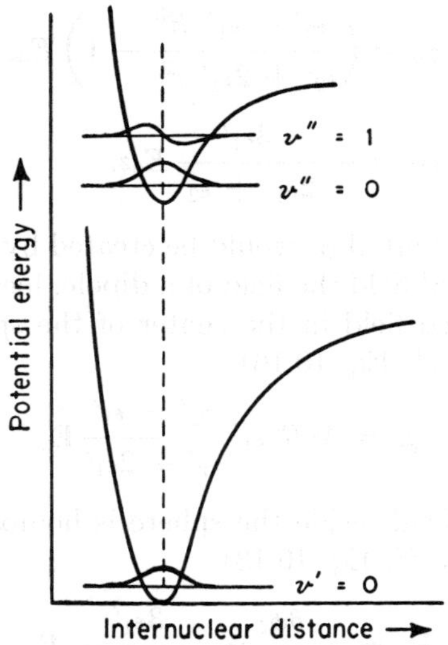

Fig. 8.2. Quantum-mechanical explanation of Franck-Condon principle. (After Herzberg.[4])

various vibration states. The former factor determines the relative intensity of the band in the band system; the latter, the relative intensity of the lines of a band. Figure 8.2 shows graphically that, in the example chosen, the line intensity, determined by the product of the wave functions, will be small for the transition $v' = 0$ to $v'' = 1$, and large for $v' = 0$ to $v'' = 0$.

For large vibrational quantum numbers the wave functions have maxima (or minima) at the classical turning points; hence quantum mechanics leads for these higher transitions to the same result as the old quantum theory.[4]

[4] See, for a more detailed discussion, G. Herzberg, *Diatomic Molecules, Molecular Spectra and Molecular Structure.* I. *Spectra of Diatomic Molecules*, Van Nostrand, New York, 1950, pp. 194 ff.

9 · Born's Lattice Theory and the Compressibility of Ionic Crystals

Born's lattice theory,[1] mentioned in II, Sec. 27, assumes that ionic crystals are held together by the elec-

[1] See M. Born, M. Göppert-Mayer, "Dynamische Gittertheorie der Kristalle," *Handbuch der Physik*, Vol. 12, Pt. 2, 1933, p. 708.

trostatic attraction between the positive and negative ions and prevented from collapsing by strong, short-range repulsive forces.

In calculating the potential energy U of the Coulomb interaction it is not sufficient, as pointed out in II, Sec. 25, to restrict ourselves to nearer neighbors. *Show that a summation over the potential contributions of any simple point lattice occupied by equal charges leads to a divergent series.* However, the fact that the unit cell of a lattice must be electrically neutral as far as its overall charge is concerned makes it clear that the assembly of these elementary cells must act like a system of multipoles. The interaction of such multipoles, as we have shown in I, Secs. 3 and 10, decreases much more rapidly with distance. Thus, by the proper lumping of lattice points before an overall summation or integration is attempted, the potential contributions of all other lattice points at a reference point P (the self-potential of the lattice at P) can be determined.[1] By multiplying this self-potential with the charge contained at P we obtain the Coulomb energy U_c which binds the reference ion to all the other ions of the ionic crystal. By summing over this Coulomb energy for all lattice points N_0 per mole of the crystal, *twice* the electrostatic lattice energy per mole is obtained, since every lattice point has entered twice into the summation process. The Coulomb energy U_c binding one lattice point of the charge ze to all the other lattice points $+ze$ and $-ze$ of the crystal can be expressed (see II, Eq. 25.8) as

$$U_c = -A\,\frac{(ze)^2}{r},$$

where r is the shortest distance to a neighboring ion of opposite polarity and the Madelung constant A characteristic for the lattice type ($A = 1.7627$ for the CsCl and 1.7476 for the NaCl type). The potential energy per mole is

$$\frac{U_c N_0}{2},$$

with N_0 representing Avogadro's number.

The repulsive energy of the lattice points on the reference ions was represented by Born by a term B/r^n and later by Born and Mayer [2] by an exponential function. The contribution of the repulsive potential energy per mole can be expressed as

$$N_0 \frac{c}{2}\frac{B}{r^n},$$

where c represents the number of nearest neighbors, the co-ordination number (see II, Table 24.1), which is 6 for the NaCl and 8 for CsCl structure.

[2] M. Born and J. E. Mayer, *Z. Physik 75*, 1 (1932).

Thus the lattice energy per mole of an NaCl crystal is represented as

$$U(r) = N_0\left(-\frac{Ae^2}{2r} + \frac{3B}{r^n}\right),$$

or the force as

$$F(r) = \frac{dU(r)}{dr} = N_0\left(\frac{Ae^2}{2r^2} - \frac{n3B}{r^{n+1}}\right).$$

To eliminate B we make use of the fact that at the experimentally observed lattice-point distance r_0 the crystal is in equilibrium, that is, $F = 0$. Hence for NaCl

$$B = r_0^{n-1}\frac{Ae^2}{6n}.$$

To determine the exponent of repulsion n, we use in general the compressibility of the crystal

$$k = \frac{1}{V}\frac{\partial V}{\partial p},$$

where V is its volume and p a uniform hydrostatic pressure acting from all sides.

Calculate the compressibility of NaCl and the lattice energy by assuming $n = 8$ and draw the potential energy characteristic $U(r)$ and the force characteristic $F(r)$.

10 · The Born-Haber Cycle

A cycle process, first introduced by Born [1] and Haber,[2] proves useful in many chemical problems to determine an unknown energy quantity from several known quantities. For example, the formation of the ionic sodium chloride crystal could be brought about by reacting a gram atom of sodium metal [Na] with one-half mole of chlorine gas $\frac{1}{2}(Cl_2)$ and obtaining a mole of sodium chloride crystal [NaCl] under development of a heat of reaction Q_{NaCl}. Alternatively, we might evaporate the sodium metal by spending the heat of sublimation S_{Na}, and dissociate the chloride gas into atoms by adding the heat of dissociation $\frac{1}{2}D_{Cl_2}$. Next we might ionize the sodium atom and attach the electrons to the chlorine atoms; thus (Na^+) and (Cl^-) in gaseous form result under loss of the ionization energy I_{Na} and gain of the electron affinity E_{Cl}. Finally, we can bring the positive and negative ions together and form one mole of the NaCl crystal, thus gaining the lattice energy U_{NaCl} (see Appendix AII, Sec. 9). By producing the crystalline phase in these

[1] M. Born, *Ber. physik. Ges. 21*, 679 (1919).

[2] F. Haber, *Ber. physik. Ges. 21*, 750 (1919).

alternative ways we have established the energy cycle:

$$\begin{array}{ccc}
[\mathrm{Na}] + \tfrac{1}{2}(\mathrm{Cl_2}) & \xrightarrow[-\frac{1}{2}D_{\mathrm{Cl_2}}]{-S_{\mathrm{Na}}} & (\mathrm{Na}) + (\mathrm{Cl}) \\
-\,Q_{\mathrm{NaCl}} \Big\uparrow & & \Big\downarrow -I_{\mathrm{Na}} \;\; +E_{\mathrm{Cl}} \\
[\mathrm{NaCl}] & \xleftarrow[+U_{\mathrm{NaCl}}]{} & (\mathrm{Na^+}) + (\mathrm{Cl^-}).
\end{array}$$

It gives the lattice energy for zero temperature by the equation

$$U_{\mathrm{NaCl}} = Q_{\mathrm{NaCl}} + S_{\mathrm{Na}} + I_{\mathrm{Na}} + \tfrac{1}{2}D_{\mathrm{Cl_2}} - E_{\mathrm{Cl}}.$$

Use this type of cycle to determine whether, in the reaction between magnesium metal and chlorine gas, the compound [MgCl] *or* [$MgCl_2$] *will form.*

11 · Electronic and Heat Conduction in Metals According to the Classical Theory

The classical theory of Drude [1] and Lorentz [2] treats the conducting electrons in metals as a gas obeying Maxwell statistics and reaching thermal equilibrium with the metal by impacts with the stationary atoms. An electron undergoes these collisions on the average every τ seconds, and each impact wipes out its directional memory. Thus, if an electric field is applied in the z-direction, the electron obeys, between collisions, the normal force law

$$m\frac{dv}{dt} = eE_z.$$

Its average velocity in the field direction, the drift velocity, is

$$\bar{v}_z = \frac{1}{\tau}\int_0^\tau \frac{eE_z}{m}\,t\,dt = \frac{eE_z}{2m}\tau,$$

or the current density

$$J = \sigma E_z = ne\bar{v}_z = \frac{ne^2\tau}{2m}E_z.$$

The collision time τ is given by the ratio of the mean free path λ and the actual average velocity of the electron $\bar{v}_0$, $\tau = \dfrac{\lambda}{\bar{v}_0}$. According to Maxwell's statistics, $\bar{v}_0$ is the mean thermal velocity

$$\bar{v}_0 = \sqrt{\frac{2kT}{\pi m}}.$$

[1] P. Drude, *Ann. Phys. 1*, 566 (1900); *3*, 369 (1900); *7*, 687 (1902).

[2] H. A. Lorentz, *The Theory of Electrons*, Stechert, New York, 1909.

Hence,

$$\sigma = \frac{ne^2\lambda}{2m}\sqrt{\frac{\pi m}{2kT}}.$$

When a temperature gradient $\dfrac{\partial T}{\partial z}$ exists, a thermal energy current is carried in the z-direction; its density is

$$Q = \kappa\frac{\partial T}{\partial z},$$

where κ designates the heat conductivity of the material. *Derive the thermal conductivity of the electron gas and the Wiedemann-Franz law,*

$$\frac{\kappa}{\sigma} = \frac{\pi^2}{3}\left(\frac{k}{e}\right)^2 T.$$

How far is the assumption justified that the thermal energy current is carried by electrons only?

12 · Space-Charge-Limited Current Flow

Whenever charge carriers are released in abundance from the cathode or anode into a dielectric material, the unidirectional current flow is limited by the space charge it creates. The potential distribution resulting must obey Poisson's equation (I, Eq. 4.12),

$$\nabla^2\phi = -\frac{\rho}{\epsilon'},$$

where the charge density ρ is a function of position. The equation of current continuity must be fulfilled (Appendix AI, Sec. 6),

$$(\nabla\cdot\mathbf{J})\,dV = -\frac{\partial\rho}{\partial t}dV,$$

for any volume element dV. Finally, the current density through any cross-sectional element dA normal to the current flow is given by the product of local charge density and drift velocity,

$$\mathbf{J}\,dA = \rho\mathbf{v}\,dA.$$

A well-known example of space-charge-limited current flow is the thermionic current in vacuum tubes. For plane geometry in the stationary case, the charge density and drift velocity depend only on the distance x between cathode ($x = 0$) and anode ($x = d$), on the voltage $\mathcal{V}(x)$, and on the dielectric medium. Poisson's equation simplifies to

$$\frac{d^2\mathcal{V}}{dx^2} = -\frac{\rho}{\epsilon_0}.$$

The continuity equation, since $\dfrac{\partial \rho}{\partial t} = 0$, prescribes constant current density; although thus

$$J = \rho v = \text{constant},$$

ρ and v separately depend on x. Finally, the dielectric medium (vacuum) can be traversed without energy loss, if radiation is neglected, that is, the potential energy of the free-falling electrons is fully converted into kinetic energy,

$$\tfrac{1}{2}mv^2 = e\mathcal{V}(x).$$

Eliminating ρ and v, we obtain the differential equation for the voltage distribution or the field strength $\dfrac{dE}{dx}$,

$$-\frac{dE}{dx} \equiv \frac{d^2\mathcal{V}}{dx^2} = -\frac{J}{\epsilon_0}\sqrt{\frac{m}{2e}}\,\mathcal{V}^{-\frac{1}{2}}.$$

Multiplying both sides with $d\mathcal{V}$,

$$\frac{dE}{dx}\,d\mathcal{V} = \frac{d\mathcal{V}}{dx}\,dE = E\,dE = \frac{J}{\epsilon_0}\sqrt{\frac{m}{2e}}\,\mathcal{V}^{-\frac{1}{2}}\,dV,$$

and integrating, we get

$$E^2 = \frac{4J}{\epsilon_0}\sqrt{\frac{m}{2e}}\,\mathcal{V}^{\frac{1}{2}} + C_1.$$

We define the cathode potential as zero ($V = 0$ for $x = 0$); furthermore, it is assumed that the emission from the cathode is so abundant that the current is independent of it, and that the electrons leave the cathode with the velocity zero. In this case, $(E)_{x=0} = 0$; hence the constant of integration $C_1 = 0$ and

$$E = \frac{d\mathcal{V}}{dx} = \sqrt{\frac{4J}{\epsilon_0}\left(\frac{m}{2e}\right)^{\frac{1}{2}}}\,\mathcal{V}^{\frac{1}{4}}.$$

Integrating again, we obtain the voltage as function of the distance in the form

$$\frac{4}{3}V^{\frac{3}{4}} = \sqrt{\frac{4J}{\epsilon_0}\left(\frac{m}{2e}\right)^{\frac{1}{2}}}\;x + C_2.$$

Since $V = 0$ for $x = 0$, the constant $C_2 = 0$. For $x = d$, the potential at the anode is the applied voltage V_0; hence the current density becomes

$$J = \frac{4\epsilon_0}{9}\sqrt{\frac{2e}{m}}\,\frac{V_0^{\frac{3}{2}}}{d^2}.$$

The law of the voltage dependence of the space-charge-limited current,

$$J = CV_0^{\frac{3}{2}},$$

is frequently known as the *Child-Langmuir law.*[1] *Show that the 3/2 power dependence is unaffected by the geometry of the vacuum tube by confirming the law for cylindrical and spherical symmetry. How does the law change if vacuum is replaced by a medium with friction in which the electrons move with constant mobility?*

[1] I. Langmuir, *Phys. Rev. 2*, 450 (1913); *21*, 419 (1923); C. D. Child, *ibid. 32*, 492 (1911).

B · Elements of Vector Analysis

Scalar. A quantity specified at any point by a *single* number, its *magnitude* (for example, temperature, mass, electrostatic potential; see I, Sec. 3).

Vector. A quantity specified at each point by its *magnitude* and *direction.* Geometrical representation: an arrow pointing in the direction, and representing by its length the magnitude (examples are velocity, momentum, force, field strength; see I, Sec. 2). A vector, completely determined by its three components along the co-ordinate directions, is specified by *three* numbers. Vectors in this treatise are designated by bold-face type.

Vector equality. Vectors of equal length and direction are equal independently of their relative position in space.

Vector addition or subtraction. A geometrical addition $\mathbf{c} = \mathbf{a} + \mathbf{b} = \mathbf{b} + \mathbf{a}$ or geometrical subtraction $(\mathbf{d} = \mathbf{a} - \mathbf{b} = -\mathbf{b} + \mathbf{a})$ (Fig. 1).

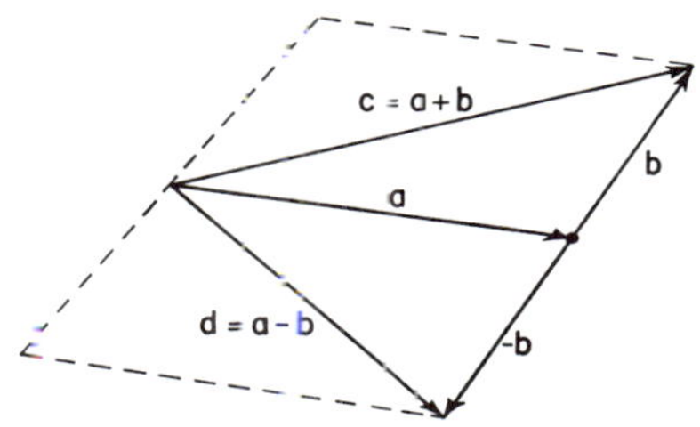

Fig. 1. Vector addition and subtraction.

Multiplication of vector by scalar; unit vectors. The magnitude is multiplied by the scalar; the direction remains unchanged. To designate explicitly a vector **A** according to magnitude and direction, we write

$$\mathbf{A} = |\mathbf{A}|\mathbf{a}^0,$$

where the symbol $|\mathbf{A}|$ represents magnitude and $\mathbf{a}^0$ a unit vector pointing in the direction of **A**. In the Cartesian (rectangular) co-ordinate system, the unit vectors are normally denoted as **i**, **j**, **k** pointing in the $+x$, $+y$, and $+z$ directions, respectively. Hence,

$$\mathbf{A} = \mathbf{A}_x\mathbf{i} + \mathbf{A}_y\mathbf{j} + \mathbf{A}_z\mathbf{k}$$

and

$$|\mathbf{A}| = \sqrt{A_x^2 + A_y^2 + A_z^2}.$$

For spherical and cylindrical co-ordinates, see I, Fig. 3.3, and I, Table 3.2. Multiplication by -1 means reversal of direction (see subtraction, Fig. 1).

Multiplication of Vector by Vector

(*a*) *Scalar product.* Defined as the product of the magnitude of two vectors **A** and **B**, multiplied by the cosine of the angle between them, the *dot product*

$$\mathbf{A}\cdot\mathbf{B} = |\mathbf{A}|\,|\mathbf{B}|\cos\theta = \mathbf{B}\cdot\mathbf{A}.$$

For Cartesian co-ordinates, since

$$\mathbf{i}\cdot\mathbf{j} = \mathbf{j}\cdot\mathbf{k} = \mathbf{k}\cdot\mathbf{i} = 0,$$

$$\mathbf{i}\cdot\mathbf{i} = \mathbf{j}\cdot\mathbf{j} = \mathbf{k}\cdot\mathbf{k} = 1,$$

$$\mathbf{A}\cdot\mathbf{B} = A_xB_x + A_yB_y + A_zB_z.$$

The scalar product is a scalar equal to the sum of the products of the components for any *orthogonal* co-ordinate system (see I, Table 3.1).

$$\mathbf{A}\cdot(\mathbf{B} + \mathbf{C} + \mathbf{D} + \cdots) = \mathbf{A}\cdot\mathbf{B} + \mathbf{A}\cdot\mathbf{C} + \mathbf{A}\cdot\mathbf{D} + \cdots,$$

$$\mathbf{A}\cdot\mathbf{A} = |\mathbf{A}|^2 \equiv \mathbf{A}^2.$$

(*b*) *Vector product.* Defined as the product of the magnitude of two vectors **A** and **B**, multiplied by the sine of the angle between them, the *cross product*

$$\mathbf{A}\times\mathbf{B} = |\mathbf{A}|\,|\mathbf{B}|\sin\theta = \mathbf{C}.$$

The magnitude of the vector **C** is equal to the area of the parallelogram defined by the two vectors **A** and **B**; the direction of **C** is normal to this plane and corresponds to the direction of advance of a right-hand screw when turned from **A** towards **B** (see I, Fig. 2.5); hence

$$\mathbf{A}\times\mathbf{B} = -(\mathbf{B}\times\mathbf{A}).$$

For Cartesian co-ordinates, since

$$\mathbf{i}\times\mathbf{i} = \mathbf{j}\times\mathbf{j} = \mathbf{k}\times\mathbf{k} = 0,$$

$$\mathbf{i}\times\mathbf{j} = \mathbf{k};\ \mathbf{j}\times\mathbf{k} = \mathbf{i};\ \mathbf{k}\times\mathbf{i} = \mathbf{j},$$

$$\mathbf{A}\times\mathbf{B} = (A_yB_z - A_zB_y)\mathbf{i} + (A_zB_x - A_xB_z)\mathbf{j} + (A_xB_y - A_yB_x)\mathbf{k},$$

or, written in determinant form,

$$\mathbf{A}\times\mathbf{B} = \begin{vmatrix} i & j & k \\ A_x & A_y & A_z \\ B_x & B_y & B_z \end{vmatrix} \quad \text{(see I, Eq. 5.9).}$$

$$\mathbf{A}\times(\mathbf{B} + \mathbf{C} + \mathbf{D} + \cdots) = \mathbf{A}\times\mathbf{B} + \mathbf{A}\times\mathbf{C} + \mathbf{A}\times\mathbf{D} + \cdots,$$

$$\mathbf{A}\times\mathbf{A} = 0.$$

Triple Products

(*a*) *Scalar triple product:* $\mathbf{A}\cdot[\mathbf{B}\times\mathbf{C}] = \mathbf{B}\cdot[\mathbf{C}\times\mathbf{A}] = \mathbf{C}\cdot[\mathbf{A}\times\mathbf{B}]$ is a scalar; geometrically it corresponds to the volume of the parallelepiped of which **A**, **B** and **C** are coterminous edges.

$$\mathbf{A}\cdot[\mathbf{B}\times\mathbf{C}] = -\mathbf{A}\cdot[\mathbf{C}\times\mathbf{B}].$$

(*b*) *Vector triple product:*

$$\mathbf{A}\times[\mathbf{B}\times\mathbf{C}] = (\mathbf{A}\cdot\mathbf{C})\mathbf{B} - (\mathbf{A}\cdot\mathbf{B})\mathbf{C};$$

$$[\mathbf{A}\times\mathbf{B}]\times\mathbf{C} = (\mathbf{A}\cdot\mathbf{C})\mathbf{B} - (\mathbf{B}\cdot\mathbf{C})\mathbf{A}.$$

Hence the triple vector product has a unique meaning only when the order in which the products are to be formed is prescribed.

Differential Operators

(*a*) The *gradient* denotes the maximum increase of some scalar function f at any point in space. The differential operation, by which this gradient is derived from f is characterized by the differential operator (*Hamiltonian operator*) ∇ (del). In Cartesian co-ordinates this is

$$\nabla = \mathbf{i}\frac{\partial}{\partial x} + \mathbf{j}\frac{\partial}{\partial y} + \mathbf{k}\frac{\partial}{\partial z},$$

and the gradient of f is

$$\nabla f = \mathbf{i}\frac{\partial f}{\partial x} + \mathbf{j}\frac{\partial f}{\partial y} + \mathbf{k}\frac{\partial f}{\partial z}.$$

(The formulation of the del operator in various co-ordinate systems is given in I, Eq. 3.18.) The change of the function f for a differential displacement ds in space is $df = \nabla f\cdot ds$. The gradient characterizes a vector *field* derived from a conservative potential field, for example, the force field of the earth gravitation potential or of the electrostatic potential (see I, Sec. 3). The del operator is handled according to the rules of vector analysis as a vector.

(*b*) The *divergence* characterizes the *sources* of a potential field (see I, Sec. 4); for example, in hydrodynamics the sources and sinks of a liquid or, in electrostatics, the positive and negative charges (I, Figs. 4.1 and 4.2). More generally speaking, the divergence of a vector field prescribes that the field pattern is consistent with the law of conservation of mass or charge (see Appendix AI, Sec. 6).

The divergence is the dot product of the del operator and the field vector in question,

$$\text{div}\,\mathbf{B} = \nabla\cdot\mathbf{B},$$

hence a scalar. It is formulated for the various coordinate systems in I, Eq. 4.9.

(*c*) The *curl* characterizes the *turbulence* of a vector field (I, Fig. 5.2) by the cross product of the del operator and the field vector in question,

$$\text{curl}\,\mathbf{B} = \nabla\times\mathbf{B}.$$

Hence, in accord with the rule of vector products, it is a vector, orientated normal to the surface element of the turbulent flow in the direction in which a right-hand screw, turned by the torque of the turbulence, would advance. The formulation of the curl in various coordinate systems is given in I, Eqs. 5.8 and 5.12.

Some Important Multiplication Operations with the Del Operator

$$\nabla(uvw) = vw\nabla u + wu\nabla v + uv\nabla w,$$

$$\nabla\cdot(u\mathbf{v}) = \mathbf{v}\nabla u + u\nabla\cdot\mathbf{v},$$

$$\nabla\times(u\mathbf{v}) = \nabla u\times\mathbf{v} + u\nabla\times\mathbf{v},$$

$$\nabla\cdot(\nabla u) = \nabla^2 u \text{ (see I, Eq. 4.16)},$$

$$\nabla\cdot\nabla(uv) = u\nabla^2 v + v\nabla^2 u + 2\nabla u\cdot\nabla v,$$

$$\nabla\times\nabla u = 0,$$

$$\nabla\cdot(\nabla\times\mathbf{u}) = 0,$$

$$\nabla\times(\nabla\times\mathbf{v}) = \nabla(\nabla\cdot\mathbf{v}) - \nabla^2\mathbf{v}.$$

C · Values of Physical Constants

c	velocity of light	2.9979×10^{8} [m sec^{-1}]
e	electronic charge	1.602×10^{-19} [coulombs]
h	Planck's constant	6.623×10^{-34} [joule sec]
k	Boltzmann's constant molecular gas constant	1.380×10^{-23} [joule deg^{-1}]
m	mass of electron	9.107×10^{-31} [kg]
m_+	mass of proton	1.6725×10^{-27} [kg]
m_B	Bohr magneton	9.27×10^{-24} [joule/weber m^{-2}]
N_L	Loschmidt's number	2.687×10^{25} [m^{-3}]
N_0	Avogadros's number	6.023×10^{23}
R	gas constant per mole	8.314 [joule mole^{-1} deg^{-1}]
R	Rydberg constant	10,967,758 [m^{-1}]
Z_0	intrinsic impedance of vacuum	376.6 [ohm]
ϵ_0	dielectric constant of vacuum	8.854×10^{-12} [farad m^{-1}]
μ_0	permeability of vacuum	1.257×10^{-6} [henry m^{-1}]

D · Table of Symbols

A	area
	Wien's displacement constant
$A_{i \leftrightarrow j}$	transition probability
a	Stefan-Boltzmann constant
	major half axis of ellipse
$\mathbf{B}$	magnetic flux density
b	minor half axis of ellipse
C	capacitance
c	velocity of light
D	dissipation factor (loss tangent)
$\mathbf{D}$	electric flux density (displacement)
d	distance
$\mathcal{E}$	total energy
$\mathbf{E}$	electric field strength
e	electric elementary charge
$\mathbf{F}$	force
F_s	radiation friction force
f	force constant
G	conductance
g	g factor (Landé splitting factor)
H	energy
$\mathbf{H}$	magnetic field strength
h	Planck constant
$\hbar$	$h/2\pi$
I	current
	light intensity
	moment of inertia
I_c	charging current
I_l	loss current
I_m	magnetization current
$\mathbf{i}$	unit vector in the x-direction
J	resultant inner quantum number
	quantum number of rotation
$\mathbf{J}$	current density
j	square root of -1
	inner quantum number
$\mathbf{j}$	unit vector in the y-direction
k	Boltzmann constant
	Bohr's azimuthal quantum number
	index of absorption (α/β)
$\mathbf{k}$	unit vector in the z-direction
	wave vector
$\mathbf{k}^0$	unit vector in direction of wave propagation
L	inductance
	resultant orbital momentum quantum number
$\mathbf{L}$	resultant orbital momentum vector
l	length
	azimuthal quantum number
$\boldsymbol{l}$	length vector
$\mathbf{l}$	orbital momentum vector
M	molecular weight
$\mathbf{M}$	magnetization (magnetic dipole moment per unit volume)
$\textcircled{M}$	resultant magnetic quantum number
m	mass
	mass of electron
m_0	rest mass
m_+	mass of proton
m_B	Bohr magneton
$\mathbf{m}$	magnetic moment
$\textcircled{m}$	magnetic quantum number
N	number of particles per unit volume
N_0	Avogadro's number
N_L	Loschmidt number
n	index of refraction
	total or principal quantum number
$\mathbf{n}$	unit vector perpendicular to surface under consideration
n'	$= n_a + n_b$, Bohr's radial quantum number
n^*	$= n(1 - jk)$, complex index of refraction
n_e	$= n_a - n_b$, electrical quantum number (linear Stark effect)
P	power
	work function
$\mathbf{P}$	polarization (electric dipole moment per unit volume)
	generalized momentum
p	pressure
	magnetic pole strength
$\mathbf{p}$	momentum
$\mathbf{p}'$	angular momentum vector
Q	electric charge
$\mathbf{Q}$	electric quality factor, Q value
$\mathbf{q}$	general space co-ordinate

R	Rydberg constant
	gas constant per mole
	energy coefficient of reflection (reflectivity)
R_s	equivalent radiation resistance of dipole antenna
r	reflection coefficient
$\mathbf{r}$	position vector, measured from origin
$\mathbf{r}^0$	unit vector in radial direction
$\mathbf{S}$	Poynting vector
	resultant spin momentum vector
$\underleftrightarrow{S}$	strain tensor
s	true (total) surface charge density
	spin quantum number
$\mathbf{s}$	spin momentum vector
T	absolute temperature
	energy coefficient of transmission (transmissibility)
$\underleftrightarrow{T}$	stress tensor
U	potential energy
u	space co-ordinate
V	volume
$\mathcal{V}$	voltage
v	velocity
	phase velocity
υ	vibrational quantum number
v_g	group velocity
W	work
x	space co-ordinate
	electronegativity
Y	admittance
Z	intrinsic impedance
	order number of atoms
Z_c	characteristic impedance
Z_0	impedance of free space
α	attenuation factor
	friction factor
	polarizability
	overlap integral
α_a	atomic polarizability
α_d	dipole (orientation) polarizability
α_e	electronic polarizability
α_m	magnetizability
α_s	space charge (interfacial) polarizability
β	phase factor
γ	complex propagation factor
	gyromagnetic ratio
δ	loss angle
$\tan\delta$	loss tangent (dissipation factor)
ϵ	numerical eccentricity of ellipse
ϵ_0	dielectric constant of free space (vacuum)
ϵ'	dielectric constant
ϵ''	loss factor
ϵ^*	(complex) permittivity
ζ	friction factor
η	viscosity
θ	angle
ϑ	angle
$\cos\theta$	power factor
κ	thermal conductivity
κ'	(relative) dielectric constant
κ''	(relative) loss factor
κ_m'	(relative) permeability
κ_m''	(relative) magnetic loss factor
κ_s'	static dielectric constant
κ_∞'	optical dielectric constant
κ^*	complex relative permittivity
κ_m^*	complex relative permeability
λ	wavelength
λ_c	cut-off wavelength
$\boldsymbol{\mu}$	electric dipole moment
$\boldsymbol{\mu}_i$	induced electric dipole moment
$\boldsymbol{\mu}_{ij}$	matrix element of the dipole moment
μ'	permeability
μ''	magnetic loss factor
μ^*	(complex) permeability
μ_0	permeability of free space (vacuum)
ν'	wave number
ν	frequency
ν_c	cut-off frequency
ν_m	Larmor frequency
ν_0	resonance frequency
Π	polarizability per mole (molar polarizability)
σ	conductivity
	dielectric conductivity $\omega\epsilon''$
τ	time constant
	relaxation time
	life time
	ringing time
Φ	magnetic flux
ϕ	angle
	phase angle
ϕ or φ	electrostatic potential
ψ	angle
	Schrödinger's wave function
ρ	density
	energy density of radiation
	space charge density
χ	electric susceptibility
χ_m	magnetic susceptibility
ω	$= 2\pi\nu$
	angular velocity
	(angular) frequency
ω_0	(angular) resonance frequency

E · Reference List of Books and Summarizing Articles

This reference list, by no means complete, has been compiled as a guide to additional literature; it may facilitate a more detailed study of special subjects.

ELECTROMAGNETIC THEORY

M. Abraham (revised by R. Becker), *The Classical Theory of Electricity and Magnetism*, Vol. I, Hafner, New York, 1951; Vol. II, *Elektronentheorie*, German edition, Teubner, Leipzig and Berlin, 1933.

M. Born, *Optik*, Springer, Berlin, 1933; Edwards Brothers, Ann Arbor, Michigan, 1943.

H. Motz, *Electromagnetic Problems of Microwave Theory*, Methuen, London, 1951.

F. Ollendorff, *Potentialfelder der Elektrotechnik*, Springer, Berlin, 1932.

H. H. Skilling, *Fundamentals of Electric Waves*, Wiley, New York, 1948.

J. A. Stratton, *Electromagnetic Theory*, McGraw-Hill, New York, 1949.

J. C. Slater and N. H. Frank, *Electromagnetism*, McGraw-Hill, New York, 1949.

TRANSMISSION LINES AND WAVE GUIDES

W. Jackson, *High Frequency Transmission Lines*, Methuen, London, 1947.

R. W. P. King, *Transmission Lines, Antennas and Wave Guides*, McGraw-Hill, New York, 1945.

C. G. Montgomery, R. H. Dicke, and E. M. Purcell, *Principles of Microwave Circuits*, McGraw-Hill, New York, 1948.

H. J. Reich, P. F. Ordung, H. L. Krauss, and J. G. Skalnik, *Microwave Theory and Techniques*, Van Nostrand, New York, 1953.

J. C. Slater, *Microwave Transmission*, McGraw-Hill, New York, 1942; *Microwave Electronics*, Van Nostrand, New York, 1950.

G. C. Southworth, *Principles and Applications of Wave-Guide Transmission*, Van Nostrand, New York, 1950.

R. I. Sarbacher and W. A. Edson, *Hyper and Ultrahigh Frequency Engineering*, Wiley, New York, 1943.

GENERAL THEORY

G. Joos, *Theoretical Physics*, Hafner, New York, 1950.

F. K. Richtmyer and E. H. Kennard, *Introduction to Modern Physics*, McGraw-Hill, New York, 1947.

DIELECTRICS

C. J. F. Böttcher, *Theory of Electric Polarization*, Elsevier Publishing Co., New York, 1952.

P. Debye, *Polar Molecules*, Chemical Catalog Co., New York, 1929.

"Dielectrics," *Ann. N. Y. Acad. Sci. 40*, 285 (1940).

"Dielectrics Conference, Bristol," *Trans. Faraday Soc. 42A* (1946).

Digest of the Literature on Dielectrics (annual volumes), Conference on Electrical Insulation, National Research Council, National Academy of Sciences, Washington, D. C.

H. Fröhlich, *Theory of Dielectrics*, Clarendon Press, Oxford, 1949.

W. Kauzmann, "Dielectric Relaxation as a Chemical Rate Process, *Revs. Mod. Phys. 14*, 12 (1942).

R. J. W. Le Fèvre, *Dipole Moments*, Methuen, London, 1953.

F. H. Müller and C. Schmelzer, "Dielektrisches Verhalten in Zusammenhang mit dem polaren Aufbau der Materie," *Ergeb. exakt. Naturw. 25*, 359 (1951).

C. P. Smyth, *Dielectric Constant and Molecular Structure*, Chemical Catalog Co., New York, 1931.

J. H. Van Vleck, *The Theory of Electric and Magnetic Susceptibilities*, Clarendon Press, Oxford, 1932.

A. von Hippel et al., *Dielectric Materials and Applications*, The Technology Press and Wiley, 1954.

L. G. Wesson, *Tables of Electric Dipole Moments*, The Technology Press, Massachusetts Institute of Technology, 1948.

QUANTUM THEORY AND ATOMIC PHYSICS

D. Bohm, *Quantum Theory*, Prentice-Hall, New York, 1952.

M. Born, *Atomic Physics*, Stechert, New York, 1936.

W. Heitler, *Elementary Wave Mechanics*, Clarendon Press, Oxford, 1945.

N. F. Mott and I. N. Sneddon, *Wave Mechanics and Its Applications*, Clarendon Press, Oxford, 1948.

K. S. Pitzer, *Quantum Chemistry*, Prentice-Hall, New York, 1953.

F. O. Rice and E. Teller, *The Structure of Matter*, Wiley, New York, 1949.

E. Schrödinger, *Collected Papers on Wave Mechanics*, Blackie and Son, London, 1929.

SPECTROSCOPY

G. Herzberg, *Atomic Spectra and Atomic Structure*, Dover Publications, New York, 1944; *Spectra of Diatomic Molecules*, Van Nostrand, New York, 1950; *Infrared and Raman Spectra of Polyatomic Molecules*, Prentice-Hall, New York, 1951.

H. M. Randall et al., *Infrared Determination of Organic Structures*, Van Nostrand, New York, 1949.

A. Sommerfeld, *Atombau und Spektrallinien*, Vieweg, Braunschweig, 1929; English edition, Dutton, New York, 1934.

H. Spooner, *Molekülspektren und ihre Aufwendung auf chemische Probleme*, Springer, Berlin, 1935.

H. E. White, *Introduction to Atomic Spectra*, McGraw-Hill, New York, 1934.

MICROWAVE SPECTROSCOPY

W. Gordy et al., *Microwave Spectroscopy*, Wiley, New York, 1953.

D. K. Coles, *Microwave Spectroscopy, Advances in Electronics* II, Academic Press, New York, 1950, pp. 300 ff.

"Microwave Spectroscopy," *Ann. N. Y. Acad. Sci. 55*, 743 (1952).

STRUCTURAL CHEMISTRY

A. E. van Arkel and J. H. de Boer, *Chemische Bindung als elektrostatische Erscheinung*, Hirzel, Leipzig, 1931.

C. S. Barrett, *Structure of Metals*, McGraw-Hill, New York, 1952.

C. W. Bunn, *Chemical Crystallography*, Clarendon Press, Oxford, 1945.

C. A. Coulson, *Valence*, Clarendon Press, Oxford, 1952.

W. Hückel, *Structural Chemistry of Inorganic Compounds*, Elsevier Publishing Co., New York, 1950.

C. K. Ingold, *Structure and Mechanism in Organic Chemistry*, Cornell University Press, Ithaca, 1953.

J. A. A. Ketelaar, *Chemical Constitution*, Elsevier Publishing Co., Amsterdam-New York, 1953.

L. Pauling, *The Nature of the Chemical Bond and the Structure of Molecules and Crystals*, Cornell University Press, Ithaca, 1940.

J. M. Robertson, *Organic Crystals and Molecules*, Cornell University Press, Ithaca, 1953.

H. A. Stuart, *Die Struktur des freien Moleküls*, Springer, Berlin, 1952.

A. F. Wells, *Structural Inorganic Chemistry*, University Press, Oxford, 1945.

SOLID STATE PHYSICS

M. Born and M. Göppert-Mayer, "Dynamische Gittertheorie der Kristalle," *Handbuch der Physik 24*, Pt. 2, Springer, Berlin, 1933, pp. 623 ff.

C. Kittel, *Introduction to Solid State Physics*, Wiley, New York, 1953.

N. F. Mott and H. Jones, *Theory of the Properties of Metals and Alloys*, Clarendon Press, Oxford, 1936.

N. F. Mott and R. W. Gurney, *Electronic Processes in Ionic Crystals*, Clarendon Press, Oxford, 1940.

W. Shockley, *Electrons and Holes in Semiconductors*, Van Nostrand, New York, 1950.

F. Seitz, *The Modern Theory of Solids*, McGraw-Hill, New York, 1940.

J. C. Slater, *Quantum Theory of Matter*, McGraw-Hill, New York, 1951.

PIEZO- AND FERROELECTRICITY

W. Voigt, *Lehrbuch der Kristallphysik*, Teubner, Leipzig and Berlin, 1910.

W. G. Cady, *Piezoelectricity*, McGraw-Hill, New York, 1946.

W. P. Mason, *Piezoelectric Crystals and Their Application to Ultrasonics*, Van Nostrand, New York, 1950.

A. von Hippel, "Ferroelectricity, Domain Structure, and Phase Transitions of Barium Titanate," *Revs. Mod. Phys. 22*, 221 (1950).

MAGNETISM

L. F. Bates, *Modern Magnetism*, Cambridge University Press, 1951.

R. Becker and W. Döring, *Ferromagnetismus*, Springer, Berlin, 1939.

F. Bitter, *Introduction to Ferromagnetism*, McGraw-Hill, New York, 1937.

R. M. Bozorth, *Ferromagnetism*, Van Nostrand, New York, 1951.

G. E. Pake, "Fundamentals of Nuclear Magnetic Resonance," *Am. J. Phys. 18*, 438, 473 (1950).

G. T. Rado, *Ferromagnetic Phenomena at Microwave Frequencies, Advances in Electronics* II, Academic Press, New York, 1950, pp. 251 ff.

"Washington Conference on Magnetism," *Revs. Mod. Phys. 25*, 1 (1953).

P. W. Selwood, *Magnetochemistry*, Interscience Publishers, New York, 1943.

E. C. Stoner, *Magnetism and Matter*, Methuen, London, 1934; *Magnetism*, Methuen, London, 1948.

CONDUCTION AND BREAKDOWN

A. von Engel and M. Steenbeck, *Elektrische Gasentladungen. Ihre Physik und Technik*, Springer, Berlin, 1932–34.

W. Franz, "Theorie des reinen elektrischen Durchschlags festen Isolatoren," *Ergeb. exakt. Naturw. 27*, 1 (1953).

H. Fröhlich and J. H. Simpson, *Intrinsic Dielectric Breakdown in Solids, Advances in Electronics* II, Academic Press, New York, 1950, pp. 185 ff.

H. Gänger, *Der elektrische Durchschlag von Gasen*, Springer, Berlin, 1953.

A. von Hippel, "Der elektrische Durchschlag in Gasen und festen Isolatoren," *Ergeb. exakt. Naturw. 14*, 79 (1935); "Electric Breakdown of Solids and Liquid Insulators," *J. Appl. Phys. 8*, 815 (1937).

F. Llewellyn Jones, "Electrical Discharges," *Reports on Progress in Physics 16*, 216 (1953).

L. B. Loeb, *Fundamental Processes of Electrical Discharge in Gases*, Wiley, New York, 1939.

H. S. W. Massey and E. H. S. Burhop, *Electronic and Ionic Impact Phenomena*, Clarendon Press, Oxford, 1952.

J. M. Meek and J. D. Craggs, *Electrical Breakdown of Gases*, Clarendon Press, Oxford, 1953.

S. Whitehead, *Dielectric Breakdown of Solids*, Clarendon Press, Oxford, 1951.

INDEX

The Artech House Microwave Library

Acoustic Charge Transport: Device Technology and Applications, R. Miller, C. Nothnick, and D. Bailey

Advanced Automated Smith Chart Software and User's Manual, Version 2.0, Leonard M. Schwab

Algorithms for Computer-Aided Design of Linear Microwave Circuits, Stanislaw Rosloniec

Analysis, Design, and Applications of Fin Lines, Bharathi Bhat and Shiban K. Koul

Analysis Methods for Electromagnetic Wave Problems, Eikichi Yamashita, editor

Automated Smith Chart Software and User's Manual, Leonard M. Schwab

C/NL2 for Windows: Linear and Nonlinear Microwave Circuit Analysis and Optimization, Software and User's Manual, Stephen A. Maas and Arthur Nichols

Capacitance, Inductance, and Crosstalk Analysis, Charles S. Walker

Design of Impedance-Matching Networks for RF and Microwave Amplifiers, Pieter L. D. Abrie

Dielectric Materials and Applications, Arthur von Hippel, editor

Dielectrics and Waves, Arthur von Hippel

Digital Microwave Receivers, James B. Tsui

Electric Filters, Martin Hasler and Jacques Neirynck

Electrical and Thermal Characterization of MESFETs, HEMTs, and HBTs, Robert Anholt

E-Plane Integrated Circuits, P. Bhartia and P. Pramanick, editors

Feedback Maximization, Boris J. Lurie

Filters with Helical and Folded Helical Resonators, Peter Vizmuller

Frequency Synthesizer Design Toolkit Software and User's Manual, Version 1.0, James A. Crawford

Fundamentals of Distributed Amplification, Thomas T. Y. Wong

GaAs MESFET Circuit Design, Robert A. Soares, editor

GASMAP: Gallium Arsenide Model Analysis Program, J. Michael Golio *et al.*

Handbook of Microwave Integrated Circuits, Reinmut K. Hoffmann

Handbook for the Mechanical Tolerancing of Waveguide Components, W. B. W. Alison

HELENA: HEMT Electrical Properties and Noise Analysis Software and User's Manual, Henri Happy and Alain Cappy

HEMTs and HBTs: Devices, Fabrication, and Circuits, Fazal Ali, Aditya Gupta, and Inder Bahl, editors

High-Power Microwave Sources, Victor Granatstein and Igor Alexeff, edtiors

High-Power GaAs FET Amplifiers, John Walker, editor

High-Power Microwaves, James Benford and John Swegle

Introduction to Microwaves, Fred E. Gardiol

Introduction to Computer Methods for Microwave Circuit Analysis and Design, Janusz A. Dobrowolski

Introduction to the Uniform Geometrical Theory of Diffraction, D. A. McNamara, C. W. I. Pistorius and J. A. G. Malherbe

LOSLIN: Lossy Line Calculation Software and User's Manual, Fred E. Gardiol

Lossy Transmission Lines, Fred E. Gardiol

Low-Angle Microwave Propagation: Physics and Modeling, Adolf Giger

Low Phase Noise Microwave Oscillator Design, Robert G. Rogers

MATCHNET: Microwave Matching Networks Synthesis, Stephen V. Sussman-Fort

Matrix Parameters for Multiconductor Transmission Lines: Software and User's Manual, A. R. Djordjevic *et al.*

Microelectronic Reliability, Volume I: Reliability, Test, and Diagnostics, Edward B. Hakim, editor

Microelectronic Reliability, Volume II: Integrity Assessment and Assurance, Emiliano Pollino, editor

Microwave and RF Circuits: Analysis, Synthesis, and Design, Max Medley

Microwave and RF Component and Subsystem Manufacturing Technology, Heriot-Watt University

Microwave Circulator Design, Douglas K. Linkhart

Microwave Engineers' Handbook, 2 Volumes, Theodore Saad, editor

Microwave Materials and Fabrication Techniques, Second Edition, Thomas S. Laverghetta

Microwave MESFETs and HEMTs, J. Michael Golio *et al.*

Microwave and Millimeter Wave Heterostructure Transistors and Applicatons, F. Ali, editor

Microwave and Millimeter Wave Phase Shifters, Volume I: Dielectric and Ferrite Phase Shifters, S. Koul and B. Bhat

Microwave and Millimeter Wave Phase Shifters, Volume II: Semiconductor and Delay Line Phase Shifters, S. Koul and B. Bhat

Microwave Mixers, Second Edition, Stephen Maas

Microwave Transmission Design Data, Theodore Moreno

Microwave Transition Design, Jamal S. Izadian and Shahin M. Izadian

Microwave Transmission Line Couplers, J. A. G. Malherbe

Microwave Tubes, A. S. Gilmour, Jr.

Microwaves: Industrial, Scientific, and Medical Applications, J. Thuery

Microwaves Made Simple: Principles and Applicatons, Stephen W. Cheung, Frederick H. Levien *et al.*

MMIC Design: GaAs FETs and HEMTs, Peter H. Ladbrooke

Modern GaAs Processing Techniques, Ralph Williams

Modern Microwave Measurements and Techniques, Thomas S. Laverghetta

Monolithic Microwave Integrated Circuits: Technology and Design, Ravender Goyal *et al.*

Nonuniform Line Microstrip Directional Couplers, Sener Uysal

PC Filter: Electronic Filter Design Software and User's Guide, Michael G. Ellis, Sr.

PLL: Linear Phase-Locked Loop Control Systems Analysis Software and User's Manual, Eric L. Unruh

RF Design Guide: Systems, Circuits, and Equations, Peter Vizmuller

Scattering Parameters of Microwave Networks with Multiconductor Transmission Lines: Software & User's Manual, A. R. Djordjevic *et al.*

Solid-State Microwave Power Oscillator Design, Eric Holzman and Ralston Robertson

Terrestrial Digital Microwave Communications, Ferdo Ivanek *et al.*

Time-Domain Response of Multiconductor Transmission Lines: Software and User's Manual, A. R. Djordjevic *et al.*

Transmission Line Design Handbook, Brian C. Waddell

Yield and Reliability in Microwave Circuit and System Design, Michael Meehan and John Purviance

For further information on these and other Artech House titles, contact:

Artech House
685 Canton Street
Norwood, MA 02062
617-769-9750
Fax: 617-769-6334
Telex: 951-659
email: artech@world.std.com

Artech House
Portland House, Stag Place
London SW1E 5XA England
+44 (0) 171-973-8077
Fax: +44 (0) 171-630-0166
Telex: 951-659
bookco@artech.demon.co.uk